Advanced Analytical Dynamics

This book provides a unique bridge between the foundations of analytical mechanics and application to multibody dynamical systems. It is intended as a textbook, particularly well suited for graduate students seeking an understanding of the theoretical underpinnings of analytical mechanics, as well as modern task space approaches for representing the resulting dynamics that can be exploited for real-world problems in areas such as biomechanics and robotics.

Established principles in mechanics are presented in a thorough and modern way. The chapters build up from general mathematical foundations to an extensive treatment of kinematics and then to a rigorous treatment of conservation and variational principles in mechanics. Parallels are drawn between the different approaches, providing the reader with insights that unify his or her understanding of analytical dynamics. Additionally, a unique treatment is presented on task space dynamical formulations that map traditional configuration space representations into more intuitive geometric spaces.

Vincent De Sapio is a research scientist at HRL Laboratories in Malibu, California. He has more than 40 publications and 16 patents (issued or pending) in the areas of multibody dynamics, robotics, biomechanics, control theory, and human motion synthesis and simulation. Dr. De Sapio is a senior member of IEEE and a member of ASME and Sigma Xi. He is a founding member of the IEEE Technical Committee on Human Movement Understanding. He received his Ph.D. and M.S. from Stanford University and his B.S. from Rensselaer Polytechnic Institute, all in mechanical engineering.

καὶ τὸ φῶς ἐν τῇ σκοτίᾳ φαίνει, καὶ ἡ σκοτία αὐτὸ οὐ κατέλαβεν.

The light shines in the darkness, and the darkness has not overcome it.

John 1:5

S. D. G.

Advanced Analytical Dynamics

Theory and Applications

VINCENT DE SAPIO
HRL Laboratories LLC

CAMBRIDGE
UNIVERSITY PRESS

CAMBRIDGE
UNIVERSITY PRESS

University Printing House, Cambridge CB2 8BS, United Kingdom

One Liberty Plaza, 20th Floor, New York, NY 10006, USA

477 Williamstown Road, Port Melbourne, VIC 3207, Australia

4843/24, 2nd Floor, Ansari Road, Daryaganj, Delhi - 110002, India

79 Anson Road, #06-04/06, Singapore 079906

Cambridge University Press is part of the University of Cambridge.

It furthers the University's mission by disseminating knowledge in the pursuit of education, learning, and research at the highest international levels of excellence.

www.cambridge.org
Information on this title: www.cambridge.org/9781107179608
DOI: 10.1017/9781316832301

First published 2017

Printed in the United States of America by Sheridan Books, Inc.

A catalogue record for this publication is available from the British Library.

Library of Congress Cataloging-in-Publication Data
Names: De Sapio, Vincent, 1968– author.
Title: Advanced analytical dynamics : theory and applications / Vincent De Sapio (HRL Laboratories LLC).
Description: Cambridge, United Kingdom ; New York, NY : Cambridge University Press, 2017. | Includes bibliographical references and index.
Identifiers: LCCN 2016036860| ISBN 9781107179608 (hardback ; alk. paper) | ISBN 1107179602 (hardback ; alk. paper)
Subjects: LCSH: Dynamics. | Mechanics, Analytic.
Classification: LCC QA845 .D42 2017 | DDC 620.1/04–dc23 LC record available at https://lccn.loc.gov/2016036860

ISBN 978-1-107-17960-8 Hardback

Contents

Illustrations

Tables

Preface

This book addresses the analytical dynamics of multibody systems and is intended for a one- to two-semester advanced graduate-level course in analytical dynamics. The emphasis is on a solid theoretical foundation with examples that concretely illustrate the theory. I have included a chapter on the fundamental mathematics that is helpful in navigating the principles of dynamics. This includes coverage of linear systems and differential geometry. A chapter on kinematics, the study of the geometry of motion, follows. The first chapter, on dynamics, addresses conservation principles, fundamentally the conservation of momentum embodied in the Newton-Euler Principles. Historically, analytical mechanics (dynamics) has referred to the so-called variational principles, rooted in the calculus of variations. Three chapters cover zeroth-, first-, and second-order variational principles, respectively. Lagrangian and Hamiltonian mechanics are among the more well-known formulations arising from variational principles covered in this book. I also cover important, but lesser known, higher-order principles, including Jourdain's Principle of Virtual Power, Gauss's Principle of Least Constraint, and Hertz's Principle of Least Curvature, as well as Kane's formulation and the Gibbs-Appell formulation.

As an aside, it is worth noting that modern theoretical physics emerged out of the classical variational principles. Einstein's general theory of relativity is commonly formulated using Lagrangian mechanics. Dirac was the first to use the Lagrangian in quantum mechanics and provided separate formulations of quantum mechanics and general relativity based on the Hamiltonian formalism. He also provided a generalized formulation of constrained Hamiltonian systems. Additionally, Feynman's path integral formulation of quantum mechanics has its classical ancestry in Hamilton's Principle of Least Action.

After the chapters on the variational principles, I have included a chapter on an alternate formulation of classical dynamics that has found significant utility in the control of robotic systems. The so-called task space formulation of dynamics was pioneered by Khatib under the name of operational space dynamics. It provides a transformation of the configuration space description of system dynamics into a more convenient task-oriented description. An applications-oriented chapter is included on biomechanical systems. This provides a basic overview of musculoskeletal and neuromuscular biomechanics with extensive coverage of application examples using actual anthropometric and muscle property data. The final chapter provides a brief survey of some analytical dynamics software. This is not intended to provide exhaustive coverage but only some examples of general purpose mathematical software with extensions for multibody dynamics, as well as dedicated multibody dynamics software, both commercial and open source. An appendix is included that touches upon the application of continuum mechanics to flexible multibody systems, which include both rigid and deformable bodies. It is my intention to extend this appendix into a fully integrated book chapter in a future edition.

I have tried to provide at least one example to illustrate each important concept covered in the book. The examples presented are thoroughly worked out in the text, so the reader should have no trouble following the methodology. In many cases, numerical simulation results are also provided. Some simple integration schemes are covered, and the reader is encouraged to explore the examples on his or her own and use any preferred mathematical software (e.g., Mathematica, Matlab) to generate simulation results. The reader would also benefit from employing mathematical software in addressing some of the exercises at the end of selected chapters. These exercises are often multipart and can involve fairly extensive mathematical computations.

I would like to thank a number of my past teachers who helped impart to me not only an understanding and insight related to the topics covered in this book but also a sense of the vast expanse of applications for analytical dynamics. These include Oussama Khatib, Scott Delp, Jean Heegaard, Bernie Roth, Ken Waldron, and others. A number of other teachers have contributed to my understanding of areas related to analytical dynamics. These include Stephen Boyd, Steve Rock, Sanjay Lall, Ron Fedkiw, Peter Pinsky, Charles Steele, Edward Goldstein, and Terry Sanger. My academic colleagues have provided productive interaction as well. These include Jaeheung Park, Luis Sentis, François Conti, Mike Zinn, James Warren, Emel Demircan, Jinsung Kwon, Dongjun Shin, Anya Petrovskaya, Peter Thaulad, Oliver Brock, Vincent Padois, Kate Saul, Rob Siston, Jeff Reinbolt, and others.

Although not directly related to the content of this book, many colleagues at HRL (formerly Hughes Research Laboratories) whom I have worked with on various projects have provided professional support and fruitful interaction. Among these individuals are Rajan Bhattacharyya, Jaehoon Choe, Yang Chen, Jose Cruz-Albrecht, Mike Daily, Son Dao, Chong Ding, Darren Earl, Karim El DeFrawy, Stephanie Goldfarb, Heiko Hoffmann, Mike Howard, Qin Jiang, Deepak Khosla, Ken Kim, Tiffany Kim, Dmitiry Korchev, Joonho Lee, Tsai-ching Lu, Charles Martin, Kevin Martin, Connie Ni, Aleksey Nogin, Yuri Owechko, Dave Payton, Matt Phillips, Praveen Pilly, Amir Rahimi, Shankar Rao, Shane Roach, Stephan Salas, Narayan Srinivasan, Nigel Stepp, Ryan Uhlenbrock, and others. Special thanks go to HRL's patent counsel, George Rapacki, my administrative assistant, Jennifer Greene, and Joonho Lee, who performed a technical proofread of part of the manuscript. In addition to working with a talented group of people, HRL's location provides me with inspiring views of the Pacific Ocean from a bluff overlooking the Malibu coastline in Southern California.

It has been a great pleasure working with Cambridge University Press on this book. I would like to thank Steve Elliot, Mark Fox, and Rebecca Rom-Frank at Cambridge. I would also like to thank Holly Monteith for copy editing the manuscript, and Vijay Bhatia for managing the book composition and typesetting. The external technical reviewers provided valuable constructive assessments of the early draft, for which I am grateful. A special extra note of thanks goes to Steve Elliot who guided me through the manuscript evaluation, approval, and contract-related processes.

A number of people in my life made this book possible, not by technical contribution, but by the consistent impact they have had on my life. First and foremost, my parents, Martin and Lucia, shaped my life in ways that were profound. As immigrants from Italy, both of them embodied the American dream and built their lives upon hard work, commitment to family, and an enduring faith. My siblings, Gaetano, Antonio, Maria, Carmine, and particularly my sisters, Rita and Sally, nurtured me throughout my life and continue to impact my life in positive ways. I can not sufficiently express my gratitude to them for the counsel they provide me in life. I would also like to thank Tom Stephen, who has provided me with his friendship and steadfast spiritual guidance, and my in-laws, Bob and Joyce Prindle, who have always encouraged me.

The greatest joys in my life are my wife, Robin, and my two boys, Robbie and Marty. Robin has stuck by me when I thought no one would. She is a source of enthusiasm, compassion, and humor, and she is my rock. Robbie and Marty inspire me in ways that only a toddler and a six-year-old can. In them I see innocence (and a little mischief), a complete passion for life, and warm hearts. They are my dearest treasures in life.

Finally, I thank God for all the blessings that He has provided in my life. It is through Him that my life has meaning.

Malibu, USA — Vincent De Sapio
November 2016 — HRL Laboratories

Notation

A specific set of notational standards is employed in this book. In the following sections we build up the notational standards that are used in the following chapters, starting with general mathematical objects and proceeding to kinematic objects, dynamical objects, and block diagram elements.

General Mathematical Objects

Sets

The following standard set notation is employed:

$\{\}$	designation of a set
$\mid$	such that
$\forall$	for all
$\in$	element of
$\perp$	orthogonal to
$\square^{\perp}$	orthogonal complement
$\mathbb{R}$	set of real numbers
$\mathbb{R}^n$	set of real n-dimensional vectors
$\mathbb{R}^{m \times n}$	set of real $m \times n$ matrices
$\mathbb{C}$	set of complex numbers
$\mathbb{H}$	set of quaternions
$\mathbb{S}^n$	set of points on an n-dimensional unit sphere

An example is as follows:

$$\begin{gathered} \boldsymbol{A}^T\boldsymbol{\lambda} \perp \boldsymbol{x}, \\ \forall \boldsymbol{\lambda} \in \mathbb{R}^m \text{ and } \forall \boldsymbol{x} \in \ker(\boldsymbol{A}), \\ \text{where } \ker(\boldsymbol{A}) = \{\boldsymbol{x} \in \mathbb{R}^n \mid \boldsymbol{A}\boldsymbol{x} = \boldsymbol{0}\}. \end{gathered}$$

This would read as follows: $\boldsymbol{A}^T\boldsymbol{\lambda}$ is orthogonal to $\boldsymbol{x}$ for all $\boldsymbol{\lambda}$ in the set of real m-dimensional vectors and for all $\boldsymbol{x}$ in the kernel of $\boldsymbol{A}$, where the kernel of $\boldsymbol{A}$ is the set of all real n-dimensional vectors, $\boldsymbol{x}$, such that $\boldsymbol{A}\boldsymbol{x} = \boldsymbol{0}$.

Scalars

Scalars (rank 0 tensors) are represented with nonbold italic characters (e.g., a). These include scalars as well as scalar components of vectors and matrices. Scalar components of vectors and matrices are denoted with a subscripted index to the right of the scalar symbol (e.g., v_i, M_{ij}). The following standard operators are employed:

δ	variation
$\frac{d}{d\square}$	derivative
$\dot{\square}$	time derivative

Complex Numbers and Quaternions

Complex numbers and quaternions are represented with nonbold lowercase (typically) italic characters. The components can be expressed as a sum of the real and imaginary parts, for example,

$$z = a + \boldsymbol{i}b$$
$$h = h_0 + h_1\boldsymbol{i} + h_2\boldsymbol{j} + h_3\boldsymbol{k}.$$

Vectors, Points, and Line Segments

Vectors (rank 1 tensors) are represented with bold lowercase (typically) italic characters. Vectors can be expressed as a 1-dimensional array or as a linear combination of basis vectors, with indexed scalar components (displayed as nonbold italic characters). Basis vectors are denoted as $\hat{\boldsymbol{e}}$. An example follows:

$$\boldsymbol{v} = \begin{pmatrix} v_1 \\ v_2 \\ v_3 \end{pmatrix} = \sum_{i=1}^{3} v_i \hat{\boldsymbol{e}}_i = v_1\hat{\boldsymbol{e}}_1 + v_2\hat{\boldsymbol{e}}_2 + v_3\hat{\boldsymbol{e}}_3.$$

Points are represented using nonbold italic characters (e.g., A). Line segments between two points are represented using an arrow (e.g., $\overrightarrow{AB}$).

Matrices and Tensors

Matrices (rank 2 tensors) are represented with bold uppercase (typically) italic characters. Matrices can be expressed as a 2-dimensional array or as a linear combination of dyads, with indexed scalar components (displayed as nonbold italic characters). Dyads consist of a pair of base vectors separated by an outer product symbol, $\otimes$, for example,

$$\boldsymbol{M} = \begin{pmatrix} M_{11} & M_{12} \\ M_{21} & M_{22} \end{pmatrix} = \sum_{i=1}^{2}\sum_{j=1}^{2} M_{ij}\hat{\boldsymbol{e}}_i \otimes \hat{\boldsymbol{e}}_j = M_{11}\hat{\boldsymbol{e}}_1 \otimes \hat{\boldsymbol{e}}_1 + M_{12}\hat{\boldsymbol{e}}_1 \otimes \hat{\boldsymbol{e}}_2$$
$$+ M_{21}\hat{\boldsymbol{e}}_2 \otimes \hat{\boldsymbol{e}}_1 + M_{22}\hat{\boldsymbol{e}}_2 \otimes \hat{\boldsymbol{e}}_2.$$

The identity matrix is denoted as **1** and the *zero* matrix is denoted as **0**.

Vector and Matrix Operators

The following standard vector and matrix operators are employed:

$\cdot$	dot product
$\langle \square, \square \rangle$	inner product
$\lVert \square \rVert$	norm
$\times$	cross-product
$\otimes$	outer product
δ	variation
$\frac{d}{ds}$	derivative with respect to a scalar, s
$\dot{\square}$	time derivative
$\frac{\partial}{\partial \square}$, ∇	partial derivative, gradient
$\mathrm{im}()$	image or range of a matrix
$\ker()$	kernel or null space of a matrix
$\mathrm{proj}_{\square}()$	projection of a vector onto a subspace
$T()$	tangent space operator
$\bar{\square}$	dynamically consistent (mass-weighted) inverse of a matrix

The partial derivative/gradient operators are overloaded for scalars and vectors. For example, given a scalar, $U \in \mathbb{R}$, and a vector, $\boldsymbol{v} \in \mathbb{R}^m$, the respective gradients are

$$\nabla U = \frac{\partial U}{\partial \boldsymbol{q}} = \sum_{i=1}^{n} \frac{\partial U}{\partial q_i} \hat{\boldsymbol{e}}_i = \begin{pmatrix} \frac{\partial U}{\partial q_1} \\ \vdots \\ \frac{\partial U}{\partial q_n} \end{pmatrix} \in \mathbb{R}^n$$

and

$$\boldsymbol{v}\nabla = \frac{\partial \boldsymbol{v}}{\partial \boldsymbol{q}} = \sum_{i=1}^{m} \sum_{j=1}^{n} \frac{\partial v_i}{\partial q_j} \hat{\boldsymbol{e}}_i \otimes \hat{\boldsymbol{e}}_j = \begin{pmatrix} \frac{\partial v_1}{\partial q_1} & \cdots & \frac{\partial v_1}{\partial q_n} \\ \vdots & \ddots & \vdots \\ \frac{\partial v_m}{\partial q_1} & \cdots & \frac{\partial v_m}{\partial q_n} \end{pmatrix} \in \mathbb{R}^{m \times n}.$$

Kinematic Objects

Objects having a kinematic meaning inherit all of the aforementioned rules with respect to their mathematical type. Additionally, they adhere to the following with regard to their physical type.

A position vector, $\boldsymbol{r}$, uses a right subscript to denote the *material point* it refers to and a left superscript to denote the basis it is expressed in. Velocity, $\boldsymbol{v}$, and acceleration, $\boldsymbol{a}$, vectors additionally denote the frame that motion is relative to using a ":" separator in the right subscript. Angular velocity, $\boldsymbol{\omega}$, and angular acceleration vectors, $\boldsymbol{\alpha}$, use a right subscript to denote the *body* they refer to and a left superscript to denote the basis they are expressed in. As with velocity, they additionally denote the frame that motion is relative to, using a ":". Any annotation can be omitted if the information conveyed by it is already clear from context. Generalized coordinates are denoted as $\boldsymbol{q}$ and operational space coordinates are denoted as $\mathbf{x}$.

Coordinate transformation matrices, including both orthogonal rotation matrices, $\boldsymbol{Q}$, and homogenous transformation matrices, $\boldsymbol{T}$, denote the frame of interest using a left subscript and the embedding frame using a left superscript. Unit quaternions, h, use similar annotation. Jacobian matrices use a right subscript to denote the object (material point, body, etc.) they refer to and a left superscript to denote the basis they are expressed in. Again, any annotation can be omitted if the information conveyed by it is already clear from context.

Cartesian Space Quantities

$\oplus$, G	center of mass point
${}^{A}\frac{d}{dt}{}_{\mathcal{O}}$	time derivative relative to $\mathcal{O}$, expressed in $\mathcal{A}$
${}^{A}\Delta_{\mathcal{O}}$	change relative to $\mathcal{O}$, expressed in $\mathcal{A}$
${}^{B}\boldsymbol{d}_{\overrightarrow{AB}}$	displacement vector between points A and B, expressed in $\mathcal{B}$
${}^{B}\boldsymbol{r}_{G_{\mathcal{A}}}$	position of center of mass, G, of body $\mathcal{A}$ expressed in $\mathcal{B}$
${}^{B}\boldsymbol{r}^{\mathcal{A}}_{\mathcal{B}}$	point on body $\mathcal{B}$ to which body $\mathcal{A}$ attaches, expressed in $\mathcal{B}$
${}^{B}\boldsymbol{v}_{G_{\mathcal{A}}:\mathcal{O}}$	velocity of center of mass, G, of body $\mathcal{A}$, relative to $\mathcal{O}$, expressed in $\mathcal{B}$
${}^{B}\boldsymbol{a}_{G_{\mathcal{A}}:\mathcal{O}}$	acceleration of center of mass, G, of body $\mathcal{A}$, relative to $\mathcal{O}$, expressed in $\mathcal{B}$
${}^{B}\boldsymbol{\omega}_{\mathcal{A}:\mathcal{O}}$	angular velocity of body $\mathcal{A}$, relative to $\mathcal{O}$, expressed in $\mathcal{B}$
${}^{B}\boldsymbol{\alpha}_{\mathcal{A}:\mathcal{O}}$	angular acceleration of body $\mathcal{A}$, relative to $\mathcal{O}$, expressed in $\mathcal{B}$
${}^{\mathcal{A}}_{\mathcal{B}}\boldsymbol{Q}$	rotation matrix of $\mathcal{B}$ with respect to $\mathcal{A}$
$\boldsymbol{Q}_k(\theta)$	rotation matrix representing a spin of θ about axis $\boldsymbol{k}$
${}^{\mathcal{A}}_{\mathcal{B}}h$	quaternion of $\mathcal{B}$ with respect to $\mathcal{A}$
$h_k(\theta)$	quaternion representing a spin of θ about axis $\boldsymbol{k}$
$\$$	screw displacement
${}^{A}_{B}\boldsymbol{T}$	homogenous transformation matrix of $\mathcal{B}$ with respect to $\mathcal{A}$
${}^{B}\boldsymbol{\Gamma}_{G_{\mathcal{A}}}$	Jacobian of position of center of mass, G, of body $\mathcal{A}$ expressed in $\mathcal{B}$
${}^{B}\boldsymbol{\Pi}_{\mathcal{A}}$	Jacobian of body $\mathcal{A}$ expressed in $\mathcal{B}$

Configuration Space Quantities

$\boldsymbol{q}$	generalized coordinate vector

Constraint Space Quantities

$\boldsymbol{\phi}$	holonomic constraint vector (general zeroth-order constraints)
$\boldsymbol{\Phi}$	holonomic constraint Jacobian matrix
$\boldsymbol{C}$	nonholonomic constraint matrix (linear first-order constraints)
$\boldsymbol{W}$	constraint null space matrix
$\boldsymbol{\psi}$	nonholonomic constraint vector (general first-order constraints)
$\boldsymbol{A}$	nonholonomic constraint matrix (linear second-order constraints)

Task Space Quantities

$\boldsymbol{x}$	task space coordinate vector
$\boldsymbol{J}$	task Jacobian matrix

Dynamic Objects

Objects having a dynamical meaning inherit all of the aforementioned rules with respect to their mathematical type. Additionally, they adhere to the following with regard to their physical type.

Translational momentum vectors, $\boldsymbol{p}$, have the same scripting as velocity vectors. Angular momentum vectors, $\boldsymbol{H}$, use the same scripting as angular velocity vectors. Additionally, angular momentum vectors denote the point about which they are evaluated using a right superscript. Inertia tensors, $\boldsymbol{I}$, have scripting similar to angular momentum vectors. Again, any annotation can be omitted if the information conveyed by it is already clear from context.

Cartesian Space Quantities

M	mass of point or body
g	acceleration due to gravity (e.g.,$\approx$ 9.8m/s^2 on earth)
${}^{B}\boldsymbol{f}_{B}^{A}$	force that body $\mathcal{A}$ exerts on body $\mathcal{B}$, expressed in $\mathcal{B}$
${}^{B}\boldsymbol{\varphi}_{B}^{A}$	moment that body $\mathcal{A}$ exerts on body $\mathcal{B}$, expressed in $\mathcal{B}$
${}^{B}\boldsymbol{p}_{G_{\mathcal{A}}:O}$	translational momentum of center of mass, G, of body $\mathcal{A}$, relative to $\mathcal{O}$, expressed in $\mathcal{B}$
${}^{B}\boldsymbol{H}_{\mathcal{A}:O}^{G_{\mathcal{A}}}$	angular momentum of body $\mathcal{A}$, relative to $\mathcal{O}$, about center of mass, G, of body $\mathcal{A}$, expressed in $\mathcal{B}$
${}^{B}\boldsymbol{I}_{\mathcal{A}:O}^{G_{\mathcal{A}}}$	inertia tensor of body $\mathcal{A}$, relative to $\mathcal{O}$, about center of mass, G, of body $\mathcal{A}$, expressed in $\mathcal{B}$

Configuration Space Quantities

$\boldsymbol{p}$	generalized momentum vector
$\boldsymbol{\tau}$	generalized force vector
$\boldsymbol{M}$	generalized mass matrix
$\boldsymbol{b}$	generalized Coriolis-centrifugal vector
$\boldsymbol{g}$	generalized gravity vector
$\mathcal{L}$	Lagrangian
$\mathcal{H}$	Hamiltonian
$\mathcal{G}$	Gauss function
$\mathcal{S}$	Gibbs function

Constraint Space Quantities

$\boldsymbol{\lambda}$	Lagrange multipliers (constraint forces)
$\mathbf{H}$	constraint space mass matrix
$\boldsymbol{\alpha}$	constraint space Coriolis-centrifugal vector
$\boldsymbol{\rho}$	constraint space gravity vector
$\boldsymbol{\Theta}^T$	constraint null space projection matrix

Task Space Quantities

$\boldsymbol{N}^T$	task null space projection matrix
$\boldsymbol{f}$	task space force vector
$\boldsymbol{\Lambda}$	task space mass matrix
$\boldsymbol{\mu}$	task space Coriolis-centrifugal vector
$\boldsymbol{p}$	task space gravity vector

Block Diagrams

Block diagrams use a number of common schematic elements, as follows. For general (nonlinear) operators, a dashed line and an unfilled arrow are used to denote the input argument into the block:

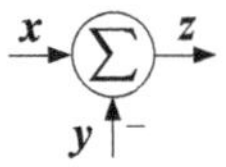

summation, $\boldsymbol{z} = \boldsymbol{x} - \boldsymbol{y}$

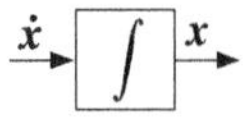

integration, $\boldsymbol{x} = \int \dot{\boldsymbol{x}}\, dt$

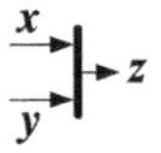

concatenation, $\boldsymbol{z} = (\boldsymbol{x}^T\ \boldsymbol{y}^T)^T$

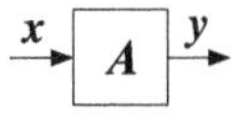

linear operator, $\boldsymbol{y} = \boldsymbol{A}\boldsymbol{x}$

general (nonlinear) operator, $\boldsymbol{y} = \boldsymbol{f}(\boldsymbol{x})$

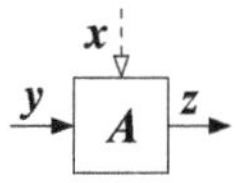

mixed operator, $\boldsymbol{z} = \boldsymbol{A}(\boldsymbol{x})\boldsymbol{y}$

1 Introduction

Dynamics is traditionally defined as the classical study of motion with respect to the physical causes of motion, that is, forces and moments. Kinematics, on the other hand, is concerned with the study of motion *without* respect to the underlying physical causes. In this sense, kinematics is really a fundamental prerequisite upon which dynamics is constructed.

For the purposes of this text, the terms *dynamics* and *mechanics* are taken to be synonymous. The choice of which term is used is based more on the academic community than on a strict technical distinction. The engineering community typically adopts the term *dynamics* and the physics and applied mathematics communities typically adopt the term *mechanics*. The term *dynamics* is predominantly used in this text.

1.1 Historical Background

Interest in the dynamics of linked multibody systems has existed throughout much of recorded human history. As an example, representations of the human form in art have included anthropomorphic constructions made up of mechanical elements like those depicted in Giovanni Braccelli's *Bizzarie di Varie Figure* published in 1624 (see Figure 1.1). Braccelli's art coincided with the birth of the mechanical philosophy of René Descartes, Pierre Gassendi, and others. The mechanical philosophy sought to describe physical phenomena in terms of intricate mechanisms. Decades after the birth of the mechanical philosophy, a systematic theory of mechanics began to flourish with Newtonian mechanics.

Analytical dynamics (historically referred to as *analytical mechanics*) is identified with a number of formulations of classical mechanics that arose after Isaac Newton published his *Philosophiae Naturalis Principia Mathematica* in 1687. The cornerstones of Newtonian mechanics are his laws of motion, which were applied to point masses. From a modern perspective, Newton's second law can be seen more generally as a conservation law, specifically, as a law of conservation of momentum. As such, Newtonian mechanics, and its extension to extended (rigid) bodies by Leonhard Euler, are based on two fundamental conservation principles: (1) the conservation of translational momentum and (2) the conservation of angular momentum.

Figure 1.1 Anthropomorphic forms adapted from plate no. 12 of *Bizzarie di Varie Figure*, 1624, by Giovanni Battista Braccelli, Livorno, Library of Congress Lessing J. Rosenwald Collection. Braccelli's Mannerist-styled work consists of a set of 50 etchings depicting anthropomorphic mechanical figures like this one. The human form has long been an inspiration in the study of mechanical systems.

In contrast to the conservation principles underlying the Newton-Euler mechanics, the analytical dynamics that followed were based on variational principles. The conservation of vector quantities like translational and angular momentum was replaced with principles rooted in the variation of scalar quantities like work and energy. An important precursor to the subsequent development of variational mechanics was the concept of the *vis viva* (living force) proposed by Gottfried Leibniz, a contemporary of Newton. The *vis viva* corresponds to our present notion of kinetic energy. In Leibniz's mechanics, momentum was replaced by kinetic energy and force was replaced by work of the force (Lanczos 1986).

Development of variational mechanics required a mathematical tool beyond the basic calculus of Newton and Leibniz. The calculus of variations, concerned with extremizing functionals (mappings of functions to scalar values), emerged as this tool. Johann Bernoulli was the first to exploit the calculus of variations in solving the brachistochrone curve problem. However, Euler is usually credited with the formal development of the calculus of variations in his 1744 *Methodus inveniendi*.

Around the time of Euler's *Methodus inveniendi*, in 1743, Jean Le Rond d'Alembert published his *Traité de dynamique*. This articulated d'Alembert's Principle of Virtual Work. Although Johann Bernoulli is credited with first proposing the Principle of Virtual Work for cases of static equilibrium, d'Alembert is credited with extending the principle to dynamic equilibrium by interpreting the acceleration terms in Newton's equation of motion as inertial forces. The principle is based on the notion of virtual displacement, defined as an infinitesimal change of the system's configuration coordinates while time is frozen. The displacement is virtual because no actual displacement occurs. Rather, a virtual displacement is used as a conceptual mechanism. D'Alembert's Principle can be viewed as a *zeroth*-order variational principle, as it is based on the *zeroth*-order derivative of displacement.

It should be noted that with the advent of analytical dynamics and variational principles like d'Alembert's, the concept of generalized coordinates became relevant. As the name implies, these coordinates are a generalization of the Cartesian coordinates used in Newton-Euler mechanics, whereby any consistent set of parameters that uniquely describe the configuration of the system can be chosen. The vector space defined by these generalized coordinates forms the configuration space of the system.

Following d'Alembert's Principle, the monumental work of Joseph-Louis Lagrange (born Giuseppe Lodovico Lagrangia) resulted in what is now known as Lagrangian mechanics. Together with contributions from Euler and William Rowan Hamilton, the Euler-Lagrange equations emerged as a logical consequence of Hamilton's Principle of Least Action. Disentangling the individual contributions of these three eminent mechanicians with respect to Lagrangian mechanics can be a bit tedious. Consequently, we will not proceed in chronological order when discussing this.

Hamilton's Principle of Least Action has been referred to as "the most direct and most natural transformation of d'Alembert's into a minimum principle" (Lanczos 1986, p. 111). The principle states that the path of a system in configuration space during a time interval is such that the action is stationary under all path variations. The action is defined as the integral of the Lagrangian over the time interval, where the Lagrangian is defined as the difference between the kinetic and potential energies of the system. The Euler-Lagrange equations of motion for the system emerge directly from this principle. Lagrange published his seminal work, *Mécanique analytique*, in 1788, formalizing these ideas into what is now known as Lagrangian mechanics.

Hamilton's reformulation of Lagrangian mechanics, published in 1833, constitutes what is now known as Hamiltonian mechanics. This reformulation involves a transformation of the Euler-Lagrange equations from a set of second-order differential equations in the generalized coordinates to a set of first-order differential equations in the generalized coordinates and generalized momenta. The Hamiltonian is defined as a new invariant corresponding to the total energy. Hamilton's equations written in terms of the Hamiltonian are known as Hamilton's canonical equations.

Up to this point, the historical flow of the variational principles of mechanics has followed the path from d'Alembert's Principle to Hamilton's Principle to the Euler-Lagrange equations and, finally, to Hamilton's canonical equations. It was mentioned that d'Alembert's Principle can be viewed as a *zeroth*-order variational principle. Holonomic constraints, which take the form of algebraic functions of the generalized coordinates and possibly time, can be inherently addressed by *zeroth*-order variational principles using the method of Lagrange multipliers.

Higher-order variational principles have also been proposed. One of the advantages of higher-order variational principles is the ability to address nonholonomic constraints, which take the form of algebraic functions of the higher-order derivatives of the

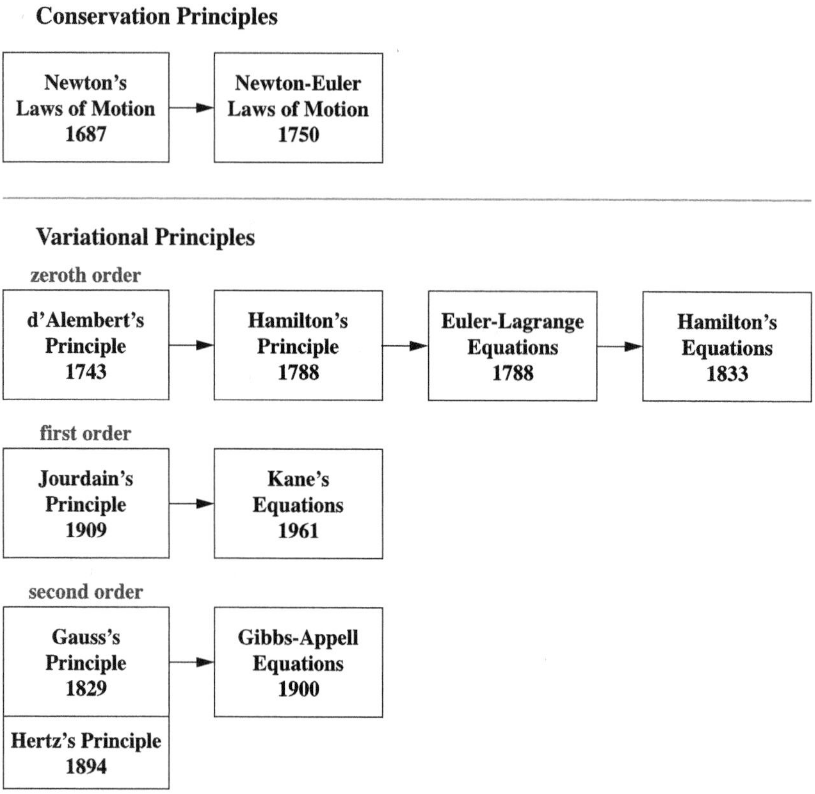

Figure 1.2 Historical progression of key principles of analytical dynamics. The variational principles, which form the basis of analytical dynamics, are comprised of a number of formulations that arose after Newton's mechanics.

generalized coordinates (generalized velocities and accelerations) and possibly time. The first-order variational principle, published by Philip E. B. Jourdain in 1909, is based on the notion of virtual velocity. Virtual power assumes the role that virtual work assumes in d'Alembert's Principle. Subsequent developments by Thomas Kane in 1961 essentially rediscovered Jourdain's Principle. Kane's approach extended Jourdain's approach to rigid bodies and introduced quasi-velocities to implicitly handle nonholonomic constraints.

Jourdain's Principle was influenced by Carl Friedrich Gauss's second-order variational principle, based on the notion of virtual accelerations, published in 1829. Gauss used the notion of virtual acceleration to establish a true minimum principle known as Gauss's Principle of Least Constraint, which minimizes the quadratic form known as the Gauss function. Heinrich Rudolf Hertz reinterpreted a special case of Gauss's Principle as the Principle of Least Curvature in 1894. Subsequent developments by Josiah Willard Gibbs (1879), Paul Appell (1900), and others led to the Gibbs-Appell equations, which have their lineage in Gauss's Principle. The Gibbs-Appell equations, like Kane's equations, make use of quasi-variables – in this case quasi-accelerations – to implicitly handle nonholonomic constraints. Although they derive from variational principles of different order, the Gibbs-Appell equations and Kane's equations can be viewed as identical. However, because the Gibbs-Appell equations are derived from a second-order variational principle, they can be stated in a concise form as the gradient of a scalar function, known as the Gibbs function, with respect to the quasi-accelerations. The Gibbs function can be related to the Gauss function.

Figure 1.2 summarizes the historical progression of some of the key principles of analytical dynamics that are covered in this book. I have by no means presented an exhaustive history, and the interested reader is referred to Dugas's excellent history of mechanics (Dugas 1988), which addresses ancient through modern developments in mechanics.

1.2 Devices That Illustrate Principles of Analytical Dynamics

Before jumping into a formal exposition of analytical dynamics, we will look at some motivating examples. Toys and other objects of amusement tend to make the most compelling examples. The reader is encouraged to refer back to these when encountering similar detailed technical examples presented in the subsequent chapters.

Figure 1.3 displays some devices composed of branching kinematic chains. A modified double pendulum (top), the Swinging Sticks Kinetic Energy Sculpture by BTS Trading GmbH, exhibits chaotic motion characterized by a sensitive dependence on initial conditions. As with all real-world mechanical systems, it dissipates energy; however, in this example, the double pendulum gives the illusion of perpetual motion through the use of electromagnetic coils mounted in the base and permanent neodymium rare-earth magnets mounted in the arms. The electromagnets measure the speed of the rotating arms and impart additional kinetic energy into the system. The gimbaled Super Precision Gyroscope distributed by Gyroscope.com (bottom left) consists of a high-speed rotor mounted, in this case, on a two-axis gimbal. The rotor on a high-speed gyroscope is precisely balanced and mounted on low-friction bearings to demonstrate the conservation of angular momentum (see the example in Section 5.3.5). Similar to the Swinging Sticks Kinetic Energy Sculpture is the Chaos Machine by Fat Brain Toys, a reconfigurable tree-structured mechanism (bottom right). This simple device illustrates the complex motion characteristic of multilink kinematic chains.

Figure 1.4 displays some devices that operate under holonomic constraints. These constraints are discussed in detail in later chapters. The holonomically constrained devices shown here all involve loop closures that can be represented as algebraic conditions on the configuration coordinates. The Falcon (top), by Novint Technologies Inc., is a haptic (force feedback) game controller based on the kinematics of the Delta parallel robot (Clavel 1991). Three translational degrees of freedom are provided by the kinematic structure of the Falcon, which uses four-bar parallelogram linkages in the three arms to maintain the fixed orientation of the end effector. Three actuators mounted in

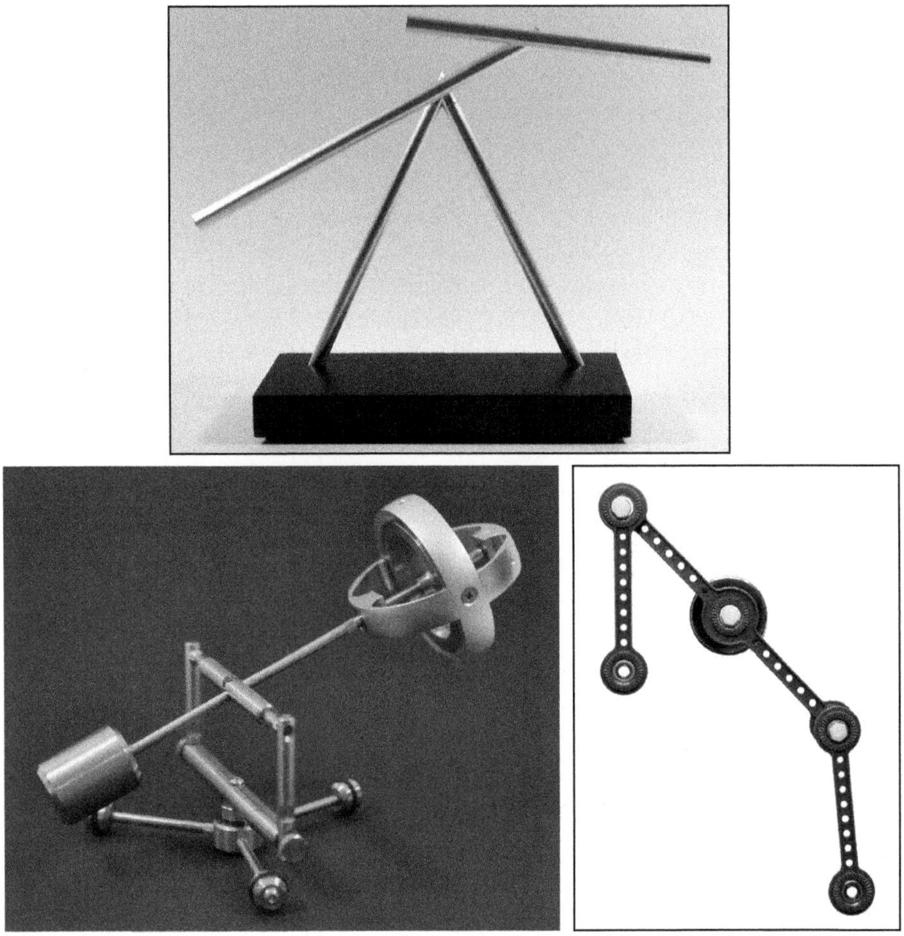

Figure 1.3 Some devices composed of branching kinematic chains. (Top) The Swinging Sticks Kinetic Energy Sculpture by BTS Trading GmbH. (Bottom Left) A gimbaled Super Precision Gyroscope distributed by Gyroscope.com. (Bottom Right) The Chaos Machine by Fat Brain Toys. All images © 2016 Vincent De Sapio.

the base allow the device to provide force feedback to the user. The Hoberman Sphere (bottom left) is a collapsing spherical structure made up of six rings, each of which comprises a series of connected four-bar parallelogram linkages that produce scissor-like motion. The overall structure has 1 degree of freedom and is able to radially expand and contract. A Stirling engine (bottom right), produced by Wiggers Stirling HeiBluft Modellbau, makes use of closed chain slider-crank and four-bar linkages to convert reciprocating piston motion into rotational motion of a flywheel.

Figure 1.5 displays some devices that operate under nonholonomic constraints. As with holonomic constraints, we discuss nonholonomic constraints in detail in later chapters. The nonholonomically constrained devices shown here all involve rolling/spinning constraints that can be represented as algebraic conditions on the configuration

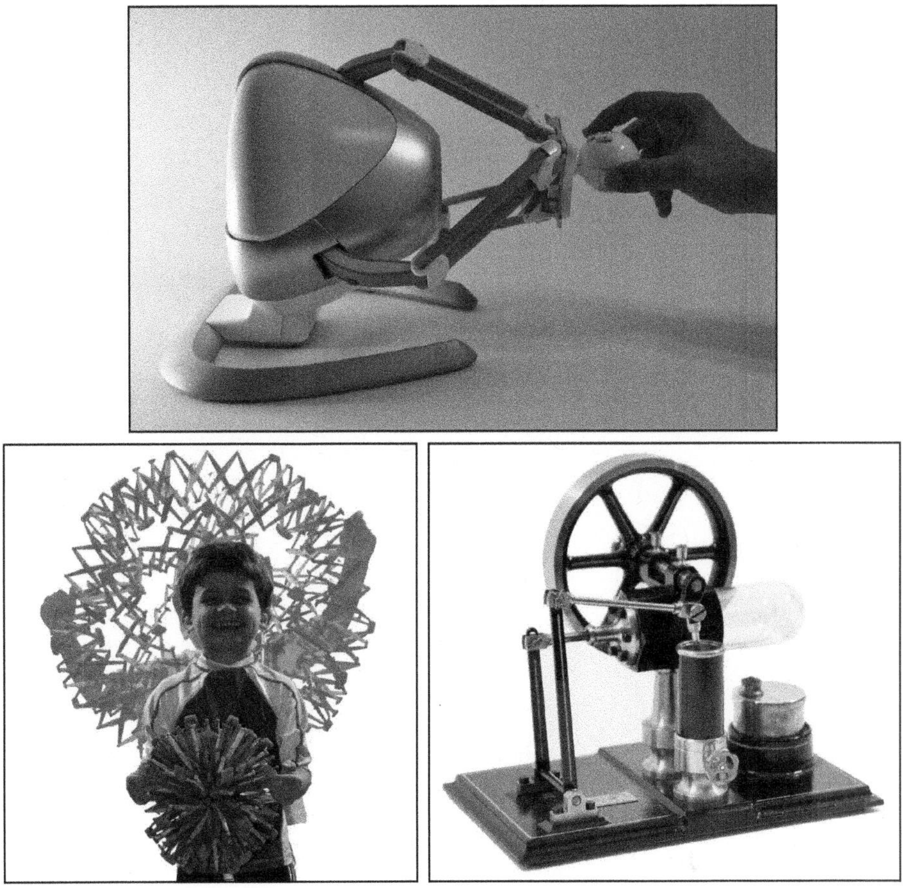

Figure 1.4 Some devices that operate under holonomic constraints. (Top) The Falcon haptic game controller by Novint Technologies Inc. (Bottom Left) The Hoberman Sphere © 1993 Charles Hoberman. (Bottom Right) A Stirling engine by Wiggers Stirling HeiBluft Modellbau. All images © 2016 Vincent De Sapio.

velocities. Specifically, the no-slip rolling/spinning condition requires *zero* velocity of the instantaneous material contact point of the device with the external surface. The Euler's Disk (top), distributed by Toysmith, illustrates the dynamics of a rolling/spinning disk on a flat surface. The complex motion of the disk produces intricate traces, which can be investigated in simulation (see an example in Section 6.2.5). Ollie (bottom left), by Sphero Inc., is a two-wheeled robot controlled by a smartphone app. The wheels are independently driven to allow steering and maneuvering. A gyroscope and accelerometer are incorporated into the control unit for inertial sensing. Sphero SPRK edition (bottom right), by Sphero Inc., is a spherical robot also controlled by a smartphone app. Internal drive wheels and a stabilizer provide forward propulsion and maneuvering. As with Ollie, a gyroscope and accelerometer are incorporated into the control unit for inertial sensing.

Figure 1.5 Some devices that operate under nonholonomic constraints. (Top) Euler's Disk by Toysmith. (Bottom Left) Ollie by Sphero Inc. (Bottom Right) Sphero SPRK edition by Sphero Inc. All images © 2016 Vincent De Sapio.

1.3 Scope of This Book

This book is intended to cover the foundations of analytical dynamics of discrete systems. By discrete systems, we mean systems made up of a discrete set of point masses or rigid bodies, as opposed to continuous systems. Continuous systems are the subject of continuum mechanics, which addresses infinite-dimensional deformable bodies, that is, bodies with infinite degrees of freedom. The study of flexible multibody systems, which include both rigid and deformable bodies, is an active area of research. Although such systems are outside the scope of the main body of this book, the appendix addresses some of the basics of the application of continuum mechanics to flexible multibody systems.

The material covered in this book is intended for an an intermediate to advanced graduate-level audience. As such, it is not intended as an introductory book on dynamics or classical mechanics. Some of the topics covered, particularly higher-order variational principles, are not commonly covered in engineering dynamics texts. Preceding the material on dynamics is a chapter providing a brief mathematical background in linear algebra, vectors and tensors, differential geometry, and optimization. A chapter on the kinematics of discrete systems directly precedes the chapters devoted to dynamics. Conservation principles are then addressed, followed by variational principles. Separate chapters cover zeroth-, first-, and second-order principles.

The variational principles are based on a configuration space description comprising generalized coordinates. After addressing these configuration space formulations, we present a chapter based on an alternate, task space formulation of dynamics using task coordinates. The following chapter presents applications to biomechanical systems. Such systems are an active area of study, and this chapter is intended to provide a brief introduction to system-level modeling of musculoskeletal and neuromuscular dynamics. The final chapter provides a short survey of some analytical dynamics software. This includes examples of general purpose mathematical software as well as dedicated multibody dynamics software.

The subsequent chapters provide example problems that are worked out in detail. The intention is to give the reader exposure to systematic approaches to applying the concepts presented to practical examples, thereby reinforcing abstract concepts with concrete applications.

2 Mathematical Preliminaries

This chapter is intended to present only a brief exposition of the mathematics helpful for understanding the material presented in the subsequent chapters. The reader is also encouraged to consult full texts in the mathematical subjects addressed here, if needed.

Linear algebra, including coverage of vectors and tensors, basic differential geometry, and static optimization, will be covered. Again, these sections are not complete expositions in these subjects but are intended to provide the reader with sufficient mathematical background.

2.1 Linear Systems

Linear systems are mathematical models of a system, physical or otherwise, that exclusively employ linear operators. While the models addressed in analytical dynamics are typically highly nonlinear, methods associated with linear systems are indispensable in the analysis of dynamical systems.

Linear algebra forms the fundamental theory for analyzing linear systems. Vector spaces and linear mappings between vector spaces are of fundamental importance. We will start by addressing the general properties of n-dimensional vector spaces, $\mathbb{R}^n$, and then address the more concrete case of vectors and tensors in $\mathbb{R}^3$.

2.1.1 Vector Spaces in $\mathbb{R}^n$

A vector or linear space is a collection of objects, namely, vectors, for which addition and multiplication by scalars is defined.

Inner Product

For our purposes, all vector spaces that we will be concerned with possess an inner product and therefore are also termed inner product spaces.

DEFINITION 2.1 The ***inner product*** $\langle \square, \square \rangle$ satisfies the following properties:

1. $\langle \boldsymbol{u} + \boldsymbol{v}, \boldsymbol{w} \rangle = \langle \boldsymbol{u}, \boldsymbol{w} \rangle + \langle \boldsymbol{v}, \boldsymbol{w} \rangle$.
2. $\langle \alpha \boldsymbol{u}, \boldsymbol{v} \rangle = \alpha \langle \boldsymbol{u}; \boldsymbol{v} \rangle$.
3. $\langle \boldsymbol{u}, \boldsymbol{v} \rangle = \langle \boldsymbol{v}, \boldsymbol{u} \rangle$.
4. $\langle \boldsymbol{u}, \boldsymbol{u} \rangle \geq 0$.

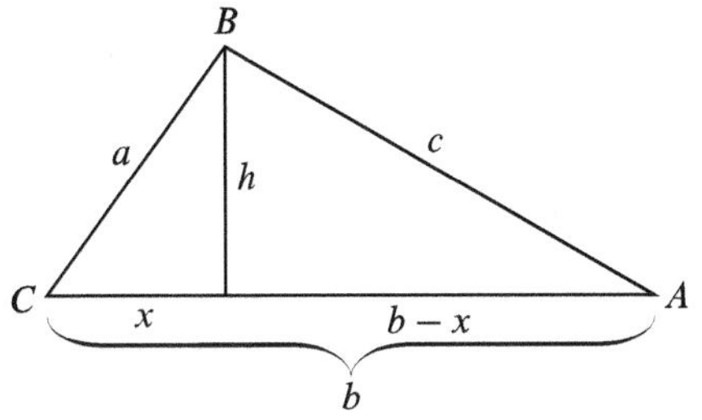

Figure 2.1 Triangle ABC demonstrating the law of cosines, $c^2 = a^2 + b^2 - 2ab\cos C$.

The inner product is a generalization of the dot product and for Euclidean space is equal to the dot product. That is,

$$\langle \boldsymbol{u}, \boldsymbol{v} \rangle = \boldsymbol{u} \cdot \boldsymbol{v} = \sum_{i=1}^{n} u_i v_i. \tag{2.1}$$

Next, we note a basic property of triangles, the *law of cosines*:

THEOREM 2.2 *Given a triangle, ABC, the following relationship holds:*

$$c^2 = a^2 + b^2 - 2ab\cos C. \tag{2.2}$$

This is known as the ***law of cosines****.*

Proof The triangle ABC can be split into two right triangles (Figure 2.1). We thus have

$$a^2 = x^2 + h^2 \tag{2.3}$$

and

$$c^2 = (b-x)^2 + h^2 = b^2 + x^2 - 2bx + h^2. \tag{2.4}$$

Substituting $h^2 = a^2 - x^2$ into (2.4), we have

$$c^2 = b^2 + x^2 - 2bx + a^2 - x^2 = a^2 + b^2 - 2bx. \tag{2.5}$$

Noting that $x = a\cos C$, we obtain the result

$$c^2 = a^2 + b^2 - 2ab\cos C. \tag{2.6}$$

□

From the *law of cosines* we can deduce the following geometric property of the inner product.

THEOREM 2.3 *The inner product of a vector* $\boldsymbol{a}$ *with a vector* $\boldsymbol{b}$ *is given by*

$$\langle \boldsymbol{a}, \boldsymbol{b} \rangle = \|\boldsymbol{a}\|\,\|\boldsymbol{b}\|\cos\theta. \tag{2.7}$$

Proof By the *law of cosines*, we have

$$\|\boldsymbol{b} - \boldsymbol{a}\|^2 = \|\boldsymbol{a}\|^2 + \|\boldsymbol{b}\|^2 - 2\,\|\boldsymbol{a}\|\,\|\boldsymbol{b}\|\cos\theta, \tag{2.8}$$

or

$$\langle \boldsymbol{b}-\boldsymbol{a}, \boldsymbol{b}-\boldsymbol{a}\rangle = \langle \boldsymbol{a}, \boldsymbol{a}\rangle + \langle \boldsymbol{b}, \boldsymbol{b}\rangle - 2\,\|\boldsymbol{a}\|\,\|\boldsymbol{b}\|\cos\theta. \tag{2.9}$$

Noting that

$$\langle \boldsymbol{b}-\boldsymbol{a}, \boldsymbol{b}-\boldsymbol{a}\rangle = \langle \boldsymbol{a}, \boldsymbol{a}\rangle + \langle \boldsymbol{b}, \boldsymbol{b}\rangle - 2\,\langle \boldsymbol{a}, \boldsymbol{b}\rangle\,, \tag{2.10}$$

we can equate (2.9) and (2.10). Eliminating terms yields the result

$$\langle \boldsymbol{a}, \boldsymbol{b}\rangle = \|\boldsymbol{a}\|\,\|\boldsymbol{b}\|\cos\theta. \tag{2.11}$$

□

Basis of a Vector Space

To describe a vector space, it is convenient to define a basis which can represent any vector in the vector space.

DEFINITION 2.4 A set of vectors, $\{\boldsymbol{v}_1, \ldots, \boldsymbol{v}_n\}$, forms a ***basis*** of a vector space, V, if the vectors span V and are linearly independent.

THEOREM 2.5 *Let a vector space, V, have a basis, $\mathcal{B}$, consisting of a set of vectors, $\{\boldsymbol{v}_1, \ldots, \boldsymbol{v}_n\}$. Any vector $\boldsymbol{x} \in V$ may be expressed as a* unique *linear combination of the basis vectors*

$$\boldsymbol{x} = c_1\boldsymbol{v}_1 + c_2\boldsymbol{v}_2 + \cdots + c_n\boldsymbol{v}_n. \tag{2.12}$$

Proof Since $\{\boldsymbol{v}_1, \ldots, \boldsymbol{v}_n\}$ is a basis of V it spans V. Consequently, there is at least one linear combination of the basis vectors that expresses $\boldsymbol{x}$, namely, $c_1\boldsymbol{v}_1 + c_2\boldsymbol{v}_2 + \cdots + c_n\boldsymbol{v}_n$. Let us assume that there is another linear combination of the same basis vectors,

$$\boldsymbol{x} = d_1\boldsymbol{v}_1 + d_2\boldsymbol{v}_2 + \cdots + d_n\boldsymbol{v}_n. \tag{2.13}$$

Then,

$$\boldsymbol{x} - \boldsymbol{x} = \boldsymbol{0} = (c_1 - d_1)\boldsymbol{v}_1 + (c_2 - d_2)\boldsymbol{v}_2 + \cdots + (c_n - d_n)\boldsymbol{v}_n. \tag{2.14}$$

Since $\{\boldsymbol{v}_1, \ldots, \boldsymbol{v}_n\}$ is a basis, it is linearly independent. As such, all of the coefficients, $(c_i - d_i)$, must be *zero*. Therefore,

$$c_1 = d_1,\; c_2 = d_2,\; \ldots,\; c_n = d_n, \tag{2.15}$$

and the set of coefficients is unique. □

For any vector, $\boldsymbol{x}$, the set of coefficients, $\{c_1, \ldots, c_2\}$, associated with the unique linear combination of basis vectors are referred to as the coordinates (or components) of $\boldsymbol{x}$ with respect to the given basis. For a basis, $\mathcal{B}$, consisting of vectors, $\{\boldsymbol{v}_1, \ldots, \boldsymbol{v}_n\}$. The coordinate vector of

$$\boldsymbol{x} = c_1\boldsymbol{v}_1 + c_2\boldsymbol{v}_2 + \cdots + c_n\boldsymbol{v}_n, \tag{2.16}$$

with respect to $\mathcal{B}$, is expressed as

$$ {}^{\mathcal{B}}\boldsymbol{x} = \begin{pmatrix} c_1 \\ \vdots \\ c_n \end{pmatrix}. \tag{2.17} $$

So, we have the following relationships:

$$ \boldsymbol{x} = \begin{pmatrix} \uparrow & \uparrow & \uparrow \\ \boldsymbol{v}_1 & \dots & \boldsymbol{v}_n \\ \downarrow & \downarrow & \downarrow \end{pmatrix} \begin{pmatrix} c_1 \\ \vdots \\ c_n \end{pmatrix} = \boldsymbol{S}\,{}^{\mathcal{B}}\boldsymbol{x} \tag{2.18} $$

and

$$ {}^{\mathcal{B}}\boldsymbol{x} = \boldsymbol{S}^{-1}\boldsymbol{x}, \tag{2.19} $$

where

$$ \boldsymbol{S} = \begin{pmatrix} \uparrow & \uparrow & \uparrow \\ \boldsymbol{v}_1 & \dots & \boldsymbol{v}_n \\ \downarrow & \downarrow & \downarrow \end{pmatrix}. \tag{2.20} $$

Example: Given a basis, $\mathcal{B}$, in $\mathbb{R}^3$ consisting of $\{\boldsymbol{v}_1, \boldsymbol{v}_2, \boldsymbol{v}_3\}$, where

$$ \boldsymbol{v}_1 = \begin{pmatrix} 2 \\ 1 \\ 1/2 \end{pmatrix}, \quad \boldsymbol{v}_2 = \begin{pmatrix} 1 \\ 2 \\ 1/2 \end{pmatrix}, \quad \boldsymbol{v}_3 = \begin{pmatrix} 1 \\ 1/2 \\ 2 \end{pmatrix}, \tag{2.21} $$

$\boldsymbol{S}$ is constructed as

$$ \boldsymbol{S} = \begin{pmatrix} \uparrow & \uparrow & \uparrow \\ \boldsymbol{v}_1 & \boldsymbol{v}_2 & \boldsymbol{v}_3 \\ \downarrow & \downarrow & \downarrow \end{pmatrix} = \begin{pmatrix} 2 & 1 & 1 \\ 1 & 2 & 1/2 \\ 1/2 & 1/2 & 2 \end{pmatrix}. \tag{2.22} $$

Given the vector $\boldsymbol{x}$ expressed as the linear combination (see Figure 2.2)

$$ \boldsymbol{x} = 2\boldsymbol{v}_1 + 3\boldsymbol{v}_2 + 1\boldsymbol{v}_3, \tag{2.23} $$

we have the following coordinate vector, with respect to $\mathcal{B}$:

$$ {}^{\mathcal{B}}\boldsymbol{x} = \begin{pmatrix} 2 \\ 3 \\ 1 \end{pmatrix}. \tag{2.24} $$

We compute $\boldsymbol{x}$ with respect to the standard basis, as

$$ \boldsymbol{x} = \boldsymbol{S}\,{}^{\mathcal{B}}\boldsymbol{x} = \begin{pmatrix} 2 & 1 & 1 \\ 1 & 2 & 1/2 \\ 1/2 & 1/2 & 2 \end{pmatrix} \begin{pmatrix} 2 \\ 3 \\ 1 \end{pmatrix} = \begin{pmatrix} 8 \\ 8\tfrac{1}{2} \\ 4\tfrac{1}{2} \end{pmatrix}. \tag{2.25} $$

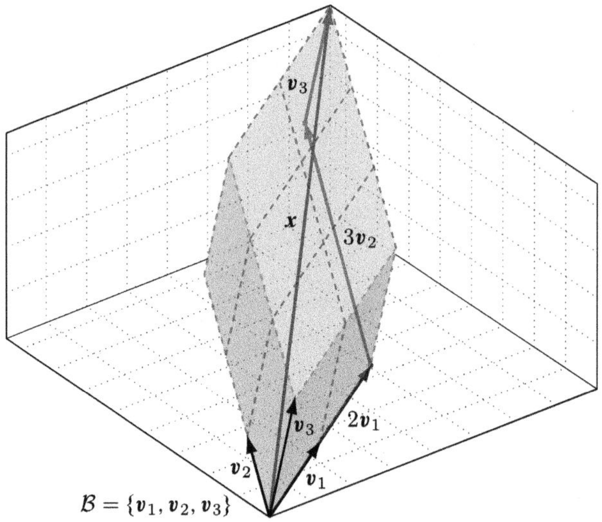

Figure 2.2 Vector coordinates with respect to a basis, $\mathcal{B}$, illustrated as a linear combination of basis vectors, $\{\boldsymbol{v}_1, \boldsymbol{v}_2, \boldsymbol{v}_3\}$.

Conversely, having started with $\boldsymbol{x}$, we would compute ${}^{\mathcal{B}}\boldsymbol{x}$ as

$$ {}^{\mathcal{B}}\boldsymbol{x} = \boldsymbol{S}^{-1}\boldsymbol{x} = \begin{pmatrix} 2 & 1 & 1 \\ 1 & 2 & 1/2 \\ 1/2 & 1/2 & 2 \end{pmatrix}^{-1} \begin{pmatrix} 8 \\ 8^1\!/_2 \\ 4^1\!/_2 \end{pmatrix} = \begin{pmatrix} 2 \\ 3 \\ 1 \end{pmatrix}. \tag{2.26} $$

Given a linear transformation

$$ \boldsymbol{\tau}(\boldsymbol{x}) = \boldsymbol{A}\boldsymbol{x}, \tag{2.27} $$

from $\mathbb{R}^n$ to $\mathbb{R}^n$ we may be interested in expressing the corresponding transformation with respect to some basis, $\mathcal{B}$:

$$ {}^{\mathcal{B}}\boldsymbol{\tau}(\boldsymbol{x}) = \boldsymbol{B}\,{}^{\mathcal{B}}\boldsymbol{x}. \tag{2.28} $$

The matrix $\boldsymbol{B}$ is determined in the following manner. First, let a vector $\boldsymbol{x}$ be expressed as a linear combination of basis vectors $\{\boldsymbol{v}_1, \ldots, \boldsymbol{v}_n\}$,

$$ \boldsymbol{x} = c_1\boldsymbol{v}_1 + \cdots + c_n\boldsymbol{v}_n. \tag{2.29} $$

The corresponding coordinate vector with respect to $\mathcal{B}$ is

$$ {}^{\mathcal{B}}\boldsymbol{x} = \begin{pmatrix} c_1 \\ \vdots \\ c_n \end{pmatrix}. \tag{2.30} $$

By linearity of $\boldsymbol{\tau}(\boldsymbol{x})$, we have

$$ \boldsymbol{\tau}(\boldsymbol{x}) = \boldsymbol{\tau}(c_1\boldsymbol{v}_1 + \cdots + c_n\boldsymbol{v}_n) = c_1\boldsymbol{\tau}(\boldsymbol{v}_1) + \cdots + c_n\boldsymbol{\tau}(\boldsymbol{v}_n). \tag{2.31} $$

Now, $\{\boldsymbol{\tau}(\boldsymbol{v}_1), \ldots, \boldsymbol{\tau}(\boldsymbol{v}_n)\}$ can be expressed as linear combinations of $\{\boldsymbol{v}_1, \ldots, \boldsymbol{v}_n\}$:

$$\begin{aligned} \boldsymbol{\tau}(\boldsymbol{v}_1) &= B_{11}\boldsymbol{v}_1 + \cdots + B_{n1}\boldsymbol{v}_n, \\ &\vdots \\ \boldsymbol{\tau}(\boldsymbol{v}_n) &= B_{1n}\boldsymbol{v}_1 + \cdots + B_{nn}\boldsymbol{v}_n, \end{aligned} \tag{2.32}$$

where the coordinate vectors, $\{{}^{\mathcal{B}}\boldsymbol{\tau}(\boldsymbol{v}_1), \ldots, {}^{\mathcal{B}}\boldsymbol{\tau}(\boldsymbol{v}_n)\}$, are

$${}^{\mathcal{B}}\boldsymbol{\tau}(\boldsymbol{v}_1) = \begin{pmatrix} B_{11} \\ \vdots \\ B_{n1} \end{pmatrix}, \ldots, {}^{\mathcal{B}}\boldsymbol{\tau}(\boldsymbol{v}_n) = \begin{pmatrix} B_{1n} \\ \vdots \\ B_{nn} \end{pmatrix}. \tag{2.33}$$

We now have the following expression for $\boldsymbol{\tau}(\boldsymbol{x})$:

$$\boldsymbol{\tau}(\boldsymbol{x}) = c_1(B_{11}\boldsymbol{v}_1 + \cdots + B_{n1}\boldsymbol{v}_n) + \cdots + c_n(B_{1n}\boldsymbol{v}_1 + \cdots + B_{nn}\boldsymbol{v}_n) \tag{2.34}$$

or

$$\boldsymbol{\tau}(\boldsymbol{x}) = (c_1B_{11} + \cdots + c_nB_{1n})\boldsymbol{v}_1 + \cdots + (c_1B_{n1} + \cdots + c_nB_{nn})\boldsymbol{v}_n. \tag{2.35}$$

So, the coordinate vector, ${}^{\mathcal{B}}\boldsymbol{\tau}(\boldsymbol{x})$, is

$${}^{\mathcal{B}}\boldsymbol{\tau}(\boldsymbol{x}) = \begin{pmatrix} c_1B_{11} + \cdots + c_nB_{1n} \\ \vdots \\ c_1B_{n1} + \cdots + c_nB_{nn} \end{pmatrix} = \begin{pmatrix} B_{11} & \ldots & B_{1n} \\ \vdots & \ddots & \vdots \\ B_{n1} & \ldots & B_{nn} \end{pmatrix} \begin{pmatrix} c_1 \\ \vdots \\ c_n \end{pmatrix} = \boldsymbol{B}\,{}^{\mathcal{B}}\boldsymbol{x}, \tag{2.36}$$

and

$$\boldsymbol{B} = \begin{pmatrix} B_{11} & \ldots & B_{1n} \\ \vdots & \ddots & \vdots \\ B_{n1} & \ldots & B_{nn} \end{pmatrix} = \begin{pmatrix} \uparrow & \uparrow & \uparrow \\ {}^{\mathcal{B}}\boldsymbol{\tau}(\boldsymbol{v}_1) & \ldots & {}^{\mathcal{B}}\boldsymbol{\tau}(\boldsymbol{v}_n) \\ \downarrow & \downarrow & \downarrow \end{pmatrix}. \tag{2.37}$$

To relate $\boldsymbol{A}$ and $\boldsymbol{B}$, we first note that

$$\boldsymbol{S}\,{}^{\mathcal{B}}\boldsymbol{\tau}(\boldsymbol{x}) = \boldsymbol{S}\boldsymbol{B}\,{}^{\mathcal{B}}\boldsymbol{x} \tag{2.38}$$

and

$${}^{\mathcal{B}}\boldsymbol{x} = \boldsymbol{S}^{-1}\boldsymbol{x}. \tag{2.39}$$

So,

$$\boldsymbol{S}\,{}^{\mathcal{B}}\boldsymbol{\tau}(\boldsymbol{x}) = \boldsymbol{S}\boldsymbol{B}\boldsymbol{S}^{-1}\boldsymbol{x}. \tag{2.40}$$

Finally,

$$\boldsymbol{\tau}(\boldsymbol{x}) = \boldsymbol{S}\,{}^{\mathcal{B}}\boldsymbol{\tau}(\boldsymbol{x}) = \boldsymbol{S}\boldsymbol{B}\boldsymbol{S}^{-1}\boldsymbol{x} = \boldsymbol{A}\boldsymbol{x}. \tag{2.41}$$

So,

$$\boldsymbol{A} = \boldsymbol{S}\boldsymbol{B}\boldsymbol{S}^{-1} \quad \text{and,} \quad \boldsymbol{B} = \boldsymbol{S}^{-1}\boldsymbol{A}\boldsymbol{S}. \tag{2.42}$$

Algorithm 1 Gram-Schmidt process

The Gram-Schmidt process allows us to construct an orthonormal basis from an arbitrary basis. Given a subspace, V, of $\mathbb{R}^n$ with some basis, $\{\boldsymbol{v}_1, \ldots, \boldsymbol{v}_m\}$, we can construct an orthonormal basis, $\{\boldsymbol{u}_1, \ldots, \boldsymbol{u}_m\}$, for V as follows:

1: $\boldsymbol{u}_1 = \frac{1}{\|\boldsymbol{v}_1\|}\boldsymbol{v}_1$ {Establish $\boldsymbol{u}_1$ by normalizing $\boldsymbol{v}_1$}
2: **for** $i = 2$ to m **do**
3: $V_{i-1} = \text{span}(\boldsymbol{u}_1, \ldots, \boldsymbol{u}_{i-1})$
4: $\text{proj}_{V_{i-1}}\boldsymbol{v}_i = (\boldsymbol{u}_1 \cdot \boldsymbol{v}_i)\boldsymbol{u}_1 + \cdots + (\boldsymbol{u}_{i-1} \cdot \boldsymbol{v}_i)\boldsymbol{u}_{i-1}$
5: $\boldsymbol{u}_i = \frac{1}{\left\|\boldsymbol{v}_i - \text{proj}_{V_{i-1}}\boldsymbol{v}_i\right\|}(\boldsymbol{v}_i - \text{proj}_{V_{i-1}}\boldsymbol{v}_i)$
6: **end for**

DEFINITION 2.6 The ***orthogonal complement***, $V^\perp$, of a vector space, V, is the set of those vectors, $\boldsymbol{x}$, in $\mathbb{R}^n$ that are orthogonal to all vectors in V:

$$V^\perp = \{\boldsymbol{x}|\boldsymbol{x} \in \mathbb{R}^n, \boldsymbol{v} \cdot \boldsymbol{x} = 0, \forall \boldsymbol{v} \in V\}. \tag{2.43}$$

Given a basis, $\{\boldsymbol{v}_1, \boldsymbol{v}_2, \ldots, \boldsymbol{v}_m\}$, $V^\perp$ can be equivalently defined as

$$V^\perp = \{\boldsymbol{x}|\boldsymbol{x} \in \mathbb{R}^n, \boldsymbol{v}_i \cdot \boldsymbol{x} = 0, \text{ for } i = 1, \ldots, m\}. \tag{2.44}$$

DEFINITION 2.7 Consider a subspace, V, of $\mathbb{R}^n$ with orthonormal basis, $\{\boldsymbol{u}_1, \ldots, \boldsymbol{u}_m\}$. For any vector, $\boldsymbol{x}$, in $\mathbb{R}^n$, there is a unique vector, $\boldsymbol{w}$, in V such that $\boldsymbol{x} - \boldsymbol{w}$ is in $V^\perp$. The vector, $\boldsymbol{w}$, is called the ***orthogonal projection*** of $\boldsymbol{x}$ onto V and is given by

$$\text{proj}_V\boldsymbol{x} = (\boldsymbol{u}_1 \cdot \boldsymbol{x})\boldsymbol{u}_1 + \cdots + (\boldsymbol{u}_m \cdot \boldsymbol{x})\boldsymbol{u}_m. \tag{2.45}$$

Given an orthonormal basis, $\{\boldsymbol{u}_1, \ldots, \boldsymbol{u}_n\}$, of $\mathbb{R}^n$,

$$\boldsymbol{x} = (\boldsymbol{u}_1 \cdot \boldsymbol{x})\boldsymbol{u}_1 + \cdots + (\boldsymbol{u}_n \cdot \boldsymbol{x})\boldsymbol{u}_n, \tag{2.46}$$

for all $\boldsymbol{x}$ in $\mathbb{R}^n$. The Gram-Schmidt process described in Algorithm 1 employs orthogonal projection to generate an orthonormal basis from an arbitrary basis.

Least Squares

We consider the over determined system

$$\boldsymbol{A}\boldsymbol{x} = \boldsymbol{y}, \tag{2.47}$$

where $\boldsymbol{A} \in \mathbb{R}^{m\times n}$ is full rank and $m > n$. In this case the vector, $\boldsymbol{y}$, is not generally in the image of $\boldsymbol{A}$, so the equality in (2.47) does not generally hold. However, if we define the vector space, V,

$$V = \text{im}(\boldsymbol{A}), \tag{2.48}$$

the projection of $\boldsymbol{y}$ onto V is in the image of $\boldsymbol{A}$. So,

$$\boldsymbol{A}\boldsymbol{x}_* = \text{proj}_V\boldsymbol{y} \tag{2.49}$$

for some vector $\boldsymbol{x}_*$. Furthermore, the residual, or difference between the observed data, $\boldsymbol{y}$, and the estimate, $\text{proj}_V \boldsymbol{y}$, is orthogonal to V:

$$(\boldsymbol{y} - \text{proj}_V \boldsymbol{y}) \perp V \tag{2.50}$$

or

$$(\boldsymbol{y} - \boldsymbol{A}\boldsymbol{x}_*) \perp \text{im}(\boldsymbol{A}). \tag{2.51}$$

Noting that

$$\ker(\boldsymbol{A}^T) = \text{im}(\boldsymbol{A})^{\perp}, \tag{2.52}$$

we have

$$(\boldsymbol{y} - \boldsymbol{A}\boldsymbol{x}_*) \in \ker(\boldsymbol{A}^T). \tag{2.53}$$

So,

$$\boldsymbol{A}^T(\boldsymbol{y} - \boldsymbol{A}\boldsymbol{x}_*) = \boldsymbol{0}, \tag{2.54}$$

and

$$\boldsymbol{A}^T\boldsymbol{A}\boldsymbol{x}_* = \boldsymbol{A}^T\boldsymbol{y}. \tag{2.55}$$

Thus our least squares solution that minimizes the residual between the observed data, $\boldsymbol{y}$, and the estimate, $\text{proj}_V \boldsymbol{y}$, is

$$\boldsymbol{x}_* = (\boldsymbol{A}^T\boldsymbol{A})^{-1}\boldsymbol{A}^T\boldsymbol{y}. \tag{2.56}$$

We define the least squares left inverse of $\boldsymbol{A}$ as

$$\boldsymbol{A}^I \triangleq (\boldsymbol{A}^T\boldsymbol{A})^{-1}\boldsymbol{A}^T. \tag{2.57}$$

Least Norm

We now consider the under determined system

$$\boldsymbol{A}\boldsymbol{x} = \boldsymbol{y}, \tag{2.58}$$

where $\boldsymbol{A} \in \mathbb{R}^{m\times n}$ is full rank and $m < n$. In this case, there is an infinite number of solutions, $\boldsymbol{x}$, for a given $\boldsymbol{y}$ in (2.58). Our solution is

$$\boldsymbol{x} = \boldsymbol{x}_P + \boldsymbol{x}_N, \tag{2.59}$$

where $\boldsymbol{x}_P$ is a particular solution and $\boldsymbol{x}_N \in \ker(\boldsymbol{A})$ are the null space solutions associated with $\boldsymbol{y} = \boldsymbol{0}$. Since there is an infinite number of particular solutions to (2.58), we can find a specific one that minimizes some quantity. For example, we may be interested in the solution that minimizes the norm of $\boldsymbol{x}$. In this case the particular solution will be orthogonal to all null space solutions,

$$\boldsymbol{x}_P \perp \boldsymbol{x}_N. \tag{2.60}$$

We note that

$$\operatorname{im}(\boldsymbol{A}^T) = \ker(\boldsymbol{A})^{\perp}. \tag{2.61}$$

So,

$$\boldsymbol{x}_P \in \operatorname{im}(\boldsymbol{A}^T), \qquad \text{or} \qquad \boldsymbol{x}_P = \boldsymbol{A}^T \boldsymbol{v}, \tag{2.62}$$

for some vector, $\boldsymbol{v}$. Thus,

$$\boldsymbol{A}\boldsymbol{x} = \boldsymbol{A}\boldsymbol{x}_P = \boldsymbol{A}\boldsymbol{A}^T \boldsymbol{v} = \boldsymbol{y}, \tag{2.63}$$

and

$$\boldsymbol{v} = (\boldsymbol{A}\boldsymbol{A}^T)^{-1}\boldsymbol{y}. \tag{2.64}$$

Our solution that minimizes the norm of $\boldsymbol{x}$ is then

$$\boldsymbol{x} = \boldsymbol{x}_P = \boldsymbol{A}^T(\boldsymbol{A}\boldsymbol{A}^T)^{-1}\boldsymbol{y}. \tag{2.65}$$

We define the least norm right inverse of $\boldsymbol{A}$ as

$$\boldsymbol{A}^{\dagger} \triangleq \boldsymbol{A}^T(\boldsymbol{A}\boldsymbol{A}^T)^{-1}. \tag{2.66}$$

All solutions to $\boldsymbol{A}\boldsymbol{x} = \boldsymbol{y}$ are given by

$$\begin{aligned} \boldsymbol{x} &= \boldsymbol{A}^{\dagger}\boldsymbol{y} + \boldsymbol{x}_N \\ \forall \boldsymbol{x}_N &\in \ker(\boldsymbol{A}). \end{aligned} \tag{2.67}$$

See Figure 2.3 for a graphical depiction of the space of solutions.

Eigenvalue Problem

DEFINITION 2.8 Consider an eigenvalue, λ, of an $n \times n$ matrix, $\boldsymbol{A}$. The kernel of $\lambda \mathbf{1}_n - \boldsymbol{A}$ is called the ***eigenspace*** of $\boldsymbol{A}$ associated with λ and is denoted by E_λ:

$$E_\lambda = \ker(\lambda \mathbf{1}_n - \boldsymbol{A}). \tag{2.68}$$

DEFINITION 2.9 Consider an $n \times n$ matrix, $\boldsymbol{A}$. A basis of $\mathbb{R}^n$ consisting of eigenvectors of $\boldsymbol{A}$ is called an ***eigenbasis*** for $\boldsymbol{A}$. The eigenbasis spans the eigenspace of $\boldsymbol{A}$.

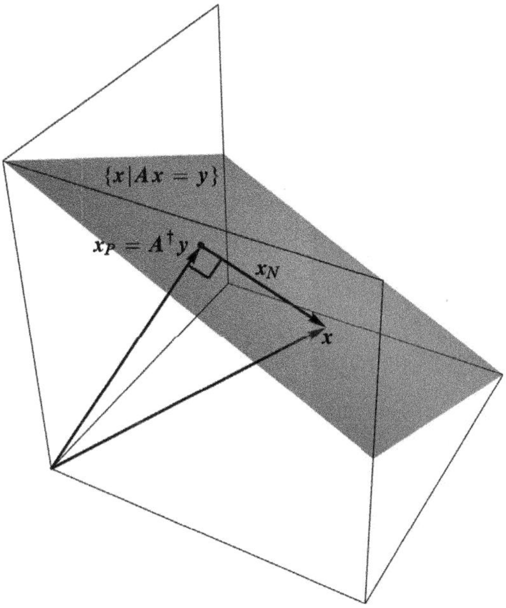

Figure 2.3 Least norm solution, $\boldsymbol{x}_P$, and space of all solutions, $\boldsymbol{x}$. All solutions are separated by a null space vector, $\boldsymbol{x}_N \in \ker(\boldsymbol{A})$.

Given an eigenbasis, $\mathcal{E} = \{\boldsymbol{v}_1, \ldots, \boldsymbol{v}_n\}$, of a matrix, $\boldsymbol{A}$, we have the following transformation with respect to $\mathcal{E}$:

$$\boldsymbol{D} = \boldsymbol{S}^{-1}\boldsymbol{A}\boldsymbol{S} = \begin{pmatrix} \uparrow & \uparrow & \uparrow \\ {}^{\mathcal{B}}\boldsymbol{\tau}(\boldsymbol{v}_1) & \cdots & {}^{\mathcal{B}}\boldsymbol{\tau}(\boldsymbol{v}_n) \\ \downarrow & \downarrow & \downarrow \end{pmatrix} = \begin{pmatrix} \lambda_1 & 0 & \cdots & 0 \\ 0 & \lambda_2 & & 0 \\ \vdots & & \ddots & \vdots \\ 0 & 0 & \cdots & \lambda_n \end{pmatrix}, \tag{2.69}$$

where $\boldsymbol{D}$ is a diagonal matrix with the eigenvalues, $\{\lambda_1, \ldots, \lambda_n\}$, as the diagonal entries. The matrix $\boldsymbol{S}$ is a matrix with the eigenvectors, $\{\boldsymbol{v}_1, \ldots, \boldsymbol{v}_n\}$, as the columns,

$$\boldsymbol{S} = \begin{pmatrix} \uparrow & \uparrow & \uparrow \\ \boldsymbol{v}_1 & \cdots & \boldsymbol{v}_n \\ \downarrow & \downarrow & \downarrow \end{pmatrix}. \tag{2.70}$$

DEFINITION 2.10 An $n \times n$ matrix, $\boldsymbol{A}$, is ***positive definite*** if all of its eigenvalues are positive. That is, $\lambda_i > 0$, for $i = 1, \ldots, n$.

2.1.2 Vectors in $\mathbb{R}^3$

Given some coordinate frame, $\mathcal{A}$, associated with an orthonormal basis, $\hat{\boldsymbol{e}}_{\mathcal{A}_i}$, we can define a vector, $\boldsymbol{v}$, as a linear combination of basis vectors:

$$\boldsymbol{v} = \sum_{i=1}^{3} v_i \hat{\boldsymbol{e}}_{\mathcal{A}_i} = v_1 \hat{\boldsymbol{e}}_{\mathcal{A}_1} + v_2 \hat{\boldsymbol{e}}_{\mathcal{A}_2} + v_3 \hat{\boldsymbol{e}}_{\mathcal{A}_3}. \tag{2.71}$$

The components of this vector, expressed in $\mathcal{A}$, can be represented as the coordinate vector

$$
\begin{aligned}
{}^{A}\boldsymbol{v} &= \sum_{i=1}^{3} v_i {}^{A}\hat{\boldsymbol{e}}_{A_i} = v_1 {}^{A}\hat{\boldsymbol{e}}_{A_1} + v_2 {}^{A}\hat{\boldsymbol{e}}_{A_2} + v_3 {}^{A}\hat{\boldsymbol{e}}_{A_3} \\
&= v_1 \begin{pmatrix} 1 \\ 0 \\ 0 \end{pmatrix} + v_2 \begin{pmatrix} 0 \\ 1 \\ 0 \end{pmatrix} + v_3 \begin{pmatrix} 0 \\ 0 \\ 1 \end{pmatrix} = \begin{pmatrix} v_1 \\ v_2 \\ v_3 \end{pmatrix}.
\end{aligned} \tag{2.72}
$$

Given another coordinate frame, $\mathcal{B}$, associated with the orthonormal basis vectors, $\{\hat{\boldsymbol{e}}_{B_1}, \hat{\boldsymbol{e}}_{B_2}, \hat{\boldsymbol{e}}_{B_3}\}$, we have the following relationship:

$$
\hat{\boldsymbol{e}}_{A_i} = \sum_{j=1}^{3} (\hat{\boldsymbol{e}}_{A_i} \cdot \hat{\boldsymbol{e}}_{B_j}) \hat{\boldsymbol{e}}_{B_j}. \tag{2.73}
$$

Vector, $\boldsymbol{v}$, can now be expressed as a linear combination of $\{\hat{\boldsymbol{e}}_{B_1}, \hat{\boldsymbol{e}}_{B_2}, \hat{\boldsymbol{e}}_{B_3}\}$:

$$
\boldsymbol{v} = \sum_{i=1}^{3} \sum_{j=1}^{3} v_i (\hat{\boldsymbol{e}}_{A_i} \cdot \hat{\boldsymbol{e}}_{B_j}) \hat{\boldsymbol{e}}_{B_j}. \tag{2.74}
$$

The components of $\hat{\boldsymbol{e}}_{A_i}$, expressed in $\mathcal{B}$, can be represented as the coordinate vector

$$
\begin{aligned}
{}^{B}\hat{\boldsymbol{e}}_{A_i} &= \sum_{j=1}^{3} (\hat{\boldsymbol{e}}_{A_i} \cdot \hat{\boldsymbol{e}}_{B_j}) {}^{B}\hat{\boldsymbol{e}}_{B_j} \\
&= (\hat{\boldsymbol{e}}_{A_i} \cdot \hat{\boldsymbol{e}}_{B_1}) \begin{pmatrix} 1 \\ 0 \\ 0 \end{pmatrix} + (\hat{\boldsymbol{e}}_{A_i} \cdot \hat{\boldsymbol{e}}_{B_2}) \begin{pmatrix} 0 \\ 1 \\ 0 \end{pmatrix} + (\hat{\boldsymbol{e}}_{A_i} \cdot \hat{\boldsymbol{e}}_{B_3}) \begin{pmatrix} 0 \\ 0 \\ 1 \end{pmatrix} \\
&= \begin{pmatrix} \hat{\boldsymbol{e}}_{A_i} \cdot \hat{\boldsymbol{e}}_{B_1} \\ \hat{\boldsymbol{e}}_{A_i} \cdot \hat{\boldsymbol{e}}_{B_2} \\ \hat{\boldsymbol{e}}_{A_i} \cdot \hat{\boldsymbol{e}}_{B_3} \end{pmatrix}.
\end{aligned} \tag{2.75}
$$

The components of $\boldsymbol{v}$ expressed in $\mathcal{B}$ are then

$$
\begin{aligned}
{}^{B}\boldsymbol{v} &= \sum_{i=1}^{3} v_i {}^{B}\hat{\boldsymbol{e}}_{A_i} = \sum_{i=1}^{3} \sum_{j=1}^{3} v_i (\hat{\boldsymbol{e}}_{A_i} \cdot \hat{\boldsymbol{e}}_{B_j}) {}^{B}\hat{\boldsymbol{e}}_{B_j} \\
&= \begin{pmatrix} \hat{\boldsymbol{e}}_{A_1} \cdot \hat{\boldsymbol{e}}_{B_1} & \hat{\boldsymbol{e}}_{A_2} \cdot \hat{\boldsymbol{e}}_{B_1} & \hat{\boldsymbol{e}}_{A_3} \cdot \hat{\boldsymbol{e}}_{B_1} \\ \hat{\boldsymbol{e}}_{A_1} \cdot \hat{\boldsymbol{e}}_{B_2} & \hat{\boldsymbol{e}}_{A_2} \cdot \hat{\boldsymbol{e}}_{B_2} & \hat{\boldsymbol{e}}_{A_3} \cdot \hat{\boldsymbol{e}}_{B_2} \\ \hat{\boldsymbol{e}}_{A_1} \cdot \hat{\boldsymbol{e}}_{B_3} & \hat{\boldsymbol{e}}_{A_2} \cdot \hat{\boldsymbol{e}}_{B_3} & \hat{\boldsymbol{e}}_{A_3} \cdot \hat{\boldsymbol{e}}_{B_3} \end{pmatrix} \begin{pmatrix} v_1 \\ v_2 \\ v_3 \end{pmatrix},
\end{aligned} \tag{2.76}
$$

or

$$
{}^{B}\boldsymbol{v} = \begin{pmatrix} \uparrow & \uparrow & \uparrow \\ {}^{B}\hat{\boldsymbol{e}}_{A_1} & {}^{B}\hat{\boldsymbol{e}}_{A_2} & {}^{B}\hat{\boldsymbol{e}}_{A_3} \\ \downarrow & \downarrow & \downarrow \end{pmatrix} \begin{pmatrix} v_1 \\ v_2 \\ v_3 \end{pmatrix} = {}^{B}_{A}\boldsymbol{Q}\, {}^{A}\boldsymbol{v}, \tag{2.77}
$$

where the rotation matrix, $\boldsymbol{Q}$, is defined as

$$ {}^{B}_{A}\boldsymbol{Q} = \begin{pmatrix} \uparrow & \uparrow & \uparrow \\ {}^{B}\hat{\boldsymbol{e}}_{\mathcal{A}_1} & {}^{B}\hat{\boldsymbol{e}}_{\mathcal{A}_2} & {}^{B}\hat{\boldsymbol{e}}_{\mathcal{A}_3} \\ \downarrow & \downarrow & \downarrow \end{pmatrix}. \tag{2.78} $$

2.1.3 Tensors in $\mathbb{R}^3$

The outer product of two vectors, $\boldsymbol{a}$ and $\boldsymbol{b}$, is denoted by $\boldsymbol{a} \otimes \boldsymbol{b}$. Given the components of $\boldsymbol{a}$ and $\boldsymbol{b}$ expressed in some coordinate frame, $\mathcal{C}$, we can compute the vector outer product as

$$ {}^{C}\boldsymbol{a} \otimes {}^{C}\boldsymbol{b} = \begin{pmatrix} a_1 \\ a_2 \\ a_3 \end{pmatrix} \begin{pmatrix} b_1 & b_2 & b_3 \end{pmatrix} = \begin{pmatrix} a_1b_1 & a_1b_2 & a_1b_3 \\ a_2b_1 & a_2b_2 & a_2b_3 \\ a_3b_1 & a_3b_2 & a_3b_3 \end{pmatrix}. \tag{2.79} $$

A dyad, $\hat{\boldsymbol{e}}_i \otimes \hat{\boldsymbol{e}}_j$, involves the vector outer product of two basis vectors. We can define a dyadic (rank 2 tensor), $\boldsymbol{T}$, as a linear combination of dyads:

$$ \boldsymbol{T} = \sum_{i=1}^{3}\sum_{j=1}^{3} T_{ij}\hat{\boldsymbol{e}}_{\mathcal{A}_i} \otimes \hat{\boldsymbol{e}}_{\mathcal{A}_j} = T_{11}\hat{\boldsymbol{e}}_{\mathcal{A}_1} \otimes \hat{\boldsymbol{e}}_{\mathcal{A}_1} + T_{12}\hat{\boldsymbol{e}}_{\mathcal{A}_1} \otimes \hat{\boldsymbol{e}}_{\mathcal{A}_2} + \cdots + T_{33}\hat{\boldsymbol{e}}_{\mathcal{A}_3} \otimes \hat{\boldsymbol{e}}_{\mathcal{A}_3}. \tag{2.80} $$

The components of this tensor, expressed in $\mathcal{A}$, can be represented in matrix form as

$$ \begin{aligned} {}^{A}\boldsymbol{T} = \sum_{i=1}^{3}\sum_{j=1}^{3} T_{ij}{}^{A}\hat{\boldsymbol{e}}_{\mathcal{A}_i} \otimes {}^{A}\hat{\boldsymbol{e}}_{\mathcal{A}_j} &= T_{11}\begin{pmatrix} 1 & 0 & 0 \\ 0 & 0 & 0 \\ 0 & 0 & 0 \end{pmatrix} + T_{12}\begin{pmatrix} 0 & 1 & 0 \\ 0 & 0 & 0 \\ 0 & 0 & 0 \end{pmatrix} \\ &+ \cdots + T_{33}\begin{pmatrix} 0 & 0 & 0 \\ 0 & 0 & 0 \\ 0 & 0 & 1 \end{pmatrix}, \end{aligned} \tag{2.81} $$

or

$$ {}^{A}\boldsymbol{T} = \begin{pmatrix} T_{11} & T_{12} & T_{13} \\ T_{21} & T_{22} & T_{23} \\ T_{13} & T_{23} & T_{33} \end{pmatrix}. \tag{2.82} $$

The components of $\boldsymbol{T}$ expressed in $\mathcal{B}$ are

$$ \begin{aligned} {}^{B}\boldsymbol{T} &= \sum_{i=1}^{3}\sum_{j=1}^{3} T_{ij}{}^{B}\hat{\boldsymbol{e}}_{\mathcal{A}_i} \otimes {}^{B}\hat{\boldsymbol{e}}_{\mathcal{A}_j} \\ &= \sum_{i=1}^{3}\sum_{j=1}^{3} T_{ij} \begin{pmatrix} \uparrow & \uparrow & \uparrow \\ {}^{B}\hat{\boldsymbol{e}}_{\mathcal{A}_1} & {}^{B}\hat{\boldsymbol{e}}_{\mathcal{A}_2} & {}^{B}\hat{\boldsymbol{e}}_{\mathcal{A}_3} \\ \downarrow & \downarrow & \downarrow \end{pmatrix} {}^{A}\hat{\boldsymbol{e}}_{\mathcal{A}_i} \otimes {}^{A}\hat{\boldsymbol{e}}_{\mathcal{A}_j} \begin{pmatrix} \leftarrow & {}^{B}\hat{\boldsymbol{e}}^{T}_{\mathcal{A}_1} & \rightarrow \\ \leftarrow & {}^{B}\hat{\boldsymbol{e}}^{T}_{\mathcal{A}_2} & \rightarrow \\ \leftarrow & {}^{B}\hat{\boldsymbol{e}}^{T}_{\mathcal{A}_3} & \rightarrow \end{pmatrix}. \end{aligned} \tag{2.83} $$

So, we have the following transformation law for rank 2 tensors:

$$ {}^{B}\boldsymbol{T} = {}^{B}_{A}\boldsymbol{Q}\,{}^{A}\boldsymbol{T}\,{}^{B}_{A}\boldsymbol{Q}^{T}. \tag{2.84} $$

2.1.4 Operations on Vectors and Tensors

The *del* operator is defined as

$$\nabla = \sum_{i=1}^{n} \frac{\partial}{\partial r_i} \hat{\boldsymbol{e}}_i \tag{2.85}$$

Given a scalar field, $\phi(\boldsymbol{r})$, a vector field, $\boldsymbol{v}(\boldsymbol{r}) = \sum_{i=1}^{n} v_i \hat{\boldsymbol{e}}_i$, and a tensor field, $\boldsymbol{T}(\boldsymbol{r}) = \sum_{i=1}^{m} \sum_{j=1}^{n} T_{ij} \hat{\boldsymbol{e}}_i \otimes \hat{\boldsymbol{e}}_j$, we define a set of operations. The *gradient* of a scalar field involves the direct product with the del operator, that is, $\nabla \otimes \phi$. The symbol, $\otimes$, will be omitted for conciseness, so

$$\text{grad}(\phi) = \nabla \phi = \sum_{i=1}^{n} \frac{\partial \phi}{\partial r_i} \hat{\boldsymbol{e}}_i \tag{2.86}$$

Similarly, the gradient of a vector field is given by

$$\text{grad}(\boldsymbol{v}) = \boldsymbol{v} \nabla = \sum_{i=1}^{m} \sum_{j=1}^{n} v_i \hat{\boldsymbol{e}}_i \otimes \frac{\partial}{\partial r_j} \hat{\boldsymbol{e}}_j = \sum_{i=1}^{m} \sum_{j=1}^{n} \frac{\partial v_i}{\partial r_j} \hat{\boldsymbol{e}}_i \otimes \hat{\boldsymbol{e}}_j \tag{2.87}$$

The *divergence* of a vector field is given by

$$\text{div}(\boldsymbol{v}) = \nabla \cdot \boldsymbol{v} = \boldsymbol{v} \cdot \nabla = \sum_{i=1}^{n} v_i \frac{\partial}{\partial r_i} = \sum_{i=1}^{n} \frac{\partial v_i}{\partial r_i} \tag{2.88}$$

The divergence of a tensor field is given by

$$\text{div}(\boldsymbol{T}) = \boldsymbol{T} \cdot \nabla = \sum_{i=1}^{m} \sum_{j=1}^{n} T_{ji} \frac{\partial}{\partial r_j} \hat{\boldsymbol{e}}_i = \sum_{i=1}^{m} \sum_{j=1}^{n} \frac{\partial T_{ji}}{\partial r_j} \hat{\boldsymbol{e}}_i \tag{2.89}$$

The *curl* is given by

$$\text{curl}(\boldsymbol{v}) = \nabla \times \boldsymbol{v} = \sum_{i=1}^{l} \sum_{j=1}^{m} \sum_{k=1}^{n} \epsilon_{ijk} \frac{\partial v_k}{\partial r_j} \hat{\boldsymbol{e}}_i \tag{2.90}$$

The *Laplacian* of a scalar field is given by

$$\text{lap}(\phi) = \nabla^2 \phi = (\nabla \cdot \nabla) \phi = \sum_{i=1}^{n} \frac{\partial \phi}{\partial r_i \partial r_i} \tag{2.91}$$

The *Laplacian* of a vector field is given by

$$\text{lap}(\boldsymbol{v}) = \nabla^2 \boldsymbol{v} = (\nabla \cdot \nabla) \boldsymbol{v} = \sum_{i=1}^{m} \sum_{j=1}^{n} \frac{\partial}{\partial r_j \partial r_j} v_i \hat{\boldsymbol{e}}_i \tag{2.92}$$

2.2 Differential Geometry

Differential geometry applies differential and integral calculus and linear algebra to the study of geometric problems. We will start off addressing the general properties of k-manifolds in $\mathbb{R}^n$ and then will address the more concrete case of curves and surfaces in $\mathbb{R}^3$.

2.2.1 k-Manifolds in $\mathbb{R}^n$

A manifold is a space of points that is locally a Euclidean space near each point. Furthermore, a parameterization, $\boldsymbol{\alpha}(\boldsymbol{u})$, of a manifold, M, is a map from a local coordinate space, U, to M. That is, $\{\boldsymbol{\alpha}|U \rightarrow M\}$, where $U = \mathbb{R}^k$ and $M \subseteq \mathbb{R}^n$ $(k \leq n)$. The parameter, $\boldsymbol{u}$, is the local coordinate vector, $\boldsymbol{u} = (u_1 \ldots u_k)^T \in U$. The dimension of the manifold is k, and the dimension of the ambient space is n.

DEFINITION 2.11 A ***diffeomorphism*** is a map between manifolds that is differentiable and has a differentiable inverse. So, the map $\{\boldsymbol{g}|M_1 \rightarrow M_2\}$ is a diffeomorphism if it is differentiable and there is a differentiable function $\{\boldsymbol{g}^{-1}|M_2 \rightarrow M_1\}$. Given a parameterization, $\boldsymbol{\alpha}_1(\boldsymbol{u})$, for M_1 and a parameterization, $\boldsymbol{\alpha}_2(\boldsymbol{u})$, for M_2, we have

$$\boldsymbol{\alpha}_2(\boldsymbol{u}) = \boldsymbol{g}(\boldsymbol{\alpha}_1(\boldsymbol{u})) = (\boldsymbol{g} \circ \boldsymbol{\alpha}_1)(\boldsymbol{u}) \tag{2.93}$$

and

$$\boldsymbol{\alpha}_1(\boldsymbol{u}) = \boldsymbol{g}^{-1}(\boldsymbol{\alpha}_2(\boldsymbol{u})) = (\boldsymbol{g}^{-1} \circ \boldsymbol{\alpha}_2)(\boldsymbol{u}). \tag{2.94}$$

2.2.2 Curves in $\mathbb{R}^3$

We now address space curves, which are 1-dimensional manifolds in $\mathbb{R}^3$. Space curves can be represented in parameterized form as

$$\boldsymbol{r}(t) = \begin{pmatrix} x(t) \\ y(t) \\ z(t) \end{pmatrix}. \tag{2.95}$$

The derivative, $\boldsymbol{r}'(t)$, is tangent to $\boldsymbol{r}(t)$. If the parameter, t, is taken to be time, then $\boldsymbol{r}'(t) = \boldsymbol{v}(t)$ is the velocity along the curve. The differential arc length is given by

$$ds = \|d\boldsymbol{r}\| = \left\|\boldsymbol{r}'(t)dt\right\| = \left\|\boldsymbol{r}'(t)\right\| dt. \tag{2.96}$$

The arc length over a region of the curve is

$$s(t) = \int ds = \int_{t_o}^{t_f} \left\|\boldsymbol{r}'(t)\right\| dt. \tag{2.97}$$

For a curve parameterized by arc length, we have

$$ds/dt = \left\|\boldsymbol{r}'(t)\right\| = 1. \tag{2.98}$$

The extrinsic curvature of a curve is given by the magnitude of the change in the unit tangent vector, $\boldsymbol{T}(t) = \boldsymbol{r}'(t)/\left\|\boldsymbol{r}'(t)\right\|$, with respect to change in arc length. That is,

$$k = \left\|\frac{d\boldsymbol{T}}{ds}\right\| = \left\|\frac{d\boldsymbol{T}}{dt}\frac{dt}{ds}\right\| = \frac{\left\|\boldsymbol{T}'(t)\right\|}{\|\boldsymbol{r}'(t)\|}. \tag{2.99}$$

If the curve is parameterized by arc length, then $\boldsymbol{T}(s) = \boldsymbol{r}'(s)$, $\left\|\boldsymbol{r}'(s)\right\| = 1$, and

$$k = \left\|\boldsymbol{r}''(s)\right\|. \tag{2.100}$$

2.2.3 Surfaces in $\mathbb{R}^3$

Surfaces are 2-dimensional manifolds in $\mathbb{R}^3$. An implicit description of a surface, S, is defined as the zero set of a function, $f(x, y, z)$,

$$S = \{x, y, z | f(x, y, z) = 0\}, \tag{2.101}$$

or $S = f^{-1}(0)$. The Gauss map, $\{\hat{\boldsymbol{n}}|S \to \mathbb{S}^2\}$, is given by

$$\hat{\boldsymbol{n}} = \frac{\nabla f}{\|\nabla f\|}. \tag{2.102}$$

Alternately, given a parameterization of S, we have

$$\boldsymbol{r}(u, v) = \begin{pmatrix} x(u, v) \\ y(u, v) \\ z(u, v) \end{pmatrix}. \tag{2.103}$$

The Gauss map is then given by

$$\hat{\boldsymbol{n}} = \frac{\boldsymbol{r}_u \times \boldsymbol{r}_v}{\|\boldsymbol{r}_u \times \boldsymbol{r}_v\|}, \tag{2.104}$$

where the right subscript is introduced as a shorthand for the partial derivative. For example,

$$\boldsymbol{r}_u = \frac{\partial \boldsymbol{r}}{\partial u}. \tag{2.105}$$

The differential of the parameterization of (2.103) is

$$d\boldsymbol{r} = \boldsymbol{r}_u du + \boldsymbol{r}_v dv. \tag{2.106}$$

Defining the basis,

$$\mathcal{R} = \{\boldsymbol{r}_u, \boldsymbol{r}_v\}, \tag{2.107}$$

the coordinate vector of $d\boldsymbol{r}$ in $\mathcal{R}$ is

$${}^{\mathcal{R}}d\boldsymbol{r} = \begin{pmatrix} du \\ dv \end{pmatrix}. \tag{2.108}$$

The differential of the Gauss map is

$$d\hat{\boldsymbol{n}} = \hat{\boldsymbol{n}}_u du + \hat{\boldsymbol{n}}_v dv. \tag{2.109}$$

The partial derivatives can be expressed in the $\mathcal{R}$ basis,

$$\hat{\boldsymbol{n}}_u = K_{11}\boldsymbol{r}_u + K_{21}\boldsymbol{r}_v \tag{2.110}$$

$$\hat{\boldsymbol{n}}_v = K_{12}\boldsymbol{r}_u + K_{22}\boldsymbol{r}_v. \tag{2.111}$$

Thus,

$$d\hat{\boldsymbol{n}} = (K_{11}\boldsymbol{r}_u + K_{21}\boldsymbol{r}_v)du + (K_{12}\boldsymbol{r}_u + K_{22}\boldsymbol{r}_v)dv, \tag{2.112}$$

or

$$d\hat{\boldsymbol{n}} = (K_{11}du + K_{12}dv)\boldsymbol{r}_u + (K_{21}du + K_{22}dv)\boldsymbol{r}_v. \tag{2.113}$$

The coordinate vector of $d\hat{\boldsymbol{n}}$ in $\mathcal{R}$ is then

$${}^{\mathcal{R}}d\hat{\boldsymbol{n}} = \begin{pmatrix} K_{11}du + K_{12}dv \\ K_{21}du + K_{22}dv \end{pmatrix} = {}^{\mathcal{R}}\boldsymbol{K}\begin{pmatrix} du \\ dv \end{pmatrix} = {}^{\mathcal{R}}\boldsymbol{K}{}^{\mathcal{R}}d\boldsymbol{r}, \tag{2.114}$$

where ${}^{\mathcal{R}}\boldsymbol{K}$ is the curvature tensor of the surface. The eigenvalues of ${}^{\mathcal{R}}\boldsymbol{K}$ are the principal curvatures, κ_1 and κ_2. The Gaussian curvature is given by

$$K = \det(\boldsymbol{K}) = \kappa_1\kappa_2, \tag{2.115}$$

and the mean curvature is given by

$$H = \frac{1}{2}\mathrm{tr}(\boldsymbol{K}) = \frac{\kappa_1 + \kappa_2}{2}. \tag{2.116}$$

We can determine the elements, K_{ij}, by manipulating (2.110) and (2.111):

$$\langle \hat{\boldsymbol{n}}_u, \boldsymbol{r}_v \rangle = K_{11}\langle \boldsymbol{r}_u, \boldsymbol{r}_v \rangle + K_{21}\langle \boldsymbol{r}_v, \boldsymbol{r}_v \rangle, \tag{2.117}$$

$$\langle \hat{\boldsymbol{n}}_v, \boldsymbol{r}_u \rangle = K_{12}\langle \boldsymbol{r}_u, \boldsymbol{r}_u \rangle + K_{22}\langle \boldsymbol{r}_v, \boldsymbol{r}_u \rangle, \tag{2.118}$$

$$\langle \hat{\boldsymbol{n}}_u, \boldsymbol{r}_u \rangle = K_{11}\langle \boldsymbol{r}_u, \boldsymbol{r}_u \rangle + K_{21}\langle \boldsymbol{r}_v, \boldsymbol{r}_u \rangle, \tag{2.119}$$

$$\langle \hat{\boldsymbol{n}}_v, \boldsymbol{r}_v \rangle = K_{12}\langle \boldsymbol{r}_u, \boldsymbol{r}_v \rangle + K_{22}\langle \boldsymbol{r}_v, \boldsymbol{r}_v \rangle. \tag{2.120}$$

Now $\langle \hat{\boldsymbol{n}}, \boldsymbol{r}_u \rangle = \langle \hat{\boldsymbol{n}}, \boldsymbol{r}_v \rangle = 0$, since $\boldsymbol{r}_u$ and $\boldsymbol{r}_v$ are orthogonal to $\hat{\boldsymbol{n}}$. So,

$$\langle \hat{\boldsymbol{n}}, \boldsymbol{r}_u \rangle_u = \langle \hat{\boldsymbol{n}}_u, \boldsymbol{r}_u \rangle + \langle \hat{\boldsymbol{n}}, \boldsymbol{r}_{uu} \rangle = 0, \tag{2.121}$$

$$\langle \hat{\boldsymbol{n}}, \boldsymbol{r}_u \rangle_v = \langle \hat{\boldsymbol{n}}_v, \boldsymbol{r}_u \rangle + \langle \hat{\boldsymbol{n}}, \boldsymbol{r}_{uv} \rangle = \langle \hat{\boldsymbol{n}}, \boldsymbol{r}_v \rangle_u = \langle \hat{\boldsymbol{n}}_u, \boldsymbol{r}_v \rangle + \langle \hat{\boldsymbol{n}}, \boldsymbol{r}_{vu} \rangle = 0, \tag{2.122}$$

$$\langle \hat{\boldsymbol{n}}, \boldsymbol{r}_v \rangle_v = \langle \hat{\boldsymbol{n}}_v, \boldsymbol{r}_v \rangle + \langle \hat{\boldsymbol{n}}, \boldsymbol{r}_{vv} \rangle = 0. \tag{2.123}$$

For conciseness we will define

$$a \triangleq \langle \hat{\boldsymbol{n}}_u, \boldsymbol{r}_u \rangle = -\langle \hat{\boldsymbol{n}}, \boldsymbol{r}_{uu} \rangle, \tag{2.124}$$

$$b \triangleq \langle \hat{\boldsymbol{n}}_v, \boldsymbol{r}_u \rangle = -\langle \hat{\boldsymbol{n}}, \boldsymbol{r}_{uv} \rangle = \langle \hat{\boldsymbol{n}}_u, \boldsymbol{r}_v \rangle = -\langle \hat{\boldsymbol{n}}, \boldsymbol{r}_{vu} \rangle, \tag{2.125}$$

$$c \triangleq \langle \hat{\boldsymbol{n}}_v, \boldsymbol{r}_v \rangle = -\langle \hat{\boldsymbol{n}}, \boldsymbol{r}_{vv} \rangle, \tag{2.126}$$

and

$$A \triangleq \langle \boldsymbol{r}_u, \boldsymbol{r}_u \rangle, \tag{2.127}$$

$$B \triangleq \langle \boldsymbol{r}_u, \boldsymbol{r}_v \rangle = \langle \boldsymbol{r}_v, \boldsymbol{r}_u \rangle, \tag{2.128}$$

$$C \triangleq \langle \boldsymbol{r}_v, \boldsymbol{r}_v \rangle. \tag{2.129}$$

Equations (2.117) through (2.120) can be written as

$$b = K_{11}B + K_{21}C, \tag{2.130}$$

$$b = K_{12}A + K_{22}B, \tag{2.131}$$

$$a = K_{11}A + K_{21}B, \tag{2.132}$$

$$c = K_{12}B + K_{22}C, \tag{2.133}$$

or

$$\begin{pmatrix} a & b \\ b & c \end{pmatrix} = \begin{pmatrix} K_{11} & K_{21} \\ K_{12} & K_{22} \end{pmatrix} \begin{pmatrix} A & B \\ B & C \end{pmatrix}. \tag{2.134}$$

So,

$$\begin{aligned} \begin{pmatrix} K_{11} & K_{21} \\ K_{12} & K_{22} \end{pmatrix} &= \begin{pmatrix} a & b \\ b & c \end{pmatrix} \begin{pmatrix} A & B \\ B & C \end{pmatrix}^{-1} \\ &= \frac{1}{AC - B^2} \begin{pmatrix} a & b \\ b & c \end{pmatrix} \begin{pmatrix} C & -B \\ -B & A \end{pmatrix} = \frac{1}{AC - B^2} \begin{pmatrix} aC - bB & bA - aB \\ bC - cB & cA - bB \end{pmatrix}, \end{aligned} \tag{2.135}$$

and

$$^{\mathcal{R}}\boldsymbol{K} = \begin{pmatrix} K_{11} & K_{12} \\ K_{21} & K_{22} \end{pmatrix} = \frac{1}{AC - B^2} \begin{pmatrix} aC - bB & bC - cB \\ bA - aB & cA - bB \end{pmatrix}. \tag{2.136}$$

Geodesics on Surfaces

A *tangent vector field*, V, on a surface, S, maps to each point on S a vector, $\boldsymbol{v}$, which is tangent to S at that point. That is,

$$\boldsymbol{v}(u, v) = v_1(u, v)\boldsymbol{r}_u + v_2(u, v)\boldsymbol{r}_v. \tag{2.137}$$

Let $\boldsymbol{v}(t)$ be the restriction of the tangent vector field, V, to a curve, $\boldsymbol{r}(t)$, sitting on the surface, S, with tangent space $T_r(S)$ at a point $\boldsymbol{r}$. The *covariant derivative* of $\boldsymbol{v}(t)$ with reference to S is given by

$$\frac{D}{dt}\boldsymbol{v}(t) = \text{proj}_T\left(\frac{d}{dt}\boldsymbol{v}(t)\right), \tag{2.138}$$

where $\text{proj}_T()$ denotes the projection of a vector onto the tangent space of S. The vector field V is said to be *parallel* if

$$\frac{D}{dt}\boldsymbol{v}(t) = \boldsymbol{0}. \tag{2.139}$$

The intrinsic geodesic curvature of a curve is

$$k_g = \left\| \frac{D}{ds}\frac{d\boldsymbol{r}}{ds} \right\| = \left\| \text{proj}_T\left(\boldsymbol{r}''(s)\right) \right\| \tag{2.140}$$

if the curve is parameterized by arc length, or

$$k_g = \left\| \text{proj}_T\left(\frac{\boldsymbol{T}'(t)}{\|\boldsymbol{r}'(t)\|}\right) \right\|, \tag{2.141}$$

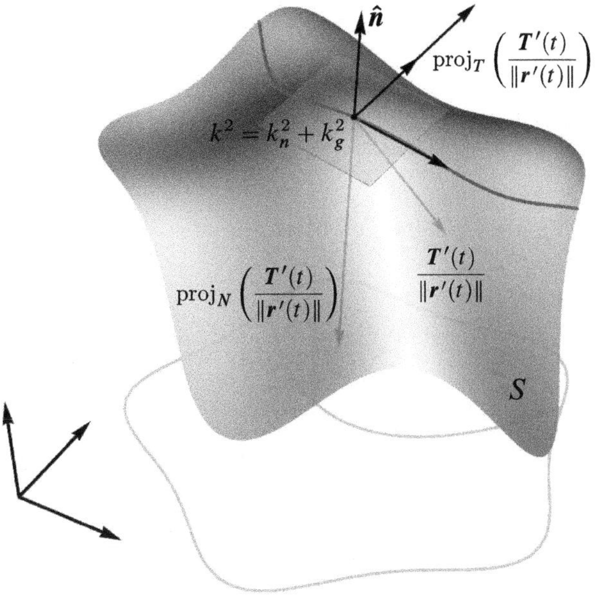

Figure 2.4 Relationship between extrinsic, geodesic, and normal curvature. The total curvature vector is projected onto the tangent and normal spaces of the surface, yielding the geodesic and normal curvature vectors, respectively.

if not. The normal curvature is given as

$$k_n = \left\| \text{proj}_N\left(\frac{T'(t)}{\|r'(t)\|}\right)\right\|. \tag{2.142}$$

We have the following vector relationship:

$$\frac{T'(t)}{\|r'(t)\|} = \text{proj}_N\left(\frac{T'(t)}{\|r'(t)\|}\right) + \text{proj}_T\left(\frac{T'(t)}{\|r'(t)\|}\right). \tag{2.143}$$

These vectors form a right triangle. The curvatures are the corresponding magnitudes in this vector relationship and are related by

$$k^2 = k_n^2 + k_g^2. \tag{2.144}$$

This is depicted in Figure 2.4.

A geodesic is the shortest path between two points on a surface. Geodesics correspond to curves with *zero* geodesic curvature ($k_g = 0$; see Figure 2.5). They can be computed by solving a set of second-order nonlinear differential equations. If we parameterize our surface as $\boldsymbol{r}(u, v) \in \mathbb{R}^3$, the differential equations are given by (Do Carmo 1976)

$$u'' + \Gamma^1_{11}u'^2 + 2\Gamma^1_{12}u'v' + \Gamma^1_{22}v'^2 = 0 \tag{2.145}$$

$$v'' + \Gamma^2_{11}u'^2 + 2\Gamma^2_{12}u'v' + \Gamma^2_{22}v'^2 = 0, \tag{2.146}$$

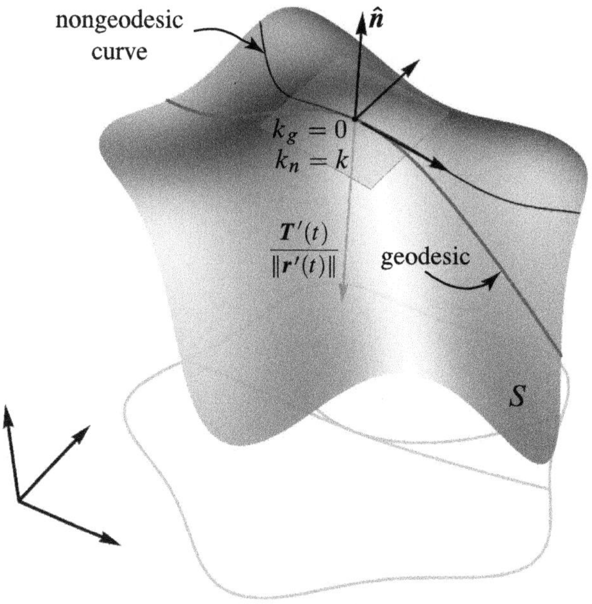

Figure 2.5 Geodesics correspond to curves with *zero* geodesic curvature. A geodesic is illustrated. Another, nongeodesic curve is shown for comparison.

where the Christoffel symbols, Γ, are found by solving the linear algebraic system

$$\Gamma^1_{11}\langle \boldsymbol{r}_u, \boldsymbol{r}_u\rangle + \Gamma^2_{11}\langle \boldsymbol{r}_u, \boldsymbol{r}_v\rangle = \langle \boldsymbol{r}_{uu}, \boldsymbol{r}_u\rangle\,, \tag{2.147}$$

$$\Gamma^1_{11}\langle \boldsymbol{r}_u, \boldsymbol{r}_v\rangle + \Gamma^2_{11}\langle \boldsymbol{r}_v, \boldsymbol{r}_v\rangle = \langle \boldsymbol{r}_{uu}, \boldsymbol{r}_v\rangle\,, \tag{2.148}$$

$$\Gamma^1_{12}\langle \boldsymbol{r}_u, \boldsymbol{r}_u\rangle + \Gamma^2_{12}\langle \boldsymbol{r}_u, \boldsymbol{r}_v\rangle = \langle \boldsymbol{r}_{uv}, \boldsymbol{r}_u\rangle\,, \tag{2.149}$$

$$\Gamma^1_{12}\langle \boldsymbol{r}_u, \boldsymbol{r}_v\rangle + \Gamma^2_{12}\langle \boldsymbol{r}_v, \boldsymbol{r}_v\rangle = \langle \boldsymbol{r}_{uv}, \boldsymbol{r}_v\rangle\,, \tag{2.150}$$

$$\Gamma^1_{22}\langle \boldsymbol{r}_u, \boldsymbol{r}_u\rangle + \Gamma^2_{22}\langle \boldsymbol{r}_u, \boldsymbol{r}_v\rangle = \langle \boldsymbol{r}_{vv}, \boldsymbol{r}_u\rangle\,, \tag{2.151}$$

$$\Gamma^1_{22}\langle \boldsymbol{r}_u, \boldsymbol{r}_v\rangle + \Gamma^2_{22}\langle \boldsymbol{r}_v, \boldsymbol{r}_v\rangle = \langle \boldsymbol{r}_{vv}, \boldsymbol{r}_v\rangle\,. \tag{2.152}$$

Equations (2.145) and (2.146) can be expressed in compact form as

$$\boldsymbol{u}'' = -\begin{pmatrix} \boldsymbol{u}'^T\boldsymbol{\Gamma}^1\boldsymbol{u}' \\ \boldsymbol{u}'^T\boldsymbol{\Gamma}^2\boldsymbol{u}' \end{pmatrix}, \tag{2.153}$$

where $\boldsymbol{u} = (\,u\ v\,)^T$ and the matrices $\boldsymbol{\Gamma}^1$ and $\boldsymbol{\Gamma}^2$ are symmetric.

2.3 Optimization

Mathematical optimization seeks a solution, within some allowable set of values (search space), that maximizes or minimizes a function. We will restrict our attention to the minimization of an objective function where the allowable search space is represented by an equality condition.

2.3.1 Constrained Static Optimization

The constrained static optimization problem seeks to find a solution, $\boldsymbol{x}$, which minimizes some objective function, $J(\boldsymbol{x})$, subject to the equality constraints

$$\boldsymbol{f}(\boldsymbol{x}) = \boldsymbol{0}. \tag{2.154}$$

We define the manifold, P, as the level set of $J(\boldsymbol{x})$,

$$P \triangleq \{\boldsymbol{x} | J(\boldsymbol{x}) = \text{constant}\}, \tag{2.155}$$

and the constraint manifold, C, as the zero set of $\boldsymbol{f}(\boldsymbol{x})$,

$$C \triangleq \{\boldsymbol{x} | \boldsymbol{f}(\boldsymbol{x}) = \boldsymbol{0}\}. \tag{2.156}$$

At a local extremum, $\boldsymbol{x}_o$, the tangent space of C lies in the tangent space of P,

$$T_{x_o}(C) \subseteq T_{x_o}(P), \tag{2.157}$$

where

$$T_{x_o}(C) = \ker(\boldsymbol{f}\nabla_{x_o}) \quad \text{and,} \quad T_{x_o}(P) = \ker(J\nabla_{x_o}). \tag{2.158}$$

Conversely, in terms of the normal spaces,

$$T_{x_o}(P)^{\perp} \subseteq T_{x_o}(C)^{\perp}, \tag{2.159}$$

or

$$N_{x_o}(P) \subseteq N_{x_o}(C). \tag{2.160}$$

So, we have

$$N_{x_o}(P) = \ker(J\nabla_{x_o})^{\perp} = \operatorname{im}(\nabla_{x_o} J) \tag{2.161}$$

and

$$N_{x_o}(C) = \ker(\boldsymbol{f}\nabla_{x_o})^{\perp} = \operatorname{im}(\nabla_{x_o}\boldsymbol{f}). \tag{2.162}$$

Thus,

$$\operatorname{im}(\nabla_{x_o} J) \subseteq \operatorname{im}(\nabla_{x_o}\boldsymbol{f}). \tag{2.163}$$

Defining the basis, $\mathcal{N}_C$, for the normal space of C,

$$\mathcal{N}_C \triangleq \{\nabla f_1, \cdots, \nabla f_m\}, \tag{2.164}$$

we can represent $\nabla_{x_o} J$ in $\mathcal{N}_C$,

$$\nabla_{x_o} J = -\sum_{i=1}^{m} \lambda_i \nabla_{x_o} f_i = -\nabla_{x_o}\boldsymbol{f}\boldsymbol{\lambda}, \tag{2.165}$$

where $\boldsymbol{\lambda}$ is a vector of Lagrange multipliers. The solution to the constrained minimization problem is then found by solving the following system:

$$\begin{aligned} \nabla J + \nabla \boldsymbol{f}\boldsymbol{\lambda} &= \boldsymbol{0} \\ \boldsymbol{f}(\boldsymbol{x}) &= \boldsymbol{0}. \end{aligned} \tag{2.166}$$

Example: Let us consider a quadratic objective function

$$J(\boldsymbol{x}) = \boldsymbol{x}^T \boldsymbol{B}\boldsymbol{x}, \tag{2.167}$$

where $\boldsymbol{B}$ is a symmetric positive definite matrix. Let us also consider a set of constraints

$$\boldsymbol{f}(\boldsymbol{x}) = \boldsymbol{A}\boldsymbol{x} - \boldsymbol{y} = \boldsymbol{0}. \tag{2.168}$$

We note that

$$J(\boldsymbol{x} + d\boldsymbol{x}) = (\boldsymbol{x} + d\boldsymbol{x})^T \boldsymbol{B}(\boldsymbol{x} + d\boldsymbol{x}) = \boldsymbol{x}^T \boldsymbol{B}\boldsymbol{x} + 2\boldsymbol{x}^T \boldsymbol{B} d\boldsymbol{x}. \tag{2.169}$$

So,

$$dJ = J(\boldsymbol{x} + d\boldsymbol{x}) - J(\boldsymbol{x}) = 2\boldsymbol{x}^T \boldsymbol{B} d\boldsymbol{x} = J\nabla d\boldsymbol{x}. \tag{2.170}$$

Thus,

$$J\nabla = 2\boldsymbol{x}^T \boldsymbol{B} \quad \text{and} \quad \nabla J = 2\boldsymbol{B}\boldsymbol{x}. \tag{2.171}$$

We also note that

$$\boldsymbol{f}\nabla = \boldsymbol{A} \quad \text{and} \quad \nabla \boldsymbol{f} = \boldsymbol{A}^T. \tag{2.172}$$

So the extremum must satisfy

$$\nabla J + \nabla \boldsymbol{f}\boldsymbol{\lambda} = 2\boldsymbol{B}\boldsymbol{x} + \boldsymbol{A}^T \boldsymbol{\lambda} = \boldsymbol{0}, \tag{2.173}$$

and

$$\boldsymbol{f}(\boldsymbol{x}) = \boldsymbol{A}\boldsymbol{x} - \boldsymbol{y} = \boldsymbol{0}. \tag{2.174}$$

Solving for $\boldsymbol{x}$ in 2.173 yields

$$\boldsymbol{x} = -\frac{1}{2}\boldsymbol{B}^{-1}\boldsymbol{A}^T\boldsymbol{\lambda}. \tag{2.175}$$

Substituting into 2.174 and solving for $\boldsymbol{\lambda}$ yields

$$\boldsymbol{\lambda} = -2(\boldsymbol{A}\boldsymbol{B}^{-1}\boldsymbol{A}^T)^{-1}\boldsymbol{y}, \tag{2.176}$$

and our solution is

$$\boldsymbol{x} = \boldsymbol{B}^{-1}\boldsymbol{A}^T(\boldsymbol{A}\boldsymbol{B}^{-1}\boldsymbol{A}^T)^{-1}\boldsymbol{y}. \tag{2.177}$$

If the weighting matrix, $\boldsymbol{B}$, is the identity matrix, then our solution is equivalent to the least norm solution from Section 2.1.1,

$$\boldsymbol{x} = \boldsymbol{A}^T(\boldsymbol{A}^{-1}\boldsymbol{A}^T)^{-1}\boldsymbol{y} = \boldsymbol{A}^{\dagger}\boldsymbol{y}. \tag{2.178}$$

Figure 2.6 provides a geometric illustration of the constrained static optimization problem.

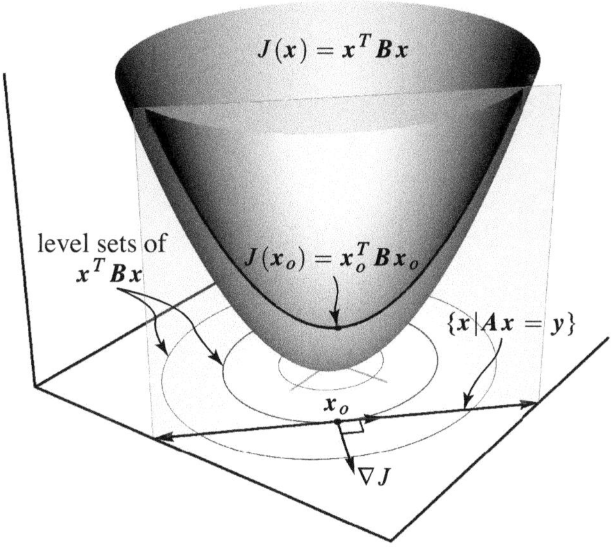

Figure 2.6 Minimization of $J(x) = x^T Bx$ subject to the constraint $Ax = y$. The solution corresponds to the point on the constraint surface (line) where the gradient of the objective function is orthogonal to the constraint tangent space.

2.4 Exercises

1. Project the vector, $x = (-3 \quad 1 \quad 4)^T$ onto the basis

$$u_1 = \begin{pmatrix}1 & 0 & 0\end{pmatrix}^T,$$
$$u_2 = \begin{pmatrix}0 & \frac{1}{\sqrt{2}} & \frac{1}{\sqrt{2}}\end{pmatrix}^T,$$
$$u_3 = \begin{pmatrix}0 & -\frac{1}{\sqrt{2}} & \frac{1}{\sqrt{2}}\end{pmatrix}^T.$$

   That is, find the coefficients, c_i, associated with the linear combination of basis vectors where $x = c_1u_1 + c_2u_2 + c_3u_3$.

2. Construct an orthonormal basis, $\{u_1, u_2, u_3\}$, from

$$v_1 = \begin{pmatrix}1 & 1 & 2 & 0\end{pmatrix}^T,$$
$$v_2 = \begin{pmatrix}3 & 0 & -2 & 1\end{pmatrix}^T,$$
$$v_3 = \begin{pmatrix}-2 & 1 & 1 & -2\end{pmatrix}^T$$

   using the Gram-Schmidt process.

3. Find the line, $y = mx + b$, that best fits the points $(x_1, y_1) = (1, .5)$, $(x_2, y_2) = (-2, -3.8)$, $(x_3, y_3) = (-1, -3)$, and $(x_4, y_4) = (3, 2.8)$, in a least squares sense.

4. Consider the system

$$\begin{pmatrix} 1 & 1 & 1 & -1 \\ -2 & 1 & 3 & 0 \end{pmatrix} \begin{pmatrix} x_1 \\ x_2 \\ x_3 \\ x_4 \end{pmatrix} = \begin{pmatrix} 3 \\ 4 \end{pmatrix}.$$

   (a) Find the least norm solution, $\boldsymbol{x}_P$.
   (b) Represent all solutions to the system by finding a basis for the null space of $\boldsymbol{A}$.

5. Consider the matrix

$$\boldsymbol{A} = \begin{pmatrix} \frac{5}{3} & \frac{5}{4} & 1 \\ \frac{5}{4} & 1 & \frac{5}{6} \\ 1 & \frac{5}{6} & \frac{3}{2} \end{pmatrix}.$$

   (a) Find the eigenvalues and eigenvectors.
   (b) Show that $\boldsymbol{D} = \boldsymbol{S}^{-1}\boldsymbol{A}\boldsymbol{S}$, where $\boldsymbol{D}$ is the diagonal matrix of eigenvalues and $\boldsymbol{S}$ is the matrix of eigenvectors.

6. Consider the time-parameterized curve (see Figure 2.7) given by

$$\boldsymbol{r}(t) = \left(2\cos(t/2)\ -3\sin(t/2)\ \cos(t)\right)^T.$$

   (a) Compute the velocity (tangent vector) of a point moving along the curve.
   (b) Compute the speed, $s'(t)$, of a point moving along the curve.

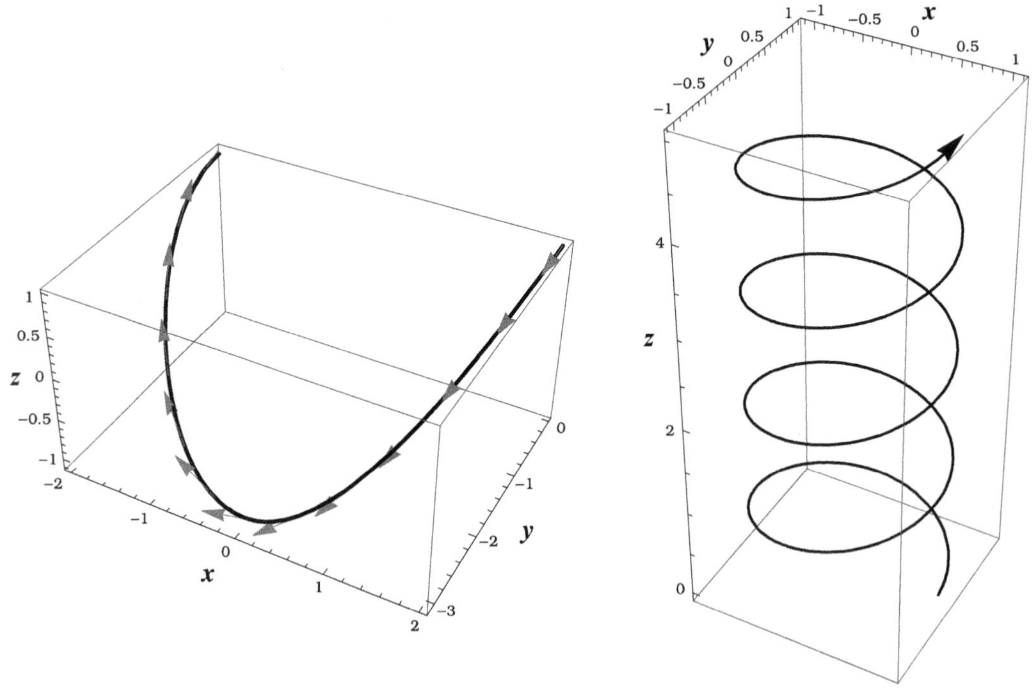

Figure 2.7 (Left) Space curve $\boldsymbol{r}(t) = (2\cos(t/2), -3\sin(t/2), \cos(t))$ (Exercise 6). (Right) Space curve $\boldsymbol{r}(t) = (\cos(t), \sin(t), t/5)$ (Exercise 7).

7. Consider the time-parameterized curve (see Figure 2.7) given by

$$\boldsymbol{r}(t) = \left(\cos(t)\ \sin(t)\ t/5\right)^T.$$

   (a) Compute the velocity (tangent vector) of a point moving along the curve.
   (b) Compute the arc length of the curve on the interval $[0, 8\pi]$.
   (c) Compute the curvature of the curve.
8. Generate the Gauss map for the surface of Figure 2.8 with implicit description

$$f = 4 + x^2 + y^2 - \left[z - \frac{1}{3}\cos(3x)\cos(3y)\right]^2 = 0.$$

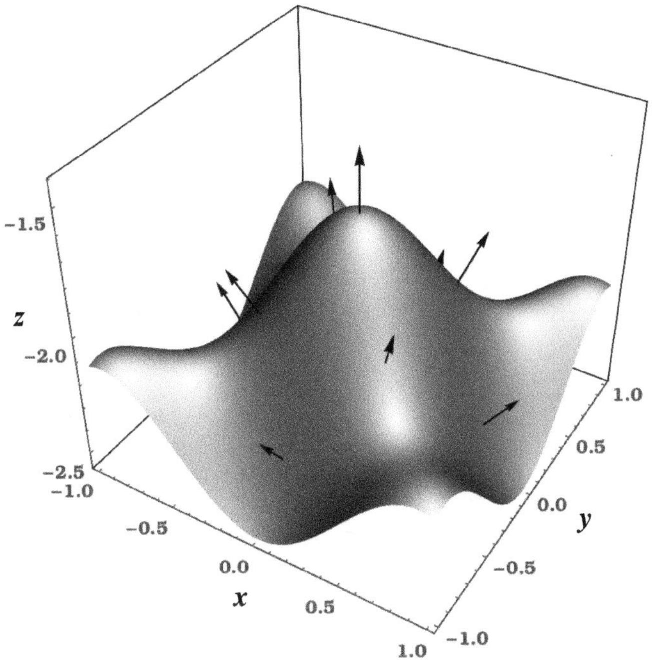

Figure 2.8 Implicit surface $f = 4 + x^2 + y^2 - \left[z - \frac{1}{3}\cos(3x)\cos(3y)\right]^2 = 0$ (Exercise 8).

9. Consider the parameterized surface (see Figure 2.9) given by

$$\boldsymbol{r}(u, v) = \left(u\ v\ -\tfrac{2}{5}(u^2 + v^2) + \cos(2u)\cos(2v)\right)^T$$

   and the parameterized curve

$$\boldsymbol{\alpha}(t) = \left(t\ t - \tfrac{3}{4}\ -\tfrac{2}{5}\left[t^2 + (t - \tfrac{3}{4})^2\right] + \cos(2t)\cos(2t - \tfrac{3}{2})\right)^T$$

   sitting on the surface.
   (a) Generate the Gauss map for the surface.
   (b) Determine the total (extrinsic) curvature, k, of $\boldsymbol{\alpha}(t)$ at point $t = 3/8$.
   (c) Determine the normal, k_n, and geodesic curvature, k_g, of $\boldsymbol{\alpha}(t)$ at the point $t = 3/8$.

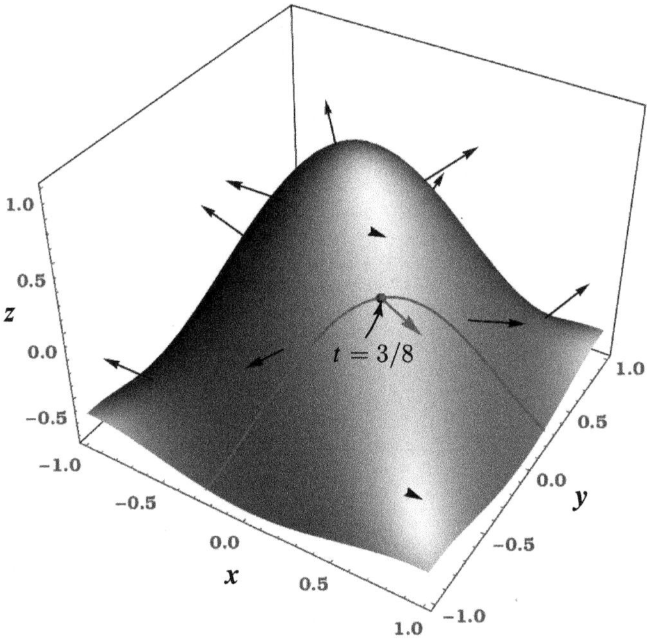

Figure 2.9 Parametric surface $\boldsymbol{r}(u, v) = \big(u, v, -\frac{2}{5}(u^2 + v^2) + \cos(2u)\cos(2v)\big)$ and parameterized curve $\boldsymbol{\alpha}(t) = \big(t, t - \frac{3}{4}, -\frac{2}{5}\left[t^2 + (t - \frac{3}{4})^2\right] + \cos(2t)\cos(2t - \frac{3}{2}\big)$ (Exercise 9).

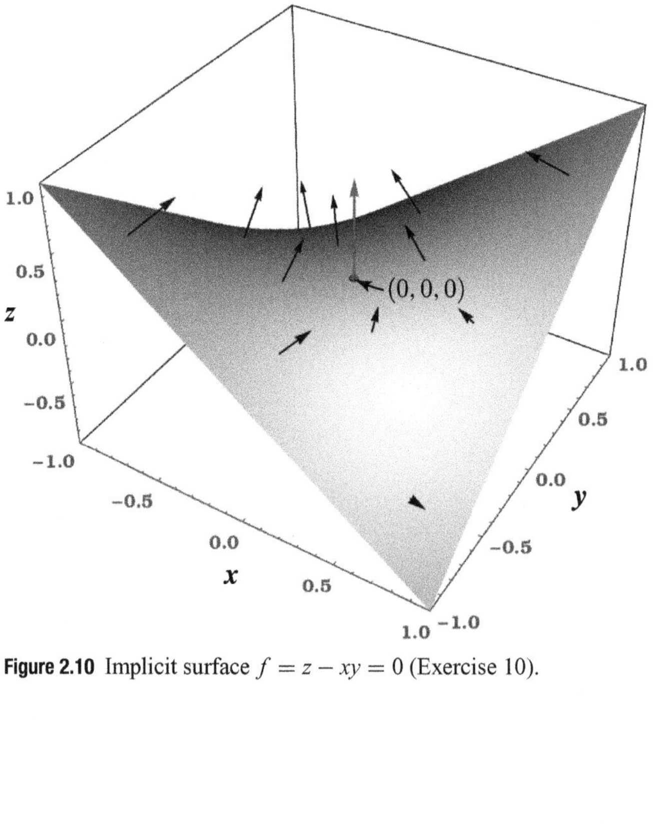

Figure 2.10 Implicit surface $f = z - xy = 0$ (Exercise 10).

10. Consider the implicit surface (see Figure 2.10) given by

$$f = z - xy = 0.$$

 (a) Generate the Gauss map for the surface.
 (b) Compute the curvature tensor, $\boldsymbol{K}$.
 (c) Determine the principal curvatures, κ_1 and κ_2, the Gaussian curvature, K, and the mean curvature, H, at the point $(x, y) = (0, 0)$.

11. Show that the geodesics of a Euclidean plane are straight lines.

3 Kinematics of Discrete Systems

Kinematics is the formal study of motion without regard to the causes of motion (i.e., forces and moments). It is a natural prerequisite to the study of dynamics, which builds on kinematics by addressing the kinetic causes of motion. The kinematics of discrete systems is concerned with systems that possess a finite number of degrees of freedom, as opposed to continuous systems which possess an infinite number of degrees of freedom. The basic element of a discrete system is the rigid body.

In this chapter we will address spherical kinematics associated with pure rotation, followed by spatial kinematics which involves both rotation and translation. Finally, we will address the kinematics of systems of rigid bodies forming linked chains. This will provide the necessary formal understanding of the geometry of motion required to address dynamical principles in the subsequent chapters.

3.1 Spherical Kinematics

Spherical kinematics is concerned with the 3-dimensional rotation group denoted as $SO(3)$. This is the group of all rotations about the origin in Euclidean space, $\mathbb{R}^3$. Formally, the group is defined as the set of all proper orthogonal matrices, $\boldsymbol{Q}$. That is,

$$SO(3) = \{\boldsymbol{Q} | \boldsymbol{Q} \in \mathbb{R}^{3\times3}, \boldsymbol{Q}^T\boldsymbol{Q} = \boldsymbol{Q}\boldsymbol{Q}^T = \mathbf{1}\}. \tag{3.1}$$

The orthogonal rotation matrix, $\boldsymbol{Q}$, will be described here, as well as some other ways to parameterize rotation. These include angle-sets, axis-angle parameters, and unit quaternions.

3.1.1 Orthogonal Rotation Matrices

Orthogonal rotation matrices encode spatial rotation by describing the orientation of one coordinate frame relative to another. The column vectors of a rotation matrix are the basis vectors of the coordinate frame of interest, expressed within an embedding frame. For example, Figure 3.1 depicts frame $\mathcal{B}$ rotated relative to frame $\mathcal{A}$.

The rotation matrix describing the orientation of frame $\mathcal{B}$ in frame $\mathcal{A}$ is

$${}^{A}_{B}\boldsymbol{Q} = \begin{pmatrix} \uparrow & \uparrow & \uparrow \\ {}^{A}\hat{\boldsymbol{e}}_{\mathcal{B}_1} & {}^{A}\hat{\boldsymbol{e}}_{\mathcal{B}_2} & {}^{A}\hat{\boldsymbol{e}}_{\mathcal{B}_3} \\ \downarrow & \downarrow & \downarrow \end{pmatrix} = \begin{pmatrix} \hat{\boldsymbol{e}}_{\mathcal{B}_1}\cdot\hat{\boldsymbol{e}}_{\mathcal{A}_1} & \hat{\boldsymbol{e}}_{\mathcal{B}_2}\cdot\hat{\boldsymbol{e}}_{\mathcal{A}_1} & \hat{\boldsymbol{e}}_{\mathcal{B}_3}\cdot\hat{\boldsymbol{e}}_{\mathcal{A}_1} \\ \hat{\boldsymbol{e}}_{\mathcal{B}_1}\cdot\hat{\boldsymbol{e}}_{\mathcal{A}_2} & \hat{\boldsymbol{e}}_{\mathcal{B}_2}\cdot\hat{\boldsymbol{e}}_{\mathcal{A}_2} & \hat{\boldsymbol{e}}_{\mathcal{B}_3}\cdot\hat{\boldsymbol{e}}_{\mathcal{A}_2} \\ \hat{\boldsymbol{e}}_{\mathcal{B}_1}\cdot\hat{\boldsymbol{e}}_{\mathcal{A}_3} & \hat{\boldsymbol{e}}_{\mathcal{B}_2}\cdot\hat{\boldsymbol{e}}_{\mathcal{A}_3} & \hat{\boldsymbol{e}}_{\mathcal{B}_3}\cdot\hat{\boldsymbol{e}}_{\mathcal{A}_3} \end{pmatrix}. \tag{3.2}$$

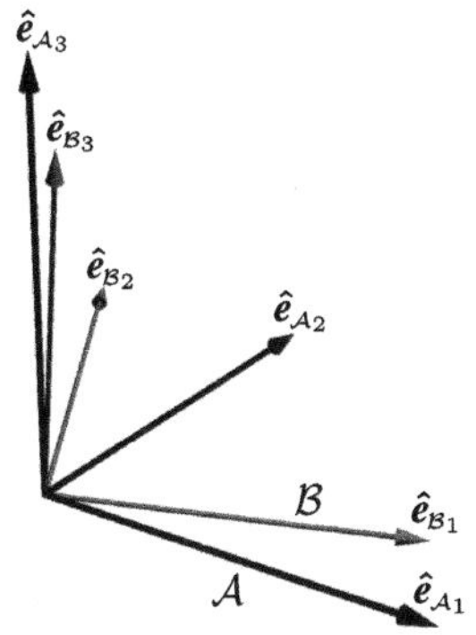

Figure 3.1 Rotation of frame $\mathcal{B}$ relative to frame $\mathcal{A}$. The basis vectors of frame $\mathcal{B}$, expressed in frame $\mathcal{A}$, are the column vectors of the rotation matrix.

For rotations about the principal axes,

$$\boldsymbol{Q}_x(\theta) = \begin{pmatrix} 1 & 0 & 0 \\ 0 & \cos\theta & -\sin\theta \\ 0 & \sin\theta & \cos\theta \end{pmatrix}, \quad \boldsymbol{Q}_y(\theta) = \begin{pmatrix} \cos\theta & 0 & \sin\theta \\ 0 & 1 & 0 \\ -\sin\theta & 0 & \cos\theta \end{pmatrix},$$
$$\boldsymbol{Q}_z(\theta) = \begin{pmatrix} \cos\theta & -\sin\theta & 0 \\ \sin\theta & \cos\theta & 0 \\ 0 & 0 & 1 \end{pmatrix}. \tag{3.3}$$

The inverse of a rotation matrix, $\boldsymbol{Q}^{-1}$, satisfies

$$\boldsymbol{Q}\boldsymbol{Q}^{-1} = \boldsymbol{Q}^{-1}\boldsymbol{Q} = \mathbf{1} = \begin{pmatrix} 1 & 0 & 0 \\ 0 & 1 & 0 \\ 0 & 0 & 1 \end{pmatrix}. \tag{3.4}$$

Since $\boldsymbol{Q}$ is an orthogonal matrix,

$${}^{A}_{B}\boldsymbol{Q}^{T}\,{}^{A}_{B}\boldsymbol{Q} = \begin{pmatrix} \leftarrow & {}^{A}\hat{\boldsymbol{e}}_{\mathcal{B}_1} & \rightarrow \\ \leftarrow & {}^{A}\hat{\boldsymbol{e}}_{\mathcal{B}_2} & \rightarrow \\ \leftarrow & {}^{A}\hat{\boldsymbol{e}}_{\mathcal{B}_3} & \rightarrow \end{pmatrix} \begin{pmatrix} \uparrow & \uparrow & \uparrow \\ {}^{A}\hat{\boldsymbol{e}}_{\mathcal{B}_1} & {}^{A}\hat{\boldsymbol{e}}_{\mathcal{B}_2} & {}^{A}\hat{\boldsymbol{e}}_{\mathcal{B}_3} \\ \downarrow & \downarrow & \downarrow \end{pmatrix} = \mathbf{1}. \tag{3.5}$$

Therefore,

$${}^{B}_{A}\boldsymbol{Q}\,{}^{A}_{B}\boldsymbol{Q} = {}^{A}_{B}\boldsymbol{Q}^{-1}\,{}^{A}_{B}\boldsymbol{Q} = \mathbf{1}, \tag{3.6}$$

and

$$\boldsymbol{Q}^{T} = \boldsymbol{Q}^{-1}. \tag{3.7}$$

Rotational transformation can be accommodated with rotation matrices using the product

$${}^{A}\boldsymbol{v} = {}^{A}_{B}\boldsymbol{Q}\,{}^{B}\boldsymbol{v}. \tag{3.8}$$

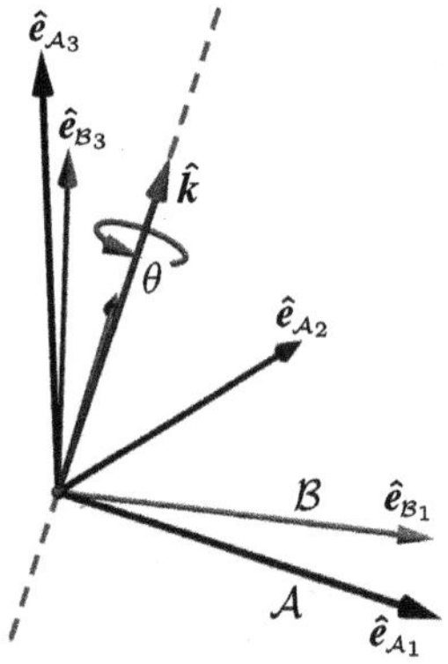

Figure 3.2 Axis-angle representation describing an arbitrary rotation using spin axis, $\hat{\boldsymbol{k}}$, and spin angle, θ, parameters.

Additionally, multiple rotations can be concatenated using multiplication:

$$ {}^{A}_{C}\boldsymbol{Q} = {}^{A}_{B}\boldsymbol{Q}\,{}^{B}_{C}\boldsymbol{Q}. \tag{3.9}$$

This process is not commutative. That is, in general,

$$ {}^{A}_{B}\boldsymbol{Q}\,{}^{B}_{C}\boldsymbol{Q} \neq {}^{B}_{C}\boldsymbol{Q}\,{}^{A}_{B}\boldsymbol{Q}. \tag{3.10}$$

Thus, the order in which successive finite (noninfinitesimal) rotations are performed affects the resulting orientation.

3.1.2 Axis-Angle Representation

We begin by noting the following:

THEOREM 3.1 *There exists a spin axis and angle for any arbitrary orientation in* $\mathbb{R}^3$. *This is known as Euler's rotation theorem.*

The axis-angle representation specifies a spin axis about which a coordinate frame is rotated by a specified angle. Figure 3.2 depicts an arbitrary rotation described by the axis-angle parameters, spin axis, $\hat{\boldsymbol{k}}$, and spin angle, θ.

We wish to determine the rotation matrix, ${}^{A}_{B}\boldsymbol{Q}$, associated with the rotation depicted in Figure 3.2, in terms of the axis-angle parameters. Let us start by defining orthonormal vectors $\hat{\boldsymbol{\imath}}$ and $\hat{\boldsymbol{\jmath}}$ to be orthogonal to unit vector $\hat{\boldsymbol{k}}$. Let coordinate frame $\mathcal{K}$ be defined by the basis vectors $\hat{\boldsymbol{\imath}}, \hat{\boldsymbol{\jmath}}, \hat{\boldsymbol{k}}$. Let $\mathcal{K}'$ be the coordinate frame that $\mathcal{K}$ is rotated into. Then,

$$ {}^{A}_{B}\boldsymbol{Q} = \boldsymbol{Q}_k(\theta) = {}^{A}_{\mathcal{K}}\boldsymbol{Q}\,{}^{\mathcal{K}}_{\mathcal{K}'}\boldsymbol{Q}\,{}^{\mathcal{K}'}_{B}\boldsymbol{Q}, \tag{3.11}$$

where

$$ {}^{A}_{\mathcal{K}}\boldsymbol{Q} = \begin{pmatrix} \uparrow & \uparrow & \uparrow \\ {}^{A}\hat{\boldsymbol{\imath}} & {}^{A}\hat{\boldsymbol{\jmath}} & {}^{A}\hat{\boldsymbol{k}} \\ \downarrow & \downarrow & \downarrow \end{pmatrix} = \begin{pmatrix} i_x & j_x & k_x \\ i_y & j_y & k_y \\ i_z & j_z & k_z \end{pmatrix}, \tag{3.12}$$

and

$$ {}^{K}_{K'}\boldsymbol{Q} = \boldsymbol{Q}_z(\theta), \tag{3.13}$$

and

$$ {}^{K'}_{B}\boldsymbol{Q} = {}^{K}_{A}\boldsymbol{Q} = {}^{A}_{K}\boldsymbol{Q}^T. \tag{3.14}$$

So,

$$ {}^{A}_{B}\boldsymbol{Q} = \boldsymbol{Q}_k(\theta) = {}^{A}_{K}\boldsymbol{Q}\boldsymbol{Q}_z(\theta){}^{K}_{A}\boldsymbol{Q}, \tag{3.15}$$

and

$$ \boldsymbol{Q}_k(\theta) = \begin{pmatrix} \uparrow & \uparrow & \uparrow \\ {}^{A}\hat{\boldsymbol{\imath}} & {}^{A}\hat{\boldsymbol{\jmath}} & {}^{A}\hat{\boldsymbol{k}} \\ \downarrow & \downarrow & \downarrow \end{pmatrix} \begin{pmatrix} \cos\theta & -\sin\theta & 0 \\ \sin\theta & \cos\theta & 0 \\ 0 & 0 & 1 \end{pmatrix} \begin{pmatrix} \leftarrow & {}^{A}\hat{\boldsymbol{\imath}}^T & \rightarrow \\ \leftarrow & {}^{A}\hat{\boldsymbol{\jmath}}^T & \rightarrow \\ \leftarrow & {}^{A}\hat{\boldsymbol{k}}^T & \rightarrow \end{pmatrix}. \tag{3.16}$$

The components of $\hat{\boldsymbol{\imath}}$ and $\hat{\boldsymbol{\jmath}}$ drop out, so

$$ \boldsymbol{Q}_k(\theta) = \begin{pmatrix} k_xk_x(1-c\theta)+c\theta & k_xk_y(1-c\theta)-k_zs\theta & k_xk_z(1-c\theta)+k_ys\theta \\ k_xk_y(1-c\theta)+k_zs\theta & k_yk_y(1-c\theta)+c\theta & k_yk_z(1-c\theta)-k_xs\theta \\ k_xk_z(1-c\theta)-k_ys\theta & k_yk_z(1-c\theta)+k_xs\theta & k_zk_z(1-c\theta)+c\theta \end{pmatrix}, \tag{3.17}$$

where $s\theta$ and $c\theta$ have been used as shorthand for $\sin(\theta)$ and $\cos(\theta)$. Conversely, the axis-angle parameters, expressed in terms of $\boldsymbol{Q}$, are

$$ \begin{aligned} \theta &= \cos^{-1}\left(\frac{Q_{11}+Q_{22}+Q_{33}-1}{2}\right), \\ k_x &= \frac{Q_{32}-Q_{23}}{2\sin\theta}, \\ k_y &= \frac{Q_{13}-Q_{31}}{2\sin\theta}, \\ k_z &= \frac{Q_{21}-Q_{12}}{2\sin\theta}. \end{aligned} \tag{3.18}$$

3.1.3 Euler Sets

The Euler angle scheme specifies a sequence of relative frame rotations about the principal axes. For example, an xyz sequence specifies a rotation of α about the x axis of the base frame, $\mathcal{A}$. Next a rotation of β is specified about the y axis of the intermediate frame associated with the completion of the first rotation, $\mathcal{A}'$. Finally a rotation of γ is specified about the z axis of the intermediate frame associated with the completion of the second rotation, $\mathcal{A}''$. Figure 3.3 depicts this sequence.

The rotation matrix, ${}^{A}_{B}\boldsymbol{Q}$, associated with this rotation sequence is given by

$$ {}^{A}_{B}\boldsymbol{Q}(\alpha,\beta,\gamma) = {}^{A}_{A'}\boldsymbol{Q}(\alpha){}^{A'}_{A''}\boldsymbol{Q}(\beta){}^{A''}_{B}\boldsymbol{Q}(\gamma). \tag{3.19}$$

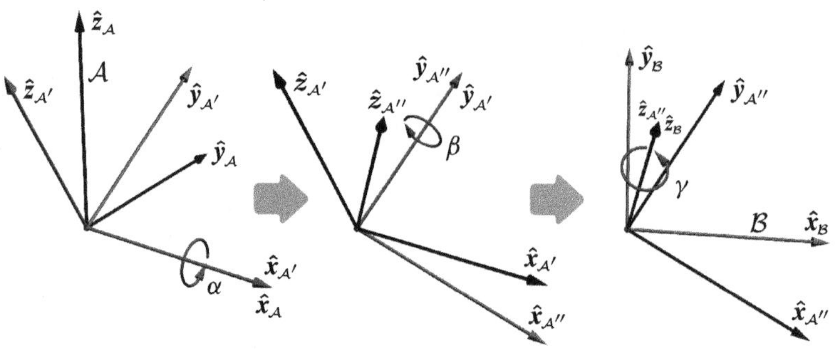

Figure 3.3 An *xyz* Euler angle sequence specifies a rotation of α about the x axis of the base frame, $\mathcal{A}$. Next, a rotation of β is specified about the y axis of the intermediate frame, $\mathcal{A}'$. Finally a rotation of γ is specified about the z axis of the intermediate frame, $\mathcal{A}''$.

Since all of these intermittent rotations are about principal axes, we have

$$
\begin{aligned}
{}^{\mathcal{A}}_{\mathcal{B}}\boldsymbol{Q}(\alpha,\beta,\gamma) &= \boldsymbol{Q}_{xyz}(\alpha,\beta,\gamma) = \boldsymbol{Q}_x(\alpha)\boldsymbol{Q}_y(\beta)\boldsymbol{Q}_z(\gamma) \\
&= \begin{pmatrix} 1 & 0 & 0 \\ 0 & \cos(\alpha) & -\sin(\alpha) \\ 0 & \sin(\alpha) & \cos(\alpha) \end{pmatrix} \begin{pmatrix} \cos(\beta) & 0 & \sin(\beta) \\ 0 & 1 & 0 \\ -\sin(\beta) & 0 & \cos(\beta) \end{pmatrix} \begin{pmatrix} \cos(\gamma) & -\sin(\gamma) & 0 \\ \sin(\gamma) & \cos(\gamma) & 0 \\ 0 & 0 & 1 \end{pmatrix}.
\end{aligned}
\tag{3.20}
$$

So,

$$
{}^{\mathcal{A}}_{\mathcal{B}}\boldsymbol{Q}(\alpha,\beta,\gamma) = \begin{pmatrix} c\beta c\gamma & -c\beta s\gamma & s\beta \\ s\alpha s\beta c\gamma + c\alpha s\gamma & -s\alpha s\beta s\gamma + c\alpha c\gamma & -s\alpha c\beta \\ -c\alpha s\beta c\gamma + s\alpha s\gamma & c\alpha s\beta s\gamma + s\alpha c\gamma & c\alpha c\beta \end{pmatrix}. \tag{3.21}
$$

In general, for an arbitrary sequence *abc*, where

$$
\begin{aligned}
a &= x, y, \text{ or } z, \\
b &= x, y, \text{ or } z, \\
c &= x, y, \text{ or } z, \\
a &\neq b, \; b \neq c,
\end{aligned}
\tag{3.22}
$$

we have

$$
\boldsymbol{Q}_{abc}(\alpha,\beta,\gamma) = \boldsymbol{Q}_a(\alpha)\boldsymbol{Q}_b(\beta)\boldsymbol{Q}_c(\gamma). \tag{3.23}
$$

The inverse problem of determining the Euler angles in terms of the rotation matrix requires a bit more effort. For the *xyz* sequence addressed earlier, we have the following solution.

If $\cos\beta \neq 0$, $(\beta \neq \pm\pi/2)$,

$$\beta = \text{Atan2}(Q_{13}, \sqrt{Q_{11}^2 + Q_{12}^2}), \tag{3.24}$$

$$\alpha = \text{Atan2}(-Q_{23}/c\beta, Q_{33}/c\beta), \tag{3.25}$$

$$\gamma = \text{Atan2}(-Q_{12}/c\beta, Q_{11}/c\beta). \tag{3.26}$$

If $\beta = \pm\pi/2$,

$$\alpha = 0 \tag{3.27}$$

$$\gamma = \text{Atan2}(Q_{21}, Q_{22}). \tag{3.28}$$

For an xzx Euler sequence we have

$$\boldsymbol{Q}_{xzx}(\alpha, \beta, \gamma) = \begin{pmatrix} c\beta & -s\beta c\gamma & s\beta s\gamma \\ c\alpha s\beta & c\alpha c\beta c\gamma - s\alpha s\gamma & -c\alpha c\beta s\gamma - s\alpha c\gamma \\ s\alpha s\beta & s\alpha c\beta c\gamma + c\alpha s\gamma & -s\alpha c\beta s\gamma + c\alpha c\gamma \end{pmatrix}, \tag{3.29}$$

and the solution to the inverse problem is

If $\sin\beta \neq 0$, $(\beta \neq 0, \pi)$,

$$\beta = \text{Atan2}(\sqrt{Q_{21}^2 + Q_{31}^2}, Q_{11}), \tag{3.30}$$

$$\alpha = \text{Atan2}(Q_{31}/s\beta, Q_{21}/s\beta), \tag{3.31}$$

$$\gamma = \text{Atan2}(Q_{13}/s\beta, -Q_{12}/s\beta). \tag{3.32}$$

If $\beta = 0$,

$$\alpha = 0 \tag{3.33}$$

$$\gamma = \text{Atan2}(Q_{32}, Q_{22}). \tag{3.34}$$

If $\beta = \pi$,

$$\alpha = 0 \tag{3.35}$$

$$\gamma = \text{Atan2}(Q_{32}, -Q_{22}). \tag{3.36}$$

3.1.4 Quaternions

A quaternion is a hypercomplex number which is the 4-dimensional analog to the 2-dimensional complex number $z = a + b\boldsymbol{i}$. The 4-dimensional space of quaternions, denoted by $\mathbb{H}$, is spanned by four orthogonal axes. These include the real axis and three principal imaginaries, $\boldsymbol{i}$, $\boldsymbol{j}$, and $\boldsymbol{k}$:

$$h = h_0 + h_1\boldsymbol{i} + h_2\boldsymbol{j} + h_3\boldsymbol{k}. \tag{3.37}$$

This can also be thought of as a scalar part, h_0, and a vector part, $\boldsymbol{h}$:

$$h = h_0 + \boldsymbol{h}. \tag{3.38}$$

Quaternion algebra is noncommutative with respect to multiplication. Specifically, the following laws apply:

$$
\begin{aligned}
i^2 = j^2 = k^2 = ijk &= -1, \\
ij = k &= -ji, \\
jk = i &= -kj, \\
ki = j &= -ik.
\end{aligned}
\tag{3.39}
$$

Multiplication of two quaternions can most easily be represented using the operations of vector cross-product and dot product:

$$
\begin{aligned}
s &= gh = s_0 + \boldsymbol{s} \\
s &= g_0 h_0 - \boldsymbol{g} \cdot \boldsymbol{h} + g_0 \boldsymbol{h} + h_0 \boldsymbol{g} + \boldsymbol{g} \times \boldsymbol{h}.
\end{aligned}
\tag{3.40}
$$

Grouping the scalar and vector parts separately, we have

$$
\begin{aligned}
s_0 &= g_0 h_0 - \boldsymbol{g} \cdot \boldsymbol{h} \\
\boldsymbol{s} &= g_0 \boldsymbol{h} + h_0 \boldsymbol{g} + \boldsymbol{g} \times \boldsymbol{h}.
\end{aligned}
\tag{3.41}
$$

It is stressed that quaternions are not vectors but rather hypercomplex numbers. Nevertheless, the components of the imaginary part of a quaternion can be used in traditional vector operations to compute quaternion products per the preceding formula. An equivalent algorithm exists for representing multiplication of two quaternions using matrix operations:

$$
\begin{pmatrix} s_0 & s_1 & s_2 & s_3 \end{pmatrix} = \begin{pmatrix} g_0 & g_1 & g_2 & g_3 \end{pmatrix} \boldsymbol{H}, \tag{3.42}
$$

where $\boldsymbol{H}$ is an antisymmetric matrix defined in terms of h as

$$
\boldsymbol{H} = \begin{pmatrix} h_0 & h_1 & h_2 & h_3 \\ -h_1 & h_0 & -h_3 & h_2 \\ -h_2 & h_3 & h_0 & -h_1 \\ -h_3 & -h_2 & h_1 & h_0 \end{pmatrix}. \tag{3.43}
$$

It is also convenient to represent a quaternion using complex matrices. First we define the following matrices in $\mathbb{C}^{2\times 2}$:

$$
\mathbf{1} = \begin{pmatrix} 1 & 0 \\ 0 & 1 \end{pmatrix}, \quad \boldsymbol{I} = \begin{pmatrix} i & 0 \\ 0 & -i \end{pmatrix}, \quad \boldsymbol{J} = \begin{pmatrix} 0 & 1 \\ -1 & 0 \end{pmatrix}, \quad \boldsymbol{K} = \begin{pmatrix} 0 & i \\ i & 0 \end{pmatrix}. \tag{3.44}
$$

These are derived from the so-called Pauli matrices. The principal imaginaries, $\boldsymbol{I}, \boldsymbol{J}$, and $\boldsymbol{K}$, adhere to the same laws stated earlier, namely,

$$
\begin{aligned}
\boldsymbol{I}^2 = \boldsymbol{J}^2 = \boldsymbol{K}^2 = \boldsymbol{IJK} &= -\mathbf{1}, \\
\boldsymbol{IJ} = \boldsymbol{K} &= -\boldsymbol{JI}, \\
\boldsymbol{JK} = \boldsymbol{I} &= -\boldsymbol{KJ}, \\
\boldsymbol{KI} = \boldsymbol{J} &= -\boldsymbol{IK}.
\end{aligned}
\tag{3.45}
$$

The quaternion is then

$$H = h_0\mathbf{1} + h_1\boldsymbol{I} + h_2\boldsymbol{J} + h_3\boldsymbol{K} = \begin{pmatrix} h_0 + ih_1 & h_2 + ih_3 \\ -h_2 + ih_3 & h_0 - ih_1 \end{pmatrix}. \tag{3.46}$$

While this is a useful representation of quaternions, we will use the conventional representation throughout the remainder of this text, namely, $h = h_0 + h_1\boldsymbol{i} + h_2\boldsymbol{j} + h_3\boldsymbol{k}$. We can define the inverse of a quaternion, h^{-1}, as satisfying

$$hh^{-1} = h^{-1}h = 1 + 0\boldsymbol{i} + 0\boldsymbol{j} + 0\boldsymbol{k} = 1. \tag{3.47}$$

Noting that the product of a quaternion and its conjugate, $\bar{h}$, is given by

$$\begin{aligned} h\bar{h} &= (h_0 + h_1\boldsymbol{i} + h_2\boldsymbol{j} + h_3\boldsymbol{k})(h_0 - h_1\boldsymbol{i} - h_2\boldsymbol{j} - h_3\boldsymbol{k}) \\ &= h_0h_0 - \boldsymbol{h}\cdot\boldsymbol{h} + h_0\boldsymbol{h} + h_0\boldsymbol{h} + \boldsymbol{h}\times\boldsymbol{h} \\ &= h_0h_0 + h_1h_1 + h_2h_2 + h_3h_3 = \|h\|^2, \end{aligned} \tag{3.48}$$

we have

$$h^{-1} = \frac{\bar{h}}{\|h\|^2}. \tag{3.49}$$

A unit quaternion $\{h | \|h\| = 1\}$ is a point on a unit hypersphere, $\mathbb{S}^3 \subset \mathbb{H}$, where $h_0^2 + h_1^2 + h_2^2 + h_3^2 = 1$. Unit quaternions are efficient at encoding spatial rotation. There is a direct relationship between the elements of a unit quaternion and the axis-angle parameters. For any unit quaternion, $h \in \mathbb{S}^3$, we have

$$h = \cos\frac{\theta}{2} + k_x\sin\frac{\theta}{2}\boldsymbol{i} + k_y\sin\frac{\theta}{2}\boldsymbol{j} + k_z\sin\frac{\theta}{2}\boldsymbol{k} = \cos\frac{\theta}{2} + \boldsymbol{u}\sin\frac{\theta}{2}, \tag{3.50}$$

where

$$\boldsymbol{u} = k_x\boldsymbol{i} + k_y\boldsymbol{j} + k_z\boldsymbol{k}. \tag{3.51}$$

In shorthand, we have

$$h = e^{\frac{\theta}{2}\boldsymbol{u}} \quad \text{and} \quad h^{-1} = \bar{h} = e^{-\frac{\theta}{2}\boldsymbol{u}}. \tag{3.52}$$

The individual elements of a unit quaternion are also referred to as Euler parameters, $\epsilon_1, \epsilon_2, \epsilon_3$, where

$$\begin{gathered} \epsilon_1 \triangleq h_1, \epsilon_2 \triangleq h_2, \epsilon_3 \triangleq h_3, \epsilon_4 \triangleq h_0, \\ \epsilon_1 = k_x\sin\frac{\theta}{2}, \epsilon_2 = k_y\sin\frac{\theta}{2}, \epsilon_3 = k_z\sin\frac{\theta}{2}, \epsilon_1 = \cos\frac{\theta}{2}. \end{gathered} \tag{3.53}$$

Rotational transformation can be accommodated with unit quaternions using a double product. To rotate a vector $\boldsymbol{v} \in \mathbb{R}^3$ about the $\boldsymbol{u}$ axis by an angle of θ, we perform the following:

$$h\boldsymbol{v}h^{-1} = h\boldsymbol{v}\bar{h} = e^{\frac{\theta}{2}\boldsymbol{u}}\boldsymbol{v}e^{-\frac{\theta}{2}\boldsymbol{u}}. \tag{3.54}$$

In this case the vector, $\boldsymbol{v}$, is represented as a pure quaternion (real part equal to *zero*):

$$\boldsymbol{v} = v_1\boldsymbol{i} + v_2\boldsymbol{j} + v_3\boldsymbol{k}. \tag{3.55}$$

In terms of frame transformations we have

$${}^A\boldsymbol{v} = {}^A_B h\,{}^B\boldsymbol{v}\,{}^B_A h = {}^A_B h\,{}^B\boldsymbol{v}\,{}^A_B\bar{h}, \tag{3.56}$$

where

$${}^A_B\bar{h} = {}^B_A h. \tag{3.57}$$

Additionally, multiple rotations can be concatenated using multiplication, for example,

$${}^A_C h = {}^A_B h\,{}^B_C h. \tag{3.58}$$

As with rotation matrices, this process is not commutative. That is, in general,

$${}^A_B h\,{}^B_C h \neq {}^B_C h\,{}^A_B h. \tag{3.59}$$

Thus, for both rotation matrices and unit quaternions, the order in which successive finite (noninfinitesimal) rotations are performed affects the resulting orientation.

It is useful to relate the unit quaternion to the other representations of orientation. For axis-angle parameters we have

$$\begin{aligned} h_0 &= \cos\tfrac{\theta}{2}, \\ h_1 &= k_x \sin\tfrac{\theta}{2}, \\ h_2 &= k_y \sin\tfrac{\theta}{2}, \\ h_3 &= k_z \sin\tfrac{\theta}{2}, \end{aligned} \qquad \text{and} \qquad \begin{aligned} \theta &= 2\cos^{-1} h_0, \\ k_x &= \tfrac{h_1}{\sqrt{1-h_0h_0}}, \\ k_y &= \tfrac{h_2}{\sqrt{1-h_0h_0}}, \\ k_z &= \tfrac{h_3}{\sqrt{1-h_0h_0}}. \end{aligned} \tag{3.60}$$

Relating a rotation matrix and unit quaternion, we have

$$\boldsymbol{Q}(h) = \begin{pmatrix} 2(h_0h_0 + h_1h_1) - 1 & 2(h_1h_2 - h_0h_3) & 2(h_1h_3 + h_0h_2) \\ 2(h_1h_2 + h_0h_3) & 2(h_0h_0 + h_2h_2) - 1 & 2(h_2h_3 - h_0h_1) \\ 2(h_1h_3 - h_0h_2) & 2(h_2h_3 + h_0h_1) & 2(h_0h_0 + h_3h_3) - 1 \end{pmatrix} \tag{3.61}$$

and

$$\begin{aligned} h_0 &= \frac{1}{2}\sqrt{1 + Q_{11} + Q_{22} + Q_{33}}, \\ h_1 &= \frac{Q_{32} - Q_{23}}{4h_0}, \\ h_2 &= \frac{Q_{13} - Q_{31}}{4h_0}, \\ h_3 &= \frac{Q_{21} - Q_{12}}{4h_0}. \end{aligned} \tag{3.62}$$

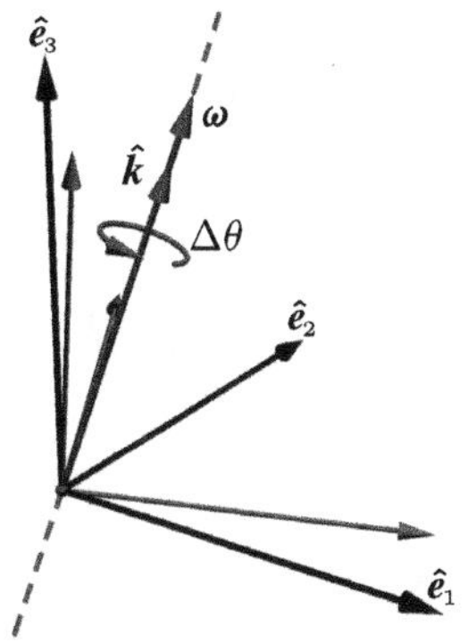

Figure 3.4 An infinitesimal rotation involving a differential spin, $\Delta\theta$, about a fixed instantaneous axis.

3.1.5 Calculus of Rotations

It will be useful to examine the rate of change for some of the orientation schemes that we have described. In particular, we can express the differential change of rotation matrices and unit quaternions as a differential spin about a fixed instantaneous axis. In doing so, we arrive at the concept of angular velocity, thereby relating the derivative of an orientation operator to angular velocity. A relationship between angle-set derivatives and angular velocity can also be derived.

We will now define the derivative of a rotation operator. Let us begin with the rotation matrix

$$\dot{\boldsymbol{Q}} = \frac{d\boldsymbol{Q}}{dt} = \lim_{\Delta t \to 0} \frac{\boldsymbol{Q}(t + \Delta t) - \boldsymbol{Q}(t)}{\Delta t}. \tag{3.63}$$

It will be convenient to express the time rate of change of a rotation operator in terms of an instantaneous spin rate and axis. This will entail applying a differential spin, $\Delta\theta$, about a fixed instantaneous axis. Figure 3.4 depicts this infinitesimal rotation. Noting that

$$\boldsymbol{Q}(t + \Delta t) = \boldsymbol{Q}_k(\Delta\theta)\boldsymbol{Q}(t), \tag{3.64}$$

the derivative can then be expressed as

$$\dot{\boldsymbol{Q}} = \lim_{\Delta t \to 0} \frac{\boldsymbol{Q}_k(\Delta\theta)\boldsymbol{Q}(t) - \boldsymbol{Q}(t)}{\Delta t} = \left(\lim_{\Delta t \to 0} \frac{\boldsymbol{Q}_k(\Delta\theta) - \boldsymbol{1}}{\Delta t} \right) \boldsymbol{Q}(t). \tag{3.65}$$

Noting small angle (infinitesimal) properties, $\boldsymbol{Q}_k(\Delta\theta)$ is

$$\boldsymbol{Q}_k(\Delta\theta) = \begin{pmatrix} 1 & -k_z\Delta\theta & k_y\Delta\theta \\ k_z\Delta\theta & 1 & -k_x\Delta\theta \\ -k_y\Delta\theta & k_x\Delta\theta & 1 \end{pmatrix}. \tag{3.66}$$

We then have

$$\dot{\boldsymbol{Q}} = \begin{pmatrix} 0 & -k_z\dot{\theta} & k_y\dot{\theta} \\ k_z\dot{\theta} & 0 & -k_x\dot{\theta} \\ -k_y\dot{\theta} & k_x\dot{\theta} & 0 \end{pmatrix} \boldsymbol{Q}(t) = \begin{pmatrix} 0 & -\omega_z & \omega_y \\ \omega_z & 0 & -\omega_x \\ -\omega_y & \omega_x & 0 \end{pmatrix} \boldsymbol{Q}(t), \tag{3.67}$$

where $\boldsymbol{\omega} = (\omega_x \ \omega_y \ \omega_z)^T$ is the angular velocity vector. Defining the antisymmetric angular velocity tensor, $\boldsymbol{\Omega}$, as

$$\boldsymbol{\Omega} = \begin{pmatrix} 0 & -\omega_z & \omega_y \\ \omega_z & 0 & -\omega_x \\ -\omega_y & \omega_x & 0 \end{pmatrix}, \tag{3.68}$$

we have

$$\dot{\boldsymbol{Q}} = \boldsymbol{\Omega}\boldsymbol{Q} = \begin{pmatrix} 0 & -\omega_z & \omega_y \\ \omega_z & 0 & -\omega_x \\ -\omega_y & \omega_x & 0 \end{pmatrix} \boldsymbol{Q}. \tag{3.69}$$

The matrix, $\boldsymbol{\Omega}$, is a rank 2 tensor and can thus be represented as the following dyadic:

$$\begin{aligned} \boldsymbol{\Omega} = \sum_{i=1}^{3}\sum_{j=1}^{3} \Omega_{ij}\hat{\boldsymbol{e}}_i \otimes \hat{\boldsymbol{e}}_j &= \omega_z(\hat{\boldsymbol{e}}_2 \otimes \hat{\boldsymbol{e}}_1 - \hat{\boldsymbol{e}}_1 \otimes \hat{\boldsymbol{e}}_2) \\ &+ \omega_y(\hat{\boldsymbol{e}}_1 \otimes \hat{\boldsymbol{e}}_3 - \hat{\boldsymbol{e}}_3 \otimes \hat{\boldsymbol{e}}_1) + \omega_x(\hat{\boldsymbol{e}}_3 \otimes \hat{\boldsymbol{e}}_2 - \hat{\boldsymbol{e}}_2 \otimes \hat{\boldsymbol{e}}_3). \end{aligned} \tag{3.70}$$

This has been formulated with the spin axis represented in the base frame. Any frame representation can be used:

$${}^{A}_{B}\dot{\boldsymbol{Q}} = {}^{A}\boldsymbol{\Omega}\,{}^{A}_{B}\boldsymbol{Q} = {}^{A}_{B}\boldsymbol{Q}\,{}^{B}\boldsymbol{\Omega}. \tag{3.71}$$

Frame transformations for the angular velocity tensor can be accommodated with the tensor double product

$${}^{A}\boldsymbol{\Omega} = {}^{A}_{B}\boldsymbol{Q}\,{}^{B}\boldsymbol{\Omega}\,{}^{B}_{A}\boldsymbol{Q}. \tag{3.72}$$

The product of the angular velocity tensor with a vector is equivalent to the cross-product of the angular velocity vector and a vector, that is,

$${}^{A}\boldsymbol{\Omega}_{A}\,{}^{A}\boldsymbol{r}_P = {}^{A}\boldsymbol{\omega}_A \times {}^{A}\boldsymbol{r}_P. \tag{3.73}$$

Defining the derivative of a unit quaternion, we have

$$\dot{h} = \frac{dh}{dt} = \lim_{\Delta t \to 0} \frac{h(t + \Delta t) - h(t)}{\Delta t}. \tag{3.74}$$

Applying a differential spin, $\Delta\theta$, about a fixed instantaneous axis (see Figure 3.4) gives us

$$h(t + \Delta t) = h_k(\Delta\theta)h(t). \tag{3.75}$$

The derivative can then be expressed as

$$\dot{h} = \lim_{\Delta t \to 0} \frac{h_k(\Delta\theta)h(t) - h(t)}{\Delta t} = \left(\lim_{\Delta t \to 0} \frac{h_k(\Delta\theta) - 1}{\Delta t} \right) h(t), \tag{3.76}$$

Table 3.1 Dualities between the rotation matrix and the unit quaternion

Property	Rotation Matrix	Quaternion
Identity	$\mathbf{1} = \begin{pmatrix} 1 & 0 & 0 \\ 0 & 1 & 0 \\ 0 & 0 & 1 \end{pmatrix}$	$1 = 1 + 0\boldsymbol{i} + 0\boldsymbol{j} + 0\boldsymbol{k}$
Inverse	$\boldsymbol{Q}^{-1} = \boldsymbol{Q}^T$	$h^{-1} = \bar{h}$
Derivative	$\dot{\boldsymbol{Q}} = \boldsymbol{\Omega Q}$	$\dot{h} = \frac{1}{2}\boldsymbol{\omega} h$
Angular velocity	$\boldsymbol{\Omega} = \begin{pmatrix} 0 & -\omega_z & \omega_y \\ \omega_z & 0 & -\omega_x \\ -\omega_y & \omega_x & 0 \end{pmatrix}$	$\boldsymbol{\omega} = 0 + \omega_x \boldsymbol{i} + \omega_y \boldsymbol{j} + \omega_z \boldsymbol{k}$
Transformation	${}^A\boldsymbol{\Omega} = {}^A_B\boldsymbol{Q}\,{}^B\boldsymbol{\Omega}\,{}^B_A\boldsymbol{Q}$	${}^A\boldsymbol{\omega} = {}^A_B h\,{}^B\boldsymbol{\omega}\,{}^B_A h$

where, in the limit, $h_k(\Delta\theta)$ is

$$h_k(\Delta\theta) = 1 + k_x \frac{\Delta\theta}{2}\boldsymbol{i} + k_y \frac{\Delta\theta}{2}\boldsymbol{j} + k_z \frac{\Delta\theta}{2}\boldsymbol{k}. \tag{3.77}$$

We then have

$$\dot{h} = \left(0 + k_x \frac{\dot{\theta}}{2}\boldsymbol{i} + k_y \frac{\dot{\theta}}{2}\boldsymbol{j} + k_z \frac{\dot{\theta}}{2}\boldsymbol{k}\right) h(t) = \left(\frac{\omega_x}{2}\boldsymbol{i} + \frac{\omega_y}{2}\boldsymbol{j} + \frac{\omega_z}{2}\boldsymbol{k}\right) h(t), \tag{3.78}$$

where $\boldsymbol{\omega}$ is the angular velocity represented as a quaternion:

$$\boldsymbol{\omega} = \omega_x \boldsymbol{i} + \omega_y \boldsymbol{j} + \omega_z \boldsymbol{k}. \tag{3.79}$$

So, we have

$$\dot{h} = \frac{1}{2}\boldsymbol{\omega} h. \tag{3.80}$$

As in the case of the derivative of the rotation matrix, this has been formulated with the spin axis represented in the base frame. Any frame representation can be used:

$${}^A_B\dot{h} = \frac{1}{2}\,{}^A\boldsymbol{\omega}\,{}^A_B h = \frac{1}{2}\,{}^A_B h\,{}^B\boldsymbol{\omega}. \tag{3.81}$$

Frame transformations for the angular velocity quaternion can be accommodated with the familiar tensor double product,

$${}^A\boldsymbol{\omega} = {}^A_B h\,{}^B\boldsymbol{\omega}\,{}^B_A h. \tag{3.82}$$

Table 3.1 lists various dualities between the rotation matrix and the unit quaternion.

We have thus far related rotation matrices and quaternions to angular velocity. We can also relate rates of change of Euler angle sets to angular velocity. The rate vector of a given angle set is

$$\dot{\boldsymbol{\phi}} = \begin{pmatrix} \dot{\alpha} \\ \dot{\beta} \\ \dot{\gamma} \end{pmatrix}. \tag{3.83}$$

Noting the relationship between a rotation matrix and angular velocity,

$$\boldsymbol{\Omega} = \begin{pmatrix} 0 & -\omega_z & \omega_y \\ \omega_z & 0 & -\omega_x \\ -\omega_y & \omega_x & 0 \end{pmatrix} = \dot{\boldsymbol{Q}}\boldsymbol{Q}^T, \tag{3.84}$$

we can determine the following:

$$\boldsymbol{\omega} = \begin{pmatrix} \omega_x \\ \omega_y \\ \omega_z \end{pmatrix} = \begin{pmatrix} \dot{Q}_{31}Q_{21} + \dot{Q}_{32}Q_{22} + \dot{Q}_{33}Q_{23} \\ \dot{Q}_{11}Q_{31} + \dot{Q}_{12}Q_{32} + \dot{Q}_{13}Q_{33} \\ \dot{Q}_{21}Q_{11} + \dot{Q}_{22}Q_{12} + \dot{Q}_{23}Q_{13} \end{pmatrix}. \tag{3.85}$$

We note that

$$\dot{Q}_{ij} = \frac{\partial Q_{ij}}{\partial \boldsymbol{\phi}}\dot{\boldsymbol{\phi}}. \tag{3.86}$$

So angular velocity is related to the angle set rate vector by the following relationship:

$$\boldsymbol{\omega} = \begin{pmatrix} \omega_x \\ \omega_y \\ \omega_z \end{pmatrix} = \begin{pmatrix} \frac{\partial Q_{31}}{\partial \boldsymbol{\phi}}Q_{21} + \frac{\partial Q_{32}}{\partial \boldsymbol{\phi}}Q_{22} + \frac{\partial Q_{33}}{\partial \boldsymbol{\phi}}Q_{23} \\ \frac{\partial Q_{11}}{\partial \boldsymbol{\phi}}Q_{31} + \frac{\partial Q_{12}}{\partial \boldsymbol{\phi}}Q_{32} + \frac{\partial Q_{13}}{\partial \boldsymbol{\phi}}Q_{33} \\ \frac{\partial Q_{21}}{\partial \boldsymbol{\phi}}Q_{11} + \frac{\partial Q_{22}}{\partial \boldsymbol{\phi}}Q_{12} + \frac{\partial Q_{23}}{\partial \boldsymbol{\phi}}Q_{13} \end{pmatrix} \dot{\boldsymbol{\phi}} = \boldsymbol{E}(\boldsymbol{\phi})\dot{\boldsymbol{\phi}}, \tag{3.87}$$

where $\boldsymbol{E}(\boldsymbol{\phi})$ is the Jacobian between angular velocity and angle set rates.

Example: We now address the motion of an atmospheric reentry body. The task of determining the orientation history of a reentry body requires a knowledge of the gyroscope data. Figure 3.5 depicts an incremental change in a body's orientation in $\mathbb{R}^3$. The incremental spin, $\Delta\theta$, about an axis, $\hat{\boldsymbol{k}}$, can be related to the instantaneous angular velocity, $\boldsymbol{\omega}$, of the body. We can use quaternions as an efficient means of representing orientation resulting from incremental motion.

Since the gyroscopes provide angular velocity in the body's local reference frame, we can approximate the orientation quaternion, ${}^{o}_{c}h(t + \Delta t)$, of the body based on the previous orientation, ${}^{o}_{c}h(t)$, and the incremental quaternion, $h_k(\Delta\theta)$, associated with a finite but small incremental rotation, $\Delta\theta$, about an axis, $\hat{\boldsymbol{k}}$, that is fixed during the rotation. This can be expressed as

$${}^{o}_{c}h(t + \Delta t) \approx {}^{o}_{c}h(t)h_k(\Delta\theta). \tag{3.88}$$

Note that the spin axis of the incremental rotation is represented in the local body frame in (3.88) as opposed to the base frame since the gyroscopes measure angular velocity in

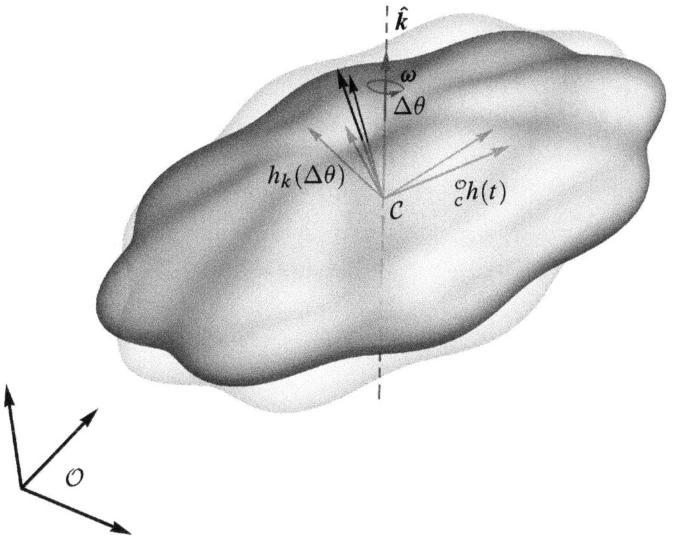

Figure 3.5 Instantaneous spin of a body about an axis. The incremental spin, $\Delta\theta$, about an axis, $\hat{\boldsymbol{k}}$, can be related to the instantaneous angular velocity, $\boldsymbol{\omega}$, of the body. Quaternions provide a numerically efficient means of integrating angular velocity to yield finite rotations.

the local body frame. Due to small angle properties,

$$h_k(\Delta\theta) \approxeq 1 + k_x \frac{\Delta\theta}{2}\boldsymbol{i} + k_y \frac{\Delta\theta}{2}\boldsymbol{j} + k_z \frac{\Delta\theta}{2}\boldsymbol{k} = 1 + \frac{1}{2}\omega\Delta t, \tag{3.89}$$

where ω is the angular velocity quaternion,

$$\omega = 0 + \omega_x \boldsymbol{i} + \omega_y \boldsymbol{j} + \omega_z \boldsymbol{k}. \tag{3.90}$$

Thus, for numerical integration, we can use the following relationship:

$${}^{O}_{C}h(t + \Delta t) \approxeq {}^{O}_{C}h(t)\left(1 + \frac{\omega_x \Delta t}{2}\boldsymbol{i} + \frac{\omega_y \Delta t}{2}\boldsymbol{j} + \frac{\omega_z \Delta t}{2}\boldsymbol{k}\right). \tag{3.91}$$

If we had implemented this integration using orthogonal rotation matrices, we would have had

$${}^{O}_{C}\boldsymbol{Q}(t + \Delta t) \approxeq {}^{O}_{C}\boldsymbol{Q}(t)\begin{pmatrix} 1 & -\omega_z & \omega_y \\ \omega_z & 1 & -\omega_x \\ -\omega_y & \omega_x & 1 \end{pmatrix}\Delta t. \tag{3.92}$$

However, the orthogonality properties of ${}^{O}_{C}\boldsymbol{Q}(t)$ would degrade with successive multiplications at finite precision, causing significant problems. In the case of the unit quaternion relationship of (3.91) the only concern would be that the length of the quaternion would deviate from unity with successive multiplications. This could be easily corrected by renormalizing the quaternion as

$${}^{O}_{C}h(t + \Delta t) \approxeq \frac{{}^{O}_{C}h(t)\left(1 + \frac{1}{2}\omega\Delta t\right)}{\left\|{}^{O}_{C}h(t)\left(1 + \frac{1}{2}\omega\Delta t\right)\right\|}. \tag{3.93}$$

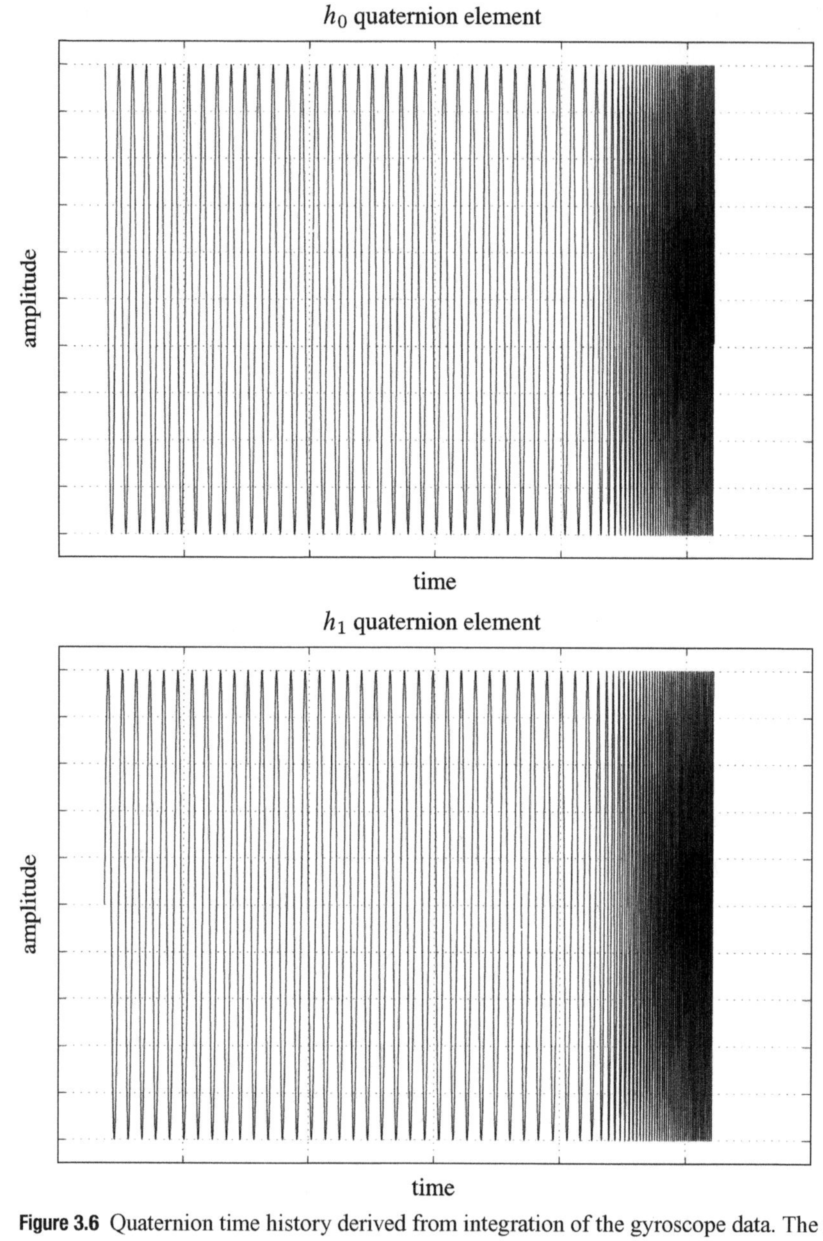

Figure 3.6 Quaternion time history derived from integration of the gyroscope data. The first two quaternion components are displayed. Units have been intentionally omitted.

It is also useful to represent the orientation in terms of Euler angles. We can convert from quaternions to rotation matrices using (3.61). We can then convert from $\boldsymbol{Q}(h)$ to an xzx Euler set, $\{\alpha, \beta, \gamma\}$, using the solution for xzx Euler angles in terms of the components of $\boldsymbol{Q}$, given by (3.30) through (3.36).

Figures 3.6 and 3.7 display plots of quaternion data calculated from gyroscope data. The data were calculated using the algorithm described here. The quaternion elements h_2 and h_3 are of special interest since they encode the k_y and k_z components of the spin

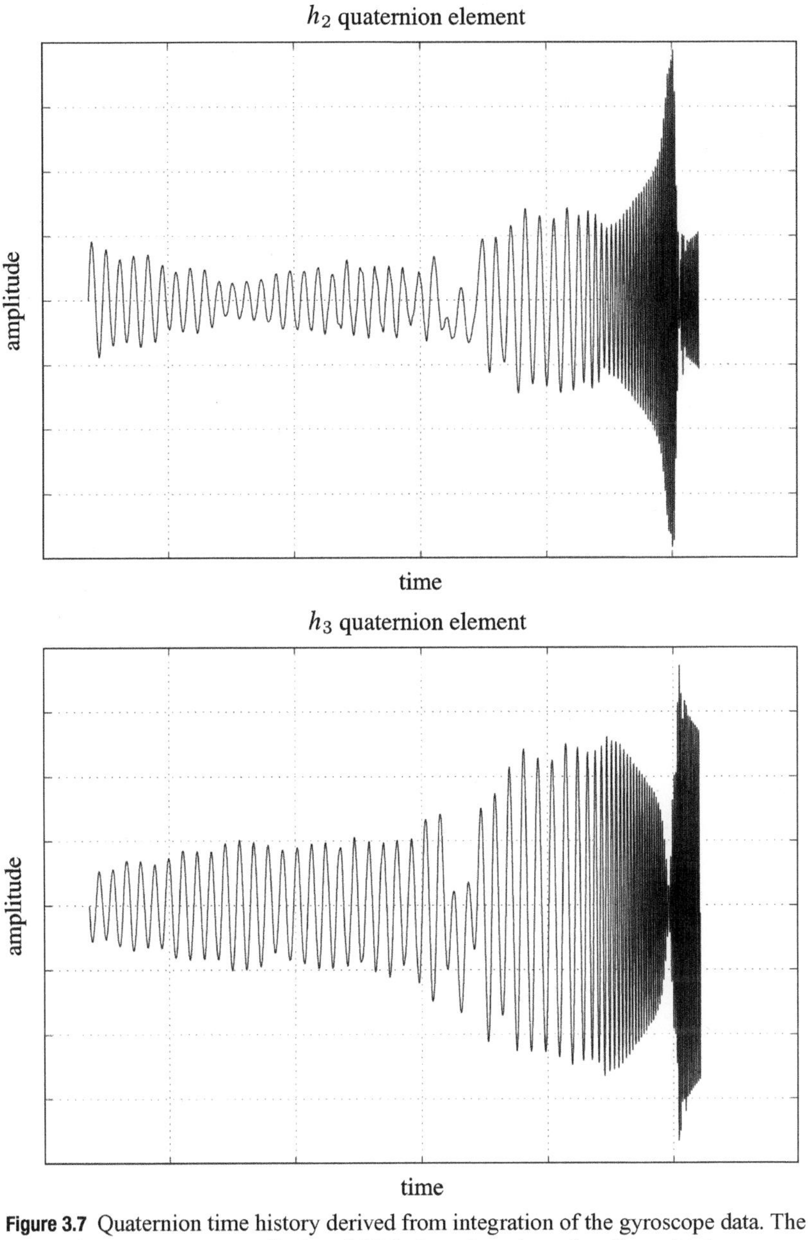

Figure 3.7 Quaternion time history derived from integration of the gyroscope data. The last two quaternion components are displayed. Units have been intentionally omitted.

axis, $\hat{\boldsymbol{k}}$. These components correspond to the lateral axes of the reentry body. While unit quaternions possess computational efficacy, Euler angles can provide an easier means of mentally decomposing the orientation of a body from 2-dimensional plots. Figures 3.8 and 3.9 display plots of the xzx Euler angles associated with the reentry body orientation. Of these angles, β is of particular interest since it indicates the angular displacement between the reentry body longitudinal axis and the base coordinate frame x-axis. The β

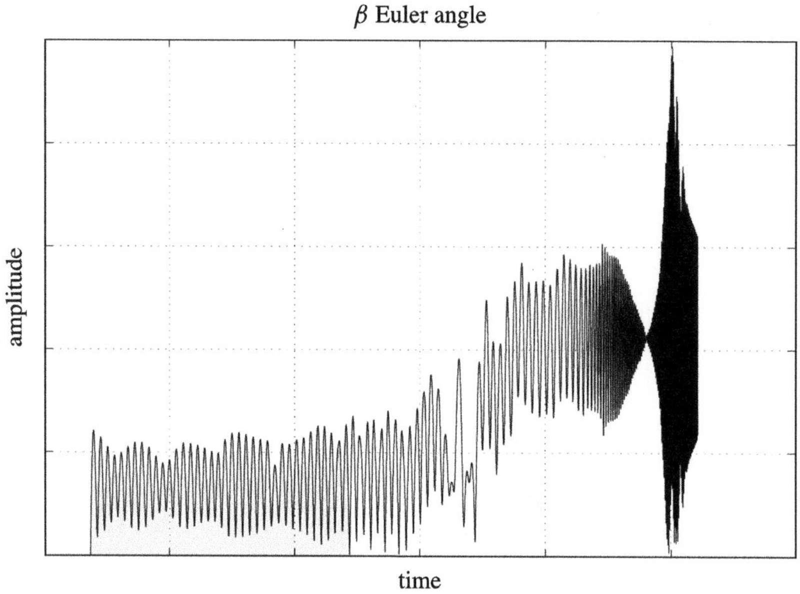

Figure 3.8 Euler angle time history derived from integration of the gyroscope data. The β Euler angle is displayed. Units have been intentionally omitted.

angle can also be related to the coning angle of the reentry body (discussed in Section 3.2.5).

Addition of Angular Velocities

We will now address the additive property of infinitesimal rotations. Consider a rotation varying over time, $\boldsymbol{Q}(t)$, which is the identity matrix at some time t. We take it to change due to two sequential infinitesimal rotations,

$$\boldsymbol{Q}(t+\Delta t) = \boldsymbol{Q}_{k_1}(\Delta\theta_1)\boldsymbol{Q}_{k_2}(\Delta\theta_2). \tag{3.94}$$

The derivative can then be expressed as

$$\dot{\boldsymbol{Q}} = \lim_{\Delta t \to 0} \frac{\boldsymbol{Q}_{k_1}(\Delta\theta_1)\boldsymbol{Q}_{k_2}(\Delta\theta_2) - \mathbf{1}}{\Delta t}. \tag{3.95}$$

Noting small angle (infinitesimal) properties, the product, $\boldsymbol{Q}_{k_1}\boldsymbol{Q}_{k_2}$, is

$$\boldsymbol{Q}_{k_1}\boldsymbol{Q}_{k_2} = \begin{pmatrix} 1-(k_{1_y}k_{2_y}+k_{1_z}k_{2_z})\Delta\theta_1\Delta\theta_2 & -k_{1_z}\Delta\theta_1 - k_{2_z}\Delta\theta_2 + k_{1_y}k_{2_x}\Delta\theta_1\Delta\theta_2 & k_{1_y}\Delta\theta_1 + k_{2_y}\Delta\theta_2 + k_{1_z}k_{2_x}\Delta\theta_1\Delta\theta_2 \\ k_{1_z}\Delta\theta_1 + k_{2_z}\Delta\theta_2 + k_{1_x}k_{2_y}\Delta\theta_1\Delta\theta_2 & 1-(k_{1_x}k_{2_x}+k_{1_z}k_{2_z})\Delta\theta_1\Delta\theta_2 & -k_{1_x}\Delta\theta_1 - k_{2_x}\Delta\theta_2 + k_{1_z}k_{2_y}\Delta\theta_1\Delta\theta_2 \\ -k_{1_y}\Delta\theta_1 - k_{2_y}\Delta\theta_2 + k_{1_x}k_{2_z}\Delta\theta_1\Delta\theta_2 & k_{1_x}\Delta\theta_1 + k_{2_x}\Delta\theta_2 + k_{1_y}k_{2_z}\Delta\theta_1\Delta\theta_2 & 1-(k_{1_x}k_{2_x}+k_{1_y}k_{2_y})\Delta\theta_1\Delta\theta_2 \end{pmatrix}. \tag{3.96}$$

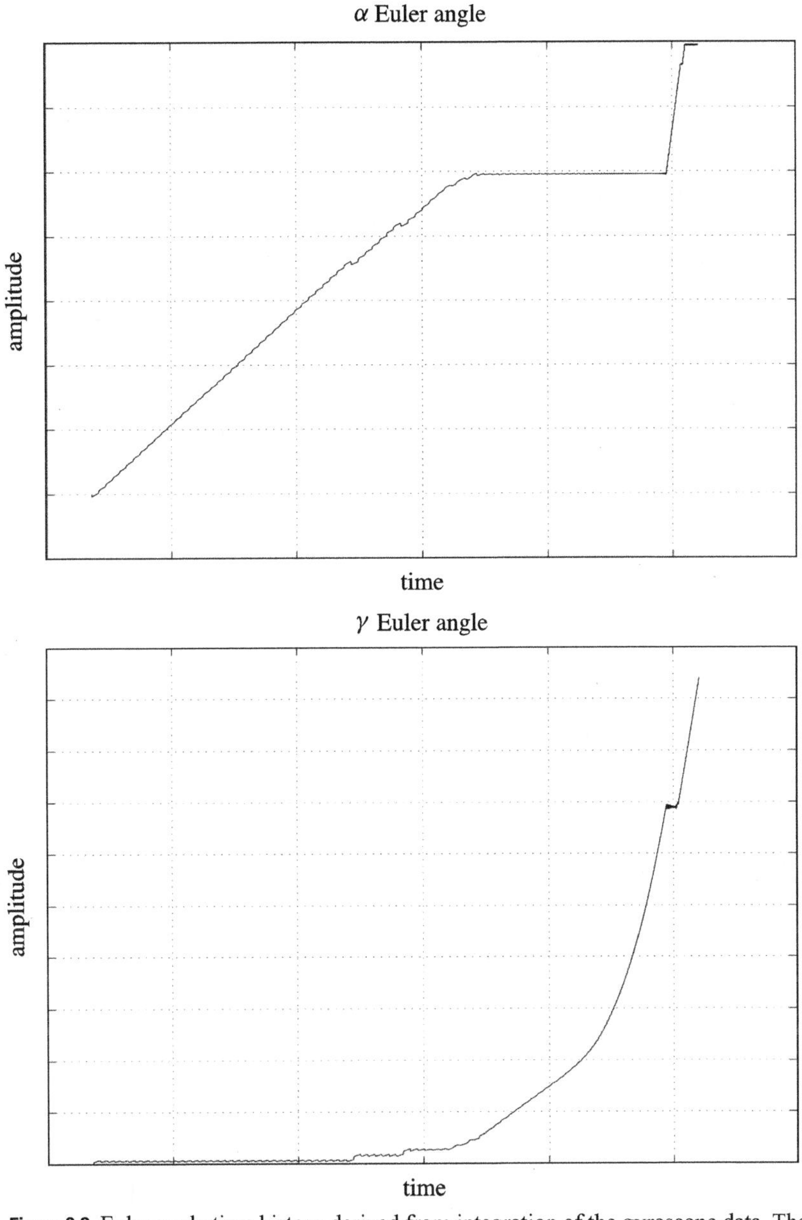

Figure 3.9 Euler angle time history derived from integration of the gyroscope data. The α and γ Euler angles are displayed. Units have been intentionally omitted.

Removing all terms involving second-order infinitesimals, $\Delta\theta_1 \Delta\theta_2$,

$$\boldsymbol{Q}_{k_1} \boldsymbol{Q}_{k_2} = \begin{pmatrix} 1 & -k_{1_z}\Delta\theta_1 - k_{2_z}\Delta\theta_2 & k_{1_y}\Delta\theta_1 + k_{2_y}\Delta\theta_2 \\ k_{1_z}\Delta\theta_1 + k_{2_z}\Delta\theta_2 & 1 & -k_{1_x}\Delta\theta_1 - k_{2_x}\Delta\theta_2 \\ -k_{1_y}\Delta\theta_1 - k_{2_y}\Delta\theta_2 & k_{1_x}\Delta\theta_1 + k_{2_x}\Delta\theta_2 & 1 \end{pmatrix}. \tag{3.97}$$

This demonstrates that, unlike finite rotations, infinitesimal rotations commute since we can observe that $\boldsymbol{Q}_{k_1}\boldsymbol{Q}_{k_2} = \boldsymbol{Q}_{k_2}\boldsymbol{Q}_{k_1}$. Thus, the order in which successive infinitesimal rotations are preformed does not affect the resulting orientation.

We then have

$$\dot{\boldsymbol{Q}} = \begin{pmatrix} 0 & -k_{1_z}\dot{\theta}_1 - k_{2_z}\dot{\theta}_2 & k_{1_y}\dot{\theta}_1 + k_{2_y}\dot{\theta}_2 \\ k_{1_z}\dot{\theta}_1 + k_{2_z}\dot{\theta}_2 & 0 & -k_{1_x}\dot{\theta}_1 - k_{2_x}\dot{\theta}_2 \\ -k_{1_y}\dot{\theta}_1 - k_{2_y}\dot{\theta}_2 & k_{1_x}\dot{\theta}_1 + k_{2_x}\dot{\theta}_2 & 0 \end{pmatrix}$$
$$= \begin{pmatrix} 0 & -\omega_{1_z} - \omega_{2_z} & \omega_{1_y} + \omega_{2_y} \\ \omega_{1_z} + \omega_{2_z} & 0 & -\omega_{1_x} - \omega_{2_x} \\ -\omega_{1_y} - \omega_{2_y} & \omega_{1_x} + \omega_{2_x} & 0 \end{pmatrix}. \tag{3.98}$$

So, the resultant angular velocity tensor is

$$\boldsymbol{\Omega} = \begin{pmatrix} 0 & -\omega_{1_z} - \omega_{2_z} & \omega_{1_y} + \omega_{2_y} \\ \omega_{1_z} + \omega_{2_z} & 0 & -\omega_{1_x} - \omega_{2_x} \\ -\omega_{1_y} - \omega_{2_y} & \omega_{1_x} + \omega_{2_x} & 0 \end{pmatrix}, \tag{3.99}$$

and the resultant angular velocity vector is

$$\boldsymbol{\omega} = \begin{pmatrix} \omega_{1_x} + \omega_{2_x} \\ \omega_{1_y} + \omega_{2_y} \\ \omega_{1_z} + \omega_{2_z} \end{pmatrix} = \boldsymbol{\omega}_1 + \boldsymbol{\omega}_2. \tag{3.100}$$

Equation (3.100) can be interpreted as the vector sum of a body's angular velocity, $\boldsymbol{\omega}_1$, relative to a given reference frame and the angular velocity, $\boldsymbol{\omega}_2$, of a second body *relative* to the first body. This yields the total angular velocity of the second body relative to the reference frame.

3.2 Spatial Kinematics

3.2.1 Homogeneous Transform Matrices

We generalize the notion of a finite displacement by considering the composition of a translation and a rotation. A position vector, ${}^{\mathcal{B}}\boldsymbol{r}_P$, represented in frame $\mathcal{B}$ can be represented in frame $\mathcal{A}$ by considering the translation and rotation of frame $\mathcal{B}$ with respect to frame $\mathcal{A}$ (see Figure 3.10). That is,

$${}^{\mathcal{A}}\boldsymbol{r}_P = {}^{\mathcal{A}}\boldsymbol{d}_{\overrightarrow{O_\mathcal{A}O_\mathcal{B}}} + {}^{\mathcal{A}}_{\mathcal{B}}\boldsymbol{Q}\,{}^{\mathcal{B}}\boldsymbol{r}_P. \tag{3.101}$$

This can be represented in homogeneous coordinates as

$$\left(\begin{array}{c} {}^{\mathcal{A}}\boldsymbol{r}_P \\ \hline 1 \end{array}\right) = \left(\begin{array}{c|c} {}^{\mathcal{A}}_{\mathcal{B}}\boldsymbol{Q} & {}^{\mathcal{A}}\boldsymbol{d}_{\overrightarrow{O_\mathcal{A}O_\mathcal{B}}} \\ \hline \boldsymbol{0} & 1 \end{array}\right) \left(\begin{array}{c} {}^{\mathcal{B}}\boldsymbol{r}_P \\ \hline 1 \end{array}\right), \tag{3.102}$$

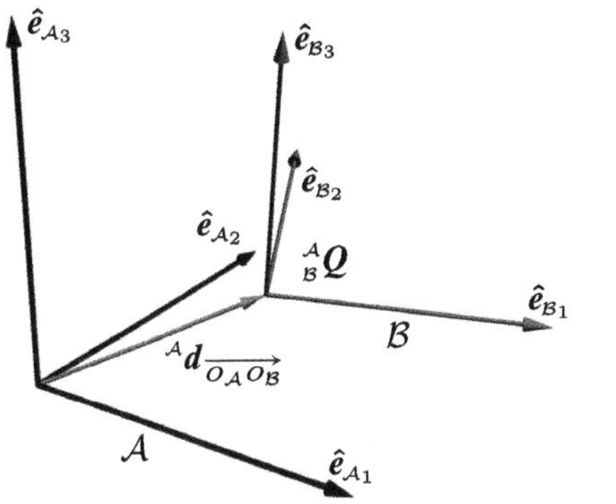

Figure 3.10 A homogenous transform generalizes the notion of a finite displacement by composing a translation and a rotation together in the same operator.

where we define the homogenous transform matrix, ${}^A_B T$, as

$$ {}^A_B T \triangleq \begin{pmatrix} {}^A_B Q & {}^A d_{\overrightarrow{O_A O_B}} \\ \mathbf{0} & 1 \end{pmatrix}. \tag{3.103} $$

As with pure rotations, multiple spatial displacements can be accommodated using matrix multiplication:

$$ {}^A_C T = {}^A_B T {}^B_C T. \tag{3.104} $$

This process is not commutative. That is, in general,

$$ {}^A_B T {}^B_C T \neq {}^B_C T {}^A_B T. \tag{3.105} $$

Thus, the order in which successive finite (noninfinitesimal) displacements are preformed affects the resulting position and orientation.

3.2.2 Screws

We define a screw, $\$$, as the set $\{s, s', \theta, d\}$, where s, s' are the Plücker vectors of the screw axis and θ, d are the displacement variables (see Figure 3.11). The moment vector, s', is given by

$$ s' = d_{\overrightarrow{OP_0}} \times s. \tag{3.106} $$

The homogenous transformation matrix associated with this screw is given by

$$ \begin{aligned} T_s(\theta, d) &= \begin{pmatrix} \mathbf{1} & d_{\overrightarrow{OP_0}} \\ 0 & 1 \end{pmatrix} \begin{pmatrix} Q_s(\theta) & \mathbf{0} \\ 0 & 1 \end{pmatrix} \begin{pmatrix} \mathbf{1} & ds \\ 0 & 1 \end{pmatrix} \begin{pmatrix} \mathbf{1} & -d_{\overrightarrow{OP_0}} \\ 0 & 1 \end{pmatrix} \\ &= \begin{pmatrix} Q_s(\theta) & [\mathbf{1} - Q_s(\theta)] d_{\overrightarrow{OP_0}} + ds \\ 0 & 1 \end{pmatrix}, \end{aligned} \tag{3.107} $$

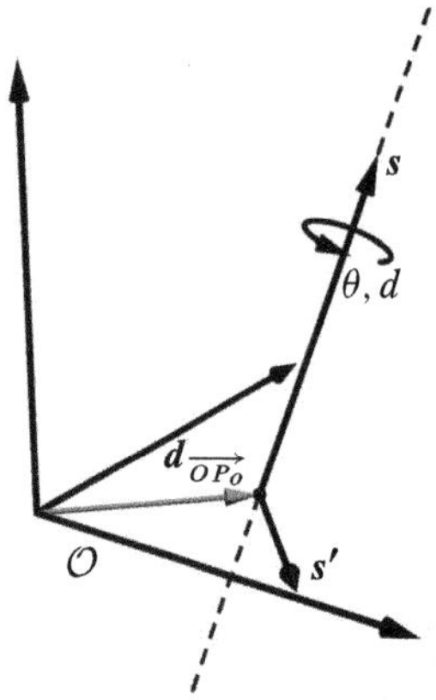

Figure 3.11 A screw description consisting of the screw axis, $\boldsymbol{s}$, the moment vector, $\boldsymbol{s}'$, and the displacement variables, θ and d.

where

$$\boldsymbol{Q}_s(\theta) = \begin{pmatrix} s_x s_x(1-c\theta)+c\theta & s_x s_y(1-c\theta)-s_z s\theta & s_x s_z(1-c\theta)+s_y s\theta \\ s_x s_y(1-c\theta)+s_z s\theta & s_y s_y(1-c\theta)+c\theta & s_y s_z(1-c\theta)-s_x s\theta \\ s_x s_z(1-c\theta)-s_y s\theta & s_y s_z(1-c\theta)+s_x s\theta & s_z s_z(1-c\theta)+c\theta \end{pmatrix}. \tag{3.108}$$

Figure 3.12 illustrates this transformation sequence.

Given a screw axis in a coordinate frame, $\mathcal{B}$,

$${}^{B}\boldsymbol{s} \text{ and, } {}^{B}\boldsymbol{s}' = \boldsymbol{d}_{\overrightarrow{O_B P_o}} \times {}^{B}\boldsymbol{s}, \tag{3.109}$$

we can represent it in coordinate frame, $\mathcal{A}$, where

$${}^{A}\boldsymbol{s} = {}^{A}_{B}\boldsymbol{Q}\,{}^{B}\boldsymbol{s}. \tag{3.110}$$

For the moment vector we note that

$${}^{A}\boldsymbol{s}' = {}^{A}\boldsymbol{d}_{\overrightarrow{O_A P_o}} \times {}^{A}\boldsymbol{s} = ({}^{A}_{B}\boldsymbol{Q}\,{}^{B}\boldsymbol{d}_{\overrightarrow{O_B P_o}} + {}^{A}\boldsymbol{d}_{\overrightarrow{O_A O_B}}) \times ({}^{A}_{B}\boldsymbol{Q}\,{}^{B}\boldsymbol{s}), \tag{3.111}$$

or

$${}^{A}\boldsymbol{s}' = ({}^{A}_{B}\boldsymbol{Q}\,{}^{B}\boldsymbol{d}_{\overrightarrow{O_B P_o}}) \times ({}^{A}_{B}\boldsymbol{Q}\,{}^{B}\boldsymbol{s}) + {}^{A}\boldsymbol{d}_{\overrightarrow{O_A O_B}} \times ({}^{A}_{B}\boldsymbol{Q}\,{}^{B}\boldsymbol{s}). \tag{3.112}$$

Simplifying, we have

$${}^{A}\boldsymbol{s}' = {}^{A}_{B}\boldsymbol{Q}(\boldsymbol{d}_{\overrightarrow{O_B P_o}} \times {}^{B}\boldsymbol{s}) + {}^{A}\boldsymbol{d}_{\overrightarrow{O_A O_B}} \times ({}^{A}_{B}\boldsymbol{Q}\,{}^{B}\boldsymbol{s}). \tag{3.113}$$

In terms of homogenous transformation matrices,

$${}^{A}\boldsymbol{T}_s = {}^{A}_{B}\boldsymbol{T}\,{}^{B}\boldsymbol{T}_s\,{}^{B}_{sA}\boldsymbol{T}. \tag{3.114}$$

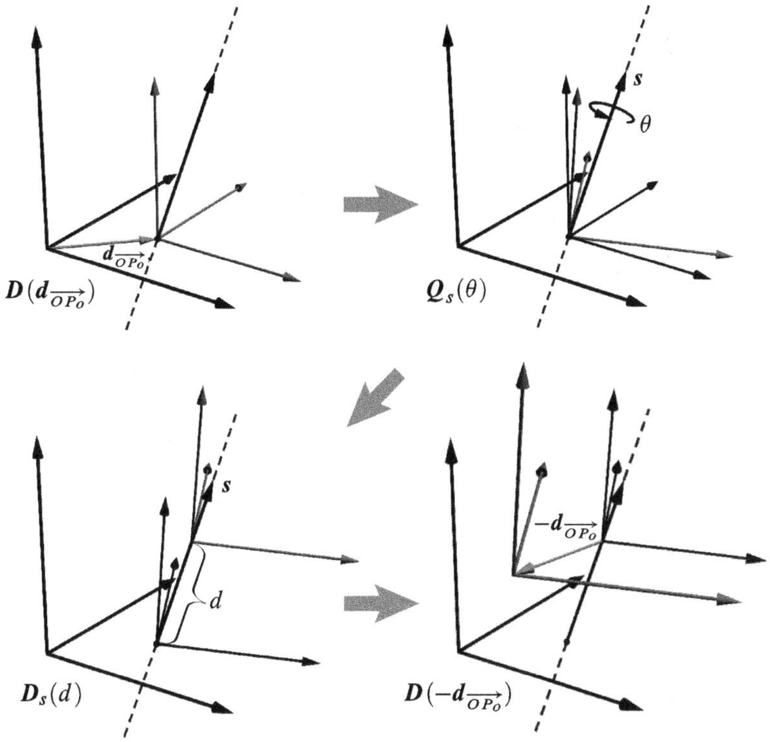

Figure 3.12 Transformation sequence of a screw displacement. First, the origin of the frame is translated by $\boldsymbol{d}_{\overrightarrow{OP_O}}$ to the screw axis, $\boldsymbol{s}$. The frame is then rotated by θ about $\boldsymbol{s}$ followed by a translation of d along $\boldsymbol{s}$. Finally, the frame is translated by $-\boldsymbol{d}_{\overrightarrow{OP_O}}$.

3.2.3 Motors

We express the spatial velocity vector of a rigid body as

$$\mathrm{v} = \begin{pmatrix} \boldsymbol{\omega} \\ \boldsymbol{v}_O \end{pmatrix}, \tag{3.115}$$

where $\boldsymbol{\omega}$ is the angular velocity of the body and $\boldsymbol{v}_O$ is the translational velocity of a point on the body coincident with the origin. This is equivalent to a motor. We can decompose this spatial vector into a line vector and a free vector:

$$\begin{pmatrix} \boldsymbol{\omega} \\ \boldsymbol{v}_O \end{pmatrix} = \begin{pmatrix} \boldsymbol{\omega} \\ \boldsymbol{d}_{\overrightarrow{OP_O}} \times \boldsymbol{\omega} \end{pmatrix} + \begin{pmatrix} \boldsymbol{0} \\ \dot{d}\frac{\boldsymbol{\omega}}{\|\boldsymbol{\omega}\|} \end{pmatrix}. \tag{3.116}$$

In terms of the screw parameters, $\{\boldsymbol{s}, \boldsymbol{s}', \dot{\theta}, \dot{d}\}$, we have

$$\begin{pmatrix} \boldsymbol{\omega} \\ \boldsymbol{v}_O \end{pmatrix} = \dot{\theta}\begin{pmatrix} \boldsymbol{s} \\ \boldsymbol{s}' \end{pmatrix} + \dot{d}\begin{pmatrix} \boldsymbol{0} \\ \boldsymbol{s} \end{pmatrix}. \tag{3.117}$$

Solving for $\boldsymbol{s}$ and $\dot{\theta}$ in terms of $\boldsymbol{\omega}$, we have

$$\boldsymbol{s} = \frac{\boldsymbol{\omega}}{\|\boldsymbol{\omega}\|} \quad \text{and} \quad \dot{\theta} = \|\boldsymbol{\omega}\|. \tag{3.118}$$

To solve for $\dot{d}$ in terms of $\boldsymbol{\omega}$ and $\boldsymbol{v}_O$, we first note that

$$\boldsymbol{v}_O = \dot{\theta}\boldsymbol{s}' + \dot{d}\boldsymbol{s}. \tag{3.119}$$

So,

$$\boldsymbol{s} \cdot \boldsymbol{v}_O = \dot{\theta}\boldsymbol{s} \cdot \boldsymbol{s}' + \dot{d}\boldsymbol{s} \cdot \boldsymbol{s}. \tag{3.120}$$

Thus,

$$\dot{d} = \boldsymbol{s} \cdot \boldsymbol{v}_O = \frac{\boldsymbol{\omega} \cdot \boldsymbol{v}_O}{\|\boldsymbol{\omega}\|}. \tag{3.121}$$

Alternately, the pitch, ρ, is

$$\rho = \frac{\dot{d}}{\dot{\theta}} = \frac{\boldsymbol{\omega} \cdot \boldsymbol{v}_O}{\|\boldsymbol{\omega}\|^2} = \frac{\boldsymbol{\omega} \cdot \boldsymbol{v}_O}{\boldsymbol{\omega} \cdot \boldsymbol{\omega}}. \tag{3.122}$$

To solve for $\boldsymbol{s}'$ in terms of $\boldsymbol{\omega}$ and $\boldsymbol{v}_O$, we note that

$$\boldsymbol{s} \times \boldsymbol{v}_O = \dot{\theta}\boldsymbol{s} \times \boldsymbol{s}' + \dot{d}\boldsymbol{s} \times \boldsymbol{s}, \tag{3.123}$$

or

$$\boldsymbol{s} \times \boldsymbol{v}_O = \dot{\theta}\boldsymbol{s} \times \boldsymbol{s}' = \dot{\theta}\boldsymbol{s} \times (\boldsymbol{d}_{\overrightarrow{OP_0}} \times \boldsymbol{s}) = \dot{\theta}\boldsymbol{d}_{\overrightarrow{OP_0}}. \tag{3.124}$$

Thus,

$$\boldsymbol{d}_{\overrightarrow{OP_0}} = \frac{\boldsymbol{s} \times \boldsymbol{v}_O}{\dot{\theta}}. \tag{3.125}$$

So,

$$\boldsymbol{s}' = \boldsymbol{d}_{\overrightarrow{OP_0}} \times \boldsymbol{s} = \frac{(\boldsymbol{s} \times \boldsymbol{v}_O) \times \boldsymbol{s}}{\dot{\theta}} = \frac{(\boldsymbol{\omega} \times \boldsymbol{v}_O) \times \boldsymbol{\omega}}{\|\boldsymbol{\omega}\|^3}. \tag{3.126}$$

3.2.4 Wrenches

We express the spatial force vector acting on a rigid body as

$$\mathbf{f} = \begin{pmatrix} \boldsymbol{f} \\ \boldsymbol{\varphi}_O \end{pmatrix}, \tag{3.127}$$

where $\boldsymbol{f}$ is the force acting on the body and $\boldsymbol{\varphi}_O$ is the moment about the origin. This is equivalent to a wrench. We can decompose this spatial vector into a line vector and a free vector:

$$\begin{pmatrix} \boldsymbol{f} \\ \boldsymbol{\varphi}_O \end{pmatrix} = \begin{pmatrix} \boldsymbol{f} \\ \boldsymbol{d}_{\overrightarrow{OP_0}} \times \boldsymbol{f} \end{pmatrix} + \begin{pmatrix} \boldsymbol{0} \\ \tau \frac{\boldsymbol{f}}{\|\boldsymbol{f}\|} \end{pmatrix}. \tag{3.128}$$

In terms of the screw parameters, $\{\boldsymbol{s}, \boldsymbol{s}', f, \tau\}$, we have

$$\begin{pmatrix} \boldsymbol{f} \\ \boldsymbol{\varphi}_O \end{pmatrix} = f \begin{pmatrix} \boldsymbol{s} \\ \boldsymbol{s}' \end{pmatrix} + \tau \begin{pmatrix} \boldsymbol{0} \\ \boldsymbol{s} \end{pmatrix}. \tag{3.129}$$

Solving for $\boldsymbol{s}$ and f in terms of $\boldsymbol{f}$, we have

$$\boldsymbol{s} = \frac{\boldsymbol{f}}{\|\boldsymbol{f}\|} \quad \text{and} \quad f = \|\boldsymbol{f}\|. \tag{3.130}$$

To solve for τ in terms of $\boldsymbol{f}$ and $\boldsymbol{\varphi}_O$, we first note that

$$\boldsymbol{\varphi}_O = f\boldsymbol{s}' + \tau\boldsymbol{s}. \tag{3.131}$$

So,

$$\boldsymbol{s} \cdot \boldsymbol{\varphi}_O = f\boldsymbol{s} \cdot \boldsymbol{s}' + \tau\boldsymbol{s} \cdot \boldsymbol{s}. \tag{3.132}$$

Thus,

$$\tau = \boldsymbol{s} \cdot \boldsymbol{\varphi}_O = \frac{\boldsymbol{f} \cdot \boldsymbol{\varphi}_O}{\|\boldsymbol{f}\|}. \tag{3.133}$$

Alternately, the pitch, ρ, is

$$\rho = \frac{\tau}{f} = \frac{\boldsymbol{f} \cdot \boldsymbol{\varphi}_O}{\|\boldsymbol{f}\|^2} = \frac{\boldsymbol{f} \cdot \boldsymbol{\varphi}_O}{\boldsymbol{f} \cdot \boldsymbol{f}}. \tag{3.134}$$

To solve for $\boldsymbol{s}'$ in terms of $\boldsymbol{f}$ and $\boldsymbol{\varphi}_O$, we note that

$$\boldsymbol{s} \times \boldsymbol{\varphi}_O = f\boldsymbol{s} \times \boldsymbol{s}' + \tau\boldsymbol{s} \times \boldsymbol{s}, \tag{3.135}$$

or

$$\boldsymbol{s} \times \boldsymbol{\varphi}_O = f\boldsymbol{s} \times \boldsymbol{s}' = f\boldsymbol{s} \times (\boldsymbol{d}_{\overrightarrow{OP_0}} \times \boldsymbol{s}) = f\boldsymbol{d}_{\overrightarrow{OP_0}}. \tag{3.136}$$

Thus,

$$\boldsymbol{d}_{\overrightarrow{OP_0}} = \frac{\boldsymbol{s} \times \boldsymbol{\varphi}_O}{f}. \tag{3.137}$$

So,

$$\boldsymbol{s}' = \boldsymbol{d}_{\overrightarrow{OP_0}} \times \boldsymbol{s} = \frac{(\boldsymbol{s} \times \boldsymbol{\varphi}_O) \times \boldsymbol{s}}{f} = \frac{(\boldsymbol{f} \times \boldsymbol{\varphi}_O) \times \boldsymbol{f}}{\cdot} \|\boldsymbol{f}\|^3. \tag{3.138}$$

3.2.5 Derivatives in Rotating Reference Frames

In taking derivatives of vector quantities we must take care to note how the quantity of interest is expressed and relative to what reference frame the derivative is sought. Because the goal is to use kinematic quantities (e.g., translational and angler velocity and acceleration) to formulate dynamical equations of motion, we are usually interested in computing derivatives relative to an inertial reference frame. A derivative representing the time rate of change of some quantity can be computed in an inertial reference frame or a noninertial reference frame, that is, one rotating relative to an inertial reference frame.

If all quantities are extrinsically expressed in the coordinates of an inertial reference frame, then differentiation is straightforward. Often, for convenience, vector quantities will be expressed in a local noninertial reference. As time elapses, the reference frame rotates with respect to the inertial frame. If we take an instantaneous quantity like translational or angular velocity that is expressed in a rotating reference frame, we need to consider how it is changing relative to the local reference frame it is expressed in, as well as how that local reference frame is itself changing.

To demonstrate, we take an arbitrary vector quantity, ${}^{A}\boldsymbol{z}$, expressed in a particular rotating reference, $\mathcal{A}$. The value of ${}^{A}\boldsymbol{z}$ at any instant is measured in $\mathcal{A}$. What we will call the simple derivative of the expression is denoted by ${}^{A}\dot{\boldsymbol{z}}$ and represents the time derivative of ${}^{A}\boldsymbol{z}$ relative to the frame it is expressed in, $\mathcal{A}$. We are interested in taking the derivative of the quantity *relative* to an inertial reference frame, $\mathcal{O}$, but expressing the coordinates in $\mathcal{A}$. We denote this operator as

$$\frac{{}^{A}d}{dt}_{\mathcal{O}}. \tag{3.139}$$

We can apply this operator to the vector, ${}^{A}\boldsymbol{z}$, where, expressing in terms of a limit,

$$\frac{{}^{A}d}{dt}_{\mathcal{O}}{}^{A}\boldsymbol{z} = \lim_{\Delta t \to 0} \frac{(\mathbf{1} + {}^{A}\Delta_{\mathcal{O}}\boldsymbol{Q})({}^{A}\boldsymbol{z} + {}^{A}\Delta_{\mathcal{A}}{}^{A}\boldsymbol{z}) - {}^{A}\boldsymbol{z}}{\Delta t}. \tag{3.140}$$

This expresses that the reference frame has rotated by ${}^{A}\Delta_{\mathcal{O}}\boldsymbol{Q}$ relative to the inertial reference frame but is still expressed in $\mathcal{A}$. Similarly, the vector, ${}^{A}\boldsymbol{z}$, has changed by ${}^{A}\Delta_{\mathcal{A}}{}^{A}\boldsymbol{z}$ relative to the rotating reference frame, $\mathcal{A}$, and is also expressed in $\mathcal{A}$. We simplify this expression:

$$\frac{{}^{A}d}{dt}_{\mathcal{O}}{}^{A}\boldsymbol{z} = \lim_{\Delta t \to 0} \left(\frac{{}^{A}\Delta_{\mathcal{A}}{}^{A}\boldsymbol{z}}{\Delta t} + \frac{{}^{A}\Delta_{\mathcal{O}}\boldsymbol{Q}}{\Delta t}{}^{A}\boldsymbol{z} \right). \tag{3.141}$$

By definition,

$$\frac{{}^{A}d}{dt}_{\mathcal{A}}{}^{A}\boldsymbol{z} = \lim_{\Delta t \to 0} \frac{{}^{A}\Delta_{\mathcal{A}}{}^{A}\boldsymbol{z}}{\Delta t} = {}^{A}\dot{\boldsymbol{z}} \tag{3.142}$$

is the simple derivative of ${}^{A}\boldsymbol{z}$, and

$$\frac{{}^{A}d}{dt}_{\mathcal{O}}\boldsymbol{Q} = \lim_{\Delta t \to 0} \frac{{}^{A}\Delta_{\mathcal{O}}\boldsymbol{Q}}{\Delta t} = \dot{\boldsymbol{Q}} = {}^{A}\boldsymbol{\Omega}_{\mathcal{A}}. \tag{3.143}$$

So,

$$\frac{{}^{A}d}{dt}_{\mathcal{O}}{}^{A}\boldsymbol{z} = \frac{{}^{A}d}{dt}_{\mathcal{A}}{}^{A}\boldsymbol{z} + \left(\frac{{}^{A}d}{dt}_{\mathcal{O}}\boldsymbol{Q}\right){}^{A}\boldsymbol{z} = {}^{A}\dot{\boldsymbol{z}} + {}^{A}\boldsymbol{\Omega}_{\mathcal{A}}{}^{A}\boldsymbol{z} = {}^{A}\dot{\boldsymbol{z}} + {}^{A}\boldsymbol{\omega}_{\mathcal{A}} \times {}^{A}\boldsymbol{z}. \tag{3.144}$$

Applying this to the differentiation of angular velocity, we have

$${}^{A}\boldsymbol{\alpha}_{\mathcal{A}\mathcal{O}} = \frac{{}^{A}d}{dt}_{\mathcal{O}}{}^{A}\boldsymbol{\omega}_{\mathcal{A}} = {}^{A}\dot{\boldsymbol{\omega}}_{\mathcal{A}} + {}^{A}\boldsymbol{\Omega}_{\mathcal{A}}{}^{A}\boldsymbol{\omega}_{\mathcal{A}} = {}^{A}\dot{\boldsymbol{\omega}}_{\mathcal{A}} + \underbrace{{}^{A}\boldsymbol{\omega}_{\mathcal{A}} \times {}^{A}\boldsymbol{\omega}_{\mathcal{A}}}_{0} = {}^{A}\dot{\boldsymbol{\omega}}_{\mathcal{A}}. \tag{3.145}$$

As a shorthand in our notation, if the frame that differentiation is relative to is omitted, we will take that to imply that it is relative to the inertial frame, for example, ${}^{A}\boldsymbol{\alpha}_{\mathcal{A}} = {}^{A}\boldsymbol{\alpha}_{\mathcal{A}\mathcal{O}}$.

Applying differentiation of translational velocity, we have

$$ {}^{A}\boldsymbol{a}_{P_A:O} = \frac{{}^{A}d}{dt}\bigg|_{O} {}^{A}\boldsymbol{v}_{P_A} = {}^{A}\dot{\boldsymbol{v}}_{P_A} + {}^{A}\boldsymbol{\Omega}_{A}\,{}^{A}\boldsymbol{v}_{P_A} = {}^{A}\dot{\boldsymbol{v}}_{P_A} + {}^{A}\boldsymbol{\omega}_{A} \times {}^{A}\boldsymbol{v}_{P_A}. \tag{3.146}$$

Example: We return to the example of an atmospheric reentry body. The task of determining the motion history of the center of mass (to be formally defined in the next chapter) of a reentry body requires a knowledge of the body's initial conditions and the gyroscope and accelerometer data. Figure 3.13 depicts a configuration where an accelerometer is mounted in each of three locations (frames 1, 2, and 3).

The following relationships exist between the acceleration of the center of mass and the accelerations of the accelerometer frames

$$ {}^{c}\boldsymbol{a}_{O_1} = {}^{c}_{1}\boldsymbol{Q}\,{}^{1}\boldsymbol{a}_{O_1} = {}^{c}\boldsymbol{a}_{G_C} + {}^{c}\boldsymbol{\omega}_{c} \times ({}^{c}\boldsymbol{\omega}_{c} \times {}^{c}\boldsymbol{d}_{1}) + {}^{c}\boldsymbol{\alpha}_{c} \times {}^{c}\boldsymbol{d}_{1}, \tag{3.147}$$

$$ {}^{c}\boldsymbol{a}_{O_2} = {}^{c}_{2}\boldsymbol{Q}\,{}^{2}\boldsymbol{a}_{O_2} = {}^{c}\boldsymbol{a}_{G_C} + {}^{c}\boldsymbol{\omega}_{c} \times ({}^{c}\boldsymbol{\omega}_{c} \times {}^{c}\boldsymbol{d}_{2}) + {}^{c}\boldsymbol{\alpha}_{c} \times {}^{c}\boldsymbol{d}_{2}, \tag{3.148}$$

$$ {}^{c}\boldsymbol{a}_{O_3} = {}^{c}_{3}\boldsymbol{Q}\,{}^{3}\boldsymbol{a}_{O_3} = {}^{c}\boldsymbol{a}_{G_C} + {}^{c}\boldsymbol{\omega}_{c} \times ({}^{c}\boldsymbol{\omega}_{c} \times {}^{c}\boldsymbol{d}_{3}) + {}^{c}\boldsymbol{\alpha}_{c} \times {}^{c}\boldsymbol{d}_{3}. \tag{3.149}$$

These three vector equations yield nine scalar equations and nine unknowns. The nine unknowns include the three center of mass acceleration components, ${}^{c}\ddot{x}_{G_C}$, ${}^{c}\ddot{y}_{G_C}$, and ${}^{c}\ddot{z}_{G_C}$, as well as the six acceleration components of frames 1, 2, and 3, which lie in directions orthogonal to the accelerometers. The other three acceleration components of frames 1, 2, and 3, are known explicitly from the accelerometer values, a_1, a_2, and a_3. For a general formulation where the accelerometers are mounted arbitrarily we have a linear system, $\boldsymbol{Ax} = \boldsymbol{b}$, at a given instant of time where $\boldsymbol{A}$ is defined as

$$ \boldsymbol{A} \triangleq \begin{pmatrix} \boldsymbol{1} & -{}^{c}_{1}\boldsymbol{Q} & \boldsymbol{0} & \boldsymbol{0} \\ \boldsymbol{1} & \boldsymbol{0} & -{}^{c}_{2}\boldsymbol{Q} & \boldsymbol{0} \\ \boldsymbol{1} & \boldsymbol{0} & \boldsymbol{0} & -{}^{c}_{3}\boldsymbol{Q} \\ \boldsymbol{0} & \mathrm{diag}(\hat{\boldsymbol{e}}_1)\boldsymbol{N} & \mathrm{diag}(\hat{\boldsymbol{e}}_2)\boldsymbol{N} & \mathrm{diag}(\hat{\boldsymbol{e}}_3)\boldsymbol{N} \end{pmatrix} \tag{3.150}$$

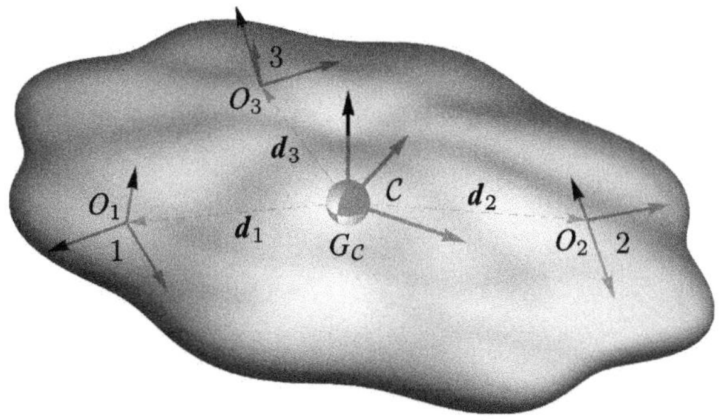

Figure 3.13 Body with acceleration components known at three different locations. Accelerometer data taken at three different locations and gyroscope data for the body can be used to determine the acceleration of the center of mass.

and $\boldsymbol{N}$ is a logical matrix defined as

$$N_{ij} \triangleq \begin{cases} 1 \text{ if accelerometer } i \text{ measures in the } j \text{ direction} \\ 0 \text{ if accelerometer } i \text{ does not measures in the } j \text{ direction.} \end{cases} \tag{3.151}$$

The vectors $\boldsymbol{x}$ and $\boldsymbol{b}$ are defined as

$$\boldsymbol{x} \triangleq \begin{pmatrix} {}^{c}\boldsymbol{a}_{G_C} \\ {}^{1}\boldsymbol{a}_{O_1} \\ {}^{2}\boldsymbol{a}_{O_2} \\ {}^{3}\boldsymbol{a}_{O_3} \end{pmatrix} \quad \text{and} \quad \boldsymbol{b} \triangleq \begin{pmatrix} -{}^{c}\boldsymbol{\omega}_c \times ({}^{c}\boldsymbol{\omega}_c \times {}^{c}\boldsymbol{d}_1) - {}^{c}\boldsymbol{\alpha}_c \times {}^{c}\boldsymbol{d}_1 \\ -{}^{c}\boldsymbol{\omega}_c \times ({}^{c}\boldsymbol{\omega}_c \times {}^{c}\boldsymbol{d}_2) - {}^{c}\boldsymbol{\alpha}_c \times {}^{c}\boldsymbol{d}_2 \\ -{}^{c}\boldsymbol{\omega}_c \times ({}^{c}\boldsymbol{\omega}_c \times {}^{c}\boldsymbol{d}_3) - {}^{c}\boldsymbol{\alpha}_c \times {}^{c}\boldsymbol{d}_3 \\ \boldsymbol{a} \end{pmatrix}, \tag{3.152}$$

where $\boldsymbol{a} \triangleq (\, a_1 \quad a_2 \quad a_3 \,)^T$ is the vector of accelerometer values. Because the accelerometers are typically aligned with the center of mass coordinate frame, the preceding system can be reduced to three scalar equations and three unknowns. For example, we will be dealing with a case where the accelerometer for frame 1 measures in the $+x$-axis of the center of mass frame, the accelerometer for frame 2 measures in the $+y$-axis, and the accelerometer for frame 3 measures in the $-z$-axis. So we have

$${}^{c}\ddot{x}_{G_C} = a_1 - \left[{}^{c}\boldsymbol{\omega}_c \times ({}^{c}\boldsymbol{\omega}_c \times {}^{c}\boldsymbol{d}_1) + {}^{c}\boldsymbol{\alpha}_c \times {}^{c}\boldsymbol{d}_1\right] \cdot \hat{\boldsymbol{e}}_1, \tag{3.153}$$

$${}^{c}\ddot{y}_{G_C} = a_2 - \left[{}^{c}\boldsymbol{\omega}_c \times ({}^{c}\boldsymbol{\omega}_c \times {}^{c}\boldsymbol{d}_2) + {}^{c}\boldsymbol{\alpha}_c \times {}^{c}\boldsymbol{d}_2\right] \cdot \hat{\boldsymbol{e}}_2, \tag{3.154}$$

$${}^{c}\ddot{z}_{G_C} = -a_3 - \left[{}^{c}\boldsymbol{\omega}_c \times ({}^{c}\boldsymbol{\omega}_c \times {}^{c}\boldsymbol{d}_3) + {}^{c}\boldsymbol{\alpha}_c \times {}^{c}\boldsymbol{d}_3\right] \cdot \hat{\boldsymbol{e}}_3. \tag{3.155}$$

Transforming into the initial world coordinate frame $\mathcal{O}$, using quaternions, we have

$${}^{\mathcal{O}}a_{G_C} = {}^{\mathcal{O}}_{c}h\,{}^{c}a_{G_C}\,{}^{c}_{\mathcal{O}}h. \tag{3.156}$$

It is noted that in the preceding case which involves quaternion operations, ${}^{\mathcal{O}}a_{G_C}$ is taken to be an acceleration quaternion. For numerical integration we can use the following trapezoidal relationships:

$${}^{\mathcal{O}}\boldsymbol{v}_{G_C}(t + \Delta t) \approxeq {}^{\mathcal{O}}\boldsymbol{v}_{G_C}(t) + \frac{1}{2}\left[{}^{\mathcal{O}}\boldsymbol{a}_{G_C}(t) + {}^{\mathcal{O}}\boldsymbol{a}_{G_C}(t + \Delta t)\right]\Delta t \tag{3.157}$$

$${}^{\mathcal{O}}\boldsymbol{r}_{G_C}(t + \Delta t) \approxeq {}^{\mathcal{O}}\boldsymbol{r}_{G_C}(t) + \frac{1}{2}\left[{}^{\mathcal{O}}\boldsymbol{v}_{G_C}(t) + {}^{\mathcal{O}}\boldsymbol{v}_{G_C}(t + \Delta t)\right]\Delta t, \tag{3.158}$$

where we have converted the acceleration quaternion, ${}^{\mathcal{O}}a_{G_C}$, back into a standard vector form. Having computed the center of mass trajectory and body orientation history, we can calculate the coning angle of the reentry body over the time series. Figure 3.14

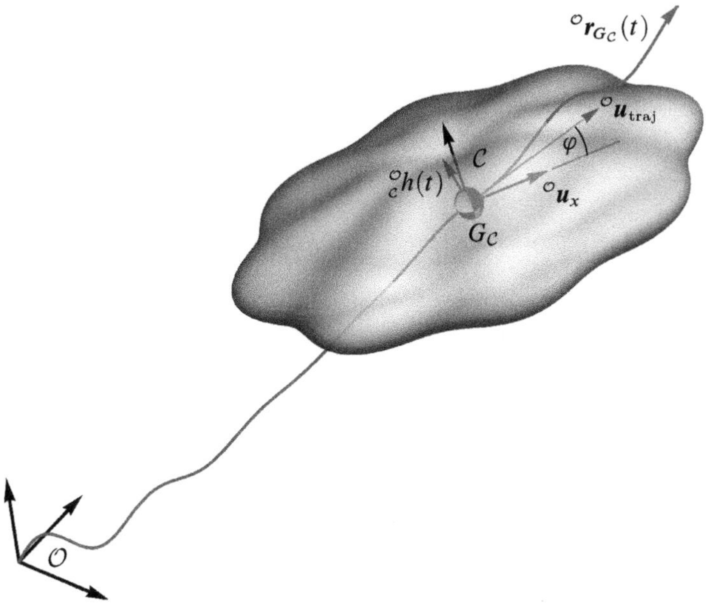

Figure 3.14 Center of mass position time history and orientation time history relative to a world frame. The coning angle, φ, is shown as the angle between body axis, $\boldsymbol{u}_x$, and the trajectory tangent, $\boldsymbol{u}_{\mathrm{traj}}$.

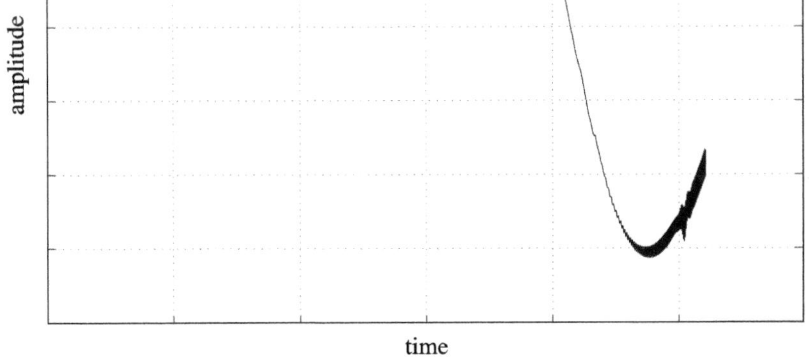

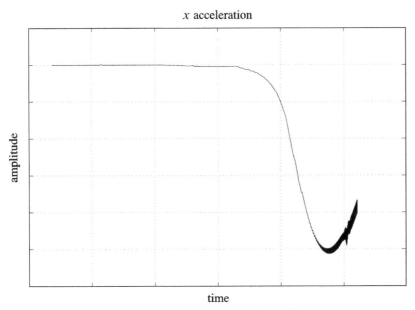

Figure 3.15 Center of mass acceleration x component derived from gyroscope and accelerometer data. Units have been intentionally omitted.

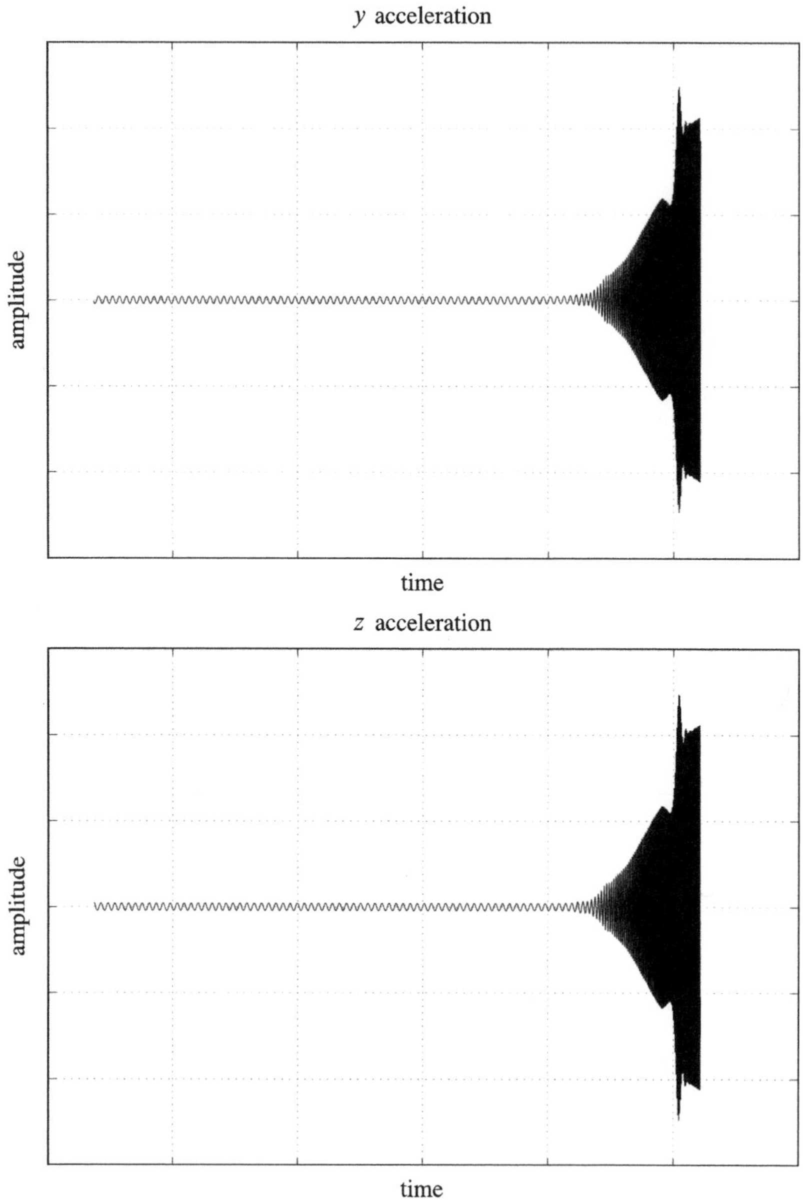

Figure 3.16 Center of mass acceleration y and z components derived from gyroscope and accelerometer data. Units have been intentionally omitted.

depicts the combined orientation and position information describing the body's motion in $\mathbb{R}^3$. The coning angle, φ, is defined as the angle between the reentry body longitudinal axis (x-axis) and the tangent to the trajectory at a given instant in time.

Using the results of Section 3.1.5, we know the α and β Euler angles of the reentry body orientation. The reentry body x-axis unit vector in the base coordinate frame is

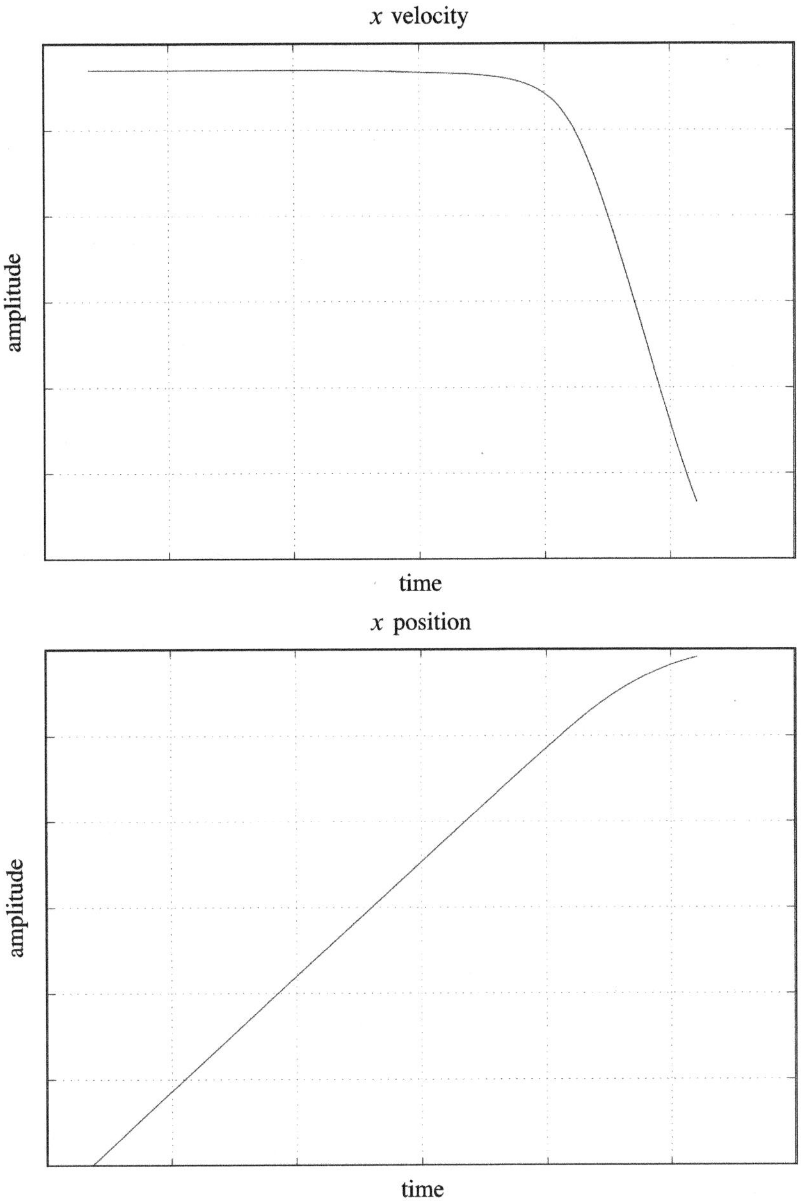

Figure 3.17 Center of mass velocity and position x components. Numerical integration was performed on the acceleration data. Units have been intentionally omitted.

thus

$$\boldsymbol{u}_x = \begin{pmatrix} \cos\beta \\ \cos\alpha \sin\beta \\ \sin\alpha \sin\beta \end{pmatrix}. \tag{3.159}$$

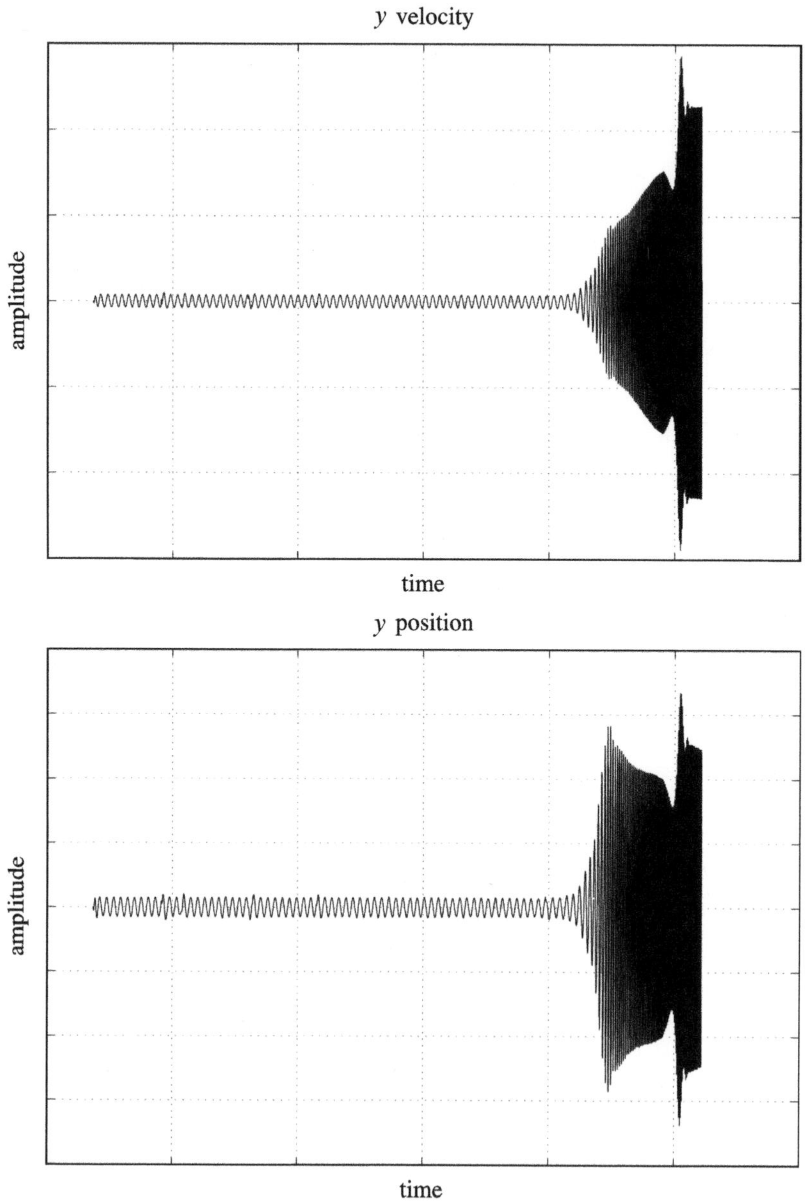

Figure 3.18 Center of mass velocity and position y component. The integrated signals were conditioned with a high-pass filter. Units have been intentionally omitted.

Alternately, this vector can be expressed as the first column of $\boldsymbol{Q}$:

$$\boldsymbol{u}_x = \boldsymbol{Q}\hat{\boldsymbol{e}}_1 = \begin{pmatrix} 2(h_0h_0 + h_1h_1) - 1 \\ 2(h_1h_2 + h_0h_3) \\ 2(h_1h_3 - h_0h_2) \end{pmatrix}. \tag{3.160}$$

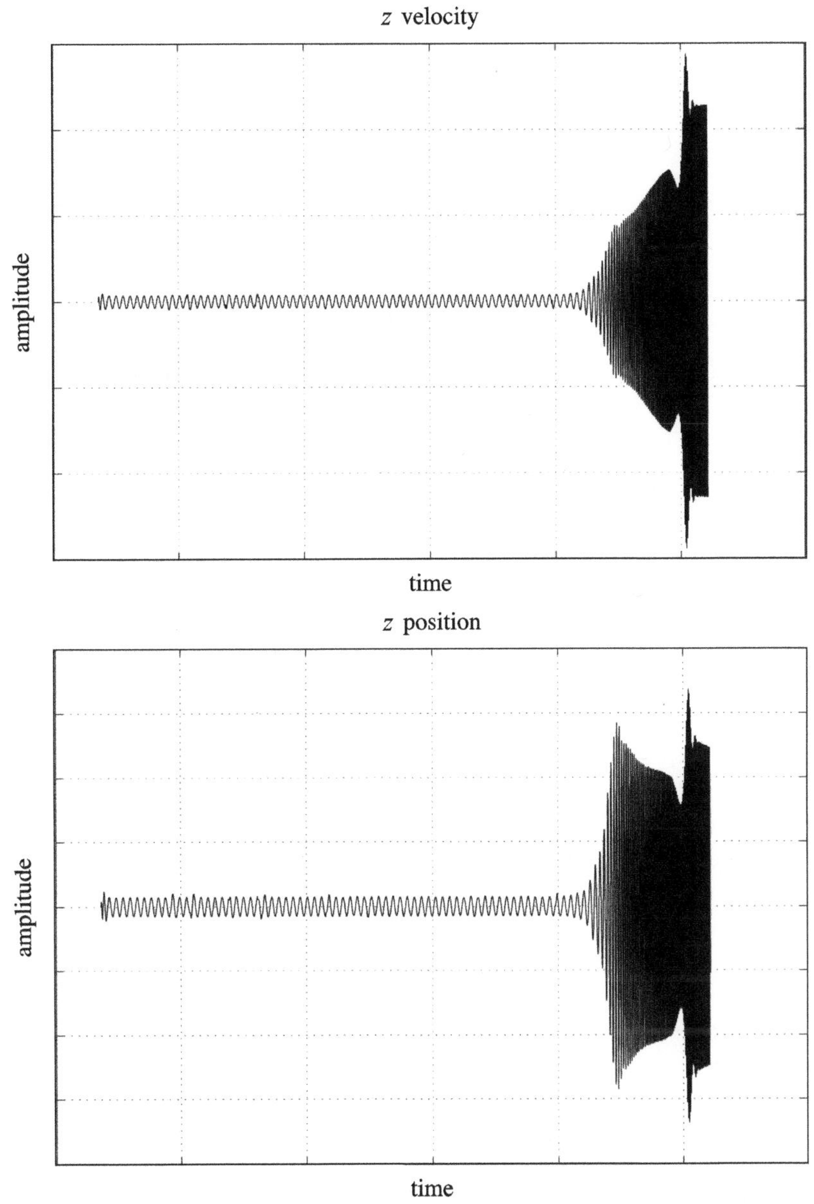

Figure 3.19 Center of mass velocity and position z component. The integrated signals were conditioned with a high-pass filter. Units have been intentionally omitted.

The tangent to the trajectory is

$$\boldsymbol{u}_{\text{traj}} = \frac{\Delta \boldsymbol{r}_{G_C}}{\left\| \Delta \boldsymbol{r}_{G_C} \right\|}. \tag{3.161}$$

So,

$$\cos \varphi = \boldsymbol{u}_x \cdot \boldsymbol{u}_{\text{traj}}. \tag{3.162}$$

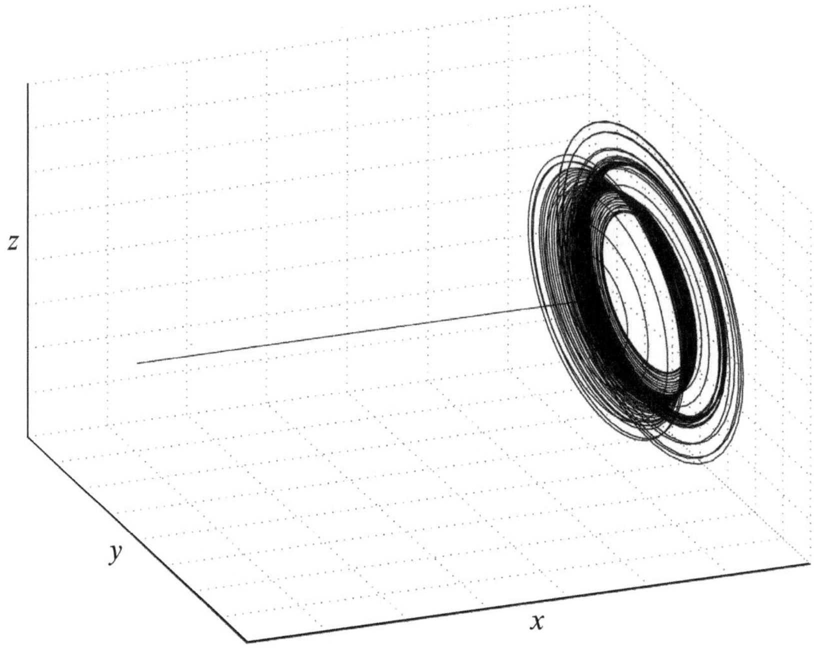

Figure 3.20 Reentry body trajectory generated after final conditioning of the integrated signals. The scale in the x direction is greatly compressed. Units have been intentionally omitted.

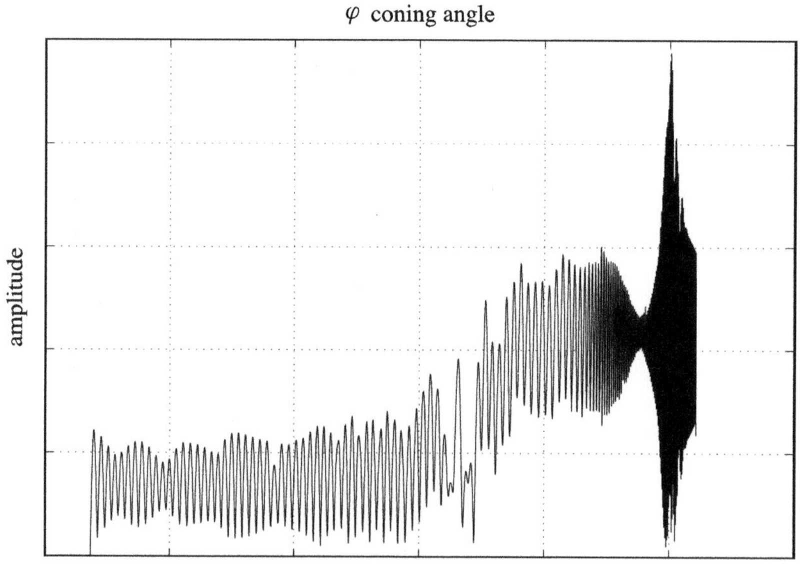

Figure 3.21 Coning angle of the reentry body determined from the quaternion orientation history and the trajectory tangent. Units have been intentionally omitted.

Figures 3.15 and 3.16 display plots of the center of mass acceleration components derived from telemetry data. Figure 3.17 displays plots of the center of mass velocity and position in the x direction. Figures 3.18 and 3.19 display plots of the center of mass velocity and position in the y and z directions. The signals were conditioned with a high-pass filter. After final conditioning of the integrated signals, a trajectory was generated. Figure 3.20 displays the center of mass trajectory of the reentry body. The scale in the x direction is greatly compressed. Figure 3.21 displays a plot of the coning angle.

3.3 Kinematic Chains

3.3.1 Denavit-Hartenberg Parameterization

The Denavit-Hartenberg parameters provide a systematic convention for assigning reference frames to the links of kinematic chains, or so-called lower-pair mechanisms (Denavit and Hartenberg 1955). Figure 3.22 depicts the Denavit-Hartenberg frame transformations. The four parameters are referred to as the link length, a_i, link twist, α_i, link offset, d_i, and joint angle, θ_i. The transformation matrix, $\boldsymbol{T}$, from link $i-1$ to link i is given by

$$ {}^{i-1}_{\ i}\boldsymbol{T} = \boldsymbol{Q}_x(\alpha_{i-1})\boldsymbol{D}_x(a_{i-1})\boldsymbol{Q}_z(\theta_i)\boldsymbol{D}_z(d_i). \qquad (3.163) $$

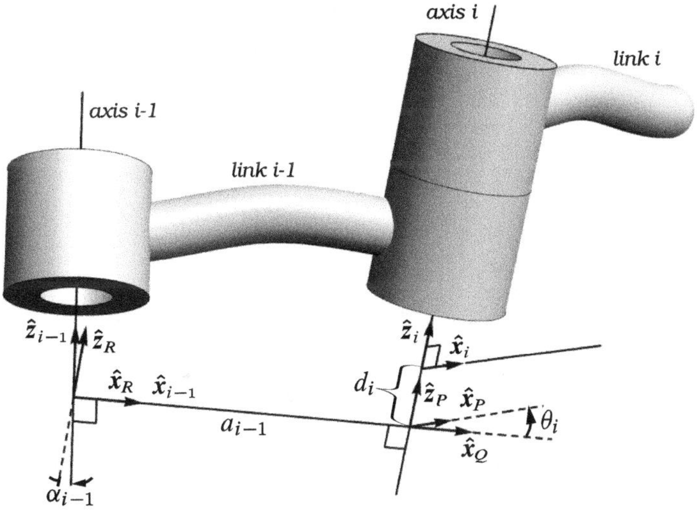

Figure 3.22 The Denavit-Hartenberg parameters provide a systematic convention for assigning reference frames to the links of lower-pair mechanisms. The four parameters, shown here, are the link length, a_i, link twist, α_i, link offset, d_i, and joint angle, θ_i.

So,

$$
{}^{i-1}_{i}\boldsymbol{T} = \begin{pmatrix} c\theta_i & -s\theta_i & 0 & a_{i-1} \\ s\theta_i c\alpha_{i-1} & c\theta_i c\alpha_{i-1} & -s\alpha_{i-1} & -s\alpha_{i-1} d_i \\ s\theta_i s\alpha_{i-1} & c\theta_i s\alpha_{i-1} & c\alpha_{i-1} & c\alpha_{i-1} d_i \\ 0 & 0 & 0 & 1 \end{pmatrix}. \tag{3.164}
$$

The overall transformation matrix, from base link 0 to link n, is given by

$$
{}^{0}_{n}\boldsymbol{T} = {}^{0}_{1}\boldsymbol{T}\,{}^{1}_{2}\boldsymbol{T} \cdots {}^{n-1}_{n}\boldsymbol{T} = \prod_{i=1}^{n} {}^{i-1}_{i}\boldsymbol{T}. \tag{3.165}
$$

3.3.2 The Jacobian Matrix

In this section we will define the kinematic Jacobians that describe the mapping from configuration space to Cartesian space. Configuration space is represented by a set of generalized coordinates, $\{q_i, \ldots, q_n\}$, corresponding to the joint variables in our kinematic chain. That is,

$$
q_i = \theta_i, \quad \text{if joint } i \text{ is revoute, and} \qquad q_i = d_i, \quad \text{if joint } i \text{ is prismatic.} \tag{3.166}
$$

We first note that displacements, relative to a fixed reference frame, are

$$
\delta\boldsymbol{r} = \frac{\partial \boldsymbol{r}}{\partial \boldsymbol{q}}\delta\boldsymbol{q} \quad \text{and} \quad \delta\boldsymbol{\theta} = \frac{\partial \boldsymbol{\theta}}{\partial \boldsymbol{q}}\delta\boldsymbol{q}. \tag{3.167}
$$

Alternately, we could write

$$
\delta\boldsymbol{r} = \frac{\partial \boldsymbol{v}}{\partial \dot{\boldsymbol{q}}}\delta\boldsymbol{q} \quad \text{and} \quad \delta\boldsymbol{\omega} = \frac{\partial \boldsymbol{\omega}}{\partial \dot{\boldsymbol{q}}}\delta\boldsymbol{q}. \tag{3.168}
$$

We will define

$$
\boldsymbol{\Gamma} \triangleq \frac{\partial \boldsymbol{r}}{\partial \boldsymbol{q}} = \frac{\partial \boldsymbol{v}}{\partial \dot{\boldsymbol{q}}} \quad \text{and} \quad \boldsymbol{\Pi} \triangleq \frac{\partial \boldsymbol{\theta}}{\partial \boldsymbol{q}} = \frac{\partial \boldsymbol{\omega}}{\partial \dot{\boldsymbol{q}}}. \tag{3.169}
$$

Consequently, the velocities, relative to a fixed reference frame, are

$$
\boldsymbol{v} = \boldsymbol{\Gamma}\dot{\boldsymbol{q}} \quad \text{and} \quad \boldsymbol{\omega} = \boldsymbol{\Pi}\dot{\boldsymbol{q}}, \tag{3.170}
$$

and the accelerations, relative to a fixed reference frame, are

$$
\boldsymbol{a} = \boldsymbol{\Gamma}\ddot{\boldsymbol{q}} + (\dot{\boldsymbol{\Gamma}} + \boldsymbol{\Omega}\boldsymbol{\Gamma})\dot{\boldsymbol{q}} \quad \text{and} \quad \boldsymbol{\alpha} = \boldsymbol{\Pi}\ddot{\boldsymbol{q}} + \dot{\boldsymbol{\Pi}}\dot{\boldsymbol{q}}, \tag{3.171}
$$

where $\boldsymbol{\Omega}$ is the angular velocity tensor of the frame in which $\boldsymbol{\Gamma}$ is expressed, relative to a fixed frame.

More generally, the Jacobian, $\boldsymbol{J}$, is a linear map between a configuration space velocity vector and a task space velocity vector. The concept of task space will be discussed in Chapter 8. For our purposes here it will suffice to interpret a task vector as a spatial vector representing both position and orientation. Given a task vector, $\boldsymbol{x}(\boldsymbol{q})$, the mapping,

$\{J|\dot{q} \mapsto \dot{x}\}$, between configuration space and task space is given by

$$\delta x = J\delta q \quad \text{or} \quad \dot{x} = J\dot{q}. \tag{3.172}$$

The Jacobian is given by

$$J = \frac{\partial x}{\partial q} = \frac{\partial \dot{x}}{\partial \dot{q}}. \tag{3.173}$$

If we take x to be the standard Cartesian vector in $\mathbb{R}^6$, then

$$\delta x = \begin{pmatrix} \delta r \\ \delta \theta \end{pmatrix} \quad \text{or} \quad \dot{x} = \begin{pmatrix} v \\ \omega \end{pmatrix}. \tag{3.174}$$

The Jacobian is then

$$J = \begin{pmatrix} \Gamma \\ \Pi \end{pmatrix}. \tag{3.175}$$

The task acceleration is given by

$$\ddot{x} = J\ddot{q} + (\dot{J} + \tilde{\Omega}J)\dot{q}, \tag{3.176}$$

where

$$\tilde{\Omega} = \text{diag}(\Omega, \Omega). \tag{3.177}$$

The term Ω is the angular velocity tensor of the frame in which the Jacobian is expressed. We note

$$\tilde{\Omega}J\dot{q} = \begin{pmatrix} \Omega\Gamma\dot{q} \\ \Omega\Pi\dot{q} \end{pmatrix} = \begin{pmatrix} \Omega\Gamma\dot{q} \\ \omega \times \omega \end{pmatrix} = \begin{pmatrix} \Omega\Gamma\dot{q} \\ 0 \end{pmatrix}. \tag{3.178}$$

Furthermore, if the Jacobian is expressed in a fixed frame, such as the base frame, then $\Omega = 0$ and

$$\ddot{x} = J\ddot{q} + \dot{J}\dot{q}. \tag{3.179}$$

If we take the task force vector, f, to be the standard Cartesian vector in $\mathbb{R}^6$, then

$$f = \begin{pmatrix} f \\ \varphi \end{pmatrix}. \tag{3.180}$$

By the Principle of Virtual Work for statics (the general principle for dynamics will be covered in Chapter 5),

$$\begin{gathered} \tau \cdot \delta q = f \cdot \delta x \\ \forall \delta q, \delta x. \end{gathered} \tag{3.181}$$

So,

$$\begin{gathered} \tau^T \delta q = f^T J \delta q \\ \forall \delta q, \end{gathered} \tag{3.182}$$

which implies

$$\tau^T = f^T J, \tag{3.183}$$

or

$$\boldsymbol{\tau} = \boldsymbol{J}^T \boldsymbol{f}. \tag{3.184}$$

Relationship between the Jacobian and Screws

We can alternately interpret the Jacobian in terms of screws. First, we recall that the spatial velocity vector is given by

$$\mathrm{v} = \begin{pmatrix} \boldsymbol{\omega} \\ \boldsymbol{v}_O \end{pmatrix} = \dot{\theta} \begin{pmatrix} \boldsymbol{s} \\ \boldsymbol{s}' \end{pmatrix} + \dot{d} \begin{pmatrix} \boldsymbol{0} \\ \boldsymbol{s} \end{pmatrix}. \tag{3.185}$$

For a revolute joint ($\dot{d} = 0$) this reduces to

$$\mathrm{v} = \dot{\theta} \begin{pmatrix} \boldsymbol{s} \\ \boldsymbol{s}' \end{pmatrix} \tag{3.186}$$

and for a prismatic joint ($\dot{\theta} = 0$) we have

$$\mathrm{v} = \dot{d} \begin{pmatrix} \boldsymbol{0} \\ \boldsymbol{s} \end{pmatrix}. \tag{3.187}$$

Defining s as

$$\mathrm{s} = \begin{pmatrix} \boldsymbol{s} \\ \boldsymbol{s}' \end{pmatrix} \tag{3.188}$$

for a revolute joint and as

$$\mathrm{s} = \begin{pmatrix} \boldsymbol{0} \\ \boldsymbol{s} \end{pmatrix} \tag{3.189}$$

for a prismatic joint, we can express the spatial velocity of link n as

$$\mathrm{v}_n = \sum_{i=1}^{n} \dot{q}_i \mathrm{s}_i = \begin{pmatrix} \uparrow & \uparrow & \uparrow \\ \mathrm{s}_1 & \cdots & \mathrm{s}_n \\ \downarrow & \downarrow & \downarrow \end{pmatrix} \begin{pmatrix} \dot{q}_1 \\ \vdots \\ \dot{q}_n \end{pmatrix}, \tag{3.190}$$

where the columns of the Jacobian,

$$\boldsymbol{J} = \begin{pmatrix} \uparrow & \uparrow & \uparrow \\ \mathrm{s}_1 & \cdots & \mathrm{s}_n \\ \downarrow & \downarrow & \downarrow \end{pmatrix}, \tag{3.191}$$

are the screw vectors, s, in the appropriate coordinate frame.

3.3.3 Propagation of Kinematic Rate Vectors

Two Rigid Bodies Connected by a Revolute Joint

We have seen that angular velocities serially propagate in an additive fashion. Given two links, $i-1$ and i, connected by a revolute joint with joint angle, θ_i, the angular velocity of link i is

$$ {}^{i}\boldsymbol{\omega}_i = {}^{i}\boldsymbol{\omega}_{i-1} + \dot{\theta}_i {}^{i}\hat{\mathfrak{z}}_i = {}^{i}_{i-1}\boldsymbol{Q}\,{}^{i-1}\boldsymbol{\omega}_{i-1} + \dot{\theta}_i {}^{i}\hat{\mathfrak{z}}_i, \tag{3.192} $$

where joint i is taken to rotate about the local $\hat{\mathfrak{z}}$ axis. Equation (3.192) simply represents the vector sum of the angular velocity of link $i-1$ and the angular velocity of link i, relative link $i-1$, to yield the total angular velocity of link i. The angular acceleration of link i is given by the inertial differentiation of ${}^{i}\boldsymbol{\omega}_i$,

$$ {}^{i}\boldsymbol{\alpha}_i = \frac{{}^{i}d}{dt}_{\mathcal{O}}\,{}^{i}\boldsymbol{\omega}_i = \underbrace{\frac{{}^{i}d}{dt}_{\mathcal{O}}\,{}^{i}\boldsymbol{\omega}_{i-1}}_{{}^{i}\boldsymbol{\alpha}_{i-1}} + \frac{{}^{i}d}{dt}_{\mathcal{O}}(\dot{\theta}_i {}^{i}\hat{\mathfrak{z}}_i). \tag{3.193} $$

Furthermore, the angular acceleration of link $i-1$ is given by the inertial differentiation of ${}^{i-1}\boldsymbol{\omega}_{i-1}$,

$$ {}^{i-1}\boldsymbol{\alpha}_{i-1} = \frac{{}^{i-1}d}{dt}_{\mathcal{O}}\,{}^{i-1}\boldsymbol{\omega}_{i-1} = {}^{i-1}\dot{\boldsymbol{\omega}}_{i-1} + {}^{i-1}\boldsymbol{\Omega}_{i-1}\,{}^{i-1}\boldsymbol{\omega}_{i-1} = {}^{i-1}\dot{\boldsymbol{\omega}}_{i-1}. \tag{3.194} $$

Transforming the frame that ${}^{i-1}\boldsymbol{\alpha}_{i-1}$ is expressed in from $i-1$ to i, we have

$$ {}^{i}\boldsymbol{\alpha}_{i-1} = {}^{i}_{i-1}\boldsymbol{Q}\,{}^{i-1}\boldsymbol{\alpha}_{i-1} = {}^{i}_{i-1}\boldsymbol{Q}\,{}^{i-1}\dot{\boldsymbol{\omega}}_{i-1}. \tag{3.195} $$

The last term in (3.193) is given by

$$ \begin{aligned} \frac{{}^{i}d}{dt}_{\mathcal{O}}(\dot{\theta}_i {}^{i}\hat{\mathfrak{z}}_i) &= \frac{{}^{i}d}{dt}_{i}(\dot{\theta}_i {}^{i}\hat{\mathfrak{z}}_i) + \left(\frac{{}^{i}d}{dt}_{\mathcal{O}}\boldsymbol{Q}\right)\dot{\theta}_i {}^{i}\hat{\mathfrak{z}}_i \\ &= \ddot{\theta}_i {}^{i}\hat{\mathfrak{z}}_i + {}^{i}\boldsymbol{\Omega}_i \dot{\theta}_i {}^{i}\hat{\mathfrak{z}}_i = \ddot{\theta}_i {}^{i}\hat{\mathfrak{z}}_i + {}^{i}\boldsymbol{\omega}_i \times \dot{\theta}_i {}^{i}\hat{\mathfrak{z}}_i. \end{aligned} \tag{3.196} $$

We further note that

$$ {}^{i}\boldsymbol{\omega}_i \times \dot{\theta}_i {}^{i}\hat{\mathfrak{z}}_i = ({}^{i}\boldsymbol{\omega}_{i-1} + \dot{\theta}_i {}^{i}\hat{\mathfrak{z}}_i) \times \dot{\theta}_i {}^{i}\hat{\mathfrak{z}}_i = {}^{i}\boldsymbol{\omega}_{i-1} \times \dot{\theta}_i {}^{i}\hat{\mathfrak{z}}_i. \tag{3.197} $$

Finally, by substituting expressions from (3.195), (3.196), and (3.197) into (3.193), we have the desired expression for ${}^{i}\boldsymbol{\alpha}_i$:

$$ \begin{aligned} {}^{i}\boldsymbol{\alpha}_i &= {}^{i}\boldsymbol{\alpha}_{i-1} + \ddot{\theta}_i {}^{i}\hat{\mathfrak{z}}_i + {}^{i}\boldsymbol{\omega}_{i-1} \times \dot{\theta}_i {}^{i}\hat{\mathfrak{z}}_i \\ &= {}^{i}_{i-1}\boldsymbol{Q}\,{}^{i-1}\boldsymbol{\alpha}_{i-1} + \ddot{\theta}_i {}^{i}\hat{\mathfrak{z}}_i + {}^{i}_{i-1}\boldsymbol{Q}\,{}^{i-1}\boldsymbol{\omega}_{i-1} \times \dot{\theta}_i {}^{i}\hat{\mathfrak{z}}_i. \end{aligned} \tag{3.198} $$

The translational velocity of the proximal end of link i can be expressed as the vector sum of the translational velocity of the proximal end of link $i-1$ and the translational velocity induced by the angular velocity of link $i-1$:

$$ \begin{aligned} {}^{i}\boldsymbol{v}_i &= {}^{i}\boldsymbol{v}_{i-1} + {}^{i}\boldsymbol{\omega}_{i-1} \times {}^{i}\boldsymbol{d}_{\overrightarrow{(i-1)i}} \\ &= {}^{i}_{i-1}\boldsymbol{Q}({}^{i-1}\boldsymbol{v}_{i-1} + {}^{i-1}\boldsymbol{\omega}_{i-1} \times {}^{i-1}\boldsymbol{d}_{\overrightarrow{(i-1)i}}), \end{aligned} \tag{3.199} $$

where ${}^{i}\boldsymbol{d}_{\overrightarrow{(i-1)i}}$ is the displacement vector from the proximal end of link $i-1$ to the proximal end of link i. The translational acceleration of the proximal end of link i is given by the inertial differentiation of ${}^{i}\boldsymbol{v}_i$,

$$ {}^{i}\boldsymbol{a}_i = \frac{{}^{i}d}{dt}_{\mathcal{O}}\,{}^{i}\boldsymbol{v}_i = \underbrace{\frac{{}^{i}d}{dt}_{\mathcal{O}}\,{}^{i}\boldsymbol{v}_{i-1}}_{{}^{i}\boldsymbol{a}_{i-1}} + \frac{{}^{i}d}{dt}_{\mathcal{O}}({}^{i}\boldsymbol{\omega}_{i-1} \times {}^{i}\boldsymbol{d}_{\overrightarrow{(i-1)i}}). \tag{3.200} $$

Furthermore,

$$\frac{{}^i d}{dt}_o ({}^i\boldsymbol{\omega}_{i-1} \times {}^i\boldsymbol{d}_{\overrightarrow{(i-1)i}}) = \frac{{}^i d}{dt}_i ({}^i\boldsymbol{\omega}_{i-1} \times {}^i\boldsymbol{d}_{\overrightarrow{(i-1)i}}) + \left(\frac{{}^i d}{dt}_o \boldsymbol{Q}\right) ({}^i\boldsymbol{\omega}_{i-1} \times {}^i\boldsymbol{d}_{\overrightarrow{(i-1)i}})$$
$$= {}^i\boldsymbol{\alpha}_{i-1} \times {}^i\boldsymbol{d}_{\overrightarrow{(i-1)i}} + {}^i\boldsymbol{\Omega}_{i-1}({}^i\boldsymbol{\omega}_{i-1} \times {}^i\boldsymbol{d}_{\overrightarrow{(i-1)i}})$$
$$= {}^i\boldsymbol{\alpha}_{i-1} \times {}^i\boldsymbol{d}_{\overrightarrow{(i-1)i}} + {}^i\boldsymbol{\omega}_{i-1} \times ({}^i\boldsymbol{\omega}_{i-1} \times {}^i\boldsymbol{d}_{\overrightarrow{(i-1)i}}). \quad (3.201)$$

This represents the tangential and centripetal acceleration terms, ${}^i\boldsymbol{\alpha}_{i-1} \times {}^i\boldsymbol{d}_{\overrightarrow{(i-1)i}}$ and ${}^i\boldsymbol{\omega}_{i-1} \times ({}^i\boldsymbol{\omega}_{i-1} \times {}^i\boldsymbol{d}_{\overrightarrow{(i-1)i}})$, respectively. Finally, substituting (3.201) into (3.200), we have

$${}^i\boldsymbol{a}_i = {}^i\boldsymbol{a}_{i-1} + {}^i\boldsymbol{\alpha}_{i-1} \times {}^i\boldsymbol{d}_{\overrightarrow{(i-1)i}} + {}^i\boldsymbol{\omega}_{i-1} \times ({}^i\boldsymbol{\omega}_{i-1} \times {}^i\boldsymbol{d}_{\overrightarrow{(i-1)i}})$$
$$= {}^i_{i-1}\boldsymbol{Q}[{}^{i-1}\boldsymbol{a}_{i-1} + {}^{i-1}\boldsymbol{\alpha}_{i-1} \times {}^{i-1}\boldsymbol{d}_{\overrightarrow{(i-1)i}} + {}^{i-1}\boldsymbol{\omega}_{i-1} \times ({}^{i-1}\boldsymbol{\omega}_{i-1} \times {}^{i-1}\boldsymbol{d}_{\overrightarrow{(i-1)i}})]. \quad (3.202)$$

The translational velocity of the center of mass of link i can be expressed as

$${}^i\boldsymbol{v}_{G_i} = {}^i\boldsymbol{v}_i + {}^i\boldsymbol{\omega}_i \times {}^i\boldsymbol{d}_{\overrightarrow{iG_i}}. \quad (3.203)$$

The translational acceleration of the center of mass of link i can be expressed as

$${}^i\boldsymbol{a}_{G_i} = {}^i\boldsymbol{a}_i + {}^i\boldsymbol{\alpha}_i \times {}^i\boldsymbol{d}_{\overrightarrow{iG_i}} + {}^i\boldsymbol{\omega}_i \times ({}^i\boldsymbol{\omega}_i \times {}^i\boldsymbol{d}_{\overrightarrow{iG_i}}). \quad (3.204)$$

Two Rigid Bodies Connected by a Prismatic Joint

Given two links, $i-1$ and i, connected by a prismatic joint with joint displacement, d_i, the angular velocity of link i is simply

$${}^i\boldsymbol{\omega}_i = {}^i\boldsymbol{\omega}_{i-1} = {}^i_{i-1}\boldsymbol{Q}\,{}^{i-1}\boldsymbol{\omega}_{i-1}. \quad (3.205)$$

The angular acceleration of link i is given by the inertial differentiation,

$${}^i\boldsymbol{\alpha}_i = \frac{{}^i d}{dt}_o {}^i\boldsymbol{\omega}_i = \underbrace{\frac{{}^i d}{dt}_o {}^i\boldsymbol{\omega}_{i-1}}_{{}^i\boldsymbol{\alpha}_{i-1}}. \quad (3.206)$$

Furthermore, the angular acceleration of link $i-1$ is given by the inertial differentiation of ${}^{i-1}\boldsymbol{\omega}_{i-1}$,

$${}^{i-1}\boldsymbol{\alpha}_{i-1} = \frac{{}^{i-1}d}{dt}_o {}^{i-1}\boldsymbol{\omega}_{i-1} = {}^{i-1}\dot{\boldsymbol{\omega}}_{i-1}. \quad (3.207)$$

The desired expression for ${}^i\boldsymbol{\alpha}_i$ is then

$${}^i\boldsymbol{\alpha}_i = {}^i\boldsymbol{\alpha}_{i-1} = {}^i_{i-1}\boldsymbol{Q}\,{}^{i-1}\boldsymbol{\alpha}_{i-1}. \quad (3.208)$$

We note that since the prismatic joint produces no relative angular velocity, there is no propagation of angular velocity and acceleration. So, clearly, $\boldsymbol{\omega}_i = \boldsymbol{\omega}_{i-1}$ and $\boldsymbol{\alpha}_i = \boldsymbol{\alpha}_{i-1}$.

The translational velocity of the proximal end of link i is expressed as the vector sum of the translational velocity of the proximal end of link $i-1$, the translational velocity induced by the angular velocity of link $i-1$, and the translational displacement rate of link i,

$$
\begin{aligned}
{}^i\boldsymbol{v}_i &= {}^i\boldsymbol{v}_{i-1} + {}^i\boldsymbol{\omega}_{i-1} \times {}^i\boldsymbol{d}_{\overrightarrow{(i-1)i}} + \dot{d}_i {}^i\hat{\boldsymbol{z}}_i \\
&= {}_{i-1}^{\;\;i}\boldsymbol{Q}({}^{i-1}\boldsymbol{v}_{i-1} + {}^{i-1}\boldsymbol{\omega}_{i-1} \times {}^{i-1}\boldsymbol{d}_{\overrightarrow{(i-1)i}}) + \dot{d}_i {}^i\hat{\boldsymbol{z}}_i,
\end{aligned} \tag{3.209}
$$

where joint i is taken to translate along the local $\hat{\boldsymbol{z}}$ axis. The translational acceleration of the proximal end of link i is given by the inertial differentiation of ${}^i\boldsymbol{v}_i$,

$$
{}^i\boldsymbol{a}_i = \frac{{}^id}{dt}_O {}^i\boldsymbol{v}_i = \underbrace{\frac{{}^id}{dt}_O {}^i\boldsymbol{v}_{i-1}}_{{}^i\boldsymbol{a}_{i-1}} + \frac{{}^id}{dt}_O ({}^i\boldsymbol{\omega}_{i-1} \times {}^i\boldsymbol{d}_{\overrightarrow{(i-1)i}}) + \frac{{}^id}{dt}_O (\dot{d}_i {}^i\hat{\boldsymbol{z}}_i). \tag{3.210}
$$

Furthermore,

$$
\begin{aligned}
\frac{{}^id}{dt}_O ({}^i\boldsymbol{\omega}_{i-1} \times {}^i\boldsymbol{d}_{\overrightarrow{(i-1)i}}) &= \frac{{}^id}{dt}_i ({}^i\boldsymbol{\omega}_{i-1} \times {}^i\boldsymbol{d}_{\overrightarrow{(i-1)i}}) + \left(\frac{{}^id}{dt}_O \boldsymbol{Q}\right) ({}^i\boldsymbol{\omega}_{i-1} \times {}^i\boldsymbol{d}_{\overrightarrow{(i-1)i}}) \\
&= {}^i\boldsymbol{\alpha}_{i-1} \times {}^i\boldsymbol{d}_{\overrightarrow{(i-1)i}} + {}^i\boldsymbol{\omega}_{i-1} \times \left(\frac{{}^id}{dt}_i {}^i\boldsymbol{d}_{\overrightarrow{(i-1)i}}\right) + {}^i\boldsymbol{\Omega}_{i-1}({}^i\boldsymbol{\omega}_{i-1} \times {}^i\boldsymbol{d}_{\overrightarrow{(i-1)i}}) \\
&= {}^i\boldsymbol{\alpha}_{i-1} \times {}^i\boldsymbol{d}_{\overrightarrow{(i-1)i}} + {}^i\boldsymbol{\omega}_{i-1} \times \dot{d}_i {}^i\hat{\boldsymbol{z}}_i + {}^i\boldsymbol{\omega}_{i-1} \times ({}^i\boldsymbol{\omega}_{i-1} \times {}^i\boldsymbol{d}_{\overrightarrow{(i-1)i}}),
\end{aligned} \tag{3.211}
$$

and,

$$
\begin{aligned}
\frac{{}^id}{dt}_O (\dot{d}_i {}^i\hat{\boldsymbol{z}}_i) &= \frac{{}^id}{dt}_i (\dot{d}_i {}^i\hat{\boldsymbol{z}}_i) + \left(\frac{{}^id}{dt}_O \boldsymbol{Q}\right) \dot{d}_i {}^i\hat{\boldsymbol{z}}_i = \ddot{d}_i {}^i\hat{\boldsymbol{z}}_i + {}^i\boldsymbol{\Omega}_{i-1} \dot{d}_i {}^i\hat{\boldsymbol{z}}_i \\
&= \ddot{d}_i {}^i\hat{\boldsymbol{z}}_i + {}^i\boldsymbol{\omega}_{i-1} \times \dot{d}_i {}^i\hat{\boldsymbol{z}}_i.
\end{aligned} \tag{3.212}
$$

Finally, substituting (3.211) and (3.212) into (3.210), we have

$$
\begin{aligned}
{}^i\boldsymbol{a}_i &= {}^i\boldsymbol{a}_{i-1} + {}^i\boldsymbol{\alpha}_{i-1} \times {}^i\boldsymbol{d}_{\overrightarrow{(i-1)i}} + {}^i\boldsymbol{\omega}_{i-1} \times ({}^i\boldsymbol{\omega}_{i-1} \times {}^i\boldsymbol{d}_{\overrightarrow{(i-1)i}}) + 2{}^i\boldsymbol{\omega}_{i-1} \times \dot{d}_i {}^i\hat{\boldsymbol{z}}_i + \ddot{d}_i {}^i\hat{\boldsymbol{z}}_i \\
&= {}_{i-1}^{\;\;i}\boldsymbol{Q}[{}^{i-1}\boldsymbol{a}_{i-1} + {}^{i-1}\boldsymbol{\alpha}_{i-1} \times {}^{i-1}\boldsymbol{d}_{\overrightarrow{(i-1)i}} + {}^{i-1}\boldsymbol{\omega}_{i-1} \times ({}^{i-1}\boldsymbol{\omega}_{i-1} \times {}^{i-1}\boldsymbol{d}_{\overrightarrow{(i-1)i}})] \\
&\quad + 2{}_{i-1}^{\;\;i}\boldsymbol{Q}{}^{i-1}\boldsymbol{\omega}_{i-1} \times \dot{d}_i {}^i\hat{\boldsymbol{z}}_i + \ddot{d}_i {}^i\hat{\boldsymbol{z}}_i.
\end{aligned} \tag{3.213}
$$

In addition to the tangential and centripetal acceleration terms, we note the Coriolis acceleration term, $2{}^i\boldsymbol{\omega}_{i-1} \times \dot{d}_i {}^i\hat{\boldsymbol{z}}_i$, and the joint acceleration term, $\ddot{d}_i {}^i\hat{\boldsymbol{z}}_i$, in (3.213).

The translational velocity of the center of mass of link i can be expressed as

$$
{}^i\boldsymbol{v}_{G_i} = {}^i\boldsymbol{v}_i + {}^i\boldsymbol{\omega}_i \times {}^i\boldsymbol{d}_{\overrightarrow{iG_i}}. \tag{3.214}
$$

The translational acceleration of the center of mass of link i can be expressed as

$$
{}^i\boldsymbol{a}_{G_i} = {}^i\boldsymbol{a}_i + {}^i\boldsymbol{\alpha}_i \times {}^i\boldsymbol{d}_{\overrightarrow{iG_i}} + {}^i\boldsymbol{\omega}_i \times ({}^i\boldsymbol{\omega}_i \times {}^i\boldsymbol{d}_{\overrightarrow{iG_i}}). \tag{3.215}
$$

Summary of Kinematic Propagation Equations

For two links, $i-1$ and i, connected by a revolute joint we have

$$ {}^{i}\boldsymbol{\omega}_i = {}^{i}_{i-1}\boldsymbol{Q}\,{}^{i-1}\boldsymbol{\omega}_{i-1} + \dot{\theta}_i\,{}^{i}\hat{\boldsymbol{z}}_i, \tag{3.216} $$

$$ {}^{i}\boldsymbol{v}_i = {}^{i}_{i-1}\boldsymbol{Q}({}^{i-1}\boldsymbol{v}_{i-1} + {}^{i-1}\boldsymbol{\omega}_{i-1} \times {}^{i-1}\boldsymbol{d}_{\overrightarrow{(i-1)i}}), \tag{3.217} $$

$$ {}^{i}\boldsymbol{v}_{G_i} = {}^{i}\boldsymbol{v}_i + {}^{i}\boldsymbol{\omega}_i \times {}^{i}\boldsymbol{d}_{\overrightarrow{iG_i}}, \tag{3.218} $$

$$ {}^{i}\boldsymbol{\alpha}_i = {}^{i}_{i-1}\boldsymbol{Q}\,{}^{i-1}\boldsymbol{\alpha}_{i-1} + \ddot{\theta}_i\,{}^{i}\hat{\boldsymbol{z}}_i + {}^{i}_{i-1}\boldsymbol{Q}\,{}^{i-1}\boldsymbol{\omega}_{i-1} \times \dot{\theta}_i\,{}^{i}\hat{\boldsymbol{z}}_i, \tag{3.219} $$

$$ {}^{i}\boldsymbol{a}_i = {}^{i}_{i-1}\boldsymbol{Q}[{}^{i-1}\boldsymbol{a}_{i-1} + {}^{i-1}\boldsymbol{\alpha}_{i-1} \times {}^{i-1}\boldsymbol{d}_{\overrightarrow{(i-1)i}} + {}^{i-1}\boldsymbol{\omega}_{i-1} \times ({}^{i-1}\boldsymbol{\omega}_{i-1} \times {}^{i-1}\boldsymbol{d}_{\overrightarrow{(i-1)i}})], \tag{3.220} $$

$$ {}^{i}\boldsymbol{a}_{G_i} = {}^{i}\boldsymbol{a}_i + {}^{i}\boldsymbol{\alpha}_i \times {}^{i}\boldsymbol{d}_{\overrightarrow{iG_i}} + {}^{i}\boldsymbol{\omega}_i \times ({}^{i}\boldsymbol{\omega}_i \times {}^{i}\boldsymbol{d}_{\overrightarrow{iG_i}}). \tag{3.221} $$

For two links, $i-1$ and i, connected by a prismatic joint we have

$$ {}^{i}\boldsymbol{\omega}_i = {}^{i}_{i-1}\boldsymbol{Q}\,{}^{i-1}\boldsymbol{\omega}_{i-1}, \tag{3.222} $$

$$ {}^{i}\boldsymbol{v}_i = {}^{i}_{i-1}\boldsymbol{Q}({}^{i-1}\boldsymbol{v}_{i-1} + {}^{i-1}\boldsymbol{\omega}_{i-1} \times {}^{i-1}\boldsymbol{d}_{\overrightarrow{(i-1)i}}) + \dot{d}_i\,{}^{i}\hat{\boldsymbol{z}}_i, \tag{3.223} $$

$$ {}^{i}\boldsymbol{v}_{G_i} = {}^{i}\boldsymbol{v}_i + {}^{i}\boldsymbol{\omega}_i \times {}^{i}\boldsymbol{d}_{\overrightarrow{iG_i}}, \tag{3.224} $$

$$ {}^{i}\boldsymbol{\alpha}_i = {}^{i}_{i-1}\boldsymbol{Q}\,{}^{i-1}\boldsymbol{\alpha}_{i-1}, \tag{3.225} $$

$$ \begin{aligned} {}^{i}\boldsymbol{a}_i = {}& {}^{i}_{i-1}\boldsymbol{Q}[{}^{i-1}\boldsymbol{a}_{i-1} + {}^{i-1}\boldsymbol{\alpha}_{i-1} \times {}^{i-1}\boldsymbol{d}_{\overrightarrow{(i-1)i}} + {}^{i-1}\boldsymbol{\omega}_{i-1} \times ({}^{i-1}\boldsymbol{\omega}_{i-1} \times {}^{i-1}\boldsymbol{d}_{\overrightarrow{(i-1)i}})] \\ & + 2\,{}^{i}_{i-1}\boldsymbol{Q}\,{}^{i-1}\boldsymbol{\omega}_{i-1} \times \dot{d}_i\,{}^{i}\hat{\boldsymbol{z}}_i + \ddot{d}_i\,{}^{i}\hat{\boldsymbol{z}}_i, \end{aligned} \tag{3.226} $$

$$ {}^{i}\boldsymbol{a}_{G_i} = {}^{i}\boldsymbol{a}_i + {}^{i}\boldsymbol{\alpha}_i \times {}^{i}\boldsymbol{d}_{\overrightarrow{iG_i}} + {}^{i}\boldsymbol{\omega}_i \times ({}^{i}\boldsymbol{\omega}_i \times {}^{i}\boldsymbol{d}_{\overrightarrow{iG_i}}). \tag{3.227} $$

3.4 Kinematic Constraints and Degrees of Freedom

The notions of constraints and degrees of freedom are important concepts that need to be formally described. The degrees of freedom of a kinematic system refer to the number of independent coordinates that are needed to fully describe the motion of the system. Constraints, on the other hand, act as restrictions to the motion of a kinematic system.

For now we will consider one type of kinematic constraint referred to as a *holonomic* constraint. Holonomic constraints can be expressed as algebraic functions of the system's generalized coordinates, exclusively. In Section 3.3 we referred to configuration space as the space of generalized coordinates corresponding to the joint variables in a kinematic chain. We can denote the number of generalized coordinates used to describe a kinematic chain as n. If the number of independent constraint equations that impose restrictions on those generalized coordinates is denoted by m, then the number of degrees of freedom, p, is given by

$$ p = n - m. \tag{3.228} $$

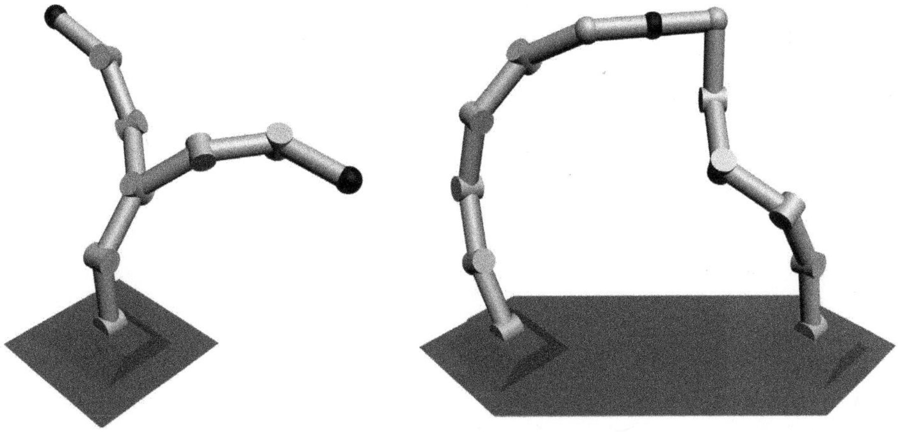

Figure 3.23 Branching and closed chain systems. (Left) Branching kinematic chains can be expressed with a minimal set of generalized coordinates, corresponding to the joint variables, and no imposed constraints. (Right) Closed kinematic chains consisting of loop closures can be represented as branching chains with imposed holonomic constraints to enforce the loop closures.

The way a kinematic system is parameterized is not unique. For example, a rigid body connected to ground by a revolute joint could be parameterized by 6 generalized coordinates ($n = 6$) with 5 imposed constraints ($m = 5$). Such a system would have 1 degree of freedom ($p = 6 - 5 = 1$). Alternately, this system could be parameterized as a kinematic chain with a single generalized coordinate ($n = 1$) associated with the revolute joint and no imposed constraints ($m = 0$). With this parameterization the system would still have 1 degree of freedom ($p = 1 - 0 = 1$). In this way the number of degrees of freedom of a kinematic system is invariant of a particular parameterization.

Branching kinematic chains like the ones addressed in Section 3.3 can be expressed with a minimal set of generalized coordinates, corresponding to the joint variables, and no imposed constraints. Therefore, for these branching chains parametrized with this minimal set of joint coordinates, the number of generalized coordinates is equivalent to the number of degrees of freedom ($p = n$).

Closed kinematic chains consisting of loop closures represent common holonomic constraints. Such systems can be represented as branching chains with imposed constraints to enforce the loop closures (see Figure 3.23). In the figure on the right, the closed chain is modeled by two open chains that meet at their terminal points (black sphere). The constraint equations, $\boldsymbol{\phi}(\boldsymbol{q}) = \mathbf{0}$, describe the enforcement of the loop closure. In the subsequent chapter we shall see how the constraints are incorporated into the dynamical equations.

3.5 Exercises

1. Given the basis representation for each frame shown in Figure 3.24, determine the rotation matrix, ${}^{A}_{B}\boldsymbol{Q}$.

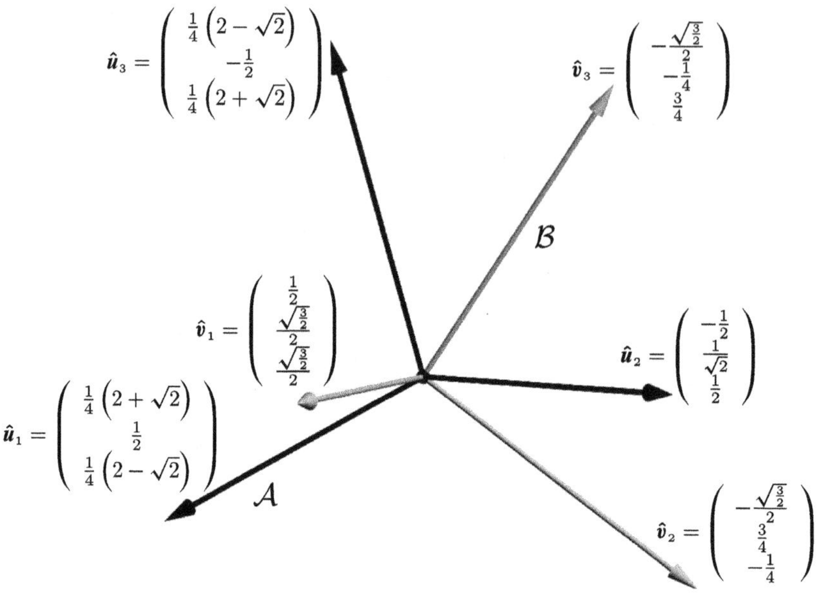

Figure 3.24 Rotation of frame $\mathcal{B}$ relative to frame $\mathcal{A}$. A basis for each frame is shown (Exercise 1).

2. Consider the rotation matrix

$$\boldsymbol{Q} = \begin{pmatrix} \frac{7}{12} & \frac{1}{12}\left(-2-3\sqrt{2}\right) & \frac{1}{12}-\frac{1}{\sqrt{2}} \\ \frac{1}{12}\left(-2+3\sqrt{2}\right) & \frac{5}{6} & \frac{1}{12}\left(-2-3\sqrt{2}\right) \\ \frac{1}{12}+\frac{1}{\sqrt{2}} & \frac{1}{12}\left(-2+3\sqrt{2}\right) & \frac{7}{12} \end{pmatrix}.$$

   (a) Compute the invariant axis (spin axis) by solving the eigenvalue problem for $\boldsymbol{Q}$. What eigenvector corresponds to the spin axis? What is the eigenvalue associated with this eigenvector?
   (b) Show that the norm of all the eigenvalues is 1. Verify that the spin axis found in (a) agrees with the solution found by using (3.18).
3. Consider the *zxz* Euler sequence.
   (a) Compute the corresponding rotation matrix, $\boldsymbol{Q}_{zxz}(\alpha, \beta, \gamma)$. That is, compute the rotation sequence, $\boldsymbol{Q}_z(\alpha)\boldsymbol{Q}_x(\beta)\boldsymbol{Q}_z(\gamma)$.
   (b) Derive the inverse relationship. That is, relate the Euler angles α , β , and γ to the elements of the rotation matrix computed in (a).
4. Consider the *yxz* Euler sequence with angles

$$\alpha = \pi/3, \quad \beta = \pi/4, \quad \text{and} \quad \gamma = \pi/6.$$

   (a) Compute the corresponding unit quaternion, $h_{yxz}(\alpha, \beta, \gamma)$. That is, compute the rotation sequence $h_y(\alpha)h_x(\beta)h_z(\gamma)$.
   (b) Find the spin axis and angle associated with the quaternion from (a).

5. What mathematical property do infinitesimal rotations possess that finite rotations do not?
6. Consider the rotation matrix $\boldsymbol{Q}(\alpha(t), \beta(t), \gamma(t)) = \boldsymbol{Q}_x(\alpha(t))\boldsymbol{Q}_y(\beta(t))\boldsymbol{Q}_z(\gamma(t))$.
   (a) Compute the angular velocity tensor in the local frame by differentiating the rotation matrix.
   (b) Express the angular velocity vector associated with the angular velocity tensor computed in (a).
   (c) Represent the angular velocity from (b) as an angular velocity quaternion.
   (d) Express the Jacobian matrix, $\boldsymbol{E}(\alpha, \beta, \gamma)$, between the angular velocity and the rates of change of α , β , and γ.
7. Consider the rotation matrix from Exercise 6.
   (a) Compute the angular velocity tensor in the base frame by differentiating the rotation matrix.
   (b) Compute the angular velocity tensor in the base frame using tensor transformation on the angular velocity tensor computed in Exercise 6(a).
8. Consider the screw displacement, $\$$, where

$$\boldsymbol{s} = \frac{1}{\sqrt{6}}\begin{pmatrix} 1 \\ -1 \\ 2 \end{pmatrix}, \quad \boldsymbol{d}_{\overrightarrow{OP_o}} = \begin{pmatrix} 1 \\ 1 \\ 0 \end{pmatrix}, \quad \theta = \pi/3, \quad \text{and} \quad d = 1.$$

   (a) Compute the moment vector $\boldsymbol{s}'$.
   (b) Compute the rotation matrix, $\boldsymbol{Q}_s(\theta)$, associated with $\$$.
   (c) Compute the transformation matrix, $\boldsymbol{T}_s(\theta, d)$, associated with $\$$.
9. Compute the spatial velocity vector $\mathbf{v} = \left(\boldsymbol{\omega}^T \ \boldsymbol{v}_O^T\right)^T$, given the motor parameters

$$\boldsymbol{s} = \frac{1}{\sqrt{11}}\begin{pmatrix} 3 \\ 1 \\ -1 \end{pmatrix}, \quad \boldsymbol{d}_{\overrightarrow{OP_o}} = \begin{pmatrix} -2 \\ 3 \\ 0 \end{pmatrix}, \quad \dot{\theta} = \pi/10, \quad \text{and} \quad \dot{d} = 1/4.$$

10. Given the frame representations of the RRR mechanism described in Figure 3.25, and the Denavit-Hartenberg parameters shown in Table 3.2, compute the kinematic transform, ${}^0_3\boldsymbol{T}$ at the joint variable values of $(\theta_1 \ \theta_2 \ \theta_3)^T = (\pi/12 \ \pi/8 \ -\pi/6)^T$.

Table 3.2 Denavit-Hartenberg parameters for RRR mechanism

i	α_{i-1}	a_{i-1}	θ_i	di
1	0	0	θ_1	0
2	$\pi/8$	3	θ_2	3/2
3	$-\pi/4$	5/2	θ_3	1

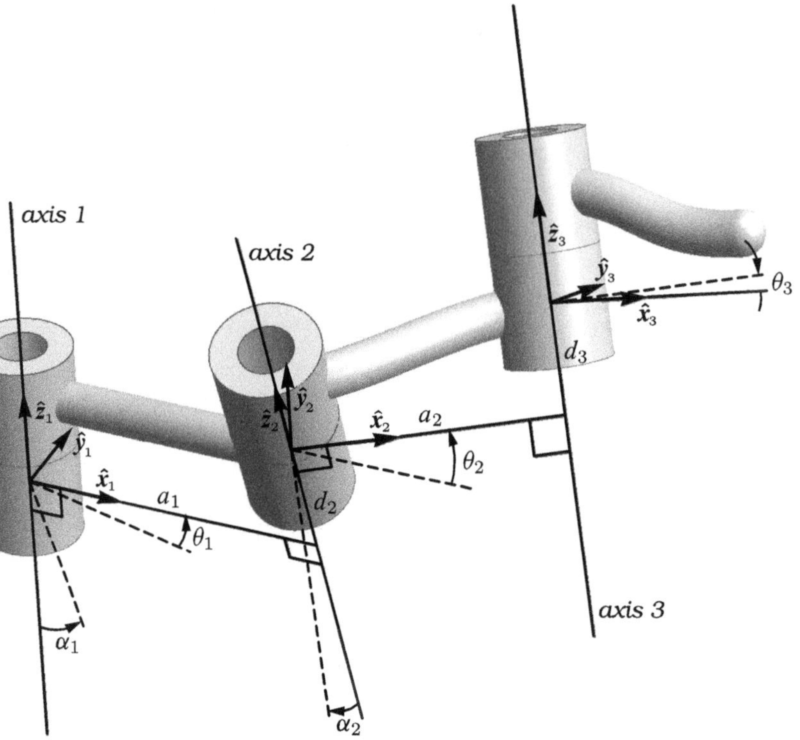

Figure 3.25 Frame representations of a RRR joint mechanism. The Denavit-Hartenberg parameters are shown (Exercises 10 and 12).

11. Given the frame representations of the PRR mechanism described in Figure 3.26, and the Denavit-Hartenberg parameters shown in Table 3.3, compute the kinematic transform, ${}^0_3\boldsymbol{T}$, at the joint variable values of $(d_1 \;\; \theta_2 \;\; \theta_3)^T = (5/4 \;\; -\pi/8 \;\; -\pi/12)^T$.

Table 3.3 Denavit-Hartenberg parameters for PRR mechanism

i	α_{i-1}	a_{i-1}	θ_i	di
1	0	0	0	d_1
2	$-\pi/2$	2	θ_2	1/2
3	$\pi/2$	3	θ_3	1

12. Consider the mechanism described in Figure 3.25 and Table 3.2, at the joint variable values of $(\theta_1 \;\; \theta_2 \;\; \theta_3)^T = (\pi/12 \;\; \pi/8 \;\; -\pi/6)^T$.
    (a) Compute the position vector of the end of link 2 in the base frame. That is, compute ${}^0\boldsymbol{r}_2^3$ (the point on link 2 to which link 3 attaches, expressed in frame 0).
    (b) Compute the Jacobian, ${}^0\boldsymbol{\Gamma}_3 = \partial\, {}^0\boldsymbol{r}_2^3 / \partial \boldsymbol{q}$, of the position vector from (a).
    (c) Compute the angular velocity vector of link 2 in the base frame. That is, compute ${}^0\boldsymbol{\omega}_2$.

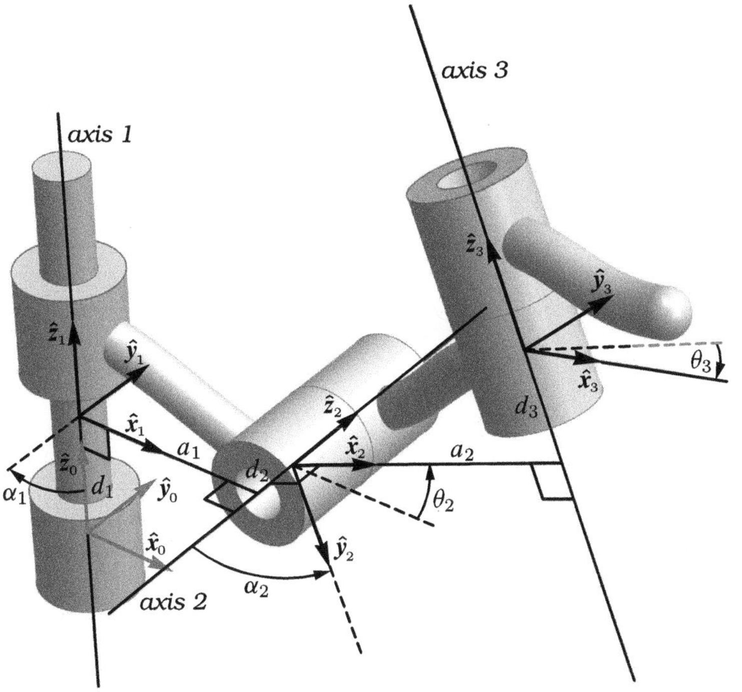

Figure 3.26 Frame representations of a PRR joint mechanism. The Denavit-Hartenberg parameters are shown (Exercises 11 and 13).

i. Compute ${}^0\boldsymbol{\omega}_2$ by taking the derivative of ${}^0_2\boldsymbol{Q}$.

ii. Compute ${}^0\boldsymbol{\omega}_2$ by forward propagation of the angular velocities.

(d) Compute the angular velocity Jacobian, ${}^0\boldsymbol{\Pi}_2 = \partial {}^0\boldsymbol{\omega}_2/\partial \dot{\boldsymbol{q}}$, associated with the angular velocity from (c).

13. Repeat Exercise 12 for the mechanism described in Figure 3.26 and Table 3.3, at the joint variable values of $(d_1 \quad \theta_2 \quad \theta_3)^T = (5/4 \quad -\pi/8 \quad -\pi/12)^T$.

14. Consider the serial chain robot of Figure 3.27. The joint axes are in the z directions of the local frames. Frame 1 is rotated by $\pi/2$ in the x direction of the base frame followed by a z-axis rotation of θ_1. Frame 2 is rotated by $-\pi/2$ in the x direction of frame 1 followed by a z-axis rotation of θ_2. The link lengths are l_1 and l_2.

(a) Compute the rotation matrices ${}^0_1\boldsymbol{Q}$, ${}^1_2\boldsymbol{Q}$, and ${}^0_2\boldsymbol{Q}$.

(b) Compute the terms ${}^i\boldsymbol{\omega}_i$, ${}^i\boldsymbol{\alpha}_i$, ${}^i\boldsymbol{a}_i$, and ${}^i\boldsymbol{a}_{G_i}$ for $i = 1, 2$ using forward kinematic propagation.

15. Determine the number of degrees of freedom for the following cases. Specify different ways of defining the generalized coordinates and imposed constraints.

(a) Two rigid bodies connected to each other by a revolute joint but otherwise free to move.

(b) Two rigid bodies connected to each other by a spherical (ball and socket) joint but otherwise free to move.

(c) Three rigid bodies connected in series by two revolute joints but otherwise free to move.

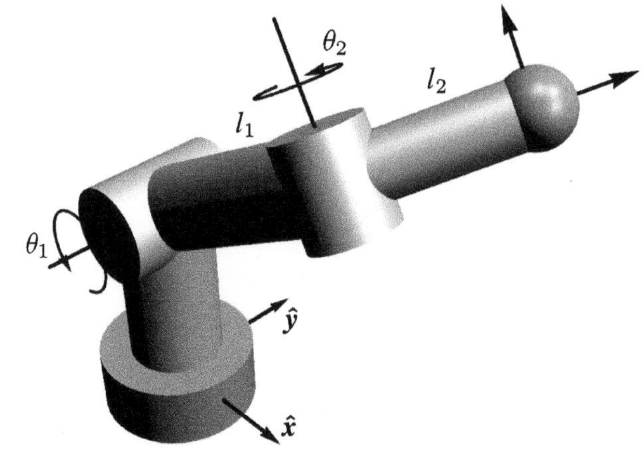

Figure 3.27 A 2 degree-of-freedom serial chain robot with joint angles θ_1 and θ_2 (Exercise 14). The link lengths are l_1 and l_2. The centers of mass are at the geometric centers of the links. The joint axes are in the z directions of the local frames. Frame 1 is rotated by $\pi/2$ in the x direction of the base frame followed by a z-axis rotation of θ_1. Frame 2 is rotated by $-\pi/2$ in the x direction of frame 1 followed by a z-axis rotation of θ_2.

(d) Three rigid bodies connected in series by one revolute joint and one spherical joint but otherwise free to move.

16. Consider a system of two particles moving in $\mathbb{R}^3$. Let the positions of the particles be represented as ${}^{o}\boldsymbol{r}_1 = (q_1 \quad q_2 \quad q_3)^T$ and ${}^{o}\boldsymbol{r}_2 = (q_4 \quad q_5 \quad q_6)^T$.
    (a) How many degrees of freedom does the system possess? How many generalized coordinates are required to parameterize the system?
    (b) Let us consider that three linear constraints are imposed on the system

$$\boldsymbol{\phi}(\boldsymbol{q}) = \begin{pmatrix} 0 & -c_3 & c_2 \\ c_3 & 0 & -c_1 \\ -c_2 & c_1 & 0 \end{pmatrix} \begin{pmatrix} q_1 \\ q_2 \\ q_3 \end{pmatrix} = \mathbf{0}.$$

    i. What type of constraints are these? Describe the constraints physically in terms of the behavior of the two particles.
    ii. How many degrees of freedom does the system have with these constraints imposed (be careful to consider the linear independence of the constraints)?

    (c) Now let us consider the following constraint imposed on the system:

$$\boldsymbol{\phi}(\boldsymbol{q}) = \left\| \begin{pmatrix} q_4 - q_1 \\ q_5 - q_2 \\ q_6 - q_3 \end{pmatrix} \right\| = 0.$$

    i. Describe the constraints physically in terms of the behavior of the two particles.
    ii. How many degrees of freedom does the system have with this constraint imposed?

4 Conservation Principles

The traditional starting point in the classical study of mechanics is with the so-called conservation principles. Newton's famous laws of motion form the basis of classical mechanics. We can consider Newton's second law as part of a more general conservation principle; specifically the conservation of translational momentum.

In this chapter we will begin with Newton's laws, which address point masses, and generalize to extended bodies (rigid bodies) using Euler's extension of Newton's laws. In both cases, point masses and rigid bodies, conservation of momentum plays the central role. We will articulate the Newton-Euler Principle as the basis for deriving the equations of motion for both a single rigid body and for systems of interconnected rigid bodies (i.e., kinematic chains). As we apply the conservation of translational and angular momentum to a single rigid body, we will define a set of fundamental inertial properties, including the center of mass and the inertia tensor. Furthermore, as we apply the conservation principles to a set of interconnected rigid bodies, we will articulate the iterative Newton-Euler method for deriving the equations of motion of kinematic chains.

A fundamental concept in Newtonian mechanics is that of an inertial (Galilean) reference frame. This is a frame of reference in which time and space are homogeneous and isotropic; that is, the laws of physics are the same for all points and orientations. All inertial references frames have constant motion (constant translational velocity) with respect to each other. Derivatives will be assumed to be taken with respect to an inertial reference frame, unless otherwise stated.

4.1 The Newton-Euler Principle

We begin by addressing a single point mass. Newton's laws of motion are as follows:

1. With respect to an inertial reference frame, a point mass at rest tends to stay at rest. A point mass in motion tends to move at a constant velocity unless acted on by external forces.
2. The sum of forces, $\boldsymbol{f}$, acting on a point mass impart to it an acceleration, $\boldsymbol{a}$, in the same direction as the resultant force vector and with a magnitude inversely proportional to the mass, M, of the body. That is, $\boldsymbol{f} = M\boldsymbol{a}$.

3. When one point mass exerts a force on another point mass, the latter point mass exerts a force of equal magnitude and opposite direction on the former point mass. This is also known as the weak law of action and reaction.

Newton's second law can be more generally stated in terms of translational momentum:

PRINCIPLE 4.1 *The translational momentum of a closed system is conserved,*

$$\frac{d\boldsymbol{p}}{dt} = 0. \tag{4.1}$$

For a system with externally applied forces we have

$$\frac{d\boldsymbol{p}}{dt} = \sum_{i=1}^{n_f} \boldsymbol{f}_i. \tag{4.2}$$

This is known as ***Newton's Principle.***

Newton's third law implies that the force that body $\mathcal{A}$ exerts on body $\mathcal{B}$ is equal and opposite to the force that body $\mathcal{B}$ exerts on body $\mathcal{A}$. That is,

$$\boldsymbol{f}_{\mathcal{B}}^{\mathcal{A}} = -\boldsymbol{f}_{\mathcal{A}}^{\mathcal{B}}. \tag{4.3}$$

Euler extended Newton's second law to rigid bodies. This extension can be generally stated as follows:

PRINCIPLE 4.2 *The translational and angular momenta of a closed system are conserved:*

$$\frac{d\boldsymbol{p}}{dt} = 0 \quad \textit{and} \quad \frac{d\boldsymbol{H}}{dt} = 0. \tag{4.4}$$

For a system with externally applied forces and moments we have

$$\frac{d\boldsymbol{p}}{dt} = \sum_{i=1}^{n_f} \boldsymbol{f}_i \quad \textit{and,} \quad \frac{d\boldsymbol{H}^{G}}{dt} = \sum_{i=1}^{n_f} \boldsymbol{d}_{\overrightarrow{GP_i}} \times \boldsymbol{f}_i + \sum_{j=1}^{n_\varphi} \boldsymbol{\varphi}_j. \tag{4.5}$$

This is known as the ***Newton-Euler Principle.***

Euler's extension of Newton's third law implies that the moment that body $\mathcal{A}$ exerts on body $\mathcal{B}$ is equal and opposite to the moment that body $\mathcal{B}$ exerts on body $\mathcal{A}$. That is,

$$\boldsymbol{\varphi}_{\mathcal{B}}^{\mathcal{A}} = -\boldsymbol{\varphi}_{\mathcal{A}}^{\mathcal{B}}. \tag{4.6}$$

4.1.1 A Single Particle

Newton's Principle for a single point mass with a discrete set of n_f external forces, $\{\boldsymbol{f}_1, \ldots, \boldsymbol{f}_{n_f}\}$, (and gravity) acting on it is expressed as

$$\frac{d\boldsymbol{p}}{dt} = \sum_{i=1}^{n_f} \boldsymbol{f}_i - Mg\hat{\boldsymbol{e}}_3, \tag{4.7}$$

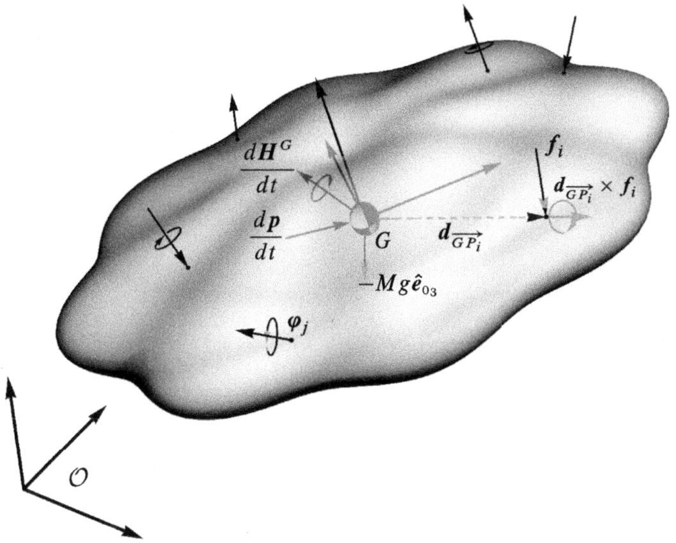

Figure 4.1 Free-body diagram showing external forces, f_i, and moments, φ_j, and the time rates of change of translational and angular momenta, $\dot{p}$ and $\dot{H}^G$, respectively. The momenta are relative to the center of mass, G.

where translational momentum is defined as

$$p \triangleq Mv. \tag{4.8}$$

Thus,

$$Ma = \sum_{i=1}^{n_f} f_i - Mg\hat{e}_3. \tag{4.9}$$

4.1.2 A Single Rigid Body

The Newton-Euler principle for a single rigid body with a discrete set of n_f external forces, $\{f_1, \ldots, f_{n_f}\}$, and n_φ external moments, $\{\varphi_1, \ldots, \varphi_{n_\varphi}\}$, acting on it is expressed as

$$\frac{dp}{dt} = \sum_{i=1}^{n_f} f_i - Mg\hat{e}_3 \quad \text{and} \quad \frac{dH^G}{dt} = \sum_{i=1}^{n_f} d_{\overrightarrow{GP_i}} \times f_i + \sum_{j=1}^{n_\varphi} \varphi_j. \tag{4.10}$$

The free-body diagram for this is illustrated in Figure 4.1.

The center of mass position vector, r_G, of a continuous rigid system is defined as

$$r_G \triangleq \frac{1}{M} \int_M r\, dM. \tag{4.11}$$

The velocity of the center of mass, $\boldsymbol{v}_G$, with respect to an inertial reference frame is then

$$\boldsymbol{v}_G = \frac{d}{dt}\boldsymbol{r}_G = \frac{1}{M}\int_M \boldsymbol{v}\, dM, \tag{4.12}$$

and the acceleration of the center of mass, $\boldsymbol{a}_G$, with respect to an inertial reference frame is

$$\boldsymbol{a}_G = \frac{d}{dt}\boldsymbol{v}_G = \frac{1}{M}\int_M \boldsymbol{a}\, dM. \tag{4.13}$$

Translational momentum is defined as

$$\boldsymbol{p} \triangleq \int_M \boldsymbol{v}\, dM = M\boldsymbol{v}_G. \tag{4.14}$$

The derivative of $\boldsymbol{p}$ with respect to an inertial reference frame can then be expressed as

$$\frac{d}{dt}\boldsymbol{p} = \frac{d}{dt}\int_M \boldsymbol{v}\, dM = \int_M \frac{d}{dt}\boldsymbol{v}\, dM = \int_M \boldsymbol{a}\, dM. \tag{4.15}$$

Alternately, we can express the derivative of $\boldsymbol{p}$ as

$$\frac{d}{dt}\boldsymbol{p} = \frac{d}{dt}M\boldsymbol{v}_G = M\boldsymbol{a}_G, \tag{4.16}$$

and by Principle 4.2,

$$M\boldsymbol{a}_G = \sum_{i=1}^{n_f} \boldsymbol{f}_i - Mg\hat{\boldsymbol{e}}_3. \tag{4.17}$$

The angular momentum of the body about an arbitrary origin point, O, is defined as

$$\boldsymbol{H}^O \triangleq \int_M \boldsymbol{r} \times \boldsymbol{v}\, dM. \tag{4.18}$$

It will be useful to represent the vector $\boldsymbol{r}$ as the sum of the vector from the origin to the center of mass, $\boldsymbol{r}_G$, and the vector from the center of mass to a point on the body, $\boldsymbol{\rho}$. Similarly, the vector $\boldsymbol{v}$ can be represented as the sum of the velocity of the center of mass, $\boldsymbol{v}_G$, and $\boldsymbol{\omega} \times \boldsymbol{\rho}$ (see Figure 4.2). Thus, the angular momentum can be expressed in

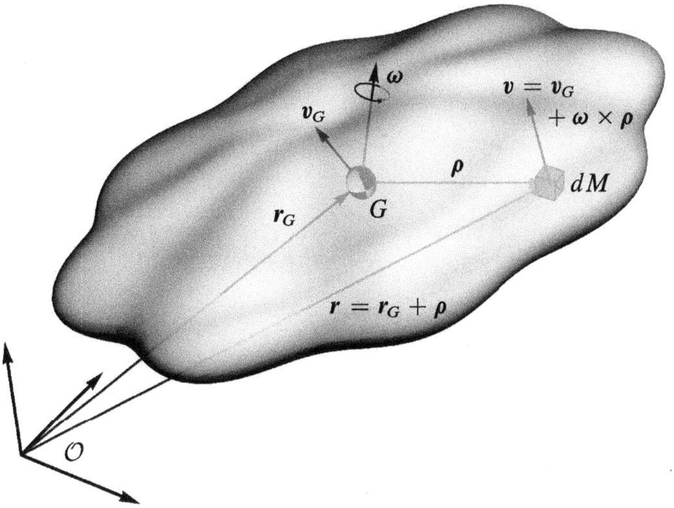

Figure 4.2 Angular momentum of a body about an arbitrary origin point is depicted. The vector from the origin to a point on the body, $\boldsymbol{r}$, is composed of the vector from the origin to the center of mass, $\boldsymbol{r}_G$, and the vector from the center of mass to a point on the body, $\boldsymbol{\rho}$. The velocity vector of a point on the body, $\boldsymbol{v}$, is composed of the velocity of the center of mass, $\boldsymbol{v}_G$, and $\boldsymbol{\omega} \times \boldsymbol{\rho}$.

terms of the center of mass and velocity of the center of mass,

$$\begin{aligned} \boldsymbol{H}^O &= \int_M (\boldsymbol{r}_G + \boldsymbol{\rho}) \times (\boldsymbol{v}_G + \boldsymbol{\omega} \times \boldsymbol{\rho})\, dM = \boldsymbol{r}_G \times \boldsymbol{v}_G \underbrace{\int_M dM}_{M} \\ &\quad + \boldsymbol{r}_G \times (\boldsymbol{\omega} \times \underbrace{\int_M \boldsymbol{\rho}\, dM}_{M\rho_G = 0}) + \underbrace{\int_M \boldsymbol{\rho}\, dM}_{M\rho_G = 0} \times \boldsymbol{v}_G + \int_M \boldsymbol{\rho} \times (\boldsymbol{\omega} \times \boldsymbol{\rho})\, dM \\ &= \boldsymbol{r}_G \times M\boldsymbol{v}_G + \int_M \boldsymbol{\rho} \times (\boldsymbol{\omega} \times \boldsymbol{\rho})\, dM. \end{aligned} \tag{4.19}$$

Noting the identity

$$\boldsymbol{a} \times (\boldsymbol{b} \times \boldsymbol{c}) = \boldsymbol{b}(\boldsymbol{a} \cdot \boldsymbol{c}) - \boldsymbol{c}(\boldsymbol{a} \cdot \boldsymbol{b}), \tag{4.20}$$

we have

$$\boldsymbol{\rho} \times (\boldsymbol{\omega} \times \boldsymbol{\rho}) = \boldsymbol{\omega}(\boldsymbol{\rho} \cdot \boldsymbol{\rho}) - \boldsymbol{\rho}(\boldsymbol{\rho} \cdot \boldsymbol{\omega}) = \|\boldsymbol{\rho}\|^2 \boldsymbol{\omega} - \left(\boldsymbol{\rho}\boldsymbol{\rho}^T\right) \boldsymbol{\omega} = \left(\|\boldsymbol{\rho}\|^2 \mathbf{1} - \boldsymbol{\rho}\boldsymbol{\rho}^T\right) \boldsymbol{\omega}. \tag{4.21}$$

So, the angular momentum can be expressed as

$$\boldsymbol{H}^O = \boldsymbol{r}_G \times M\boldsymbol{v}_G + \left[\int_M \left(\|\boldsymbol{\rho}\|^2 \mathbf{1} - \boldsymbol{\rho}\boldsymbol{\rho}^T\right) dM \right] \boldsymbol{\omega}. \tag{4.22}$$

We define the inertia tensor evaluated about the center of mass:

$$\boldsymbol{I}^G \triangleq \int_M \left(\|\boldsymbol{\rho}\|^2 \mathbf{1} - \boldsymbol{\rho}\boldsymbol{\rho}^T\right) dM = \int_M [(\boldsymbol{\rho} \cdot \boldsymbol{\rho})\mathbf{1} - \boldsymbol{\rho} \otimes \boldsymbol{\rho}]\, dM. \tag{4.23}$$

The inertia tensor is symmetric positive definite, a fact that we will not show here. We have

$$\boldsymbol{H}^O = \boldsymbol{r}_G \times M\boldsymbol{v}_G + \boldsymbol{I}^G\boldsymbol{\omega}, \tag{4.24}$$

or, noting that $\boldsymbol{v}_G = \boldsymbol{v}_O + \boldsymbol{\omega} \times \boldsymbol{r}_G$, we have

$$\begin{aligned}\boldsymbol{H}^O &= M\boldsymbol{r}_G \times \boldsymbol{v}_O + M\boldsymbol{r}_G \times (\boldsymbol{\omega} \times \boldsymbol{r}_G) + \boldsymbol{I}^G\boldsymbol{\omega} \\ &= M\boldsymbol{r}_G \times \boldsymbol{v}_O + M\left[(\boldsymbol{r}_G \cdot \boldsymbol{r}_G)\mathbf{1} - \boldsymbol{r}_G \otimes \boldsymbol{r}_G\right]\boldsymbol{\omega} + \boldsymbol{I}^G\boldsymbol{\omega} \\ &= M\boldsymbol{r}_G \times \boldsymbol{v}_O + \left(\boldsymbol{I}^G + M[(\boldsymbol{r}_G \cdot \boldsymbol{r}_G)\mathbf{1} - \boldsymbol{r}_G \otimes \boldsymbol{r}_G]\right)\boldsymbol{\omega}.\end{aligned} \tag{4.25}$$

If we evaluate the angular momentum about the center of mass, then

$$\boldsymbol{H}^G = \int_M \boldsymbol{\rho} \times (\boldsymbol{v}_G + \boldsymbol{\omega} \times \boldsymbol{\rho})\, dM = \underbrace{\int_M \boldsymbol{\rho}\, dM}_{M\rho_G = 0} \times \boldsymbol{v}_G + \int_M \boldsymbol{\rho} \times (\boldsymbol{\omega} \times \boldsymbol{\rho})\, dM = \boldsymbol{I}^G\boldsymbol{\omega}. \tag{4.26}$$

We note that

$$\boldsymbol{I}^O = \boldsymbol{I}^G + M[(\boldsymbol{r}_G \cdot \boldsymbol{r}_G)\mathbf{1} - \boldsymbol{r}_G \otimes \boldsymbol{r}_G]. \tag{4.27}$$

This relationship is known as the parallel axis theorem. Furthermore,

$$\begin{aligned}\boldsymbol{H}^O &= \boldsymbol{r}_G \times M\boldsymbol{v}_G + \boldsymbol{H}^G \\ &= M\boldsymbol{r}_G \times \boldsymbol{v}_O + \boldsymbol{I}^O\boldsymbol{\omega}.\end{aligned} \tag{4.28}$$

Frame transformations for the inertia tensor can be accommodated with the familiar tensor double product. That is,

$$^A\boldsymbol{I}^G = {}^A_B\boldsymbol{Q}\,{}^B\boldsymbol{I}^{GB}_A\boldsymbol{Q}. \tag{4.29}$$

This can be demonstrated by noting that

$$^B\boldsymbol{H}^G = {}^B\boldsymbol{I}^{GB}\boldsymbol{\omega}. \tag{4.30}$$

Therefore,

$$^A\boldsymbol{H}^G = {}^A_B\boldsymbol{Q}\,{}^B\boldsymbol{H}^G = {}^A_B\boldsymbol{Q}({}^B\boldsymbol{I}^{GB}\boldsymbol{\omega}) = ({}^A_B\boldsymbol{Q}\,{}^B\boldsymbol{I}^{GB}_A\boldsymbol{Q})^A\boldsymbol{\omega} = {}^A\boldsymbol{I}^{GA}\boldsymbol{\omega}. \tag{4.31}$$

Using the expression for angular momentum given in (4.18), the derivative of $\boldsymbol{H}^O$ with respect to an inertial reference frame can be expressed as

$$\frac{d}{dt}\boldsymbol{H}^O = \frac{d}{dt}\int_M \boldsymbol{r} \times \boldsymbol{v}\, dM = \int_M \frac{d}{dt}\boldsymbol{r} \times \boldsymbol{v}\, dM = \int_M (\boldsymbol{v} \times \boldsymbol{v} + \boldsymbol{r} \times \boldsymbol{a})\, dM \tag{4.32}$$

$$= \int_M \boldsymbol{r} \times \boldsymbol{a}\, dM. \tag{4.33}$$

More rigorously, we can see that

$$\frac{d}{dt}(\boldsymbol{r} \times \boldsymbol{v}) = \lim_{\Delta t \to 0} \frac{(\boldsymbol{r} + \Delta \boldsymbol{r}) \times (\boldsymbol{v} + \Delta \boldsymbol{v}) - \boldsymbol{r} \times \boldsymbol{v}}{\Delta t} = \boldsymbol{r} \times \boldsymbol{a}. \tag{4.34}$$

Expanding this expression for the derivative of $\boldsymbol{H}^o$, we can express it in terms of the center of mass and the acceleration of the center of mass:

$$\begin{aligned}
\frac{d}{dt}\boldsymbol{H}^o &= \int_M \boldsymbol{r} \times \boldsymbol{a}\, dM = \int_M (\boldsymbol{r}_G + \boldsymbol{\rho}) \times [\boldsymbol{a}_G + \boldsymbol{\alpha} \times \boldsymbol{\rho} + \boldsymbol{\omega} \times (\boldsymbol{\omega} \times \boldsymbol{\rho})]\, dM \\
&= \boldsymbol{r}_G \times \boldsymbol{a}_G \underbrace{\int_M dM}_{M} + \boldsymbol{r}_G \times (\boldsymbol{\alpha} \times \underbrace{\int_M \boldsymbol{\rho}\, dM}_{M\boldsymbol{\rho}_G = \boldsymbol{0}}) + \boldsymbol{r}_G \times [\boldsymbol{\omega} \times (\boldsymbol{\omega} \times \underbrace{\int_M \boldsymbol{\rho}\, dM}_{M\boldsymbol{\rho}_G = \boldsymbol{0}})] \\
&\quad + \underbrace{\int_M \boldsymbol{\rho}\, dM}_{M\boldsymbol{\rho}_G = \boldsymbol{0}} \times \boldsymbol{a}_G + \int_M \boldsymbol{\rho} \times (\boldsymbol{\alpha} \times \boldsymbol{\rho})\, dM + \int_M \boldsymbol{\rho} \times [\boldsymbol{\omega} \times (\boldsymbol{\omega} \times \boldsymbol{\rho})] dM \\
&= \boldsymbol{r}_G \times M\boldsymbol{a}_G + \int_M \boldsymbol{\rho} \times (\boldsymbol{\alpha} \times \boldsymbol{\rho})\, dM + \int_M \boldsymbol{\rho} \times [\boldsymbol{\omega} \times (\boldsymbol{\omega} \times \boldsymbol{\rho})] dM.
\end{aligned} \tag{4.35}$$

We have

$$\boldsymbol{\rho} \times (\boldsymbol{\alpha} \times \boldsymbol{\rho}) = \boldsymbol{\alpha}(\boldsymbol{\rho} \cdot \boldsymbol{\rho}) - \boldsymbol{\rho}(\boldsymbol{\rho} \cdot \boldsymbol{\alpha}) = \|\boldsymbol{\rho}\|^2 \boldsymbol{\alpha} - \left(\boldsymbol{\rho}\boldsymbol{\rho}^T\right) \boldsymbol{\alpha} = \left(\|\boldsymbol{\rho}\|^2 \mathbf{1} - \boldsymbol{\rho}\boldsymbol{\rho}^T\right) \boldsymbol{\alpha}, \tag{4.36}$$

and

$$\begin{aligned}
\boldsymbol{\rho} \times [\boldsymbol{\omega} \times (\boldsymbol{\omega} \times \boldsymbol{\rho})] &= \boldsymbol{\rho} \times [\boldsymbol{\omega}(\boldsymbol{\omega} \cdot \boldsymbol{\rho}) - \boldsymbol{\rho}(\boldsymbol{\omega} \cdot \boldsymbol{\omega})] = \boldsymbol{\rho} \times \boldsymbol{\omega}(\boldsymbol{\omega} \cdot \boldsymbol{\rho}) = -\boldsymbol{\omega} \times \boldsymbol{\rho}(\boldsymbol{\omega} \cdot \boldsymbol{\rho}) \\
&= \boldsymbol{\omega} \times (-\boldsymbol{\rho}^T \boldsymbol{\rho})\boldsymbol{\omega} = \boldsymbol{\omega} \times (-\boldsymbol{\rho}^T \boldsymbol{\rho})\boldsymbol{\omega} + \boldsymbol{\omega} \times \|\boldsymbol{\rho}\|^2 \boldsymbol{\omega} \\
&= \boldsymbol{\omega} \times [(-\boldsymbol{\rho}^T \boldsymbol{\rho})\boldsymbol{\omega} + \|\boldsymbol{\rho}\|^2 \boldsymbol{\omega}] = \boldsymbol{\omega} \times (\|\boldsymbol{\rho}\|^2 \mathbf{1} - \boldsymbol{\rho}^T \boldsymbol{\rho})\boldsymbol{\omega}.
\end{aligned} \tag{4.37}$$

So,

$$\begin{aligned}
\frac{d}{dt}\boldsymbol{H}^o = \boldsymbol{r}_G \times M\boldsymbol{a}_G &+ \left[\int_M \left(\|\boldsymbol{\rho}\|^2 \mathbf{1} - \boldsymbol{\rho}\boldsymbol{\rho}^T\right) dM\right] \boldsymbol{\alpha} \\
&+ \boldsymbol{\omega} \times \left[\int_M \left(\|\boldsymbol{\rho}\|^2 \mathbf{1} - \boldsymbol{\rho}\boldsymbol{\rho}^T\right) dM\right] \boldsymbol{\omega},
\end{aligned} \tag{4.38}$$

or

$$\frac{d}{dt}\boldsymbol{H}^o = \boldsymbol{r}_G \times M\boldsymbol{a}_G + \boldsymbol{I}^G\boldsymbol{\alpha} + \boldsymbol{\omega} \times \boldsymbol{I}^G\boldsymbol{\omega}. \tag{4.39}$$

Alternately, by using the expression for angular momentum from (4.24), we can express the derivative of $\boldsymbol{H}^o$ as

$$\frac{d}{dt}\boldsymbol{H}^o = \frac{d}{dt}(\boldsymbol{r}_G \times M\boldsymbol{v}_G + \boldsymbol{I}^G\boldsymbol{\omega}) = \frac{d}{dt}(\boldsymbol{r}_G \times M\boldsymbol{v}_G) + \frac{d}{dt}\boldsymbol{I}^G\boldsymbol{\omega}, \tag{4.40}$$

where

$$\frac{d}{dt}(\boldsymbol{r}_G \times M\boldsymbol{v}_G) = \lim_{\Delta t \to 0} \frac{(\boldsymbol{r}_G + \Delta\boldsymbol{r}_G) \times M(\boldsymbol{v}_G + \Delta\boldsymbol{v}_G) - \boldsymbol{r}_G \times M\boldsymbol{v}_G}{\Delta t} \tag{4.41}$$

$$= \boldsymbol{r}_G \times M\boldsymbol{a}_G \tag{4.42}$$

and

$$\frac{d}{dt}\boldsymbol{I}^G\boldsymbol{\omega} = \lim_{\Delta t \to 0} \frac{(\boldsymbol{1} + \Delta\boldsymbol{Q})\boldsymbol{I}^G(\boldsymbol{\omega} + \Delta\boldsymbol{\omega}) - \boldsymbol{I}^G\boldsymbol{\omega}}{\Delta t} = \boldsymbol{I}^G\boldsymbol{\alpha} + \boldsymbol{\omega} \times \boldsymbol{I}^G\boldsymbol{\omega}. \tag{4.43}$$

So,

$$\frac{d}{dt}\boldsymbol{H}^O = \boldsymbol{r}_G \times M\boldsymbol{a}_G + \boldsymbol{I}^G\boldsymbol{\alpha} + \boldsymbol{\omega} \times \boldsymbol{I}^G\boldsymbol{\omega}. \tag{4.44}$$

Evaluated about the center of mass ($\boldsymbol{r}_G = 0$), we have

$$\frac{d}{dt}\boldsymbol{H}^G = \boldsymbol{I}^G\boldsymbol{\alpha} + \boldsymbol{\omega} \times \boldsymbol{I}^G\boldsymbol{\omega}, \tag{4.45}$$

and by *Principle* 4.2,

$$\boldsymbol{I}^G\boldsymbol{\alpha} + \boldsymbol{\omega} \times \boldsymbol{I}^G\boldsymbol{\omega} = \sum_{i=1}^{n_f} \boldsymbol{d}_{\overrightarrow{GP_i}} \times \boldsymbol{f}_i + \sum_{j=1}^{n_\varphi} \boldsymbol{\varphi}_j. \tag{4.46}$$

Example: We consider the mass properties of the cone depicted in Figure 4.3. The radius of the base is R and the height is h in the z direction. Using cylindrical coordinates, the infinitesimal volume and mass elements are

$$dV = r\,d\theta\,dr\,dz \quad \text{and} \quad dM = \rho\,r\,d\theta\,dr\,dz, \tag{4.47}$$

where ρ is the material density (assumed to be a constant throughout the body). The position vector of a point in the body can be represented as

$$\boldsymbol{r} = \begin{pmatrix} r\cos(\theta) & r\sin(\theta) & z \end{pmatrix}^T. \tag{4.48}$$

Using the coordinate frame at the vertex of the cone, we can integrate to compute the total mass. The integration limits in the radial direction vary linearly with the z coordinate. Specifically, the outer radius of the cone is $R_o = zR/h$. The mass integral is then

$$\begin{aligned} M = \int dM &= \int_0^h \int_0^{zR/h} \int_0^{2\pi} \rho\,r\,d\theta\,dr\,dz = \int_0^h \int_0^{zR/h} 2\rho\,\pi\,r\,dr\,dz \\ &= \int_0^h \frac{\rho\,\pi\,z^2 R^2}{h^2}\,dz = \frac{1}{3}\rho\,h\,\pi\,R^2. \end{aligned} \tag{4.49}$$

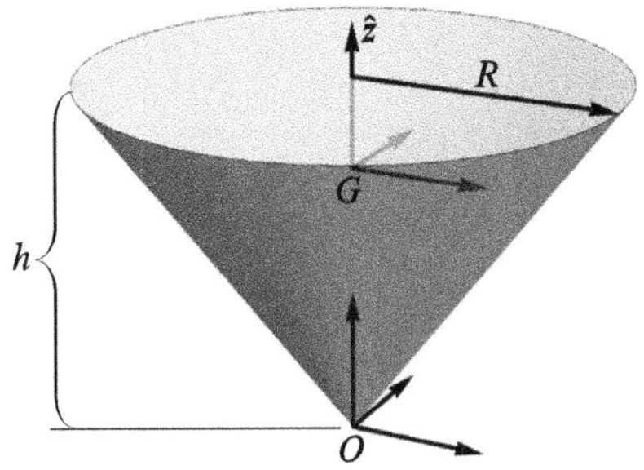

Figure 4.3 A cone with a base radius of R and height of h in the z direction.

The center of mass computed with respect to the vertex is

$$\begin{aligned} \boldsymbol{r}_G &= \frac{1}{M}\int \boldsymbol{r}\, dM = \frac{1}{M}\int_0^h \int_0^{zR/h} \int_0^{2\pi} \begin{pmatrix} r\cos(\theta) & r\sin(\theta) & z \end{pmatrix}^T \rho\, r\, d\theta\, dr\, dz \\ &= \frac{1}{M}\int_0^h \int_0^{zR/h} \begin{pmatrix} 0 & 0 & \rho\, 2\pi\, r\, z \end{pmatrix}^T dr\, dz \\ &= \frac{1}{M}\int_0^h \begin{pmatrix} 0 & 0 & \frac{\rho\,\pi\, R^2 z^3}{h^2} \end{pmatrix}^T dz = \frac{1}{M}\begin{pmatrix} 0 & 0 & \frac{1}{4}\rho\,\pi\, R^2 h^2 \end{pmatrix}^T. \end{aligned} \tag{4.50}$$

Since $M = \frac{1}{3}\rho\, h\,\pi\, R^2$, we have

$$\boldsymbol{r}_G = \begin{pmatrix} 0 & 0 & \frac{3}{4}h \end{pmatrix}^T. \tag{4.51}$$

The inertia tensor computed about the vertex is

$$\boldsymbol{I}^O = \int_M [(\boldsymbol{r}\cdot\boldsymbol{r})\mathbf{1} - \boldsymbol{r}\otimes\boldsymbol{r}]\, dM = \int_0^h \int_0^{zR/h} \int_0^{2\pi} [(\boldsymbol{r}\cdot\boldsymbol{r})\mathbf{1} - \boldsymbol{r}\otimes\boldsymbol{r}]\, \rho\, r\, d\theta\, dr\, dz. \tag{4.52}$$

We note that

$$\begin{aligned} (\boldsymbol{r}\cdot\boldsymbol{r})\mathbf{1} - \boldsymbol{r}\otimes\boldsymbol{r} &= (r^2+z^2)\mathbf{1} - \begin{pmatrix} r^2\cos^2(\theta) & r^2\cos(\theta)\sin(\theta) & rz\cos(\theta) \\ r^2\cos(\theta)\sin(\theta) & r^2\sin^2(\theta) & rz\sin(\theta) \\ rz\cos(\theta) & rz\sin(\theta) & z^2 \end{pmatrix} \\ &= \begin{pmatrix} z^2 + r^2\sin^2(\theta) & -r^2\cos(\theta)\sin(\theta) & -rz\cos(\theta) \\ -r^2\cos(\theta)\sin(\theta) & z^2 + r^2\cos^2(\theta) & -rz\sin(\theta) \\ -rz\cos(\theta) & -rz\sin(\theta) & r^2 \end{pmatrix}. \end{aligned} \tag{4.53}$$

So,

$$I^O = \int_0^h \int_0^{zR/h} \int_0^{2\pi} \begin{pmatrix} z^2 + r^2\sin^2(\theta) & -r^2\cos(\theta)\sin(\theta) & -rz\cos(\theta) \\ -r^2\cos(\theta)\sin(\theta) & z^2 + r^2\cos^2(\theta) & -rz\sin(\theta) \\ -rz\cos(\theta) & -rz\sin(\theta) & r^2 \end{pmatrix} \rho\, r\, d\theta\, dr\, dz$$

$$= \int_0^h \int_0^{zR/h} \begin{pmatrix} \rho\,\pi\, r(r^2+2z^2) & 0 & 0 \\ 0 & \rho\,\pi\, r(r^2+2z^2) & 0 \\ 0 & 0 & \rho\, 2\pi\, r^3 \end{pmatrix} dr\, dz$$

$$= \int_0^h \begin{pmatrix} \frac{\rho\,\pi\,(4h^2R^2+R^4)z^4}{4h^4} & 0 & 0 \\ 0 & \frac{\rho\,\pi\,(4h^2R^2+R^4)z^4}{4h^4} & 0 \\ 0 & 0 & \frac{\rho\,\pi\, R^4 z^4}{2h^4} \end{pmatrix} dz$$

$$= \begin{pmatrix} \frac{1}{20}\rho\, h\,\pi\, R^2(4h^2+R^2) & 0 & 0 \\ 0 & \frac{1}{20}\rho\, h\,\pi\, R^2(4h^2+R^2) & 0 \\ 0 & 0 & \frac{1}{10}\rho\, h\,\pi\, R^4 \end{pmatrix}, \tag{4.54}$$

or

$$I^O = \begin{pmatrix} \frac{3}{20}M(4h^2+R^2) & 0 & 0 \\ 0 & \frac{3}{20}M(4h^2+R^2) & 0 \\ 0 & 0 & \frac{3}{10}R^2 \end{pmatrix}. \tag{4.55}$$

We can compute the inertia tensor about the center of mass. The limits on the middle and outer integrals change to reflect the change in coordinates,

$$I^G = \int_{-3h/4}^{h/4} \int_0^{zR/h+3R/4} \int_0^{2\pi} \begin{pmatrix} z^2 + r^2\sin^2(\theta) & -r^2\cos(\theta)\sin(\theta) & -rz\cos(\theta) \\ -r^2\cos(\theta)\sin(\theta) & z^2 + r^2\cos^2(\theta) & -rz\sin(\theta) \\ -rz\cos(\theta) & -rz\sin(\theta) & r^2 \end{pmatrix} \rho\, r\, d\theta\, dr\, dz$$

$$= \begin{pmatrix} \frac{1}{80}\rho\, h\,\pi\, R^2(h^2+4R^2) & 0 & 0 \\ 0 & \frac{1}{80}\rho h\,\pi\, R^2(h^2+4R^2) & 0 \\ 0 & 0 & \frac{1}{10}\rho\, h\,\pi\, R^4 \end{pmatrix}, \tag{4.56}$$

or

$$I^G = \begin{pmatrix} \frac{3}{80}M(h^2+4R^2) & 0 & 0 \\ 0 & \frac{3}{80}M(h^2+4R^2) & 0 \\ 0 & 0 & \frac{3}{10}R^2 \end{pmatrix}. \tag{4.57}$$

We note that

$$M[(\boldsymbol{r}_G \cdot \boldsymbol{r}_G)\mathbf{1} - \boldsymbol{r}_G \otimes \boldsymbol{r}_G] = \begin{pmatrix} \frac{9}{16}h^2M & 0 & 0 \\ 0 & \frac{9}{16}h^2M & 0 \\ 0 & 0 & 0 \end{pmatrix} \tag{4.58}$$

and verify that $I^O = I^G + M[(\boldsymbol{r}_G \cdot \boldsymbol{r}_G)\mathbf{1} - \boldsymbol{r}_G \otimes \boldsymbol{r}_G]$.

4.1.3 A System of Particles

Newton's Principle applied to a system of n_p unconstrained particles with mutual forces of interaction between them yields

$$\frac{d\boldsymbol{p}_i}{dt} = \sum_{j=1}^{n_p} \boldsymbol{f}_i^j - M_i g \hat{\boldsymbol{e}}_3 \tag{4.59}$$

for $i = 1, \ldots, n_p$. The term $\boldsymbol{f}_i^j$ is the force that particle j exerts on particle i. Thus,

$$M_i \boldsymbol{a}_i = \sum_{j=1}^{n_p} \boldsymbol{f}_i^j - M_i g \hat{\boldsymbol{e}}_3, \tag{4.60}$$

where $\boldsymbol{f}_i^i = \boldsymbol{0}$ and, by Newton's third law, $\boldsymbol{f}_j^i = -\boldsymbol{f}_i^j$.

4.1.4 A System of Rigid Bodies

We now turn our attention to a system of rigid bodies forming a serial chain. Performing a force balance on link i, we have

$${}^i\boldsymbol{f}_i^{i-1} + {}^i\boldsymbol{f}_i^{i+1} = M_i({}^i\boldsymbol{a}_{G_i} + g\,{}^i\hat{\boldsymbol{e}}_{0_3}). \tag{4.61}$$

By Newton's third law, the reaction forces between links are equal and opposite. Thus, ${}^i\boldsymbol{f}_i^{i+1} = -{}^i\boldsymbol{f}_{i+1}^i$, so

$${}^i\boldsymbol{f}_i^{i-1} - {}^i\boldsymbol{f}_{i+1}^i = M_i({}^i\boldsymbol{a}_{G_i} + g\,{}^i\hat{\boldsymbol{e}}_{0_3}), \tag{4.62}$$

or

$${}^i\boldsymbol{f}_i^{i-1} = M_i({}^i\boldsymbol{a}_{G_i} + g\,{}^i\hat{\boldsymbol{e}}_{0_3}) + {}^i_{i+1}\boldsymbol{Q}\,{}^{i+1}\boldsymbol{f}_{i+1}^i. \tag{4.63}$$

Performing a moment balance on link i about the center of mass, we have

$${}^i\boldsymbol{\varphi}_i^{i-1} + {}^i\boldsymbol{\varphi}_i^{i+1} + {}^i\boldsymbol{d}_{\overrightarrow{G_i i}} \times {}^i\boldsymbol{f}_i^{i-1} + {}^i\boldsymbol{d}_{\overrightarrow{G_i(i+1)}} \times {}^i\boldsymbol{f}_i^{i+1} = {}^i\boldsymbol{I}_i^{G_i}\,{}^i\boldsymbol{\alpha}_i + {}^i\boldsymbol{\omega}_i \times {}^i\boldsymbol{I}_i^{G_i}\,{}^i\boldsymbol{\omega}_i. \tag{4.64}$$

Based on Euler's extension of Newton's third law to moments, the reaction moments between links are equal and opposite. Thus, ${}^i\boldsymbol{\varphi}_i^{i+1} = -{}^i\boldsymbol{\varphi}_{i+1}^i$, so

$${}^i\boldsymbol{\varphi}_i^{i-1} - {}^i\boldsymbol{\varphi}_{i+1}^i + {}^i\boldsymbol{d}_{\overrightarrow{G_i i}} \times {}^i\boldsymbol{f}_i^{i-1} - {}^i\boldsymbol{d}_{\overrightarrow{G_i(i+1)}} \times {}^i\boldsymbol{f}_{i+1}^i = {}^i\boldsymbol{I}_i^{G_i}\,{}^i\boldsymbol{\alpha}_i + {}^i\boldsymbol{\omega}_i \times {}^i\boldsymbol{I}_i^{G_i}\,{}^i\boldsymbol{\omega}_i, \tag{4.65}$$

or

$${}^i\boldsymbol{\varphi}_i^{i-1} = {}^i\boldsymbol{I}_i^{G_i}\,{}^i\boldsymbol{\alpha}_i + {}^i\boldsymbol{\omega}_i \times {}^i\boldsymbol{I}_i^{G_i}\,{}^i\boldsymbol{\omega}_i + {}^i_{i+1}\boldsymbol{Q}\,{}^{i+1}\boldsymbol{\varphi}_{i+1}^i - {}^i\boldsymbol{d}_{\overrightarrow{G_i i}} \times {}^i\boldsymbol{f}_i^{i-1} + {}^i\boldsymbol{d}_{\overrightarrow{G_i(i+1)}} \times {}^i_{i+1}\boldsymbol{Q}\,{}^{i+1}\boldsymbol{f}_{i+1}^i. \tag{4.66}$$

Alternately, performing a moment balance on link i about the proximal point of link i, we have

$${}^i\boldsymbol{\varphi}_i^{i-1} - {}^i\boldsymbol{\varphi}_{i+1}^i - {}^i\boldsymbol{d}_{\overrightarrow{iG_i}} \times M_i({}^i\boldsymbol{a}_{G_i} + g\,{}^i\hat{\boldsymbol{e}}_{0_3}) + {}^i\boldsymbol{d}_{\overrightarrow{i(i+1)}} \times {}^i\boldsymbol{f}_i^{i+1} = {}^i\boldsymbol{I}_i^{G_i}\,{}^i\boldsymbol{\alpha}_i + {}^i\boldsymbol{\omega}_i \times {}^i\boldsymbol{I}_i^{G_i}\,{}^i\boldsymbol{\omega}_i, \tag{4.67}$$

Algorithm 2 Iterative Newton-Euler method

$^{0}\boldsymbol{\omega}_0 = \mathbf{0}$ {initialization}

$^{0}\boldsymbol{\alpha}_0 = \mathbf{0}$ {initialization}

$^{0}\boldsymbol{a}_0 = \mathbf{0}$ {initialization}

for $i = 1$ to n **do**

  if $j_i = 1$ {joint is revolute} **then**

$$^{i}\boldsymbol{\omega}_i = {}^{i}_{i-1}\boldsymbol{Q}\,{}^{i-1}\boldsymbol{\omega}_{i-1} + \dot{\theta}_i\,{}^{i}\hat{\boldsymbol{z}}_i$$

$$^{i}\boldsymbol{\alpha}_i = {}^{i}_{i-1}\boldsymbol{Q}\,{}^{i-1}\boldsymbol{\alpha}_{i-1} + \ddot{\theta}_i\,{}^{i}\hat{\boldsymbol{z}}_i + {}^{i}_{i-1}\boldsymbol{Q}\,{}^{i-1}\boldsymbol{\omega}_{i-1} \times \dot{\theta}_i\,{}^{i}\hat{\boldsymbol{z}}_i$$

$$\begin{aligned} ^{i}\boldsymbol{a}_i = {}^{i}_{i-1}\boldsymbol{Q}[&{}^{i-1}\boldsymbol{a}_{i-1} + {}^{i-1}\boldsymbol{\alpha}_{i-1} \times {}^{i-1}\boldsymbol{d}_{\overrightarrow{(i-1)i}} \\ &+ {}^{i-1}\boldsymbol{\omega}_{i-1} \times ({}^{i-1}\boldsymbol{\omega}_{i-1} \times {}^{i-1}\boldsymbol{d}_{\overrightarrow{(i-1)i}})] \end{aligned}$$

$$^{i}\boldsymbol{a}_{G_i} = {}^{i}\boldsymbol{a}_i + {}^{i}\boldsymbol{\alpha}_i \times {}^{i}\boldsymbol{d}_{\overrightarrow{iG_i}} + {}^{i}\boldsymbol{\omega}_i \times ({}^{i}\boldsymbol{\omega}_i \times {}^{i}\boldsymbol{d}_{\overrightarrow{iG_i}})$$

  else

$$^{i}\boldsymbol{\omega}_i = {}^{i}_{i-1}\boldsymbol{Q}\,{}^{i-1}\boldsymbol{\omega}_{i-1}$$

$$^{i}\boldsymbol{\alpha}_i = {}^{i}_{i-1}\boldsymbol{Q}\,{}^{i-1}\boldsymbol{\alpha}_{i-1}$$

$$\begin{aligned} ^{i}\boldsymbol{a}_i = {}^{i}_{i-1}\boldsymbol{Q}[&{}^{i-1}\boldsymbol{a}_{i-1} + {}^{i-1}\boldsymbol{\alpha}_{i-1} \times {}^{i-1}\boldsymbol{d}_{\overrightarrow{(i-1)i}} \\ &+ {}^{i-1}\boldsymbol{\omega}_{i-1} \times ({}^{i-1}\boldsymbol{\omega}_{i-1} \times {}^{i-1}\boldsymbol{d}_{\overrightarrow{(i-1)i}})] + 2\,{}^{i}_{i-1}\boldsymbol{Q}\,{}^{i-1}\boldsymbol{\omega}_{i-1} \times \dot{d}_i\,{}^{i}\hat{\boldsymbol{z}}_i + \ddot{d}_i\,{}^{i}\hat{\boldsymbol{z}}_i \end{aligned}$$

$$^{i}\boldsymbol{a}_{G_i} = {}^{i}\boldsymbol{a}_i + {}^{i}\boldsymbol{\alpha}_i \times {}^{i}\boldsymbol{d}_{\overrightarrow{iG_i}} + {}^{i}\boldsymbol{\omega}_i \times ({}^{i}\boldsymbol{\omega}_i \times {}^{i}\boldsymbol{d}_{\overrightarrow{iG_i}})$$

  end if

end for

$^{n+1}\boldsymbol{f}^{n}_{n+1} = \mathbf{0}$ {initialization}

$^{n+1}\boldsymbol{\varphi}^{n}_{n+1} = \mathbf{0}$ {initialization}

for $i = n$ to 1 **do**

  if $j_i = 1$ {joint is revolute} **then**

$$^{i}\boldsymbol{f}^{i-1}_{i} = M_i({}^{i}\boldsymbol{a}_{G_i} + g\,{}^{i}\hat{\boldsymbol{e}}_{0_3}) + {}^{i}_{i+1}\boldsymbol{Q}\,{}^{i+1}\boldsymbol{f}^{i}_{i+1}$$

$$\begin{aligned} ^{i}\boldsymbol{\varphi}^{i-1}_{i} = {}^{i}\boldsymbol{I}^{G_i}_{i}\,{}^{i}\boldsymbol{\alpha}_i + {}^{i}\boldsymbol{\omega}_i \times {}^{i}\boldsymbol{I}^{G_i}_{i}\,{}^{i}\boldsymbol{\omega}_i + {}^{i}_{i+1}\boldsymbol{Q}\,{}^{i+1}\boldsymbol{\varphi}^{i}_{i+1} - {}^{i}\boldsymbol{d}_{\overrightarrow{G_i i}} \times {}^{i}\boldsymbol{f}^{i-1}_{i} \\ + {}^{i}\boldsymbol{d}_{\overrightarrow{G_i(i+1)}} \times {}^{i}_{i+1}\boldsymbol{Q}\,{}^{i+1}\boldsymbol{f}^{i}_{i+1} \end{aligned}$$

$$\tau_i = {}^{i}\hat{\boldsymbol{z}}_i \cdot {}^{i}\boldsymbol{\varphi}^{i-1}_{i}$$

  else

$$^{i}\boldsymbol{f}^{i-1}_{i} = M_i({}^{i}\boldsymbol{a}_{G_i} + g\,{}^{i}\hat{\boldsymbol{e}}_{0_3}) + {}^{i}_{i+1}\boldsymbol{Q}\,{}^{i+1}\boldsymbol{f}^{i}_{i+1}$$

$$\tau_i = {}^{i}\hat{\boldsymbol{z}}_i \cdot {}^{i}\boldsymbol{f}^{i-1}_{i}$$

  end if

end for

or

$$ {}^{i}\boldsymbol{\varphi}_{i}^{i-1} = {}^{i}\boldsymbol{I}_{i}^{G_i}\,{}^{i}\boldsymbol{\alpha}_{i} + {}^{i}\boldsymbol{\omega}_{i} \times {}^{i}\boldsymbol{I}_{i}^{G_i}\,{}^{i}\boldsymbol{\omega}_{i} + {}_{i+1}^{\ i}\boldsymbol{Q}\,{}^{i+1}\boldsymbol{\varphi}_{i+1}^{i} + {}^{i}\boldsymbol{d}_{\overrightarrow{iG_i}} \times M_i({}^{i}\boldsymbol{a}_{G_i} + g\,{}^{i}\hat{\boldsymbol{e}}_{0_3}) + {}^{i}\boldsymbol{d}_{\overrightarrow{i(i+1)}} \times {}_{i+1}^{\ i}\boldsymbol{Q}\,{}^{i+1}\boldsymbol{f}_{i+1}^{i}. \tag{4.68} $$

The iterative Newton-Euler method consists of n outward iterations that propagate the kinematic quantities of the system along the chain. These kinematic quantities are given by (3.216), (3.219) through (3.222), and (3.225) through (3.227). This is followed by n inward iterations that propagate the kinetic quantities of the system along the chain. These kinetic quantities are given by (4.63) and (4.66).

The applied joint forces and moments in the local ${}^{i}\hat{\boldsymbol{x}}_i$ and ${}^{i}\hat{\boldsymbol{y}}_i$ directions constitute reaction forces and moments. The applied joint forces and moments in the local ${}^{i}\hat{\boldsymbol{z}}_i$ direction constitute forces/moments, τ_i, associated with joint actuators. The joint forces/moments are given by

$$ \tau_i = {}^{i}\hat{\boldsymbol{z}}_i \cdot {}^{i}\boldsymbol{\varphi}_{i}^{i-1} \tag{4.69} $$

for revolute joints and by,

$$ \tau_i = {}^{i}\hat{\boldsymbol{z}}_i \cdot {}^{i}\boldsymbol{f}_{i}^{i-1} \tag{4.70} $$

for prismatic joints. The iterative Newton-Euler method is summarized in Algorithm 2.

Example: We consider a serial chain robot. We parameterize the system using three joint angles, as shown in Figure 4.4. The rotation matrices are

$$ {}_{1}^{0}\boldsymbol{Q} = \boldsymbol{Q}_z(\theta_1) = \begin{pmatrix} \cos(\theta_1) & -\sin(\theta_1) & 0 \\ \sin(\theta_1) & \cos(\theta_1) & 0 \\ 0 & 0 & 1 \end{pmatrix}, \tag{4.71} $$

$$ {}_{2}^{1}\boldsymbol{Q} = \boldsymbol{Q}_x(\pi/2)\boldsymbol{Q}_z(\theta_2) = \begin{pmatrix} \cos(\theta_2) & -\sin(\theta_2) & 0 \\ 0 & 0 & -1 \\ \sin(\theta_2) & \cos(\theta_2) & 0 \end{pmatrix}, \tag{4.72} $$

$$ {}_{3}^{2}\boldsymbol{Q} = \boldsymbol{Q}_z(\theta_3) = \begin{pmatrix} \cos(\theta_3) & -\sin(\theta_3) & 0 \\ \sin(\theta_3) & \cos(\theta_3) & 0 \\ 0 & 0 & 1 \end{pmatrix}. \tag{4.73} $$

The outward kinematic propagation equations are computed. For link 1 we have

$$ {}^{1}\boldsymbol{\omega}_1 = \dot{\theta}_1\,{}^{1}\hat{\boldsymbol{z}}_1 = \begin{pmatrix} 0 \\ 0 \\ \dot{\theta}_1 \end{pmatrix} \tag{4.74} $$

$$ {}^{1}\boldsymbol{\alpha}_1 = \ddot{\theta}_i\,{}^{1}\hat{\boldsymbol{z}}_1 = \begin{pmatrix} 0 \\ 0 \\ \ddot{\theta}_1 \end{pmatrix} \tag{4.75} $$

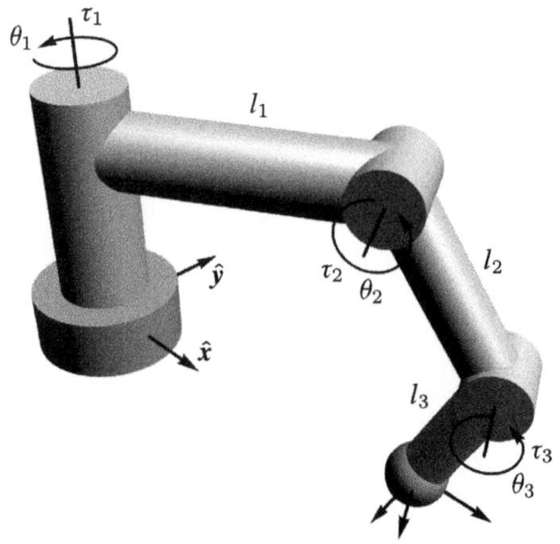

Figure 4.4 A 3 degree-of-freedom serial chain robot parameterized by the joint angles θ_1, θ_2, and θ_3. The link lengths are l_1, l_2, and l_3, the radii are r_1, r_2, and r_3, and the masses of the links are M_1, M_2, and M_3.

and

$$^1\boldsymbol{a}_1 = \boldsymbol{0} \tag{4.76}$$

$$^1\boldsymbol{a}_{G_1} = {}^1\boldsymbol{a}_1 + {}^1\boldsymbol{\alpha}_1 \times \frac{l_1}{2}\hat{\boldsymbol{e}}_1 + {}^1\boldsymbol{\omega}_1 \times \left({}^1\boldsymbol{\omega}_1 \times \frac{l_1}{2}\hat{\boldsymbol{e}}_1\right) = \frac{l_1}{2}\begin{pmatrix} -\dot{\theta}_1^2 \\ \ddot{\theta}_1 \\ 0 \end{pmatrix}. \tag{4.77}$$

For link 2 we have

$$^2\boldsymbol{\omega}_2 = {}^2_1\boldsymbol{Q}\,{}^1\boldsymbol{\omega}_1 + \dot{\theta}_2\,{}^2\hat{\boldsymbol{z}}_2 = \begin{pmatrix} \sin(\theta_2)\dot{\theta}_1 \\ \cos(\theta_2)\dot{\theta}_1 \\ \dot{\theta}_2 \end{pmatrix} \tag{4.78}$$

$$^2\boldsymbol{\alpha}_2 = {}^2_1\boldsymbol{Q}\,{}^1\boldsymbol{\alpha}_1 + \ddot{\theta}_2\,{}^2\hat{\boldsymbol{z}}_2 + {}^2_1\boldsymbol{Q}\,{}^1\boldsymbol{\omega}_1 \times \dot{\theta}_2\,{}^2\hat{\boldsymbol{z}}_2 = \begin{pmatrix} \cos(\theta_2)\dot{\theta}_1\dot{\theta}_2 + \sin(\theta_2)\ddot{\theta}_2 \\ -\sin(\theta_2)\dot{\theta}_1\dot{\theta}_2 + \cos(\theta_2)\ddot{\theta}_2 \\ \ddot{\theta}_2 \end{pmatrix} \tag{4.79}$$

and

$$^2\boldsymbol{a}_2 = {}^2_1\boldsymbol{Q}[{}^1\boldsymbol{a}_1 + {}^1\boldsymbol{\alpha}_1 \times l_1\hat{\boldsymbol{e}}_1 + {}^1\boldsymbol{\omega}_1 \times ({}^1\boldsymbol{\omega}_1 \times l_1\hat{\boldsymbol{e}}_1)] = l_1\begin{pmatrix} -\cos(\theta_2)\dot{\theta}_1^2 \\ \sin(\theta_2)\dot{\theta}_1^2 \\ -\ddot{\theta}_1 \end{pmatrix}, \tag{4.80}$$

$$\begin{aligned} ^2\boldsymbol{a}_{G_2} &= {}^2\boldsymbol{a}_2 + {}^2\boldsymbol{\alpha}_2 \times \frac{l_2}{2}\hat{\boldsymbol{e}}_1 + {}^2\boldsymbol{\omega}_2 \times \left({}^2\boldsymbol{\omega}_2 \times \frac{l_2}{2}\hat{\boldsymbol{e}}_1\right), \\ &= \frac{1}{2}\begin{pmatrix} -\cos(\theta_2)[2l_1 + l_2\cos(\theta_2)]\dot{\theta}_1^2 - l_2\dot{\theta}_2^2 \\ \sin(\theta_2)[2l_1 + l_2\cos(\theta_2)]\dot{\theta}_1^2 + l_2\ddot{\theta}_2 \\ 2l_2\sin(\theta_2)\dot{\theta}_1\dot{\theta}_2 - l_2[2l_1 + l_2\cos(\theta_2)]\ddot{\theta}_1 \end{pmatrix}. \end{aligned} \tag{4.81}$$

For link 3 we have

$$ {}^3\boldsymbol{\omega}_3 = {}^3_2\boldsymbol{Q}\,{}^2\boldsymbol{\omega}_2 + \dot{\theta}_3\,{}^3\hat{\boldsymbol{z}}_3 = \begin{pmatrix} \sin(\theta_2+\theta_3)\dot{\theta}_1 \\ \cos(\theta_2+\theta_3)\dot{\theta}_1 \\ \dot{\theta}_2+\dot{\theta}_3 \end{pmatrix} \tag{4.82} $$

$$ \begin{aligned} {}^3\boldsymbol{\alpha}_3 &= {}^3_2\boldsymbol{Q}\,{}^2\boldsymbol{\alpha}_2 + \ddot{\theta}_3\,{}^3\hat{\boldsymbol{z}}_3 + {}^3_2\boldsymbol{Q}\,{}^2\boldsymbol{\omega}_2 \times \dot{\theta}_3\,{}^3\hat{\boldsymbol{z}}_3 \\ &= \begin{pmatrix} \cos(\theta_2+\theta_3)\dot{\theta}_1(\dot{\theta}_2+\dot{\theta}_3) + \sin(\theta_2+\theta_3)\ddot{\theta}_1 \\ -\sin(\theta_2+\theta_3)\dot{\theta}_1(\dot{\theta}_2+\dot{\theta}_3) + \cos(\theta_2+\theta_3)\ddot{\theta}_1 \\ \ddot{\theta}_2+\ddot{\theta}_3 \end{pmatrix} \end{aligned} \tag{4.83} $$

and

$$ \begin{aligned} {}^3\boldsymbol{a}_3 &= {}^3_2\boldsymbol{Q}[{}^2\boldsymbol{a}_2 + {}^2\boldsymbol{\alpha}_2 \times l_2\hat{\boldsymbol{e}}_1 + {}^2\boldsymbol{\omega}_2 \times ({}^2\boldsymbol{\omega}_2 \times l_2\hat{\boldsymbol{e}}_1)] \\ &= \begin{pmatrix} -[l_1 + l_2\cos(\theta_2)]\cos(\theta_2+\theta_3)\dot{\theta}_1^2 - l_2[\cos(\theta_3)\dot{\theta}_2^2 - \sin(\theta_3)\ddot{\theta}_2] \\ [l_1 + l_2\cos(\theta_2)]\sin(\theta_2+\theta_3)\dot{\theta}_1^2 + l_2[\sin(\theta_3)\dot{\theta}_2^2 + \cos(\theta_3)\ddot{\theta}_2] \\ 2l_2\sin(\theta_2)\dot{\theta}_1\dot{\theta}_2 - [l_1 + l_2\cos(\theta_2)]\ddot{\theta}_1 \end{pmatrix} \end{aligned} \tag{4.84} $$

$$ \begin{aligned} {}^3\boldsymbol{a}_{G_3} &= {}^3\boldsymbol{a}_3 + {}^3\boldsymbol{\alpha}_3 \times \frac{l_3}{2}\hat{\boldsymbol{e}}_1 + {}^3\boldsymbol{\omega}_3 \times \left({}^3\boldsymbol{\omega}_3 \times \frac{l_3}{2}\hat{\boldsymbol{e}}_1\right) \\ &= \frac{1}{2}\begin{pmatrix} \frac{1}{2}[-2\cos(\theta_2+\theta_3)[2l_1 + 2l_2\cos(\theta_2)] + l_3\cos(\theta_2+\theta_3)\dot{\theta}_1^2 \\ [2l_1 + 2l_2\cos(\theta_2) + l_3\cos(\theta_2+\theta_3)]\sin(\theta_2+\theta_3)\dot{\theta}_1^2 \\ 2\dot{\theta}_1[[2l_2\sin(\theta_2) + l_3\sin(\theta_2+\theta_3)]\dot{\theta}_2 + l_3\sin(\theta_2+\theta_3)\dot{\theta}_3] \end{pmatrix. \\ &\qquad \begin{matrix} -2[[l_3 + 2l_2\cos(\theta_3)]\dot{\theta}_2^2 + l_3\dot{\theta}_3^2 - 2l_2\sin(\theta_3)\ddot{\theta}_2]] \\ +2l_2\sin(\theta_3)\dot{\theta}_2^2 + l_3\ddot{\theta}_2 + 2l_2\cos(\theta_3)\ddot{\theta}_2 + l_3\ddot{\theta}_3 \\ -[2l_1 + 2l_2\cos(\theta_2) + l_3\cos(\theta_2+\theta_3)]\ddot{\theta}_1 \end{matrix}\Bigg) . \end{aligned} \tag{4.85} $$

Now we can compute the inward kinetic propagation equations. First, the inertia tensors are

$$ {}^i\boldsymbol{I}_i^{G_i} = \begin{pmatrix} \frac{1}{2}M_i r_i^2 & 0 & 0 \\ 0 & \frac{1}{12}l_i^2 M_i + \frac{1}{4}M_i r_i^2 & 0 \\ 0 & 0 & \frac{1}{12}l_i^2 M_i + \frac{1}{4}M_i r_i^2 \end{pmatrix}. \tag{4.86} $$

For link 3 we have

$$ {}^3\boldsymbol{f}_3^2 = M_3({}^3\boldsymbol{a}_{G_3} + g\,{}^3\hat{\boldsymbol{e}}_{0_3}), \tag{4.87} $$

$$ {}^3\boldsymbol{\varphi}_3^2 = {}^3\boldsymbol{I}_3^{G_3}\,{}^3\boldsymbol{\alpha}_3 + {}^3\boldsymbol{\omega}_3 \times {}^3\boldsymbol{I}_3^{G_3}\,{}^3\boldsymbol{\omega}_3 + \frac{l_3}{2}\hat{\boldsymbol{e}}_1 \times {}^3\boldsymbol{f}_3^2, \tag{4.88} $$

$$ \begin{aligned} \tau_3 &= {}^3\hat{\boldsymbol{z}}_3 \cdot {}^3\boldsymbol{\varphi}_3^2 = \frac{1}{12}M_3\,[6gl_3\cos(\theta_2+\theta_3) \\ &\quad +[6l_1l_3 + 6l_2l_3\cos(\theta_2) + (4l_3^2 - 3r_3^2)\cos(\theta_2+\theta_3)]\sin(\theta_2+\theta_3)\dot{\theta}_1^2 \\ &\quad +6l_2l_3\sin(\theta_3)\dot{\theta}_2^2 + [4l_3^2 + 3r_3^2 + 6l_2l_3\cos(\theta_3)]\ddot{\theta}_2 + (4l_3^2 + 3r_3^2)\ddot{\theta}_3\big]. \end{aligned} \tag{4.89} $$

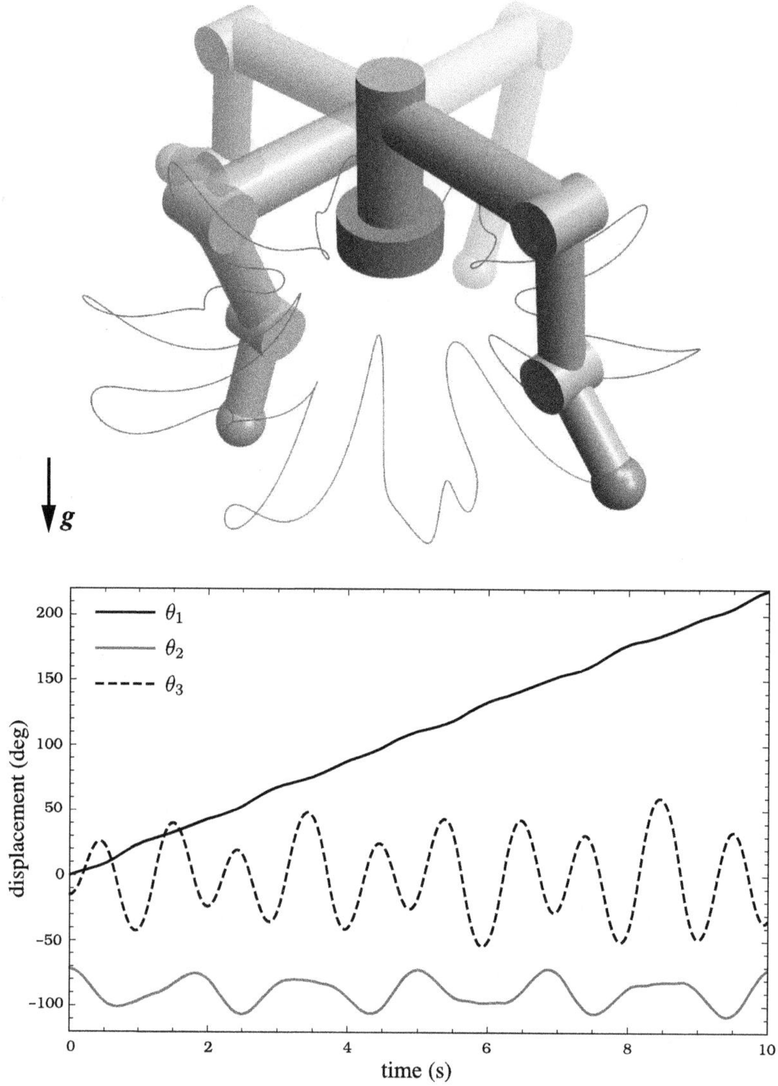

Figure 4.5 (Top) Animation frames from the simulation of a serial chain robot. No actuator torque was specified, but an initial velocity condition of $\pi/10$ rad/s (18 deg/s) was specified on joint 1 to provide the spin about the vertical axis. The trace records the trajectory of the endpoint of the chain. (Bottom) Time history of the joint angles θ_1, θ_2, and θ_3, parameterizing the configuration of the mechanism.

For link 2 we have

$$ {}^2\boldsymbol{f}_2^1 = M_2({}^2\boldsymbol{a}_{G_2} + g\,{}^2\hat{\boldsymbol{e}}_{0_3}) + {}^2_3\boldsymbol{Q}\,{}^3\boldsymbol{f}_3^2, \tag{4.90} $$

$$ {}^2\boldsymbol{\varphi}_2^1 = {}^2\boldsymbol{I}_2^{G_2}\,{}^2\boldsymbol{\alpha}_2 + {}^2\boldsymbol{\omega}_2 \times {}^2\boldsymbol{I}_2^{G_2}\,{}^2\boldsymbol{\omega}_2 + {}^2_3\boldsymbol{Q}\,{}^3\boldsymbol{\varphi}_3^2 + \frac{l_2}{2}\hat{\boldsymbol{e}}_1 \times {}^2\boldsymbol{f}_2^1 + \frac{l_2}{2}\hat{\boldsymbol{e}}_1 \times {}^2_3\boldsymbol{Q}\,{}^3\boldsymbol{f}_3^2, \tag{4.91} $$

$$ \tau_2 = {}^2\hat{\boldsymbol{z}}_2 \cdot {}^2\boldsymbol{\varphi}_2^1, \tag{4.92} $$

and for link 1 we have

$$ {}^{1}\boldsymbol{f}_{1}^{0} = M_1({}^{1}\boldsymbol{a}_{G_1} + g\,{}^{1}\hat{\boldsymbol{e}}_{0_3}) + {}^{1}_{2}\boldsymbol{Q}^{2}\boldsymbol{f}_{2}^{1}, \tag{4.93} $$

$$ {}^{1}\boldsymbol{\varphi}_{1}^{0} = {}^{1}\boldsymbol{I}_{1}^{G_1}\,{}^{1}\boldsymbol{\alpha}_{1} + {}^{1}\boldsymbol{\omega}_{1} \times {}^{1}\boldsymbol{I}_{1}^{G_1}\,{}^{1}\boldsymbol{\omega}_{1} + {}^{1}_{2}\boldsymbol{Q}^{2}\boldsymbol{\varphi}_{2}^{1} + \frac{l_1}{2}\hat{\boldsymbol{e}}_{1} \times {}^{1}\boldsymbol{f}_{1}^{0} + \frac{l_1}{2}\hat{\boldsymbol{e}}_{1} \times {}^{1}_{2}\boldsymbol{Q}^{2}\boldsymbol{f}_{2}^{1}, \tag{4.94} $$

$$ \tau_1 = {}^{1}\hat{\boldsymbol{z}}_{1} \cdot {}^{1}\boldsymbol{\varphi}_{1}^{0}. \tag{4.95} $$

The number of terms quickly grows for links 2 and 1, so we will not express them here.

The simulation results are displayed in Figure 4.5. The following values were used for the constants:

$$ \begin{aligned} r_1 &= 0.125, & r_2 &= 0.1, & r_3 &= 0.09, \\ l_1 &= 1, & l_2 &= 0.75, & l_3 &= 0.5, \\ M_1 &= 1, & M_2 &= 0.75, & M_3 &= 0.5. \end{aligned} \tag{4.96} $$

The initial conditions used were

$$ \boldsymbol{\theta}_o = \begin{pmatrix} 0 & -\frac{\pi}{2.5} & -\frac{\pi}{12} \end{pmatrix}^T \tag{4.97} $$

$$ \dot{\boldsymbol{\theta}}_o = \begin{pmatrix} \frac{\pi}{10} & 0 & 0 \end{pmatrix}^T. \tag{4.98} $$

4.2 Exercises

1. Consider the cylindrical tube with an inside radius of r_1, an outside radius of r_2, and a height of h in the z direction (see Figure 4.6).
   (a) Compute the mass using a uniform material density of ρ. Use the coordinate system associated with the point O.
   (b) Compute the center of mass. Again, use the coordinate system associated with the point O.
   (c) Compute the inertia tensor about point O, $\boldsymbol{I}^{O}$.

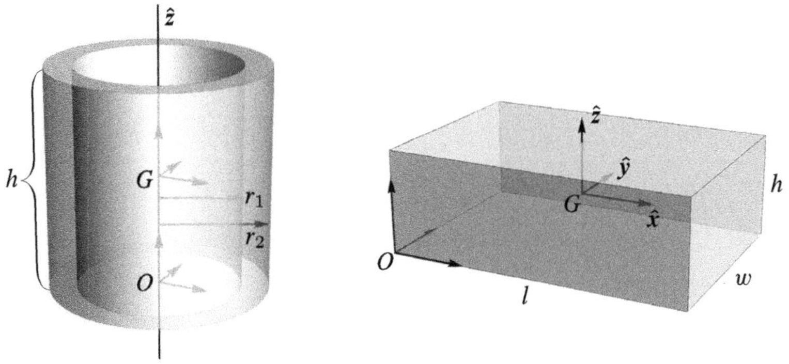

Figure 4.6 (Left) Cylindrical tube with an inside diameter of r_1, an outside diameter of r_2, and a height of h in the z direction (Exercise 1). (Right) Cuboid with a length of l in the x direction, a width of w in the y direction, and a height of h in the z direction (Exercise 2).

(d) Compute the inertia tensor about the center of mass, $\boldsymbol{I}^G$.

(e) Verify that $\boldsymbol{I}^O = \boldsymbol{I}^G + M[(\boldsymbol{r}_G \cdot \boldsymbol{r}_G)\mathbf{1} - \boldsymbol{r}_G \otimes \boldsymbol{r}_G]$.

2. Repeat Exercise 1 for the cuboid with a length of l in the x direction, a width of w in the y direction, and a height of h in the z direction (see Figure 4.6).
3. Consider the serial chain robot from Section 3.5, Exercise 14 (shown in Figure 4.7). The link lengths are l_1 and l_2. The link radii are r_1 and r_2, the link masses are M_1 and M_2, and the link inertia tensors are

$$
{}^i\boldsymbol{I}_i^{G_i} = \begin{pmatrix} \frac{1}{2}M_i r_i^2 & 0 & 0 \\ 0 & \frac{1}{12}l_i^2 M_i + \frac{1}{4}M_i r_i^2 & 0 \\ 0 & 0 & \frac{1}{12}l_i^2 M_i + \frac{1}{4}M_i r_i^2 \end{pmatrix}
$$

for links $i = 1, 2$.

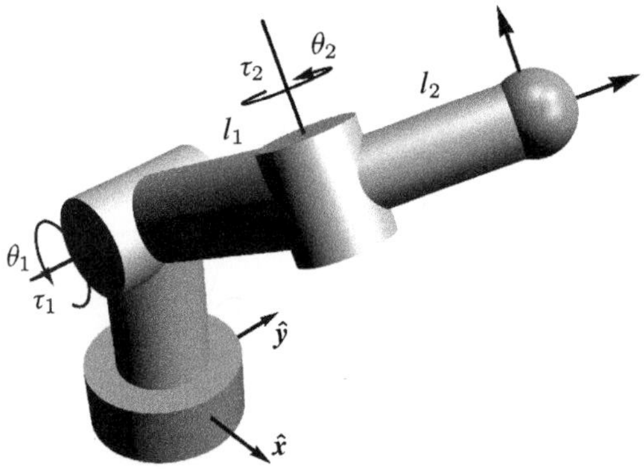

Figure 4.7 A 2 degree-of-freedom serial chain robot with joint angles θ_1 and θ_2 and joint torques τ_1 and τ_2 (Exercise 3). The link lengths are l_1 and l_2. The centers of mass are at the geometric centers of the links. The link radii are r_1 and r_2, the link masses are M_1 and M_2, and the link principal inertia components are ${}^iI_{i_{11}}^{G_i} = \frac{1}{2}M_i r_i^2$ and ${}^iI_{i_{22}}^{G_i} = {}^iI_{i_{33}}^{G_i} = \frac{1}{12}l_i^2 M_i + \frac{1}{4}M_i r_i^2$ for links $i = 1, 2$.

(a) If not already completed, compute the terms ${}^i\boldsymbol{\omega}_i$, ${}^i\boldsymbol{\alpha}_i$, ${}^i\boldsymbol{a}_i$, and ${}^i\boldsymbol{a}_{G_i}$ for $i = 1, 2$ using forward kinematic propagation.

(b) Compute the terms ${}^i\boldsymbol{f}_i^{i-1}$, ${}^i\boldsymbol{\varphi}_i^{i-1}$, and τ_i for $i = 1, 2$ using backward propagation.

(c) Solve the τ_1 and τ_2 expressions for $\ddot{\theta}_1$ and $\ddot{\theta}_2$.

5 Zeroth-Order Variational Principles

In the previous chapter we addressed principles based on the conservation of vector quantities, specifically translational and angular momentum. In this chapter we will address principles rooted in the variation of scalar quantities like work and energy. We will begin with d'Alembert's Principle of Virtual Work, which is an extension of Bernoulli's static principle to dynamics. The principle is based on the notion of virtual displacement. We will refer to d'Alembert's Principle as a zeroth-order variational principle to denote that it is based on the variation of the zeroth-order derivative of displacement. This is in contrast to higher-order variations related to velocity and acceleration, which will be discussed in subsequent chapters.

5.1 Virtual Displacements

Virtual displacements refer to all displacements of a system that satisfy the scleronomic constraints of the system. Scleronomic constraints refer to constraints that are not explicitly dependent on time, as opposed to rheonomic constraints, which are explicitly dependent on time. In the case of virtual displacements, time is frozen or stationary.

5.2 D'Alembert's Principle of Virtual Work

PRINCIPLE 5.1 *The virtual work of a system is stationary. That is,*

$$\delta W = 0. \tag{5.1}$$

Additionally, the constraints of the system perform no virtual work:

$$\delta W_c = 0. \tag{5.2}$$

This is known as ***d'Alembert's Principle****.*

It is noted that while d'Alembert's Principle can be seen as providing an alternate statement of Newton's second law, for interacting bodies, a law of action and reaction (Newton's third law) is still needed. Therefore, when we use d'Alembert's Principle to derive the equations of motion for systems of particles/bodies, we will invoke the law of action and reaction.

5.2.1 A Single Particle

D'Alembert's Principle for a single point mass with a discrete set of n_f external forces, $\{\boldsymbol{f}_1, \ldots, \boldsymbol{f}_{n_f}\}$, acting on it is expressed as

$$\delta W = \sum_{i=1}^{n_f} \boldsymbol{f}_i \cdot \delta\boldsymbol{r} - Mg\hat{\boldsymbol{e}}_3 \cdot \delta\boldsymbol{r} - M\boldsymbol{a} \cdot \delta\boldsymbol{r} = 0, \quad \forall \delta\boldsymbol{r} \in \mathbb{R}^3, \tag{5.3}$$

where $\delta\boldsymbol{r}$ represents the displacement variations. During these variations time is stationary. That is, $\delta t = 0$. More concisely, we can express

$$\delta W = \left(\sum_{i=1}^{n_f} \boldsymbol{f}_i - Mg\hat{\boldsymbol{e}}_3 - M\boldsymbol{a} \right) \cdot \delta\boldsymbol{r} = 0, \quad \forall \delta\boldsymbol{r} \in \mathbb{R}^3, \tag{5.4}$$

which implies

$$\sum_{i=1}^{n_f} \boldsymbol{f}_i - Mg\hat{\boldsymbol{e}}_3 - M\boldsymbol{a} = \boldsymbol{0}. \tag{5.5}$$

5.2.2 A Single Rigid Body

D'Alembert's Principle for a single rigid body with a discrete set of n_f external forces, $\{\boldsymbol{f}_1, \ldots, \boldsymbol{f}_{n_f}\}$, and n_φ external moments, $\{\boldsymbol{\varphi}_1, \ldots, \boldsymbol{\varphi}_{n_\varphi}\}$, acting on it is expressed as

$$\delta W = \sum_{i=1}^{n_f} \boldsymbol{f}_i \cdot \delta\boldsymbol{r}_{P_i} + \sum_{i=1}^{n_\varphi} \boldsymbol{\varphi}_j \cdot \delta\boldsymbol{\theta} - Mg\hat{\boldsymbol{e}}_3 \cdot \delta\boldsymbol{r}_G - M\boldsymbol{a}_G \cdot \delta\boldsymbol{r}_G - (\boldsymbol{I}^G\boldsymbol{\alpha} + \boldsymbol{\omega} \times \boldsymbol{I}^G\boldsymbol{\omega}) \cdot \delta\boldsymbol{\theta} = 0,$$
$$\forall \delta\boldsymbol{r}_G \in \mathbb{R}^3, \quad \text{and} \quad \forall \delta\boldsymbol{\theta} \in \mathbb{R}^3, \tag{5.6}$$

where the $\delta\boldsymbol{r}$ and $\delta\boldsymbol{\theta}$ terms represent all displacement variations consistent with the rigid-body constraint. This is depicted in Figure 5.1. During these variations, time is stationary. That is, $\delta t = 0$. D'Alembert's Principle states that the virtual work associated with all internal forces and moments consistent with the rigid body constraint is *zero* ($W_c = 0$). We further note that

$$\delta\boldsymbol{r}_{P_i} = \delta\boldsymbol{r}_G + \delta\boldsymbol{\theta} \times \boldsymbol{d}_{\overrightarrow{GP_i}}. \tag{5.7}$$

Therefore,

$$\sum_{i=1}^{n_f} \boldsymbol{f}_i \cdot \delta\boldsymbol{r}_{P_i} = \sum_{i=1}^{n_f} \boldsymbol{f}_i \cdot (\delta\boldsymbol{r}_G + \delta\boldsymbol{\theta} \times \boldsymbol{d}_{\overrightarrow{GP_i}}) = \left(\sum_{i=1}^{n_f} \boldsymbol{f}_i \right) \cdot \delta\boldsymbol{r}_G + \sum_{i=1}^{n_f} \boldsymbol{f}_i \cdot (\delta\boldsymbol{\theta} \times \boldsymbol{d}_{\overrightarrow{GP_i}}). \tag{5.8}$$

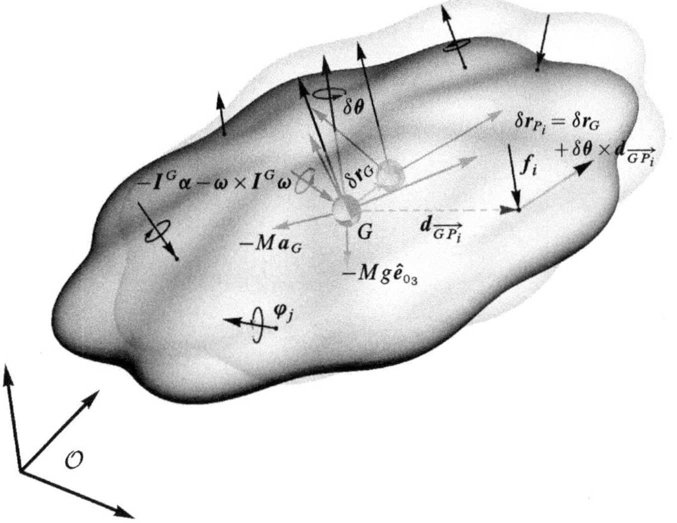

Figure 5.1 A virtual displacement of a rigid body consisting of a translational, $\delta\boldsymbol{r}_G$, and rotational, $\delta\boldsymbol{\theta}$, displacement. We are concerned with all such displacement variations consistent with the rigid-body constraint. During these variations, time is stationary.

Since $\boldsymbol{a}\cdot(\boldsymbol{b}\times\boldsymbol{c}) = (\boldsymbol{c}\times\boldsymbol{a})\cdot\boldsymbol{b}$,

$$\sum_{i=1}^{n_f} \boldsymbol{f}_i\cdot(\delta\boldsymbol{\theta}\times\boldsymbol{d}_{\overrightarrow{GP_i}}) = \sum_{i=1}^{n_f}(\boldsymbol{d}_{\overrightarrow{GP_i}}\times\boldsymbol{f}_i)\cdot\delta\boldsymbol{\theta} = \left(\sum_{i=1}^{n_f}\boldsymbol{d}_{\overrightarrow{GP_i}}\times\boldsymbol{f}_i\right)\cdot\delta\boldsymbol{\theta}. \tag{5.9}$$

So,

$$\sum_{i=1}^{n_f}\boldsymbol{f}_i\cdot\delta\boldsymbol{r}_{P_i} = \left(\sum_{i=1}^{n_f}\boldsymbol{f}_i\right)\cdot\delta\boldsymbol{r}_G + \left(\sum_{i=1}^{n_f}\boldsymbol{d}_{\overrightarrow{GP_i}}\times\boldsymbol{f}_i\right)\cdot\delta\boldsymbol{\theta}. \tag{5.10}$$

Substituting (5.10) into (5.6), we get

$$\begin{aligned}\delta W = &\left(\sum_{i=1}^{n_f}\boldsymbol{f}_i - Mg\hat{\boldsymbol{e}}_3 - M\boldsymbol{a}_G\right)\cdot\delta\boldsymbol{r}_G\\ &+\left(\sum_{i=1}^{n_f}\boldsymbol{d}_{\overrightarrow{GP_i}}\times\boldsymbol{f}_i + \sum_{i=1}^{n_\varphi}\boldsymbol{\varphi}_j - \boldsymbol{I}^G\boldsymbol{\alpha} - \boldsymbol{\omega}\times\boldsymbol{I}^G\boldsymbol{\omega}\right)\cdot\delta\boldsymbol{\theta} = 0,\end{aligned} \tag{5.11}$$

or

$$\begin{pmatrix}\sum_{i=1}^{n_f}\boldsymbol{f}_i - Mg\hat{\boldsymbol{e}}_3 - M\boldsymbol{a}_G\\ \sum_{i=1}^{n_f}\boldsymbol{d}_{\overrightarrow{GP_i}}\times\boldsymbol{f}_i + \sum_{i=1}^{n_\varphi}\boldsymbol{\varphi}_j - \boldsymbol{I}^G\boldsymbol{\alpha} - \boldsymbol{\omega}\times\boldsymbol{I}^G\boldsymbol{\omega}\end{pmatrix}\cdot\begin{pmatrix}\delta\boldsymbol{r}_G\\ \delta\boldsymbol{\theta}\end{pmatrix} = 0, \tag{5.12}$$

$$\forall\delta\boldsymbol{r}_G\in\mathbb{R}^3, \quad \text{and} \quad \forall\delta\boldsymbol{\theta}\in\mathbb{R}^3,$$

which implies

$$\sum_{i=1}^{n_f} \boldsymbol{f}_i - Mg\hat{\boldsymbol{e}}_3 - M\boldsymbol{a}_G = \boldsymbol{0} \tag{5.13}$$

$$\sum_{i=1}^{n_f} \boldsymbol{d}_{\overrightarrow{GP_i}} \times \boldsymbol{f}_i + \sum_{i=1}^{n_\varphi} \boldsymbol{\varphi}_j - \boldsymbol{I}^G\boldsymbol{\alpha} - \boldsymbol{\omega} \times \boldsymbol{I}^G\boldsymbol{\omega} = \boldsymbol{0}. \tag{5.14}$$

5.2.3 A System of Particles

We now apply d'Alembert's Principle to a system of n_p particles with a discrete set of forces acting on them. The virtual work associated with a given particle i is given by

$$\delta W_i = \left(\sum_{j=0}^{n_p} \boldsymbol{f}_i^j\right) \cdot \delta\boldsymbol{r}_i - M_i(\boldsymbol{a}_i + g\hat{\boldsymbol{e}}_3) \cdot \delta\boldsymbol{r}_i = 0, \quad \forall \delta\boldsymbol{r}_i \in \mathbb{R}^3, \tag{5.15}$$

for $i = 1, \ldots, n_p$. The term $\boldsymbol{f}_i^j$ is the force that particle j exerts on particle i, where $\boldsymbol{f}_i^0$ is the force exerted by ground (inertial reference frame) on the ith particle. Equation (5.15) is true for any particle i in the system, under all independent variations, $\delta\boldsymbol{r}_i$.

If we sum (5.15) over all particles, we obtain

$$\delta W = \sum_{i=1}^{n_p} \delta W_i = \sum_{i=1}^{n_p} \left(\sum_{j=0}^{n_p} \boldsymbol{f}_i^j\right) \cdot \delta\boldsymbol{r}_i - \sum_{i=1}^{n_p} M_i(\boldsymbol{a}_i + g\hat{\boldsymbol{e}}_3) \cdot \delta\boldsymbol{r}_i = 0. \tag{5.16}$$

The first summation is associated with the virtual work performed by the interparticle reaction forces. It will be useful to rearrange the summation so as to pair up the equal and opposite reaction forces, $\boldsymbol{f}_i^j$ and $\boldsymbol{f}_j^i$. Doing this, we have

$$\sum_{i=1}^{n_p} \left(\sum_{j=0}^{n_p} \boldsymbol{f}_i^j\right) \cdot \delta\boldsymbol{r}_i = \sum_{i=0}^{n_p-1} \sum_{j=i+1}^{n_p} (\boldsymbol{f}_i^j \cdot \delta\boldsymbol{r}_i + \boldsymbol{f}_j^i \cdot \delta\boldsymbol{r}_j) + \sum_{i=1}^{n_p} \boldsymbol{f}_i^i \cdot \delta\boldsymbol{r}_i - \sum_{j=1}^{n_p} \boldsymbol{f}_0^j \cdot \delta\boldsymbol{r}_0. \tag{5.17}$$

Since $\boldsymbol{f}_i^i = \boldsymbol{0}$ and $\delta\boldsymbol{r}_0 = \boldsymbol{0}$, we have

$$\sum_{i=1}^{n_p} \left(\sum_{j=0}^{n_p} \boldsymbol{f}_i^j\right) \cdot \delta\boldsymbol{r}_i = \sum_{i=0}^{n_p-1} \sum_{j=i+1}^{n_p} (\boldsymbol{f}_i^j \cdot \delta\boldsymbol{r}_i + \boldsymbol{f}_j^i \cdot \delta\boldsymbol{r}_j). \tag{5.18}$$

We note that $\boldsymbol{f}_j^i = -\boldsymbol{f}_i^j$. Thus,

$$\sum_{i=1}^{n_p} \left(\sum_{j=0}^{n_p} \boldsymbol{f}_i^j\right) \cdot \delta\boldsymbol{r}_i = \sum_{i=0}^{n_p-1} \sum_{j=i+1}^{n_p} \boldsymbol{f}_i^j \cdot (\delta\boldsymbol{r}_i - \delta\boldsymbol{r}_j). \tag{5.19}$$

This reflects the virtual work done by interparticle reaction forces, summed over all particles.

If we consider that the particles are subject to a set of holonomic constraints, the positions, $\boldsymbol{r}_i$, are not independent. In this case we will assume that the positions can be expressed in terms of a set of n_q independent generalized coordinates, $\boldsymbol{q}$. If we now consider only the variations, $\delta\boldsymbol{r}_i$, that are consistent with the kinematic constraints, then (5.19) reflects a projection of the interparticle forces on the direction of the interparticle motion. D'Alembert's Principle states that the virtual work associated with all forces orthogonal to the interparticle motion (reaction forces) is *zero* ($W_c = 0$) and only the generalized force acting in the direction of interparticle motion produces virtual work. Thus,

$$\sum_{i=0}^{n_p-1}\sum_{j=i+1}^{n_p} \boldsymbol{f}_i^j \cdot (\delta\boldsymbol{r}_i - \delta\boldsymbol{r}_j) = \sum_{i=1}^{n_q} \tau_i \cdot \delta q_i, \tag{5.20}$$

and (5.16) can be expressed as

$$\sum_{i=1}^{n_q} \tau_i \cdot \delta q_i - \sum_{i=0}^{n_p} M_i(\boldsymbol{a}_i + g\hat{\boldsymbol{e}}_3) \cdot \delta\boldsymbol{r}_i = 0 \tag{5.21}$$

for all variations, $\delta\boldsymbol{r}_i$, that are consistent with the kinematic constraints.

Expressing the variations in terms of variations in the generalized coordinates, we have

$$\delta\boldsymbol{r}_i = \boldsymbol{\Gamma}_i \delta\boldsymbol{q}. \tag{5.22}$$

So, (5.21) can be expressed as

$$\boldsymbol{\tau} \cdot \delta\boldsymbol{q} - \sum_{i=1}^{n_p} M_i(\boldsymbol{a}_i + g\hat{\boldsymbol{e}}_3) \cdot (\boldsymbol{\Gamma}_i \delta\boldsymbol{q}) = 0, \quad \forall \delta\boldsymbol{q} \in \mathbb{R}^n, \tag{5.23}$$

or

$$\left[\boldsymbol{\tau} - \sum_{i=1}^{n_p} \boldsymbol{\Gamma}_i^T (M_i \boldsymbol{a}_i + M_i g\hat{\boldsymbol{e}}_3)\right] \cdot \delta\boldsymbol{q} = 0 \quad \forall \delta\boldsymbol{q} \in \mathbb{R}^n, \tag{5.24}$$

which implies

$$\boldsymbol{\tau} = \sum_{i=1}^{n_p} \boldsymbol{\Gamma}_i^T (M_i \boldsymbol{a}_i + M_i g\hat{\boldsymbol{e}}_3). \tag{5.25}$$

Noting that

$$\boldsymbol{a}_i = \boldsymbol{\Gamma}_i \ddot{\boldsymbol{q}}, \tag{5.26}$$

we have

$$\boldsymbol{\tau} = \left(\sum_{i=1}^{n_p} M_i \boldsymbol{\Gamma}_i^T \boldsymbol{\Gamma}_i\right) \ddot{\boldsymbol{q}} + g \sum_{i=1}^{n_q} M_i \boldsymbol{\Gamma}_i^T \hat{\boldsymbol{e}}_3. \tag{5.27}$$

Defining the system mass matrix, $\boldsymbol{M}(\boldsymbol{q})$, and the vector of generalized gravity forces, $\boldsymbol{g}(\boldsymbol{q})$,

$$\boldsymbol{M}(\boldsymbol{q}) \triangleq \sum_{i=1}^{n_p} M_i \boldsymbol{\Gamma}_i^T \boldsymbol{\Gamma}_i \tag{5.28}$$

$$\boldsymbol{g}(\boldsymbol{q}) \triangleq g \sum_{i=1}^{n_p} M_i \boldsymbol{\Gamma}_i^T \hat{\boldsymbol{e}}_3, \tag{5.29}$$

we have

$$\boldsymbol{\tau} = \boldsymbol{M}(\boldsymbol{q})\ddot{\boldsymbol{q}} + \boldsymbol{g}(\boldsymbol{q}). \tag{5.30}$$

5.2.4 A System of Rigid Bodies

We now apply d'Alembert's Principle to a system of n_b rigid bodies forming a serial chain, with a discrete set of forces acting on them. The virtual work associated with a given body i is given by

$$\begin{aligned} \delta W_i = \boldsymbol{f}_i^{i-1} \cdot \delta \boldsymbol{r}_i^{i-1} + \boldsymbol{f}_i^{i+1} \cdot \delta \boldsymbol{r}_i^{i+1} - M_i(\boldsymbol{a}_{G_i} + g\hat{\boldsymbol{e}}_{0_3}) \cdot \delta \boldsymbol{r}_{G_i} + \boldsymbol{\varphi}_i^{i-1} \cdot \delta \boldsymbol{\theta}_i \\ + \boldsymbol{\varphi}_i^{i+1} \cdot \delta \boldsymbol{\theta}_i - (\boldsymbol{I}_i^{G_i} \boldsymbol{\alpha}_i + \boldsymbol{\omega}_i \times \boldsymbol{I}_i^{G_i} \boldsymbol{\omega}_i) \cdot \delta \boldsymbol{\theta}_i = 0, \\ \forall \delta \boldsymbol{r}_{G_i} \in \mathbb{R}^3, \quad \text{and} \quad \forall \delta \boldsymbol{\theta}_i \in \mathbb{R}^3, \end{aligned} \tag{5.31}$$

for $i = 1, \ldots, n_b$. The term $\boldsymbol{f}_i^{i-1}$ is the force that body $i-1$ exerts on body i. Likewise, $\boldsymbol{\varphi}_i^{i-1}$ is the moment that body $i-1$ exerts on body i. The term $\boldsymbol{r}_i^{i-1}$ is the point on body i to which body $i-1$ attaches. Equation (5.31) is true for any rigid body i in the system, under all independent variations, $\delta \boldsymbol{r}_{G_i}$ and $\delta \boldsymbol{\theta}_i$ (we note that $\delta \boldsymbol{r}_i^{i-1}$ and $\delta \boldsymbol{r}_i^{i+1}$ are functions of $\delta \boldsymbol{r}_{G_i}$ and $\delta \boldsymbol{\theta}_i$ due to the rigid-body constraint). If we sum (5.31) over all rigid bodies, we obtain

$$\begin{aligned} \delta W = \sum_{i=1}^{n_b} \delta W_i = \sum_{i=1}^{n_b} (\boldsymbol{f}_i^{i-1} \cdot \delta \boldsymbol{r}_i^{i-1} + \boldsymbol{f}_i^{i+1} \cdot \delta \boldsymbol{r}_i^{i+1}) \\ - \sum_{i=1}^{n_b} M_i(\boldsymbol{a}_{G_i} + g\hat{\boldsymbol{e}}_{0_3}) \cdot \delta \boldsymbol{r}_{G_i} + \sum_{i=1}^{n_b} (\boldsymbol{\varphi}_i^{i-1} \cdot \delta \boldsymbol{\theta}_i + \boldsymbol{\varphi}_i^{i+1} \cdot \delta \boldsymbol{\theta}_i) \\ - \sum_{i=1}^{n_b} (\boldsymbol{I}_i^{G_i} \boldsymbol{\alpha}_i + \boldsymbol{\omega}_i \times \boldsymbol{I}_i^{G_i} \boldsymbol{\omega}_i) \cdot \delta \boldsymbol{\theta}_i = 0. \end{aligned} \tag{5.32}$$

The first summation is associated with the virtual work performed by the interlink reaction forces. It will be useful to rearrange the summation so as to pair up the equal and opposite reaction forces, $\boldsymbol{f}_i^{i-1}$ and $\boldsymbol{f}_{i-1}^i$, acting through each joint i. Doing this, we have

$$\sum_{i=1}^{n_b}(\boldsymbol{f}_i^{i-1}\cdot\delta\boldsymbol{r}_i^{i-1}+\boldsymbol{f}_i^{i+1}\cdot\delta\boldsymbol{r}_i^{i+1})=\sum_{i=1}^{n_b}(\boldsymbol{f}_i^{i-1}\cdot\delta\boldsymbol{r}_i^{i-1}+\boldsymbol{f}_{i-1}^{i}\cdot\delta\boldsymbol{r}_{i-1}^{i})+\boldsymbol{f}_{n_b}^{n_b+1}\cdot\delta\boldsymbol{r}_{n_b}^{n_b+1}-\boldsymbol{f}_0^1\cdot\delta\boldsymbol{r}_0^1. \tag{5.33}$$

Since $\boldsymbol{f}_{n_b}^{n_b+1}=\boldsymbol{0}$ and $\delta\boldsymbol{r}_0^1=\boldsymbol{0}$, we have

$$\sum_{i=1}^{n_b}(\boldsymbol{f}_i^{i-1}\cdot\delta\boldsymbol{r}_i^{i-1}+\boldsymbol{f}_i^{i+1}\cdot\delta\boldsymbol{r}_i^{i+1})=\sum_{i=1}^{n_b}(\boldsymbol{f}_i^{i-1}\cdot\delta\boldsymbol{r}_i^{i-1}+\boldsymbol{f}_{i-1}^{i}\cdot\delta\boldsymbol{r}_{i-1}^{i}). \tag{5.34}$$

We note that $\boldsymbol{f}_{i-1}^i=-\boldsymbol{f}_i^{i-1}$. Thus,

$$\sum_{i=1}^{n_b}(\boldsymbol{f}_i^{i-1}\cdot\delta\boldsymbol{r}_i^{i-1}+\boldsymbol{f}_i^{i+1}\cdot\delta\boldsymbol{r}_i^{i+1})=\sum_{i=1}^{n_b}\boldsymbol{f}_i^{i-1}\cdot(\delta\boldsymbol{r}_i^{i-1}-\delta\boldsymbol{r}_{i-1}^{i}). \tag{5.35}$$

This reflects the virtual work done by reaction forces at each joint, summed over all joints.

As with the reaction forces, it will be useful to rearrange the summation associated with the virtual work performed by the interlink reaction moments so as to pair up the equal and opposite reaction moments, $\boldsymbol{\varphi}_i^{i-1}$ and $\boldsymbol{\varphi}_{i-1}^i$, acting about each joint. Using the same procedure as with the reaction forces, we have

$$\begin{aligned}\sum_{i=1}^{n_b}(\boldsymbol{\varphi}_i^{i-1}\cdot\delta\boldsymbol{\theta}_i+\boldsymbol{\varphi}_i^{i+1}\cdot\delta\boldsymbol{\theta}_i)&=\sum_{i=1}^{n_b}(\boldsymbol{\varphi}_i^{i-1}\cdot\delta\boldsymbol{\theta}_i+\boldsymbol{\varphi}_{i-1}^{i}\cdot\delta\boldsymbol{\theta}_{i-1})\\&=\sum_{i=1}^{n_b}\boldsymbol{\varphi}_i^{i-1}\cdot(\delta\boldsymbol{\theta}_i-\delta\boldsymbol{\theta}_{i-1}).\end{aligned} \tag{5.36}$$

This reflects the virtual work done by reaction moments at each joint, summed over all joints. So, the total virtual work performed by the interlink reaction forces and moments can be expressed compactly as

$$\begin{aligned}&\sum_{i=1}^{n_b}(\boldsymbol{f}_i^{i-1}\cdot\delta\boldsymbol{r}_i^{i-1}+\boldsymbol{f}_i^{i+1}\cdot\delta\boldsymbol{r}_i^{i+1})+\sum_{i=1}^{n_b}(\boldsymbol{\varphi}_i^{i-1}\cdot\delta\boldsymbol{\theta}_i+\boldsymbol{\varphi}_i^{i+1}\cdot\delta\boldsymbol{\theta}_i)\\&\qquad=\sum_{i=1}^{n_b}\begin{pmatrix}\boldsymbol{f}_i^{i-1}\\ \boldsymbol{\varphi}_i^{i-1}\end{pmatrix}\cdot\begin{pmatrix}\delta\boldsymbol{r}_i^{i-1}-\delta\boldsymbol{r}_{i-1}^{i}\\ \delta\boldsymbol{\theta}_i-\delta\boldsymbol{\theta}_{i-1}\end{pmatrix}.\end{aligned} \tag{5.37}$$

If we now consider only the variations $\delta\boldsymbol{r}_i^{i-1}$, $\delta\boldsymbol{r}_{i-1}^i$, $\delta\boldsymbol{\theta}_i$, and $\delta\boldsymbol{\theta}_{i-1}$ that are consistent with the kinematic constraints, then (5.37) reflects a projection of the interlink forces and moments on the direction of the joint motion. D'Alembert's Principle states that the virtual work associated with all forces and moments orthogonal to the joint motion (reaction forces/moments) is *zero* ($W_c=0$) and only the generalized force acting in the direction of joint motion produces virtual work. Thus,

$$\sum_{i=1}^{n_b} \begin{pmatrix} \boldsymbol{f}_i^{i-1} \\ \boldsymbol{\varphi}_i^{i-1} \end{pmatrix} \cdot \begin{pmatrix} \delta \boldsymbol{r}_i^{i-1} - \delta \boldsymbol{r}_{i-1}^{i} \\ \delta \boldsymbol{\theta}_i - \delta \boldsymbol{\theta}_{i-1} \end{pmatrix} = \sum_{i=1}^{n_q} \tau_i \cdot \delta q_i, \tag{5.38}$$

and (5.32) can be expressed as

$$\sum_{i=1}^{n_q} \tau_i \cdot \delta q_i - \sum_{i=1}^{n_b} M_i(\boldsymbol{a}_{G_i} + g\hat{\boldsymbol{e}}_{0_3}) \cdot \delta \boldsymbol{r}_{G_i} - \sum_{i=1}^{n_b} (\boldsymbol{I}_i^{G_i} \boldsymbol{\alpha}_i + \boldsymbol{\omega}_i \times \boldsymbol{I}_i^{G_i} \boldsymbol{\omega}_i) \cdot \delta \boldsymbol{\theta}_i = 0 \tag{5.39}$$

for all variations, $\delta \boldsymbol{r}_{G_i}$ and $\delta \boldsymbol{\theta}_i$, that are consistent with the kinematic constraints.

Using d'Alembert's Principle, we can address a system of n_b rigid bodies forming a branching chain in a similar manner. With a serial chain, the parent/child structure is implicit to the numbering scheme. Every link i has a single parent link, $i - 1$, at its proximal end and a single child link, $i + 1$, at its distal end (except for the nth link). The ith joint is at the proximal end of the ith link. With a branching chain the numbering of links is more arbitrary, without a parent/child structure implicit to the numbering scheme. We can explicitly capture the parent/child structure, however, by defining three additional parameters, λ_i, c_i, μ_{ij}. The term λ_i is the parent link number of the ith link, c_i is the number of child links for the ith link, and $\mu_{i1} \cdots \mu_{ic_i}$ are the child link numbers of the ith link. Given these parameters, the virtual work associated with a given body i is given by

$$\begin{aligned} \delta W_i = \boldsymbol{f}_i^{\lambda_i} \cdot \delta \boldsymbol{r}_i^{\lambda_i} + \sum_{j=1}^{c_i} \boldsymbol{f}_i^{\mu_{ij}} \cdot \delta \boldsymbol{r}_i^{\mu_{ij}} - M_i(\boldsymbol{a}_{G_i} + g\hat{\boldsymbol{e}}_{0_3}) \cdot \delta \boldsymbol{r}_{G_i} + \boldsymbol{\varphi}_i^{\lambda_i} \cdot \delta \boldsymbol{\theta}_i \\ + \sum_{j=1}^{c_i} \boldsymbol{\varphi}_i^{\mu_{ij}} \cdot \delta \boldsymbol{\theta}_i - (\boldsymbol{I}_i^{G_i} \boldsymbol{\alpha}_i + \boldsymbol{\omega}_i \times \boldsymbol{I}_i^{G_i} \boldsymbol{\omega}_i) \cdot \delta \boldsymbol{\theta}_i = 0, \\ \forall \delta \boldsymbol{r}_{G_i} \in \mathbb{R}^3, \quad \text{and} \quad \forall \delta \boldsymbol{\theta}_i \in \mathbb{R}^3, \end{aligned} \tag{5.40}$$

for $i = 1, \ldots, n_b$. The term $\boldsymbol{r}_i^{\lambda_i}$ is the point on body i to which body (parent) λ_i attaches, and likewise $\boldsymbol{r}_i^{\mu_{i1}} \cdots \boldsymbol{r}_i^{\mu_{ic_i}}$ are the points on body i to which bodies (children) $\mu_{i1} \cdots \mu_{ic_i}$ attach. We note that $\delta \boldsymbol{r}_i^{\lambda_i}$ and $\delta \boldsymbol{r}_i^{\mu_{i1}} \cdots \delta \boldsymbol{r}_i^{\mu_{ic_i}}$ are functions of $\delta \boldsymbol{r}_{G_i}$ and $\delta \boldsymbol{\theta}_i$ due to the rigid-body constraint. If we sum (5.40) over all rigid bodies, we obtain

$$\begin{aligned} \delta W = \sum_{i=1}^{n_b} \delta W_i = \sum_{i=1}^{n_b} \left(\boldsymbol{f}_i^{\lambda_i} \cdot \delta \boldsymbol{r}_i^{\lambda_i} + \sum_{j=1}^{c_i} \boldsymbol{f}_i^{\mu_{ij}} \cdot \delta \boldsymbol{r}_i^{\mu_{ij}} \right) \\ - \sum_{i=1}^{n_b} M_i(\boldsymbol{a}_{G_i} + g\hat{\boldsymbol{e}}_{0_3}) \cdot \delta \boldsymbol{r}_{G_i} + \sum_{i=1}^{n_b} \left(\boldsymbol{\varphi}_i^{\lambda_i} \cdot \delta \boldsymbol{\theta}_i + \sum_{j=1}^{c_i} \boldsymbol{\varphi}_i^{\mu_{ij}} \cdot \delta \boldsymbol{\theta}_i \right) \\ - \sum_{i=1}^{n_b} (\boldsymbol{I}_i^{G_i} \boldsymbol{\alpha}_i + \boldsymbol{\omega}_i \times \boldsymbol{I}_i^{G_i} \boldsymbol{\omega}_i) \cdot \delta \boldsymbol{\theta}_i = 0. \end{aligned} \tag{5.41}$$

The term associated with the virtual work performed by the interlink reaction forces is

$$\sum_{i=1}^{n_b}\left(\boldsymbol{f}_i^{\lambda_i}\cdot\delta\boldsymbol{r}_i^{\lambda_i}+\sum_{j=1}^{c_i}\boldsymbol{f}_i^{\mu_{ij}}\cdot\delta\boldsymbol{r}_i^{\mu_{ij}}\right). \tag{5.42}$$

It will be useful to rearrange the summation so as to pair up the equal and opposite reaction forces, $\boldsymbol{f}_i^{\lambda_i}$ and $\boldsymbol{f}_{\lambda_i}^{i}$, acting through each joint i. We can rewrite the summation of (5.42) based on considering the virtual work at each joint. The sum over all joints then gives us

$$\sum_{i=1}^{n_b}(\boldsymbol{f}_i^{\lambda_i}\cdot\delta\boldsymbol{r}_i^{\lambda_i}+\boldsymbol{f}_{\lambda_i}^{i}\cdot\delta\boldsymbol{r}_{\lambda_i}^{i})=\sum_{i=1}^{n_b}\boldsymbol{f}_i^{\lambda_i}\cdot(\delta\boldsymbol{r}_i^{\lambda_i}-\delta\boldsymbol{r}_{\lambda_i}^{i}). \tag{5.43}$$

The term associated with the virtual work performed by the interlink reaction moments is

$$\sum_{i=1}^{n_b}\left(\boldsymbol{\varphi}_i^{\lambda_i}\cdot\delta\boldsymbol{\theta}_i+\sum_{j=1}^{c_i}\boldsymbol{\varphi}_i^{\mu_{ij}}\cdot\delta\boldsymbol{\theta}_i\right). \tag{5.44}$$

In a similar manner as before, we can rewrite this summation based on considering the virtual work at each joint. The sum over all joints then gives us

$$\sum_{i=1}^{n_b}(\boldsymbol{\varphi}_i^{\lambda_i}\cdot\delta\boldsymbol{\theta}_i+\boldsymbol{\varphi}_{\lambda_i}^{i}\cdot\delta\boldsymbol{\theta}_{\lambda_i})=\sum_{i=1}^{n_b}\boldsymbol{\varphi}_i^{\lambda_i}\cdot(\delta\boldsymbol{\theta}_i-\delta\boldsymbol{\theta}_{\lambda_i}). \tag{5.45}$$

So, the total virtual work performed by the interlink reaction forces and moments can be expressed compactly as

$$\begin{aligned}\sum_{i=1}^{n_b}\left(\boldsymbol{f}_i^{\lambda_i}\cdot\delta\boldsymbol{r}_i^{\lambda_i}+\sum_{j=1}^{c_i}\boldsymbol{f}_i^{\mu_{ij}}\cdot\delta\boldsymbol{r}_i^{\mu_{ij}}\right)&+\sum_{i=1}^{n_b}\left(\boldsymbol{\varphi}_i^{\lambda_i}\cdot\delta\boldsymbol{\theta}_i+\sum_{j=1}^{c_i}\boldsymbol{\varphi}_i^{\mu_{ij}}\cdot\delta\boldsymbol{\theta}_i\right)\\&=\sum_{i=1}^{n_b}\begin{pmatrix}\boldsymbol{f}_i^{\lambda_i}\\\boldsymbol{\varphi}_i^{\lambda_i}\end{pmatrix}\cdot\begin{pmatrix}\delta\boldsymbol{r}_i^{\lambda_i}-\delta\boldsymbol{r}_{\lambda_i}^{i}\\\delta\boldsymbol{\theta}_i-\delta\boldsymbol{\theta}_{\lambda_i}\end{pmatrix}.\end{aligned} \tag{5.46}$$

If we now consider only the variations $\delta\boldsymbol{r}_i^{\lambda_i}$, $\delta\boldsymbol{r}_{\lambda_i}^{i}$, $\delta\boldsymbol{\theta}_i$, and $\delta\boldsymbol{\theta}_{\lambda_i}$ that are consistent with the kinematic constraints, then (5.46) reflects a projection of the interlink forces and moments on the direction of the joint motion. D'Alembert's Principle states that the virtual work associated with all forces and moments orthogonal to the joint motion (reaction forces/moments) is *zero* ($W_c = 0$) and only the generalized force acting in the direction of joint motion produces virtual work. Thus,

$$\sum_{i=1}^{n_b}\begin{pmatrix}\boldsymbol{f}_i^{\lambda_i}\\\boldsymbol{\varphi}_i^{\lambda_i}\end{pmatrix}\cdot\begin{pmatrix}\delta\boldsymbol{r}_i^{\lambda_i}-\delta\boldsymbol{r}_{\lambda_i}^{i}\\\delta\boldsymbol{\theta}_i-\delta\boldsymbol{\theta}_{\lambda_i}\end{pmatrix}=\sum_{i=1}^{n_q}\tau_i\cdot\delta q_i, \tag{5.47}$$

and (5.41) can be expressed as

$$\sum_{i=1}^{n_q}\tau_i\cdot\delta q_i-\sum_{i=1}^{n_b}M_i(\boldsymbol{a}_{G_i}+g\hat{\boldsymbol{e}}_{0_3})\cdot\delta\boldsymbol{r}_{G_i}-\sum_{i=1}^{n_b}(\boldsymbol{I}_i^{G_i}\boldsymbol{\alpha}_i+\boldsymbol{\omega}_i\times\boldsymbol{I}_i^{G_i}\boldsymbol{\omega}_i)\cdot\delta\boldsymbol{\theta}_i=0, \tag{5.48}$$

for all variations, $\delta \boldsymbol{r}_{G_i}$ and $\delta \boldsymbol{\theta}_i$, that are consistent with the kinematic constraints. Expressing the variations in terms of variations in the generalized coordinates, we have

$$ {}^i\delta \boldsymbol{r}_{G_i} = {}^i\boldsymbol{\Gamma}_{G_i}\delta \boldsymbol{q} \quad \text{and} \quad {}^i\delta \boldsymbol{\theta}_i = {}^i\boldsymbol{\Pi}_i \delta \boldsymbol{q}, \tag{5.49} $$

where the terms are expressed in the local link frame i for convenience. So, (5.48) can be expressed as

$$ \boldsymbol{\tau} \cdot \delta \boldsymbol{q} - \sum_{i=1}^{n_b} M_i({}^i\boldsymbol{a}_{G_i} + g{}^i\hat{\boldsymbol{e}}_{0_3}) \cdot ({}^i\boldsymbol{\Gamma}_{G_i}\delta \boldsymbol{q}) - \sum_{i=1}^{n_b} ({}^i\boldsymbol{I}_i^{G_i}{}^i\boldsymbol{\alpha}_i + {}^i\boldsymbol{\omega}_i \times {}^i\boldsymbol{I}_i^{G_i}{}^i\boldsymbol{\omega}_i) \cdot ({}^i\boldsymbol{\Pi}_i \delta \boldsymbol{q}) = 0, \quad \forall \delta \boldsymbol{q} \in \mathbb{R}^n, \tag{5.50} $$

or

$$ \boldsymbol{\tau} \cdot \delta \boldsymbol{q} - \sum_{i=1}^{n_b} \left[M_i {}^i\boldsymbol{\Gamma}_{G_i}^T {}^i\boldsymbol{a}_{G_i} + M_i g {}^i\boldsymbol{\Gamma}_{G_i}^T {}^i\hat{\boldsymbol{e}}_{0_3} + {}^i\boldsymbol{\Pi}_i^T ({}^i\boldsymbol{I}_i^{G_i}{}^i\boldsymbol{\alpha}_i + {}^i\boldsymbol{\omega}_i \times {}^i\boldsymbol{I}_i^{G_i}{}^i\boldsymbol{\omega}_i) \right] \cdot \delta \boldsymbol{q} = 0 \quad \forall \delta \boldsymbol{q} \in \mathbb{R}^n. \tag{5.51} $$

In matrix form we have

$$ \left[\boldsymbol{\tau} - \sum_{i=1}^{n_b} \begin{pmatrix} {}^i\boldsymbol{\Gamma}_{G_i}^T & {}^i\boldsymbol{\Pi}_i^T \end{pmatrix} \begin{pmatrix} M_i {}^i\boldsymbol{a}_{G_i} + M_i g {}^i\hat{\boldsymbol{e}}_{0_3} \\ {}^i\boldsymbol{I}_i^{G_i}{}^i\boldsymbol{\alpha}_i + {}^i\boldsymbol{\omega}_i \times {}^i\boldsymbol{I}_i^{G_i}{}^i\boldsymbol{\omega}_i \end{pmatrix} \right] \cdot \delta \boldsymbol{q} = 0, \quad \forall \delta \boldsymbol{q} \in \mathbb{R}^n, \tag{5.52} $$

which implies

$$ \boldsymbol{\tau} = \sum_{i=1}^{n_b} \begin{pmatrix} {}^i\boldsymbol{\Gamma}_{G_i}^T & {}^i\boldsymbol{\Pi}_i^T \end{pmatrix} \begin{pmatrix} M_i {}^i\boldsymbol{a}_{G_i} + M_i g {}^i\hat{\boldsymbol{e}}_{0_3} \\ {}^i\boldsymbol{I}_i^{G_i}{}^i\boldsymbol{\alpha}_i + {}^i\boldsymbol{\Omega}_i {}^i\boldsymbol{I}_i^{G_i}{}^i\boldsymbol{\omega}_i \end{pmatrix}. \tag{5.53} $$

Noting that

$$ {}^i\boldsymbol{\omega}_i = {}^i\boldsymbol{\Pi}_i \dot{\boldsymbol{q}}, \quad {}^i\boldsymbol{a}_{G_i} = {}^i\boldsymbol{\Gamma}_{G_i}\ddot{\boldsymbol{q}} + ({}^i\dot{\boldsymbol{\Gamma}}_{G_i} + {}^i\boldsymbol{\Omega}_i {}^i\boldsymbol{\Gamma}_{G_i})\dot{\boldsymbol{q}}, \quad \text{and} \quad {}^i\boldsymbol{\alpha}_i = {}^i\boldsymbol{\Pi}_i\ddot{\boldsymbol{q}} + {}^i\dot{\boldsymbol{\Pi}}_i\dot{\boldsymbol{q}}, \tag{5.54} $$

we have

$$ \begin{aligned} \boldsymbol{\tau} = & \left[\sum_{i=1}^{n_b} (M_i {}^i\boldsymbol{\Gamma}_{G_i}^T {}^i\boldsymbol{\Gamma}_{G_i} + {}^i\boldsymbol{\Pi}_i^T {}^i\boldsymbol{I}_i^{G_i} {}^i\boldsymbol{\Pi}_i) \right] \ddot{\boldsymbol{q}} \\ & + \left[\sum_{i=1}^{n_b} \left(M_i {}^i\boldsymbol{\Gamma}_{G_i}^T ({}^i\dot{\boldsymbol{\Gamma}}_{G_i} + {}^i\boldsymbol{\Omega}_i {}^i\boldsymbol{\Gamma}_{G_i}) + {}^i\boldsymbol{\Pi}_i^T ({}^i\boldsymbol{I}_i^{G_i} {}^i\dot{\boldsymbol{\Pi}}_i + {}^i\boldsymbol{\Omega}_i {}^i\boldsymbol{I}_i^{G_i} {}^i\boldsymbol{\Pi}_i) \right) \right] \dot{\boldsymbol{q}} + g \sum_{i=1}^{n_b} M_i {}^i\boldsymbol{\Gamma}_{G_i}^T {}^i\hat{\boldsymbol{e}}_{0_3}. \end{aligned} \tag{5.55} $$

Algorithm 3 Second-order method for integrating the equations of motion

1: $\boldsymbol{q}_0 = \boldsymbol{q}_o$ {initialization}
2: $\dot{\boldsymbol{q}}_0 = \dot{\boldsymbol{q}}_o$ {initialization}
3: **for** $i = 0$ to $n_s - 1$ **do**
4: $\quad \ddot{\boldsymbol{q}}_i = \boldsymbol{M}_i^{-1}(\boldsymbol{\tau}_i - \boldsymbol{b}_i - \boldsymbol{g}_i)$
5: $\quad \dot{\boldsymbol{q}}_{i+1} = \dot{\boldsymbol{q}}_i + \ddot{\boldsymbol{q}}_i \Delta t$
6: $\quad \boldsymbol{q}_{i+1} = \boldsymbol{q}_i + \dot{\boldsymbol{q}}_i \Delta t + \frac{1}{2}\ddot{\boldsymbol{q}}_i \Delta t^2$
7: **end for**

Defining the symmetric positive definite mass matrix, $\boldsymbol{M}(\boldsymbol{q})$, the centrifugal and Coriolis force vector, $\boldsymbol{b}(\boldsymbol{q}, \dot{\boldsymbol{q}})$, and the gravity force vector, $\boldsymbol{g}(\boldsymbol{q})$,

$$\boldsymbol{M}(\boldsymbol{q}) \triangleq \sum_{i=1}^{n_b} (M_i {}^i\boldsymbol{\Gamma}_{G_i}^T {}^i\boldsymbol{\Gamma}_{G_i} + {}^i\boldsymbol{\Pi}_i^T {}^i\boldsymbol{I}_i^{G_i} {}^i\boldsymbol{\Pi}_i), \tag{5.56}$$

$$\boldsymbol{b}(\boldsymbol{q}, \dot{\boldsymbol{q}}) \triangleq \left[\sum_{i=1}^{n_b} \left(M_i {}^i\boldsymbol{\Gamma}_{G_i}^T ({}^i\dot{\boldsymbol{\Gamma}}_{G_i} + {}^i\boldsymbol{\Omega}_i {}^i\boldsymbol{\Gamma}_{G_i}) + {}^i\boldsymbol{\Pi}_i^T ({}^i\boldsymbol{I}_i^{G_i} {}^i\dot{\boldsymbol{\Pi}}_i + {}^i\boldsymbol{\Omega}_i {}^i\boldsymbol{I}_i^{G_i} {}^i\boldsymbol{\Pi}_i) \right) \right] \dot{\boldsymbol{q}}, \tag{5.57}$$

$$\boldsymbol{g}(\boldsymbol{q}) \triangleq g \sum_{i=1}^{n_b} M_i {}^i\boldsymbol{\Gamma}_{G_i}^T {}^i\hat{\boldsymbol{e}}_{0_3}, \tag{5.58}$$

we have

$$\boldsymbol{\tau} = \boldsymbol{M}(\boldsymbol{q})\ddot{\boldsymbol{q}} + \boldsymbol{b}(\boldsymbol{q}, \dot{\boldsymbol{q}}) + \boldsymbol{g}(\boldsymbol{q}). \tag{5.59}$$

Numerical Integration

We can numerically integrate (5.59) using a second-order method. Solving for the generalized accelerations, we have

$$\ddot{\boldsymbol{q}} = \boldsymbol{M}^{-1}(\boldsymbol{\tau} - \boldsymbol{b} - \boldsymbol{g}). \tag{5.60}$$

The second-order method for integrating this system can be summarized as shown in Algorithm 3.

Example: We consider a serial chain robot and parameterize the system using four generalized coordinates as shown in Figure 5.2. The centers of mass of each link can be

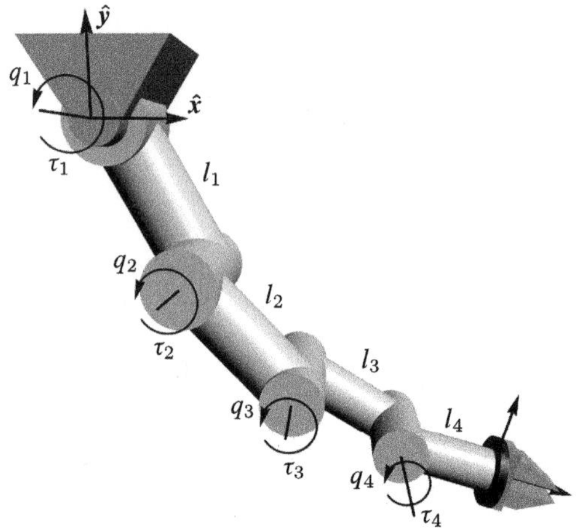

Figure 5.2 A 4 degree-of-freedom serial chain robot parameterized using four generalized coordinates.

computed in a straightforward fashion:

$$\boldsymbol{r}_{G_1} = \begin{pmatrix} \frac{l_1}{2}\cos(q_1) \\ \frac{l_1}{2}\sin(q_1) \end{pmatrix}, \tag{5.61}$$

$$\boldsymbol{r}_{G_2} = \begin{pmatrix} l_1\cos(q_1) + \frac{l_2}{2}\cos(q_1+q_2) \\ l_1\sin(q_1) + \frac{l_2}{2}\sin(q_1+q_2) \end{pmatrix}, \tag{5.62}$$

$$\boldsymbol{r}_{G_3} = \begin{pmatrix} l_1\cos(q_1) + l_2\cos(q_1+q_2) + \frac{l_3}{2}\cos(q_1+q_2+q_3) \\ l_1\sin(q_1) + l_2\sin(q_1+q_2) + \frac{l_3}{2}\sin(q_1+q_2+q_3) \end{pmatrix}, \tag{5.63}$$

$$\boldsymbol{r}_{G_4} = \left(\begin{matrix} l_1\cos(q_1) + l_2\cos(q_1+q_2) + l_3\cos(q_1+q_2+q_3) \\ l_1\sin(q_1) + l_2\sin(q_1+q_2) + l_3\sin(q_1+q_2+q_3) \end{matrix} \right. \tag{5.64}$$

$$\left. \dots \begin{matrix} + \frac{l_4}{2}\cos(q_1+q_2+q_3+q_4) \\ + \frac{l_4}{2}\sin(q_1+q_2+q_3+q_4) \end{matrix} \right). \tag{5.65}$$

Similarly, the angular velocities of each link can be computed easily:

$$\boldsymbol{\omega}_1 = \dot{q}_1, \tag{5.66}$$

$$\boldsymbol{\omega}_2 = \dot{q}_1 + \dot{q}_2, \tag{5.67}$$

$$\boldsymbol{\omega}_3 = \dot{q}_1 + \dot{q}_2 + \dot{q}_3, \tag{5.68}$$

$$\boldsymbol{\omega}_4 = \dot{q}_1 + \dot{q}_2 + \dot{q}_3 + \dot{q}_4. \tag{5.69}$$

The translational Jacobians are computed as

$$\boldsymbol{\Gamma}_{G_i} = \frac{\partial \boldsymbol{r}_{G_i}}{\partial \boldsymbol{q}}, \qquad \text{for } i = 1, \dots, 3, \tag{5.70}$$

and the angular velocity Jacobians as

$$\boldsymbol{\Pi}_i = \frac{\partial \boldsymbol{\omega}_i}{\partial \dot{\boldsymbol{q}}}, \qquad \text{for } i = 1, \dots, 3. \tag{5.71}$$

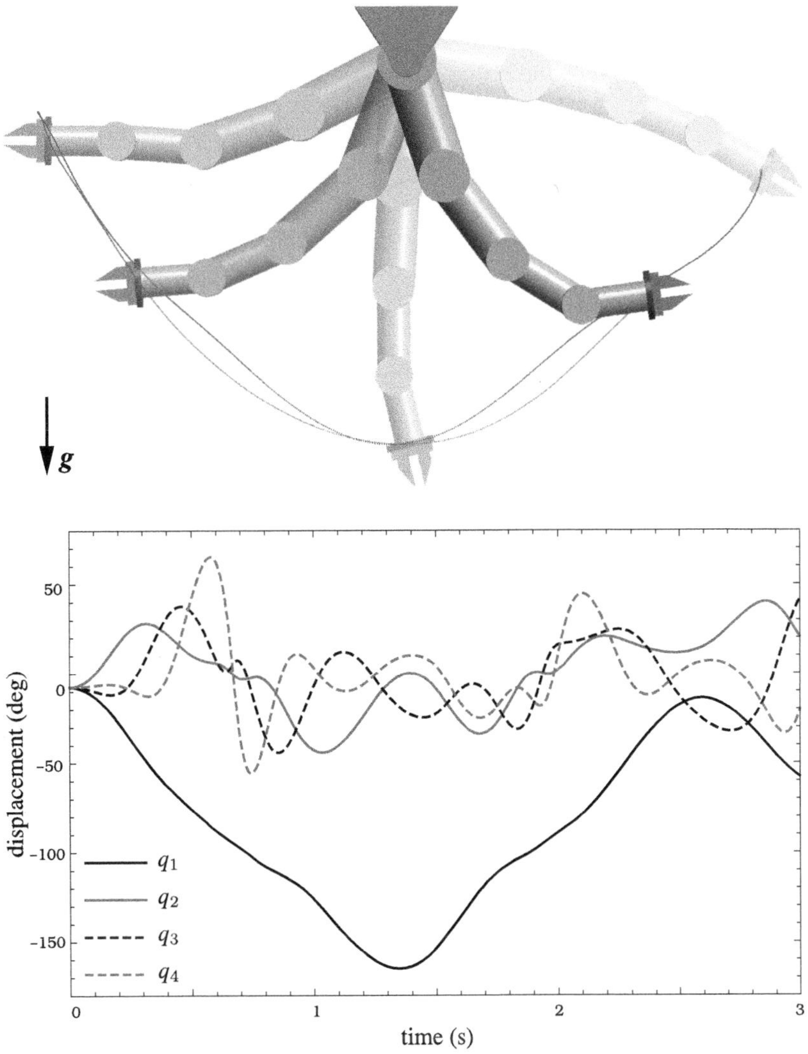

Figure 5.3 (Top) Animation frames from the simulation of a serial chain robot. (Bottom) Time history of the robot generalized coordinates (q_1, q_2, q_3, q_4).

The dynamical terms are

$$\boldsymbol{M}(\boldsymbol{q}) = \sum_{i=1}^{4} (M_i \boldsymbol{\Gamma}_{G_i}^T \boldsymbol{\Gamma}_{G_i} + \boldsymbol{\Pi}_i^T \boldsymbol{I}_i^{G_i} \boldsymbol{\Pi}_i), \tag{5.72}$$

$$\boldsymbol{b}(\boldsymbol{q}, \dot{\boldsymbol{q}}) = \sum_{i=1}^{4} M_i \boldsymbol{\Gamma}_{G_i}^T \dot{\boldsymbol{\Gamma}}_{G_i} \dot{\boldsymbol{q}}, \tag{5.73}$$

$$\boldsymbol{g}(\boldsymbol{q}) = g \sum_{i=1}^{4} M_i \boldsymbol{\Gamma}_{G_i}^T \hat{\boldsymbol{e}}_2, \tag{5.74}$$

where we take the rotational inertias to be

$$I_i^{G_i} = \frac{M_i r_i^2}{4} + \frac{M_i l_i^2}{12}. \tag{5.75}$$

The simulation results are displayed in Figure 5.3. The following values were used for the constants:

$$\begin{aligned} &r_1 = 0.15, \quad r_2 = 0.125, \quad r_3 = 0.1, \quad r_4 = 0.0875, \\ &l_1 = 0.7, \quad l_2 = 0.6, \quad l_3 = 0.5, \quad l_4 = 0.4, \\ &M_1 = 2.5, \quad M_2 = 1.5, \quad M_3 = 1.0, \quad M_4 = 0.75. \end{aligned} \tag{5.76}$$

The initial conditions used were $q_i = -\pi/24$ and $\dot{q}_i = 0$.

5.2.5 Auxiliary Constraints

Holonomic Constraints

The standard taxonomy of multibody systems by kinematic topology consists of branching or tree-like structures and graph or closed-loop structures. In the case of branching structures a set of independent generalized coordinates is chosen. The kinematic constraints between the bodies are implied by the choice of these generalized coordinates, as we saw in Section 5.2.4. Since these generalized coordinates are independent (unconstrained), we will refer to branching structures as unconstrained with respect to configuration space.

Analysis of graph or closed-loop structures typically involves breaking the loop(s) and deriving the dynamics of the resulting branching structures. The last step involves the imposition of a set of holonomic constraint equations, $\boldsymbol{\phi}(\boldsymbol{q}) = \boldsymbol{0}$, to enforce the loop closures. A closed-loop topology is depicted in Figure 5.4 (left), with the nodes representing the bodies and the edges representing the joints. Closed-loop structures represent a subset of the larger class of holonomically constrained multibody systems. Such systems involve holonomic constraints in the form of general algebraic dependencies between generalized coordinates, as depicted in Figure 5.4 (right). Again, a set of holonomic constraint equations is imposed in conjunction with the unconstrained equations of motion. Since the systems of Figure 5.4 involve explicit, or auxilliary, constraints between the generalized coordinates (in addition to the implicit constraints between the bodies suggested by the choice of generalized coordinates), we will refer to these structures as constrained with respect to configuration space. In Section 3.4 we considered the number of degrees of freedom of holonomically constrained systems. The number of degrees of freedom, p, is given by

$$p = n - m, \tag{5.77}$$

where n is the number of generalized coordinates and m is the number of independent constraint equations.

The constrained systems described thus far exclusively involve holonomic constraints. Nonholonomic systems, which involve nonintegrable constraints on the generalized velocities, will be addressed in the subsequent chapters. They can be handled with

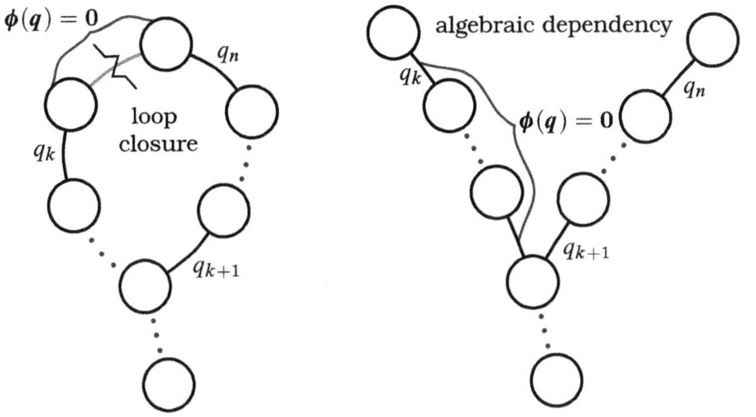

Figure 5.4 Constrained structures involving closed loops (Left) and general algebraic dependencies (Right). The dynamics of both systems can be derived from the unconstrained system dynamics through the imposition of a set of holonomic constraint equations, $\boldsymbol{\phi}(\boldsymbol{q}) = \mathbf{0}$.

higher-order variational principles (Flannery 2005) and methods like Jourdain's Principle, Kane's method, Gauss's Principle, or the Gibbs-Appell method. Jourdain's Principle and Gauss's Principle deal with the constraints explicitly, while Kane's method and the Gibbs-Appell method deal with the costraints implicitly by defining a set of independent quasi-velocities (generalized speeds) and/or quasi-accelerations.

We now consider the general case of auxiliary holonomic constraints. Given the multibody system

$$\boldsymbol{\tau} = \boldsymbol{M}(\boldsymbol{q})\ddot{\boldsymbol{q}} + \boldsymbol{b}(\boldsymbol{q}, \dot{\boldsymbol{q}}) + \boldsymbol{g}(\boldsymbol{q}), \tag{5.78}$$

subject to the holonomic constraints

$$\boldsymbol{\phi}(\boldsymbol{q}) = \mathbf{0}, \tag{5.79}$$

we begin by first expressing the zeroth-order variational equation associated with d'Alembert's Principle:

$$\boldsymbol{\tau}_C \cdot \delta \boldsymbol{q} + (\boldsymbol{\tau} - \boldsymbol{M}\boldsymbol{q} - \boldsymbol{b} - \boldsymbol{g}) \cdot \delta \boldsymbol{q} = 0, \tag{5.80}$$

where $\boldsymbol{\tau}_C$ is the vector of generalized constraint forces. The virtual displacements, $\delta \boldsymbol{q}$, refer to all displacement variations which satisfy the constraints, while time is fixed. With $\delta t = 0$ the variation of the constraint equation yields

$$\delta \boldsymbol{\phi} = \frac{\partial \boldsymbol{\phi}}{\partial \boldsymbol{q}} \delta \boldsymbol{q} = \boldsymbol{\Phi} \delta \boldsymbol{q} = \mathbf{0}, \tag{5.81}$$

which implies that $\delta \boldsymbol{q} \in \ker(\boldsymbol{\Phi})$, where $\boldsymbol{\Phi} = \frac{\partial \boldsymbol{\phi}}{\partial \boldsymbol{q}}$ is the constraint matrix. Under this condition, (5.80) can be restricted to constraint-consistent virtual displacements:

$$\begin{gathered} \boldsymbol{\tau}_C \cdot \delta \boldsymbol{q} + (\boldsymbol{\tau} - \boldsymbol{M}\boldsymbol{q} - \boldsymbol{b} - \boldsymbol{g}) \cdot \delta \boldsymbol{q} = 0 \\ \forall \delta \boldsymbol{q} \in \ker(\boldsymbol{\Phi}). \end{gathered} \tag{5.82}$$

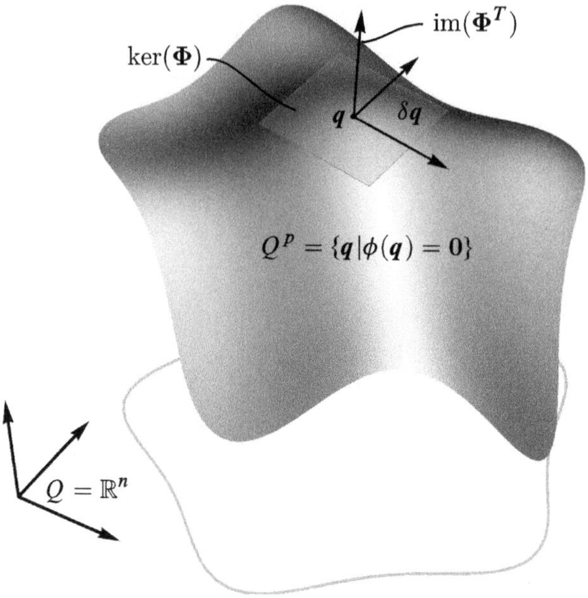

Figure 5.5 The configuration space constrained-motion manifold, Q^p. All constraint-consistent virtual variations, $\delta \boldsymbol{q}$, lie in the tangent space of Q^p and are orthogonal to the constraint forces.

We have

$$\begin{gathered} \boldsymbol{\tau}_C \perp \delta \boldsymbol{q} \\ \forall \delta \boldsymbol{q} \in \ker(\boldsymbol{\Phi}). \end{gathered} \tag{5.83}$$

The $\ker(\boldsymbol{\Phi})$ represents the tangent space of the constrained-motion manifold, Q^p, at a point, $\boldsymbol{q}$, in configuration space. The constraint-consistent virtual displacements, $\delta \boldsymbol{q}$, lie in this tangent space, and the generalized constraint forces, $\boldsymbol{\tau}_C$, are orthogonal to it. This is illustrated in Figure 5.5. Based on this, the following is implied:

$$\boldsymbol{\tau}_C \in \ker(\boldsymbol{\Phi})^{\perp} = \operatorname{im}(\boldsymbol{\Phi}^T). \tag{5.84}$$

Thus, the generalized constraint force, $\boldsymbol{\tau}_C$, can be represented as a linear combination of the columns of $\boldsymbol{\Phi}^T$. That is, $\boldsymbol{\tau}_C = \boldsymbol{\Phi}^T \boldsymbol{\lambda}$, where $\boldsymbol{\lambda}$ is a vector of unknown Lagrange multipliers. The term $\boldsymbol{\tau}_C \cdot \delta \boldsymbol{q}$ vanishes from (5.82) and we have the orthogonality relation

$$\begin{gathered} (\boldsymbol{M}\boldsymbol{q} + \boldsymbol{b} + \boldsymbol{g} - \boldsymbol{\tau}) \cdot \delta \boldsymbol{q} = 0 \\ \forall \delta \boldsymbol{q} \in \ker(\boldsymbol{\Phi}). \end{gathered} \tag{5.85}$$

The constrained multibody equations of motion are then

$$\boldsymbol{\tau} = \boldsymbol{M}\ddot{\boldsymbol{q}} + \boldsymbol{b} + \boldsymbol{g} - \boldsymbol{\Phi}^T \boldsymbol{\lambda}, \tag{5.86}$$

subject to

$$\boldsymbol{\phi}(\boldsymbol{q}) = \boldsymbol{0} \quad \Rightarrow \quad \dot{\boldsymbol{\phi}} = \boldsymbol{0}, \ddot{\boldsymbol{\phi}} = \boldsymbol{0}. \tag{5.87}$$

Solution of the Constrained Dynamics Problem

We can arrive at an explicit solution of the constrained dynamics problem,

$$\boldsymbol{M}\ddot{\boldsymbol{q}} = \boldsymbol{\tau} - \boldsymbol{b} - \boldsymbol{g} + \boldsymbol{\Phi}^T\boldsymbol{\lambda} \tag{5.88}$$

$$\boldsymbol{\Phi}\ddot{\boldsymbol{q}} = -\dot{\boldsymbol{\Phi}}\dot{\boldsymbol{q}}. \tag{5.89}$$

It is useful to define the $m_C \times m_C$ constraint space mass matrix, which reflects the system inertia projected at the constraint:

$$\mathbf{H} \triangleq (\boldsymbol{\Phi}\boldsymbol{M}^{-1}\boldsymbol{\Phi}^T)^{-1}. \tag{5.90}$$

We now express the mass-weighted (right) inverse of $\boldsymbol{\Phi}$:

$$\bar{\boldsymbol{\Phi}} = \boldsymbol{M}^{-1}\boldsymbol{\Phi}^T(\boldsymbol{\Phi}\boldsymbol{M}^{-1}\boldsymbol{\Phi}^T)^{-1}, \tag{5.91}$$

where $\boldsymbol{\Phi}\bar{\boldsymbol{\Phi}} = \mathbf{1}$ and, equivalently, $\bar{\boldsymbol{\Phi}}^T\boldsymbol{\Phi}^T = \mathbf{1}$. Defining the $n \times n$ constraint null space matrix, $\boldsymbol{\Theta} \triangleq \mathbf{1} - \bar{\boldsymbol{\Phi}}\boldsymbol{\Phi}$, we can solve the system. This yields

$$\ddot{\boldsymbol{q}} = -\bar{\boldsymbol{\Phi}}\dot{\boldsymbol{\Phi}}\dot{\boldsymbol{q}} + \boldsymbol{\Theta}\boldsymbol{M}^{-1}(\boldsymbol{\tau} - \boldsymbol{b} - \boldsymbol{g}) \tag{5.92}$$

$$\boldsymbol{\lambda} = -\bar{\boldsymbol{\Phi}}^T(\boldsymbol{\tau} - \boldsymbol{b} - \boldsymbol{g}) - \mathbf{H}\dot{\boldsymbol{\Phi}}\dot{\boldsymbol{q}}. \tag{5.93}$$

It is noted that $\boldsymbol{\Phi}$ and $\boldsymbol{\Theta}$ satisfy the condition $\boldsymbol{\Phi}\boldsymbol{\Theta} = \mathbf{0}$ and, equivalently, $\boldsymbol{\Theta}^T\boldsymbol{\Phi}^T = \mathbf{0}$. Furthermore, if we form the projection matrix $\boldsymbol{P}^T = \boldsymbol{P}$ which projects any vector in $\mathbb{R}^n$ onto the null space of $\boldsymbol{\Phi}$, we have

$$\boldsymbol{P}^T = \boldsymbol{W}\boldsymbol{W}^T = \mathbf{1} - \boldsymbol{\Phi}^T\boldsymbol{\Phi}^{+T} = \mathbf{1} - \boldsymbol{\Phi}^T(\boldsymbol{\Phi}\boldsymbol{\Phi}^T)^{-1}\boldsymbol{\Phi}, \tag{5.94}$$

where $\boldsymbol{W}$ spans the null space of $\boldsymbol{\Phi}$, and $\boldsymbol{\Phi}^+ = \boldsymbol{\Phi}^T(\boldsymbol{\Phi}\boldsymbol{\Phi}^T)^{-1}$ is the pseudoinverse (right inverse) of $\boldsymbol{\Phi}$. The expression for $\boldsymbol{P}^T$ in (5.94) has a similar form as the expression

$$\boldsymbol{\Theta}^T = \mathbf{1} - \boldsymbol{\Phi}^T\bar{\boldsymbol{\Phi}}^T = \mathbf{1} - \boldsymbol{\Phi}^T(\boldsymbol{\Phi}\boldsymbol{M}^{-1}\boldsymbol{\Phi}^T)^{-1}\boldsymbol{\Phi}\boldsymbol{M}^{-1}. \tag{5.95}$$

Consequently $\boldsymbol{P}^T = \boldsymbol{W}\boldsymbol{W}^T$ can be regarded as a *kinematic* constraint null space projection matrix and $\boldsymbol{\Theta}^T$ can be regarded as a *mass-weighted* constraint null space projection matrix. The physical and geometric meaning of $\boldsymbol{\Phi}$ and $\boldsymbol{\Theta}$ will be discussed further in Section 8.1.3.

Numerical Integration

In the previous section we solved the constrained dynamical system for the generalized accelerations and the Lagrange multipliers. In practice the integration of the forward dynamics would also require constraint stabilization to mitigate drift in the constraints. Baumgarte stabilization (Baumgarte 1972) involves replacing our original acceleration constraint equation with a linear combination of acceleration, velocity, and position constraint terms:

$$\ddot{\boldsymbol{\phi}} + \beta\dot{\boldsymbol{\phi}} + \alpha\boldsymbol{\phi} = \mathbf{0} \tag{5.96}$$

or

$$\boldsymbol{\Phi}\ddot{\boldsymbol{q}} + \dot{\boldsymbol{\Phi}}\dot{\boldsymbol{q}} + \beta\boldsymbol{\Phi}\dot{\boldsymbol{q}} + \alpha\boldsymbol{\phi} = \mathbf{0}, \tag{5.97}$$

Algorithm 4 Second-order method for integrating the zeroth-order holonomically constrained equations of motion

$q_0 = q_o$ {initialization}

$\dot{q}_0 = \dot{q}_o$ {initialization}

for $i = 0$ to $n_s - 1$ **do**

$\quad \ddot{q}_i = -\bar{\Phi}_i \dot{\Phi}_i \dot{q}_i + \Theta_i M_i^{-1}(\tau_i - b_i - g_i) - \bar{\Phi}_i(\alpha \phi_i + \beta \Phi_i \dot{q}_i)$

$\quad \lambda_i = -\bar{\Phi}_i^T(\tau_i - b_i - g_i) - \mathbf{H}_i \dot{\Phi}_i \dot{q}_i - \mathbf{H}_i(\alpha \phi_i + \beta \Phi_i \dot{q}_i)$

$\quad \dot{q}_{i+1} = \dot{q}_i + \ddot{q}_i \Delta t$

$\quad q_{i+1} = q_i + \dot{q}_i \Delta t + \frac{1}{2} \ddot{q}_i \Delta t^2$

end for

where α and β are constant parameters chosen to drive the first- and zeroth-order derivatives of the holonomic constraint equations to *zero*, thereby compensating for position and velocity drift in the constraints. The constraint stabilized equations of motion are then

$$\begin{pmatrix} M & -\Phi^T \\ -\Phi & 0 \end{pmatrix} \begin{pmatrix} \ddot{q} \\ \lambda \end{pmatrix} = \begin{pmatrix} \tau - b - g \\ \dot{\Phi}\dot{q} + \beta \Phi \dot{q} + \alpha \phi \end{pmatrix}, \tag{5.98}$$

and the solution of this system is

$$\ddot{q} = -\bar{\Phi}\dot{\Phi}\dot{q} + \Theta M^{-1}(\tau - b - g) - \bar{\Phi}(\alpha \phi + \beta \Phi \dot{q}) \tag{5.99}$$

$$\lambda = -\bar{\Phi}^T(\tau - b - g) - \mathbf{H}\dot{\Phi}\dot{q} - \mathbf{H}(\alpha \phi + \beta \Phi \dot{q}). \tag{5.100}$$

The term $-\mathbf{H}(\alpha \phi + \beta \Phi \dot{q})$ in the expression for λ can be physically interpreted as a corrective constraint force term used to compensate for any drift in the constraints. This is analogous to a proportional-derivative (PD) control law in a feedback system.

A second-order method for integrating this system can be summarized as shown in Algorithm 4.

Example: A Stewart platform parallel mechanism can be described by a set of 24 generalized coordinates, as shown in Figure 5.6.

The constraint equations associated with the loop closures are given by

$$r_{l_i} = r_{p_i} \qquad \text{for } i = 1, \cdots, 6, \tag{5.101}$$

where r_{l_i} is the terminal point of the ith strut subsystem which connects to r_{p_i}, the ith position on the platform subsystem. In vector form we have

$$\phi = \begin{pmatrix} r_{l_1} - r_{p_1} \\ \vdots \\ r_{l_6} - r_{p_6} \end{pmatrix} = \mathbf{0}. \tag{5.102}$$

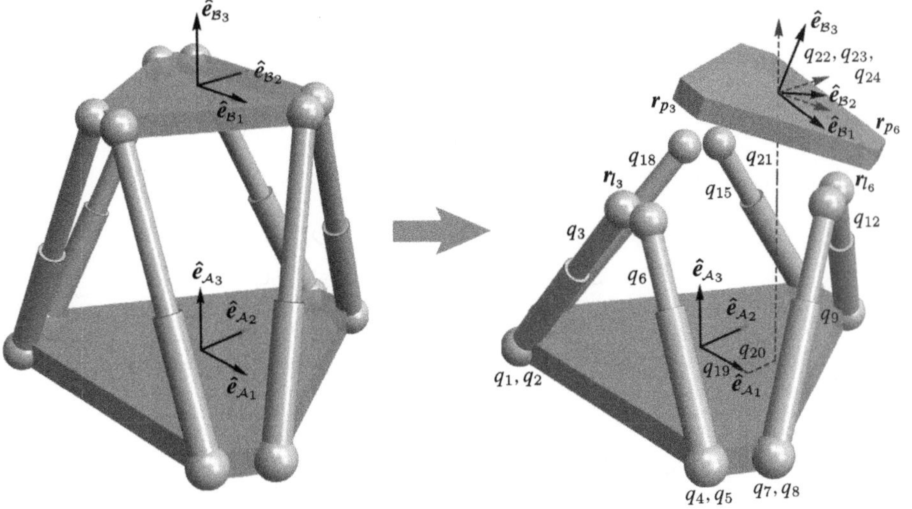

Figure 5.6 (Left) Stewart platform actuated by six prismatic struts (remaining joints are passive). (Right) The closed-loop mechanism is cut at various locations to create serial chains described by the set of generalized coordinates, $q_1, \ldots, q_{24}$.

Taking the derivative yields

$$\dot{\boldsymbol{\phi}} = \begin{pmatrix} \dot{\boldsymbol{r}}_{l_1} - \dot{\boldsymbol{r}}_{p_1} \\ \vdots \\ \dot{\boldsymbol{r}}_{l_6} - \dot{\boldsymbol{r}}_{p_6} \end{pmatrix} = \boldsymbol{\Phi}\dot{\boldsymbol{q}} = \boldsymbol{0}, \tag{5.103}$$

where

$$\dot{\boldsymbol{r}}_{l_i} = \boldsymbol{\Gamma}_{l_i} \begin{pmatrix} \dot{q}_{3i-2} \\ \dot{q}_{3i-1} \\ \dot{q}_{3i} \end{pmatrix} \quad \text{and,} \quad \dot{\boldsymbol{r}}_{p_i} = \boldsymbol{\Gamma}_{p_i} \begin{pmatrix} \dot{q}_{19} \\ \vdots \\ \dot{q}_{24} \end{pmatrix}, \quad \text{for } i = 1, \cdots, 6. \tag{5.104}$$

The terms, $\boldsymbol{\Gamma}_{l_i}$ and $\boldsymbol{\Gamma}_{p_i}$, are the corresponding Jacobians of $\boldsymbol{r}_{l_i}$ and $\boldsymbol{r}_{p_i}$, respectively. So,

$$\boldsymbol{\Phi}\dot{\boldsymbol{q}} = \begin{pmatrix} \boldsymbol{\Gamma}_{l_1} & \ldots & \boldsymbol{0} & -\boldsymbol{\Gamma}_{p_1} \\ \vdots & \ddots & \vdots & \vdots \\ \boldsymbol{0} & \ldots & \boldsymbol{\Gamma}_{l_6} & -\boldsymbol{\Gamma}_{p_6} \end{pmatrix} \begin{pmatrix} \dot{q}_1 \\ \vdots \\ \dot{q}_{24} \end{pmatrix} = \boldsymbol{0}, \tag{5.105}$$

where

$$\boldsymbol{\Phi} = \begin{pmatrix} \boldsymbol{\Gamma}_{l_1} & \ldots & \boldsymbol{0} & -\boldsymbol{\Gamma}_{p_1} \\ \vdots & \ddots & \vdots & \vdots \\ \boldsymbol{0} & \ldots & \boldsymbol{\Gamma}_{l_6} & -\boldsymbol{\Gamma}_{p_6} \end{pmatrix}. \tag{5.106}$$

The constraint forces, $\boldsymbol{\lambda}$, are shown in Figure 5.7.

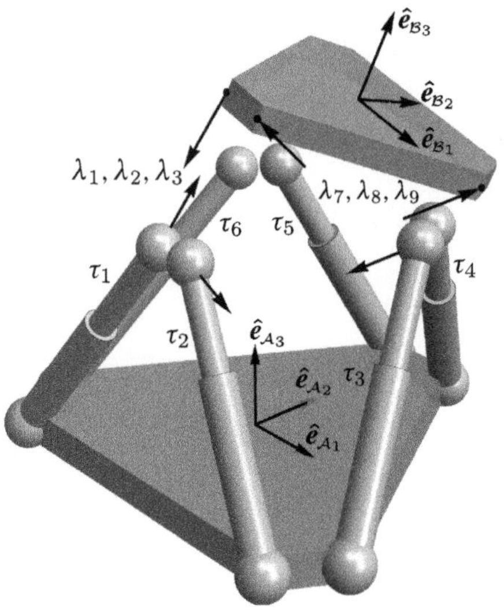

Figure 5.7 Constraint forces associated with loop closures. Lagrange multipliers, $\boldsymbol{\lambda}$, represent constraint forces at various locations between the serial chains to enforce the constraints associated with the loop closures.

The unconstrained equations of motion for the six strut serial subsystems are

$$\begin{pmatrix} 0 \\ 0 \\ \tau_1 \end{pmatrix} = \boldsymbol{M}_{l_1} \begin{pmatrix} \ddot{q}_1 \\ \ddot{q}_2 \\ \ddot{q}_3 \end{pmatrix} + \boldsymbol{b}_{l_1}(q_1, \ldots, \dot{q}_3) + \boldsymbol{g}_{l_1}(q_1, \ldots, q_3), \tag{5.107}$$

$$\vdots$$

$$\begin{pmatrix} 0 \\ 0 \\ \tau_6 \end{pmatrix} = \boldsymbol{M}_{l_6} \begin{pmatrix} \ddot{q}_{16} \\ \ddot{q}_{17} \\ \ddot{q}_{18} \end{pmatrix} + \boldsymbol{b}_{l_6}(q_{16}, \ldots, \dot{q}_{18}) + \boldsymbol{g}_{l_6}(q_{16}, \ldots, q_{18}), \tag{5.108}$$

and for the platform subsystem we have

$$\boldsymbol{0} = \boldsymbol{M}_p \begin{pmatrix} \ddot{q}_{19} \\ \vdots \\ \ddot{q}_{24} \end{pmatrix} + \boldsymbol{b}_p(q_{19}, \ldots, \dot{q}_{24}) + \boldsymbol{g}_p(q_{19}, \ldots, q_{24}). \tag{5.109}$$

The entire unconstrained system is described by

$$\begin{pmatrix} \boldsymbol{0} \\ \tau_1 \\ \vdots \\ \boldsymbol{0} \\ \tau_6 \\ \boldsymbol{0} \end{pmatrix} = \begin{pmatrix} \boldsymbol{M}_{l_1} & \boldsymbol{0} & \cdots & \boldsymbol{0} \\ \boldsymbol{0} & \ddots & & \vdots \\ \vdots & & \boldsymbol{M}_{l_6} & \boldsymbol{0} \\ \boldsymbol{0} & \cdots & \boldsymbol{0} & \boldsymbol{M}_p \end{pmatrix} \begin{pmatrix} \ddot{q}_1 \\ \ddot{q}_2 \\ \vdots \\ \ddot{q}_{24} \end{pmatrix} + \begin{pmatrix} \boldsymbol{b}_{l_1} \\ \vdots \\ \boldsymbol{b}_{l_6} \\ \boldsymbol{b}_p \end{pmatrix} + \begin{pmatrix} \boldsymbol{g}_{l_1} \\ \vdots \\ \boldsymbol{g}_{l_6} \\ \boldsymbol{g}_p \end{pmatrix}, \tag{5.110}$$

and the constrained system is

$$\begin{pmatrix} \mathbf{0} \\ \tau_1 \\ \vdots \\ \mathbf{0} \\ \tau_6 \\ \mathbf{0} \end{pmatrix} = \begin{pmatrix} \boldsymbol{M}_{l_1} & \mathbf{0} & \cdots & \mathbf{0} \\ \mathbf{0} & \ddots & & \vdots \\ \vdots & & \boldsymbol{M}_{l_6} & \mathbf{0} \\ \mathbf{0} & \cdots & \mathbf{0} & \boldsymbol{M}_p \end{pmatrix} \begin{pmatrix} \ddot{q}_1 \\ \ddot{q}_2 \\ \vdots \\ \ddot{q}_{24} \end{pmatrix} + \begin{pmatrix} \boldsymbol{b}_{l_1} \\ \vdots \\ \boldsymbol{b}_{l_6} \\ \boldsymbol{b}_p \end{pmatrix} + \begin{pmatrix} \boldsymbol{g}_{l_1} \\ \vdots \\ \boldsymbol{g}_{l_6} \\ \boldsymbol{g}_p \end{pmatrix}$$

$$+ \begin{pmatrix} \boldsymbol{\Gamma}^T_{l_1} & \cdots & \mathbf{0} \\ \vdots & \ddots & \vdots \\ \mathbf{0} & \cdots & \boldsymbol{\Gamma}^T_{l_6} \\ -\boldsymbol{\Gamma}^T_{p_1} & \cdots & -\boldsymbol{\Gamma}^T_{p_6} \end{pmatrix} \begin{pmatrix} \lambda_1 \\ \vdots \\ \lambda_{18} \end{pmatrix}, \tag{5.111}$$

where the dimensions of the terms are,

$$\boldsymbol{M} \in \mathbb{R}^{n \times n}, \qquad \boldsymbol{b}, \boldsymbol{g}, \boldsymbol{\tau} \in \mathbb{R}^n, \qquad \boldsymbol{\Phi} \in \mathbb{R}^{m \times n}, \qquad \boldsymbol{\lambda} \in \mathbb{R}^m, \tag{5.112}$$

and $n = 24$ and $m = 18$. The constrained system has $p = n - m = 6$ degrees of freedom.

Noting that the generalized constraint force $\boldsymbol{\tau}_C$ is given by

$$\boldsymbol{\tau}_C = \boldsymbol{\Phi}^T \boldsymbol{\lambda}, \tag{5.113}$$

the virtual work of the constraint forces is given by

$$\boldsymbol{\tau}_C \cdot \delta \boldsymbol{q} = \boldsymbol{\tau}_C^T \delta \boldsymbol{q} = \boldsymbol{\lambda}^T \boldsymbol{\Phi} \delta \boldsymbol{q}. \tag{5.114}$$

So,

$$\begin{aligned} &\boldsymbol{\tau}_C \cdot \delta \boldsymbol{q} = \mathbf{0} \\ &\forall \delta \boldsymbol{q} \in \delta Q^p, \end{aligned} \tag{5.115}$$

where $\delta Q^p = \ker(\boldsymbol{\Phi})$. Furthermore, we note that

$$\boldsymbol{\Phi} \delta \boldsymbol{q} = \begin{pmatrix} \delta \boldsymbol{r}_{l_1} - \delta \boldsymbol{r}_{p_1} \\ \vdots \\ \delta \boldsymbol{r}_{l_6} - \delta \boldsymbol{r}_{p_6} \end{pmatrix}. \tag{5.116}$$

So,

$$\boldsymbol{\tau}_C \cdot \delta \boldsymbol{q} = \boldsymbol{\lambda}^T \boldsymbol{\Phi} \delta \boldsymbol{q} = \begin{pmatrix} \lambda_1 \\ \vdots \\ \lambda_{18} \end{pmatrix}^T \begin{pmatrix} \delta \boldsymbol{r}_{l_1} - \delta \boldsymbol{r}_{p_1} \\ \vdots \\ \delta \boldsymbol{r}_{l_6} - \delta \boldsymbol{r}_{p_6} \end{pmatrix}. \tag{5.117}$$

Thus, the virtual work of the constraint forces can also be expressed by

$$\begin{pmatrix} \lambda_1 \\ \vdots \\ \lambda_{18} \end{pmatrix} \cdot \begin{pmatrix} \delta \boldsymbol{r}_{l_1} - \delta \boldsymbol{r}_{p_1} \\ \vdots \\ \delta \boldsymbol{r}_{l_6} - \delta \boldsymbol{r}_{p_6} \end{pmatrix} = \boldsymbol{0}, \tag{5.118}$$

$$\forall \delta \boldsymbol{r}_{l_i}, \delta \boldsymbol{r}_{p_i} | \delta \boldsymbol{r}_{l_i} = \delta \boldsymbol{r}_{p_i}, \quad i = 1, \dots, 6.$$

Example: A parallel mechanism is depicted in Figure 5.8. The constraint equations describe the loop closures and are given by

$$\boldsymbol{\phi}(\boldsymbol{q}) = \begin{pmatrix} \boldsymbol{r}_{l_1} - \boldsymbol{r}_{p_1} \\ \boldsymbol{r}_{l_2} - \boldsymbol{r}_{p_2} \\ \boldsymbol{r}_{l_3} - \boldsymbol{r}_{p_3} \end{pmatrix}, \tag{5.119}$$

where $\boldsymbol{r}_{l_i}$ is the terminal point of the ith elbow chain subsystem which connects to $\boldsymbol{r}_{p_i}$, the ith position on the platform subsystem. Taking the derivative yields

$$\dot{\boldsymbol{\phi}} = \begin{pmatrix} \dot{\boldsymbol{r}}_{l_1} - \dot{\boldsymbol{r}}_{p_1} \\ \dot{\boldsymbol{r}}_{l_2} - \dot{\boldsymbol{r}}_{p_2} \\ \dot{\boldsymbol{r}}_{l_3} - \dot{\boldsymbol{r}}_{p_3} \end{pmatrix} = \boldsymbol{\Phi}\dot{\boldsymbol{q}} = \boldsymbol{0}, \tag{5.120}$$

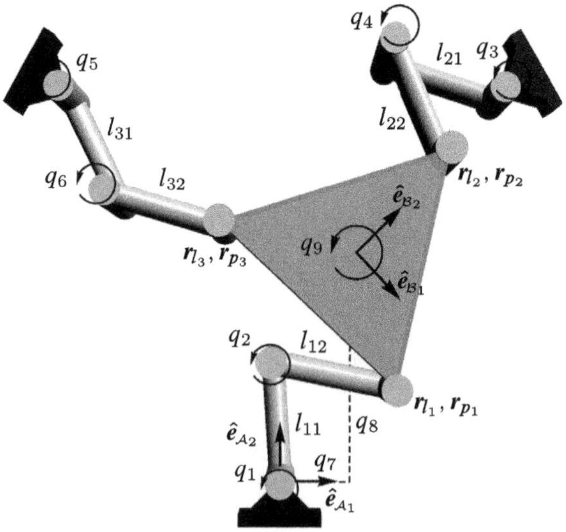

Figure 5.8 Parallel mechanism consisting of serial chains with loop closures. The three elbow joints are actuated, while the remaining joints are passive.

where

$$\dot{\boldsymbol{r}}_{l_i} = \boldsymbol{\Gamma}_{l_i} \begin{pmatrix} \dot{q}_{2i-1} \\ \dot{q}_{2i} \end{pmatrix} \quad \text{and,} \quad \dot{\boldsymbol{r}}_{p_i} = \boldsymbol{\Gamma}_{p_i} \begin{pmatrix} \dot{q}_7 \\ \dot{q}_8 \\ \dot{q}_9 \end{pmatrix}, \tag{5.121}$$

$$\text{for } i = 1, \ldots, 3.$$

The terms, $\boldsymbol{\Gamma}_{l_i}$ and $\boldsymbol{\Gamma}_{p_i}$ are the corresponding Jacobians of $\boldsymbol{r}_{l_i}$ and $\boldsymbol{r}_{p_i}$, respectively. So,

$$\boldsymbol{\Phi}\dot{\boldsymbol{q}} = \begin{pmatrix} \boldsymbol{\Gamma}_{l_1} & \mathbf{0} & \mathbf{0} & -\boldsymbol{\Gamma}_{p_1} \\ \mathbf{0} & \boldsymbol{\Gamma}_{l_2} & \mathbf{0} & -\boldsymbol{\Gamma}_{p_2} \\ \mathbf{0} & \mathbf{0} & \boldsymbol{\Gamma}_{l_3} & -\boldsymbol{\Gamma}_{p_3} \end{pmatrix} \begin{pmatrix} \dot{q}_1 \\ \vdots \\ \dot{q}_9 \end{pmatrix} = \mathbf{0}, \tag{5.122}$$

where

$$\boldsymbol{\Phi} = \begin{pmatrix} \boldsymbol{\Gamma}_{l_1} & \mathbf{0} & \mathbf{0} & -\boldsymbol{\Gamma}_{p_1} \\ \mathbf{0} & \boldsymbol{\Gamma}_{l_2} & \mathbf{0} & -\boldsymbol{\Gamma}_{p_2} \\ \mathbf{0} & \mathbf{0} & \boldsymbol{\Gamma}_{l_3} & -\boldsymbol{\Gamma}_{p_3} \end{pmatrix}. \tag{5.123}$$

For the three elbow link chains we compute

$$\boldsymbol{\Gamma}_{l_i} = \begin{pmatrix} -l_{i1}\sin(q_{2i-1}) - l_{i2}\sin(q_{2i-1}+q_{2i}) & -l_{i2}\sin(q_{2i-1}+q_{2i}) \\ l_{i1}\cos(q_{2i-1}) + l_{i2}\cos(q_{2i-1}+q_{2i}) & l_{i2}\cos(q_{2i-1}+q_{2i}) \end{pmatrix}, \tag{5.124}$$

$$\text{for } i = 1, \ldots, 3,$$

and for the platform we compute

$$\boldsymbol{\Gamma}_{p_1} = \begin{pmatrix} -1 & 0 & -1.08\sin(0.47 - q_9) \\ 0 & -1 & -1.08\cos(0.47 - q_9) \end{pmatrix}, \tag{5.125}$$

$$\boldsymbol{\Gamma}_{p_2} = \begin{pmatrix} -1 & 0 & 1.08\sin(1.62 + q_9) \\ 0 & -1 & -1.08\cos(1.62 + q_9) \end{pmatrix}, \tag{5.126}$$

$$\boldsymbol{\Gamma}_{p_3} = \begin{pmatrix} -1 & 0 & 1.08\sin(3.72 + q_9) \\ 0 & -1 & -1.08\cos(3.72 + q_9) \end{pmatrix}, \tag{5.127}$$

$$\text{for } i = 1, \ldots, 3.$$

For the six elbow link centers of mass (one proximal and one distal link for each chain) we compute

$$\boldsymbol{\Gamma}_{G_{l_{i1}}} = \begin{pmatrix} -\frac{l_{i1}}{2}\sin(q_{2i-1}) & 0 \\ \frac{l_{i1}}{2}\cos(q_{2i-1}) & 0 \end{pmatrix}, \tag{5.128}$$

$$\text{for } i = 1, \ldots, 3,$$

and

$$\boldsymbol{\Gamma}_{G_{l_{i2}}} = \begin{pmatrix} -l_{i1}\sin(q_{2i-1}) - \frac{l_{i2}}{2}\sin(q_{2i-1}+q_{2i}) & -\frac{l_{i2}}{2}\sin(q_{2i-1}+q_{2i}) \\ l_{i1}\cos(q_{2i-1}) + \frac{l_{i2}}{2}\cos(q_{2i-1}+q_{2i}) & \frac{l_{i2}}{2}\cos(q_{2i-1}+q_{2i}) \end{pmatrix}, \tag{5.129}$$

$$\text{for } i = 1, \ldots, 3,$$

The terms, $\boldsymbol{\Gamma}_{G_{l_{ij}}}$ and l_{ij}, are the center of mass Jacobian and link length, respectively, of the jth link of the ith elbow chain subsystem. The unconstrained equations of motion for the three elbow chain subsystems are

$$\begin{pmatrix} 0 \\ \tau_i \end{pmatrix} = \boldsymbol{M}_{l_i} \begin{pmatrix} \ddot{q}_{2i-1} \\ \ddot{q}_{2i} \end{pmatrix} + \boldsymbol{b}_{l_i}(q_{2i-1}, \ldots, \dot{q}_{2i}) + \boldsymbol{g}_{l_1}(q_{2i-1}, q_{2i}), \quad \text{for } i = 1, \ldots, 3, \tag{5.130}$$

where, taking the rotational inertia of the individual links to be *zero*, we have

$$\boldsymbol{M}_{l_i} = \sum_{j=1}^{2} M_{l_{ij}} \boldsymbol{\Gamma}^T_{G_{l_{ij}}} \boldsymbol{\Gamma}_{G_{l_{ij}}}, \tag{5.131}$$

$$\boldsymbol{b}_{l_i} = \left(\sum_{j=1}^{2} M_{l_{ij}} \boldsymbol{\Gamma}^T_{G_{l_{ij}}} \dot{\boldsymbol{\Gamma}}_{G_{l_{ij}}} \right) \dot{\boldsymbol{q}}, \tag{5.132}$$

$$\boldsymbol{g}_{l_i} = g \sum_{j=1}^{2} M_{l_{ij}} \boldsymbol{\Gamma}^T_{G_{l_{ij}}} \hat{\boldsymbol{e}}_2, \quad \text{for } i = 1, \ldots, 3. \tag{5.133}$$

The term, $M_{l_{ij}}$, is the mass of the jth link of the ith elbow chain subsystem. The terms are computed as

$$\boldsymbol{M}_{l_i} = \begin{pmatrix} \frac{1}{4}[l_{i2}^2 M_{l_{i2}} + l_{i1}^2(M_{l_{i1}} + 4M_{l_{i2}}) + 4 l_{i1} l_{i2} M_{l_{i2}} \cos(q_{2i})] & \cdots \frac{1}{4} l_{i2} M_{l_{i2}} [l_{i2} + 2 l_{i1} \cos(q_{2i})] \\ \frac{1}{4} l_{i2} M_{l_{i2}} [l_{i2} + 2 l_{i1} \cos(q_{2i})] & \frac{1}{4} l_{i2}^2 M_{l_{i2}} \end{pmatrix}, \tag{5.134}$$

$$\boldsymbol{b}_{l_i} = \begin{pmatrix} -\frac{1}{2} l_{i1} l_{i2} M_{l_{i2}} \sin(q_{2i}) \dot{q}_{2i} (2\dot{q}_{2i-1} + \dot{q}_{2i}) \\ \frac{1}{2} l_{i1} l_{i2} M_{l_{i2}} \sin(q_{2i}) \dot{q}_{2i-1}^2 \end{pmatrix}, \tag{5.135}$$

$$\boldsymbol{g}_{l_i} = \begin{pmatrix} \frac{1}{2} g [l_{i1}(M_{l_{i1}} + 2M_{l_{i2}}) \cos(q_{2i-1}) + l_{i2} M_{l_{i2}} \cos(q_{2i-1} + q_{2i})] \\ \frac{1}{2} g l_{i2} M_{l_{i2}} \cos(q_{2i-1} + q_{2i}) \end{pmatrix}, \quad \text{for } i = 1, \ldots, 3. \tag{5.136}$$

For the platform subsystem we have

$$\boldsymbol{0} = \boldsymbol{M}_p \begin{pmatrix} \ddot{q}_7 \\ \ddot{q}_8 \\ \ddot{q}_9 \end{pmatrix} + \boldsymbol{b}_p(q_7, \ldots, \dot{q}_9) + \boldsymbol{g}_p(q_7, q_8, q_9), \tag{5.137}$$

where

$$\boldsymbol{M}_p = M_p \boldsymbol{\Gamma}^T_{G_p} \boldsymbol{\Gamma}_{G_p} + I_p^{G_p} \boldsymbol{\Pi}_p^T \boldsymbol{\Pi}_p, \tag{5.138}$$

$$\boldsymbol{b}_p = \left(M_p \boldsymbol{\Gamma}^T_{G_p} \dot{\boldsymbol{\Gamma}}_{G_p} + I_p^{G_p} \boldsymbol{\Pi}_p^T \dot{\boldsymbol{\Pi}}_p \right) \dot{\boldsymbol{q}}, \tag{5.139}$$

$$\boldsymbol{g}_p = g M_p \boldsymbol{\Gamma}^T_{G_p} \hat{\boldsymbol{e}}_2. \tag{5.140}$$

The Jacobians are simply $\boldsymbol{\Gamma}_{G_p} = \mathbf{1}_{2\times 2}$ and $\boldsymbol{\Pi}_p = 1$. The dynamical terms are then computed as

$$\boldsymbol{M}_p = \begin{pmatrix} M_p & 0 & 0 \\ 0 & M_p & 0 \\ 0 & 0 & I_p^{G_p} \end{pmatrix}, \tag{5.141}$$

$$\boldsymbol{b}_p = \mathbf{0}, \tag{5.142}$$

$$\boldsymbol{g}_p = \begin{pmatrix} 0 \\ M_p g \\ 0 \end{pmatrix}. \tag{5.143}$$

The entire unconstrained system is composed as

$$\begin{pmatrix} 0 \\ \tau_1 \\ 0 \\ \tau_2 \\ 0 \\ \tau_3 \\ \mathbf{0} \end{pmatrix} = \begin{pmatrix} \boldsymbol{M}_{l_1} & \mathbf{0} & \mathbf{0} & \mathbf{0} \\ \mathbf{0} & \boldsymbol{M}_{l_2} & \mathbf{0} & \vdots \\ \mathbf{0} & \mathbf{0} & \boldsymbol{M}_{l_3} & \mathbf{0} \\ \mathbf{0} & \mathbf{0} & \mathbf{0} & \boldsymbol{M}_p \end{pmatrix} \begin{pmatrix} \ddot{q}_1 \\ \ddot{q}_2 \\ \vdots \\ \ddot{q}_9 \end{pmatrix} + \begin{pmatrix} \boldsymbol{b}_{l_1} \\ \boldsymbol{b}_{l_2} \\ \boldsymbol{b}_{l_3} \\ \boldsymbol{b}_p \end{pmatrix} + \begin{pmatrix} \boldsymbol{g}_{l_1} \\ \boldsymbol{g}_{l_2} \\ \boldsymbol{g}_{l_3} \\ \boldsymbol{g}_p \end{pmatrix}, \tag{5.144}$$

and the constrained system is

$$\begin{aligned} \begin{pmatrix} 0 \\ \tau_1 \\ 0 \\ \tau_2 \\ 0 \\ \tau_3 \\ \mathbf{0} \end{pmatrix} &= \begin{pmatrix} \boldsymbol{M}_{l_1} & \mathbf{0} & \mathbf{0} & \mathbf{0} \\ \mathbf{0} & \boldsymbol{M}_{l_2} & \mathbf{0} & \vdots \\ \mathbf{0} & \mathbf{0} & \boldsymbol{M}_{l_3} & \mathbf{0} \\ \mathbf{0} & \mathbf{0} & \mathbf{0} & \boldsymbol{M}_p \end{pmatrix} \begin{pmatrix} \ddot{q}_1 \\ \ddot{q}_2 \\ \vdots \\ \ddot{q}_9 \end{pmatrix} + \begin{pmatrix} \boldsymbol{b}_{l_1} \\ \boldsymbol{b}_{l_2} \\ \boldsymbol{b}_{l_3} \\ \boldsymbol{b}_p \end{pmatrix} + \begin{pmatrix} \boldsymbol{g}_{l_1} \\ \boldsymbol{g}_{l_2} \\ \boldsymbol{g}_{l_3} \\ \boldsymbol{g}_p \end{pmatrix} \\ &+ \begin{pmatrix} \boldsymbol{\Gamma}_{l_1}^T & \mathbf{0} & \mathbf{0} \\ \mathbf{0} & \boldsymbol{\Gamma}_{l_2}^T & \mathbf{0} \\ \mathbf{0} & \mathbf{0} & \boldsymbol{\Gamma}_{l_3}^T \\ -\boldsymbol{\Gamma}_{p_1}^T & -\boldsymbol{\Gamma}_{p_2}^T & -\boldsymbol{\Gamma}_{p_3}^T \end{pmatrix} \begin{pmatrix} \lambda_1 \\ \vdots \\ \lambda_6 \end{pmatrix}, \end{aligned} \tag{5.145}$$

where the dimensions of the terms are

$$\boldsymbol{M} \in \mathbb{R}^{n\times n}, \qquad \boldsymbol{b}, \boldsymbol{g}, \boldsymbol{\tau} \in \mathbb{R}^n, \qquad \boldsymbol{\Phi} \in \mathbb{R}^{m\times n}, \qquad \boldsymbol{\lambda} \in \mathbb{R}^m, \tag{5.146}$$

and $n = 9$ and $m = 6$. The constrained system has $p = n - m = 3$ degrees of freedom.

Results from simulation of the system under gravity are shown in Figures 5.9 and 5.10. All geometric and inertial constants were chosen to be 1:

$$\begin{gathered} l_{ij} = 1, \ M_{ij} = 1, \ M_p = 1, \ I_p^{G_p} = 1, \\ \text{for } i = 1, \ldots, 3, \quad \text{and, } j = 1, 2. \end{gathered} \tag{5.147}$$

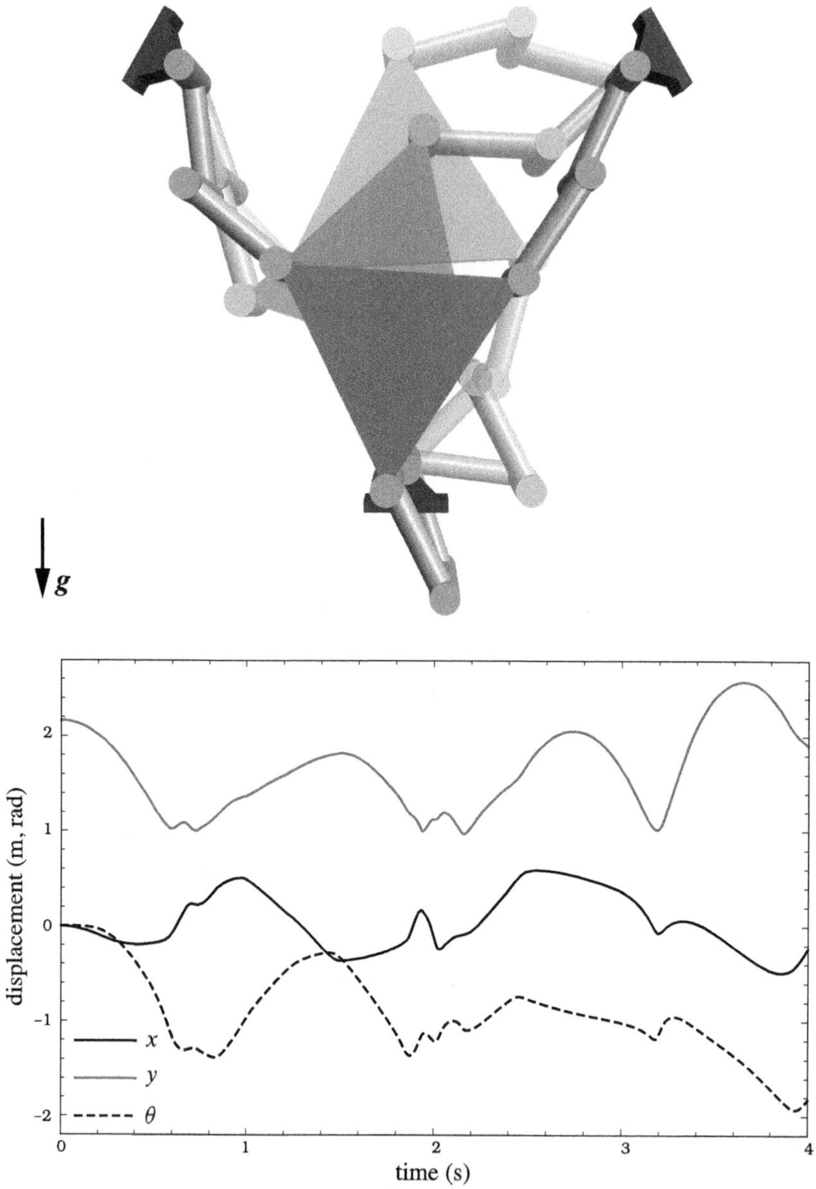

Figure 5.9 (Top) Animation frames from the simulation of the parallel mechanism falling under gravity. (Bottom) Time history of the motion of the platform (x, y, θ).

The initial conditions used were

$$\boldsymbol{q}_o = \begin{pmatrix} \pi/4 & \pi/6 & 11\pi/12 & \pi/6 & -5\pi/12 & \pi/6 & 0 & 2.165 & 0 \end{pmatrix}^T \tag{5.148}$$

$$\dot{\boldsymbol{q}}_o = \begin{pmatrix} 0 & 0 & 0 & 0 & 0 & 0 & 0 & 0 & 0 \end{pmatrix}^T. \tag{5.149}$$

It can be verified that these satisfy the constraints.

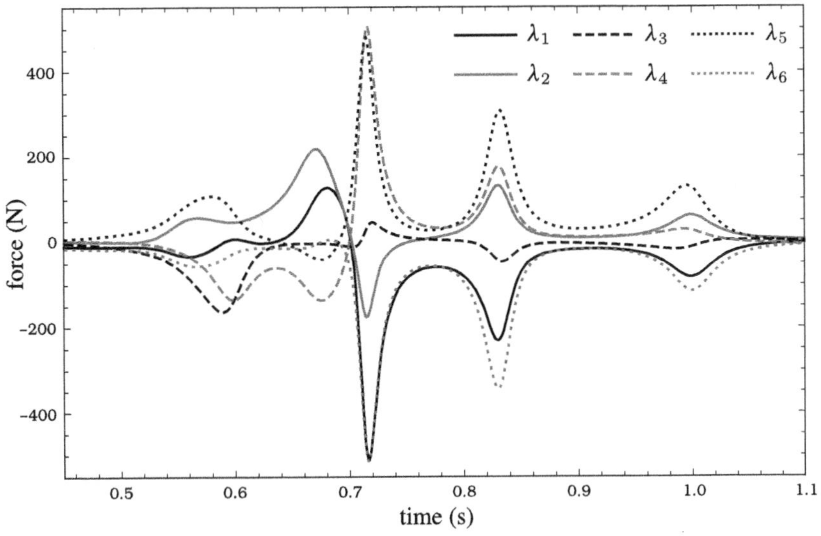

Figure 5.10 Lagrange multipliers (constraint forces) for the parallel mechanism falling under gravity.

5.2.6 Generalized Constrained Equation of Motion

Given the explicit solution of the constrained dynamics problem from Section 5.2.5, we wish to express an alternate form of the constrained dynamical equations of motion. We begin by recalling (5.93):

$$\boldsymbol{\lambda} = -\bar{\boldsymbol{\Phi}}^T(\boldsymbol{\tau} - \boldsymbol{b} - \boldsymbol{g}) - \mathbf{H}\dot{\boldsymbol{\Phi}}\dot{\boldsymbol{q}}. \tag{5.150}$$

Substituting (5.150) into (5.86) yields

$$\boldsymbol{M}\ddot{\boldsymbol{q}} + \boldsymbol{b} + \boldsymbol{g} = -\boldsymbol{\Phi}^T\mathbf{H}\dot{\boldsymbol{\Phi}}\dot{\boldsymbol{q}} + (\mathbf{1} - \boldsymbol{\Phi}^T\bar{\boldsymbol{\Phi}}^T)\boldsymbol{\tau} + \boldsymbol{\Phi}^T\bar{\boldsymbol{\Phi}}^T(\boldsymbol{b} + \boldsymbol{g}). \tag{5.151}$$

We now define the $m_C \times 1$ vector of centrifugal and Coriolis forces projected at the constraint,

$$\boldsymbol{\alpha} \triangleq \bar{\boldsymbol{\Phi}}^T\boldsymbol{b} - \mathbf{H}\dot{\boldsymbol{\Phi}}\dot{\boldsymbol{q}}, \tag{5.152}$$

and the $m_C \times 1$ vector of gravity forces projected at the constraint,

$$\boldsymbol{\rho} \triangleq \bar{\boldsymbol{\Phi}}^T\boldsymbol{g}. \tag{5.153}$$

Thus, we have the concise expression, which we will refer to as the *generalized constrained equation of motion* (De Sapio, Khatib, and Delp 2006):

$$\boldsymbol{\Theta}^T\boldsymbol{\tau} = \boldsymbol{M}\ddot{\boldsymbol{q}} + \boldsymbol{b} + \boldsymbol{g} - \boldsymbol{\Phi}^T(\boldsymbol{\alpha} + \boldsymbol{\rho}). \tag{5.154}$$

An alternative means of deriving this equation involves directly mapping the configuration space equation (5.86) into the constraint null space using $\boldsymbol{\Theta}^T$,

$$\boldsymbol{\Theta}^T\boldsymbol{\tau} = \boldsymbol{\Theta}^T\boldsymbol{M}\ddot{\boldsymbol{q}} + \boldsymbol{\Theta}^T\boldsymbol{b} + \boldsymbol{\Theta}^T\boldsymbol{g} - \boldsymbol{\Theta}^T\boldsymbol{\Phi}^T\boldsymbol{\lambda}. \tag{5.155}$$

Noting that $\boldsymbol{\Theta}^T \boldsymbol{\Phi}^T = \mathbf{0}$ and manipulating, we have

$$\begin{aligned} \boldsymbol{\Theta}^T \boldsymbol{\tau} &= \boldsymbol{M}\ddot{\boldsymbol{q}} + \boldsymbol{b} + \boldsymbol{g} - \boldsymbol{\Phi}^T \bar{\boldsymbol{\Phi}}^T \boldsymbol{M}\ddot{\boldsymbol{q}} - \boldsymbol{\Phi}^T \bar{\boldsymbol{\Phi}}^T \boldsymbol{b} - \boldsymbol{\Phi}^T \bar{\boldsymbol{\Phi}}^T \boldsymbol{g} \\ &= \boldsymbol{M}\ddot{\boldsymbol{q}} + \boldsymbol{b} + \boldsymbol{g} - \boldsymbol{\Phi}^T \mathbf{H} \boldsymbol{\Phi}\ddot{\boldsymbol{q}} - \boldsymbol{\Phi}^T (\boldsymbol{p} + \boldsymbol{\alpha}) - \boldsymbol{\Phi}^T \mathbf{H} \dot{\boldsymbol{\Phi}}\dot{\boldsymbol{q}}. \end{aligned} \tag{5.156}$$

Substituting in our constraint condition, $\dot{\boldsymbol{\Phi}}\dot{\boldsymbol{q}} = -\boldsymbol{\Phi}\ddot{\boldsymbol{q}}$, yields

$$\boldsymbol{\Theta}^T \boldsymbol{\tau} = \boldsymbol{M}\ddot{\boldsymbol{q}} + \boldsymbol{b} + \boldsymbol{g} - \boldsymbol{\Phi}^T (\boldsymbol{p} + \boldsymbol{\alpha}). \tag{5.157}$$

5.3 Hamilton's Principle of Least Action

An alternate way of arriving at the the equations of motion is through a least action principle. The fundamental theme associated with least action principles is that the evolution of a dynamical system can be revealed by examining the stationary condition for an appropriately defined action integral.

While least action can refer to a general family of variational principles, perhaps the most significant among these is Hamilton's *Principle of Least Action*. This principle states that the path, $\boldsymbol{q}(t)$, of a system in configuration space over an interval, $[t_o, t_f]$, is such that the action is *stationary* under all path variations that vanish at the endpoints, $\boldsymbol{q}(t_o)$ and $\boldsymbol{q}(t_f)$. It is noted that this does not strictly imply a minimization of the action, as the name of the principle suggests, but rather an extremization of the action.

PRINCIPLE 5.2 *For scleronomic systems (no explicit time dependence) the path of a system in configuration space during an interval, $[t_o, t_f]$, is such that the action*

$$I = \int_{t_o}^{t_f} \mathcal{L}(\boldsymbol{q}, \dot{\boldsymbol{q}}) dt \tag{5.158}$$

is stationary under all path variations. The scalar term, $\mathcal{L}$, is the Lagrangian, defined as

$$\mathcal{L} \triangleq T - V, \tag{5.159}$$

where T and V are the system kinetic and potential energies, respectively. Furthermore

$$\begin{gathered} \delta I = 0 \\ \forall \delta | \delta \boldsymbol{q}(t_o) = \delta \boldsymbol{q}(t_f) = \mathbf{0}. \end{gathered} \tag{5.160}$$

*This is known as **Hamilton's Principle of Least Action**.*

For forced systems the Principle of Least Action is modified such that the variation in the action is given by

$$\delta I = \delta \int_{t_o}^{t_f} \mathcal{L}\, dt + \int_{t_o}^{t_f} \boldsymbol{\tau} \cdot \delta \boldsymbol{q}\, dt. \tag{5.161}$$

5.3.1 Euler-Lagrange Equations

It is straightforward to apply calculus of variations to this problem (Goldstein, Poole, and Safko 2002). We can express the first term in (5.161) as

$$\delta \int_{t_o}^{t_f} \mathcal{L}\, dt = \int_{t_o}^{t_f} \left(\frac{\partial \mathcal{L}}{\partial \boldsymbol{q}} \cdot \delta \boldsymbol{q} + \frac{\partial \mathcal{L}}{\partial \dot{\boldsymbol{q}}} \cdot \delta \dot{\boldsymbol{q}} \right) dt = \int_{t_o}^{t_f} \frac{\partial \mathcal{L}}{\partial \boldsymbol{q}} \cdot \delta \boldsymbol{q}\, dt + \int_{t_o}^{t_f} \frac{\partial \mathcal{L}}{\partial \dot{\boldsymbol{q}}} \cdot \frac{d(\delta \boldsymbol{q})}{dt}\, dt. \tag{5.162}$$

Noting that

$$\frac{\partial \mathcal{L}}{\partial \dot{\boldsymbol{q}}} \cdot \frac{d(\delta \boldsymbol{q})}{dt} = \frac{d}{dt} \left(\frac{\partial \mathcal{L}}{\partial \dot{\boldsymbol{q}}} \cdot \delta \boldsymbol{q} \right) - \frac{d}{dt} \frac{\partial \mathcal{L}}{\partial \dot{\boldsymbol{q}}} \cdot \delta \boldsymbol{q}, \tag{5.163}$$

we have

$$\int_{t_o}^{t_f} \left(\frac{\partial \mathcal{L}}{\partial \boldsymbol{q}} \cdot \delta \boldsymbol{q} + \frac{\partial \mathcal{L}}{\partial \dot{\boldsymbol{q}}} \cdot \frac{d(\delta \boldsymbol{q})}{dt} \right) dt = \int_{t_o}^{t_f} \frac{d}{dt} \left(\frac{\partial \mathcal{L}}{\partial \dot{\boldsymbol{q}}} \cdot \delta \boldsymbol{q} \right) dt + \int_{t_o}^{t_f} \left(\frac{\partial \mathcal{L}}{\partial \boldsymbol{q}} \cdot \delta \boldsymbol{q} - \frac{d}{dt} \frac{\partial \mathcal{L}}{\partial \dot{\boldsymbol{q}}} \cdot \delta \boldsymbol{q} \right) dt. \tag{5.164}$$

Since the variations, $\delta \boldsymbol{q}$, vanish at the endpoints,

$$\int_{t_o}^{t_f} \frac{d}{dt} \left(\frac{\partial \mathcal{L}}{\partial \dot{\boldsymbol{q}}} \cdot \delta \boldsymbol{q} \right) dt = \left. \frac{\partial \mathcal{L}}{\partial \dot{\boldsymbol{q}}} \cdot \delta \boldsymbol{q} \right|_{t_o}^{t_f} = 0. \tag{5.165}$$

Thus,

$$\delta \int_{t_o}^{t_f} \mathcal{L}\, dt = \int_{t_o}^{t_f} \left(\frac{\partial \mathcal{L}}{\partial \boldsymbol{q}} - \frac{d}{dt} \frac{\partial \mathcal{L}}{\partial \dot{\boldsymbol{q}}} \right) \cdot \delta \boldsymbol{q}\, dt \tag{5.166}$$

and

$$\delta I = \int_{t_o}^{t_f} \left(\frac{\partial \mathcal{L}}{\partial \boldsymbol{q}} - \frac{d}{dt} \frac{\partial \mathcal{L}}{\partial \dot{\boldsymbol{q}}} + \boldsymbol{\tau} \right) \cdot \delta \boldsymbol{q}\, dt. \tag{5.167}$$

The condition

$$\begin{gathered} \delta I = 0 \\ \forall \delta \boldsymbol{q}, \end{gathered} \tag{5.168}$$

implies the following *Euler-Lagrange* equations:

$$\boldsymbol{\tau} = \frac{d}{dt} \frac{\partial \mathcal{L}}{\partial \dot{\boldsymbol{q}}} - \frac{\partial \mathcal{L}}{\partial \boldsymbol{q}}. \tag{5.169}$$

We can express the Euler-Lagrange equations as

$$\boldsymbol{\tau} = \frac{d}{dt}\frac{\partial T}{\partial \dot{\boldsymbol{q}}} - \frac{\partial T}{\partial \boldsymbol{q}} + \frac{\partial V}{\partial \boldsymbol{q}}. \tag{5.170}$$

5.3.2 A Single Particle

For a single point mass with a discrete set of n_f external forces, $\{\boldsymbol{f}_1, \ldots, \boldsymbol{f}_{n_f}\}$, acting on it, the Euler-Lagrange equations are

$$\sum_{i=1}^{n_f} \boldsymbol{f}_i = \frac{d}{dt}\frac{\partial T}{\partial \dot{\boldsymbol{r}}} - \frac{\partial T}{\partial \boldsymbol{r}} + \frac{\partial V}{\partial \boldsymbol{r}}. \tag{5.171}$$

The kinetic energy, T, is given by

$$T = \frac{1}{2} M \boldsymbol{v}^T \boldsymbol{v} = \frac{1}{2} M \dot{\boldsymbol{r}}^T \dot{\boldsymbol{r}}. \tag{5.172}$$

So,

$$\frac{\partial T}{\partial \dot{\boldsymbol{r}}} = M\dot{\boldsymbol{r}} \tag{5.173}$$

and

$$\frac{d}{dt}\frac{\partial T}{\partial \dot{\boldsymbol{r}}} = M\ddot{\boldsymbol{r}} = M\boldsymbol{a}. \tag{5.174}$$

We note that

$$\frac{\partial T}{\partial \boldsymbol{r}} = 0. \tag{5.175}$$

The potential energy, V, is given by

$$V = M\langle \boldsymbol{r}, g\hat{\boldsymbol{e}}_3 \rangle = Mg\boldsymbol{r}^T \hat{\boldsymbol{e}}_3. \tag{5.176}$$

So,

$$\frac{\partial V}{\partial \boldsymbol{r}} = Mg\hat{\boldsymbol{e}}_3. \tag{5.177}$$

Thus,

$$\sum_{i=1}^{n_f} \boldsymbol{f}_i = M\boldsymbol{a} + Mg\hat{\boldsymbol{e}}_3. \tag{5.178}$$

5.3.3 A Single Rigid Body

For a single rigid body with a discrete set of n_f external forces, $\{\boldsymbol{f}_1, \ldots, \boldsymbol{f}_{n_f}\}$, and n_φ external moments, $\{\boldsymbol{\varphi}_1, \ldots, \boldsymbol{\varphi}_{n_\varphi}\}$, acting on it, the Euler-Lagrange equations are

$$\begin{pmatrix} \sum_{i=1}^{n_f} \boldsymbol{f}_i \\ \sum_{i=1}^{n_f} \boldsymbol{d}_{\overrightarrow{GP_i}} \times \boldsymbol{f}_i + \sum_{i=1}^{n_\varphi} \boldsymbol{\varphi}_j \end{pmatrix} = \frac{d}{dt}\frac{\partial T}{\partial \dot{\mathbf{x}}} - \frac{\partial T}{\partial \mathbf{x}} + \frac{\partial V}{\partial \mathbf{x}}, \tag{5.179}$$

where

$$\mathbf{x} = \begin{pmatrix} \boldsymbol{r}_G \\ \boldsymbol{\theta} \end{pmatrix} \quad \text{and} \quad \dot{\mathbf{x}} = \begin{pmatrix} \boldsymbol{v}_G \\ \boldsymbol{\omega} \end{pmatrix}. \tag{5.180}$$

The kinetic energy, T, is given by

$$T = \frac{1}{2}(M\boldsymbol{v}_G^T\boldsymbol{v}_G + \boldsymbol{\omega}^T\boldsymbol{I}^G\boldsymbol{\omega}). \tag{5.181}$$

So,

$$\frac{\partial T}{\partial \dot{\mathbf{x}}} = \begin{pmatrix} M\boldsymbol{v}_G \\ \boldsymbol{I}^G\boldsymbol{\omega} \end{pmatrix} \tag{5.182}$$

and

$$\frac{d}{dt}\frac{\partial T}{\partial \dot{\mathbf{x}}} = \begin{pmatrix} M\boldsymbol{a}_G \\ \boldsymbol{I}^G\boldsymbol{\alpha} + \boldsymbol{\omega} \times \boldsymbol{I}^G\boldsymbol{\omega} \end{pmatrix}. \tag{5.183}$$

The potential energy, V, is given by

$$V = M\langle \boldsymbol{r}, g\hat{\boldsymbol{e}}_3 \rangle = Mg\boldsymbol{r}^T\hat{\boldsymbol{e}}_3. \tag{5.184}$$

So,

$$\frac{\partial V}{\partial \mathbf{x}} = \begin{pmatrix} Mg\hat{\boldsymbol{e}}_3 \\ \mathbf{0} \end{pmatrix}. \tag{5.185}$$

Thus,

$$\begin{pmatrix} \sum_{i=1}^{n_f} \boldsymbol{f}_i \\ \sum_{i=1}^{n_f} \boldsymbol{d}_{\overrightarrow{GP_i}} \times \boldsymbol{f}_i + \sum_{i=1}^{n_\varphi} \boldsymbol{\varphi}_j \end{pmatrix} = \begin{pmatrix} M\boldsymbol{a}_G + Mg\hat{\boldsymbol{e}}_3 \\ \boldsymbol{I}^G\boldsymbol{\alpha} + \boldsymbol{\omega} \times \boldsymbol{I}^G\boldsymbol{\omega} \end{pmatrix}. \tag{5.186}$$

The Kinetic Energy Ellipsoid

Rotational kinetic energy is given by

$$T = \frac{1}{2}\boldsymbol{\omega}^T\boldsymbol{I}^G\boldsymbol{\omega}. \tag{5.187}$$

We will assume this expression is represented in the base frame for convenience. Since $\boldsymbol{I}^G$ is symmetric positive definite, it has positive eigenvalues and an orthogonal eigenbasis (principal axes), $\mathcal{E}$. Therefore,

$$T = \frac{1}{2} {}^{\mathcal{O}}\boldsymbol{\omega}^{T\,\mathcal{O}}\boldsymbol{I}^{G\mathcal{O}}\boldsymbol{\omega} = \frac{1}{2} {}^{\mathcal{E}}\boldsymbol{\omega}^{T\,\mathcal{O}}_{\ \mathcal{E}}\boldsymbol{Q}^{T\,\mathcal{O}}\boldsymbol{I}^{G\mathcal{O}}_{\ \mathcal{E}}\boldsymbol{Q}^{\mathcal{E}}\boldsymbol{\omega}, \tag{5.188}$$

where the columns of ${}^{\mathcal{O}}_{\mathcal{E}}\boldsymbol{Q}$ are the eigenvectors, $\{\boldsymbol{v}_1, \boldsymbol{v}_2, \boldsymbol{v}_3\}$, of ${}^{\mathcal{O}}\boldsymbol{I}^G$,

$${}^{\mathcal{O}}_{\mathcal{E}}\boldsymbol{Q} = \begin{pmatrix} \uparrow & \uparrow & \uparrow \\ \boldsymbol{v}_1 & \boldsymbol{v}_2 & \boldsymbol{v}_3 \\ \downarrow & \downarrow & \downarrow \end{pmatrix}. \tag{5.189}$$

We then have

$${}^{\mathcal{O}}_{\mathcal{E}}\boldsymbol{Q}^{T\,\mathcal{O}}\boldsymbol{I}^{G\mathcal{O}}_{\ \mathcal{E}}\boldsymbol{Q} = {}^{\mathcal{E}}\boldsymbol{I}^G, \tag{5.190}$$

where ${}^{\mathcal{E}}\boldsymbol{I}^G$ is a diagonal matrix of eigenvalues, $\{\lambda_1, \lambda_2, \lambda_3\}$, of ${}^{\mathcal{O}}\boldsymbol{I}^G$:

$${}^{\mathcal{E}}\boldsymbol{I}^G = \begin{pmatrix} \lambda_1 & 0 & 0 \\ 0 & \lambda_2 & 0 \\ 0 & 0 & \lambda_3 \end{pmatrix}. \tag{5.191}$$

In terms of the eigenbasis,

$$T = \frac{1}{2} {}^{\mathcal{E}}\boldsymbol{\omega}^{T\,\mathcal{E}}\boldsymbol{I}^{G\mathcal{E}}\boldsymbol{\omega}. \tag{5.192}$$

For constant values of the kinetic energy T, this represents an ellipsoid expressed with respect to the principal axes. In scalar form,

$$T = \frac{1}{2}\left(\lambda_1 {}^{\mathcal{E}}\omega_1^2 + \lambda_2 {}^{\mathcal{E}}\omega_2^2 + \lambda_3 {}^{\mathcal{E}}\omega_3^2\right). \tag{5.193}$$

The semiaxes of the ellipsoid in frame $\mathcal{E}$ are then

$$a_i = \sqrt{2T/\lambda_i}. \tag{5.194}$$

Example: We consider the cone addressed in Section 4.1.2. The inertia tensor about the center of mass is

$$\boldsymbol{I}^G = \begin{pmatrix} \frac{3}{80}M(h^2 + 4R^2) & 0 & 0 \\ 0 & \frac{3}{80}M(h^2 + 4R^2) & 0 \\ 0 & 0 & \frac{3}{10}R^2 \end{pmatrix}. \tag{5.195}$$

Specifying $M = 1$, $h = 1$, and $R = .75$, the eigenvalues are $\lambda_1 = 0.684$, $\lambda_2 = 0.684$, and $\lambda_3 = 0.169$. The ellipsoid associated with a constant kinetic energy, $T = 1$, has semiaxes $a_1 = 1.71$, $a_2 = 1.71$, and $a_3 = 3.443$, aligned with the cone axes depicted in Figure 4.3.

5.3.4 A System of Particles

For a system of particles with generalized forces, $\boldsymbol{\tau}$, acting on the system the Euler-Lagrange equations are

$$\boldsymbol{\tau} = \frac{d}{dt}\frac{\partial T}{\partial \dot{\boldsymbol{q}}} - \frac{\partial T}{\partial \boldsymbol{q}} + \frac{\partial V}{\partial \boldsymbol{q}}. \tag{5.196}$$

The kinetic energy, T, is given by

$$T = \frac{1}{2}\sum_{i=1}^{n_p} M_i \boldsymbol{v}_i^T \boldsymbol{v}_i = \frac{1}{2}\dot{\boldsymbol{q}}^T \left(\sum_{i=1}^{n_p} M_i \boldsymbol{\Gamma}_i^T \boldsymbol{\Gamma}_i\right)\dot{\boldsymbol{q}}. \tag{5.197}$$

We note that

$$\frac{\partial T}{\partial \dot{\boldsymbol{q}}} = \left(\sum_{i=1}^{n_p} M_i \boldsymbol{\Gamma}_i^T \boldsymbol{\Gamma}_i\right)\dot{\boldsymbol{q}}. \tag{5.198}$$

So,

$$\frac{d}{dt}\frac{\partial T}{\partial \dot{\boldsymbol{q}}} = \left(\sum_{i=1}^{n_p} M_i \boldsymbol{\Gamma}_i^T \boldsymbol{\Gamma}_i\right)\ddot{\boldsymbol{q}}. \tag{5.199}$$

The potential energy, V, is given by

$$V = \sum_{i=1}^{n_p} M_i \langle \boldsymbol{r}_i, g\hat{\boldsymbol{e}}_3\rangle = g\sum_{i=1}^{n_p} M_i \boldsymbol{r}_i^T \hat{\boldsymbol{e}}_3. \tag{5.200}$$

So

$$\frac{\partial V}{\partial \boldsymbol{q}} = g\sum_{i=1}^{n_p} M_i \left(\frac{\partial \boldsymbol{r}_i}{\partial \boldsymbol{q}}\right)^T \hat{\boldsymbol{e}}_3 = g\sum_{i=1}^{n_p} M_i \boldsymbol{\Gamma}_i^T \hat{\boldsymbol{e}}_3 \tag{5.201}$$

and

$$\boldsymbol{\tau} = \sum_{i=1}^{n_p} \boldsymbol{\Gamma}_i^T (M_i \boldsymbol{a}_i + M_i g\hat{\boldsymbol{e}}_3). \tag{5.202}$$

Thus,

$$\boldsymbol{\tau} = \frac{d}{dt}\frac{\partial T}{\partial \dot{\boldsymbol{q}}} - \frac{\partial T}{\partial \boldsymbol{q}} + \frac{\partial V}{\partial \boldsymbol{q}} = \left(\sum_{i=1}^{n_p} M_i \boldsymbol{\Gamma}_i^T \boldsymbol{\Gamma}_i\right)\ddot{\boldsymbol{q}} + g\sum_{i=1}^{n_q} M_i \boldsymbol{\Gamma}_i^T \hat{\boldsymbol{e}}_3. \tag{5.203}$$

For a system of particles, we had previously defined

$$\boldsymbol{M}(\boldsymbol{q}) \triangleq \sum_{i=1}^{n_p} M_i \boldsymbol{\Gamma}_i^T \boldsymbol{\Gamma}_i \tag{5.204}$$

$$\boldsymbol{g}(\boldsymbol{q}) \triangleq g\sum_{i=1}^{n_p} M_i \boldsymbol{\Gamma}_i^T \hat{\boldsymbol{e}}_3. \tag{5.205}$$

So,

$$\boldsymbol{\tau} = \boldsymbol{M}(\boldsymbol{q})\ddot{\boldsymbol{q}} + \boldsymbol{g}(\boldsymbol{q}). \tag{5.206}$$

5.3.5 A System of Rigid Bodies

For a system of rigid bodies with generalized forces, $\boldsymbol{\tau}$, acting on the system, the Euler-Lagrange equations are

$$\boldsymbol{\tau} = \frac{d}{dt}\frac{\partial T}{\partial \dot{\boldsymbol{q}}} - \frac{\partial T}{\partial \boldsymbol{q}} + \frac{\partial V}{\partial \boldsymbol{q}}. \tag{5.207}$$

The kinetic energy, T, is given by

$$\begin{aligned} T &= \frac{1}{2}\sum_{i=1}^{n_b}\left(M_i {}^i\boldsymbol{v}_{G_i}^T {}^i\boldsymbol{v}_{G_i} + {}^i\boldsymbol{\omega}_i^T {}^i\boldsymbol{I}_i^{G_i} {}^i\boldsymbol{\omega}_i\right) \\ &= \frac{1}{2}\dot{\boldsymbol{q}}^T\left[\sum_{i=1}^{n_b}\left(M_i {}^i\boldsymbol{\Gamma}_{G_i}^T {}^i\boldsymbol{\Gamma}_{G_i} + {}^i\boldsymbol{\Pi}_i^T {}^i\boldsymbol{I}_i^{G_i} {}^i\boldsymbol{\Pi}_i\right)\right]\dot{\boldsymbol{q}}. \end{aligned} \tag{5.208}$$

We note that

$$\frac{\partial T}{\partial \dot{\boldsymbol{q}}} = \left[\sum_{i=1}^{n_b}\left(M_i {}^i\boldsymbol{\Gamma}_{G_i}^T {}^i\boldsymbol{\Gamma}_{G_i} + {}^i\boldsymbol{\Pi}_i^T {}^i\boldsymbol{I}_i^{G_i} {}^i\boldsymbol{\Pi}_i\right)\right]\dot{\boldsymbol{q}}. \tag{5.209}$$

For a system of rigid bodies, we had previously defined

$$\boldsymbol{M}(\boldsymbol{q}) \triangleq \sum_{i=1}^{n_b}\left(M_i {}^i\boldsymbol{\Gamma}_{G_i}^T {}^i\boldsymbol{\Gamma}_{G_i} + {}^i\boldsymbol{\Pi}_i^T {}^i\boldsymbol{I}_i^{G_i} {}^i\boldsymbol{\Pi}_i\right). \tag{5.210}$$

So,

$$\frac{d}{dt}\frac{\partial T}{\partial \dot{\boldsymbol{q}}} = \boldsymbol{M}\ddot{\boldsymbol{q}} + \dot{\boldsymbol{M}}\dot{\boldsymbol{q}} \tag{5.211}$$

and

$$\frac{\partial T}{\partial \boldsymbol{q}} = \frac{1}{2}\begin{pmatrix} \dot{\boldsymbol{q}}^T \frac{\partial \boldsymbol{M}}{\partial q_1}\dot{\boldsymbol{q}} \\ \vdots \\ \dot{\boldsymbol{q}}^T \frac{\partial \boldsymbol{M}}{\partial q_n}\dot{\boldsymbol{q}} \end{pmatrix}. \tag{5.212}$$

The potential energy, V, is given by

$$V = \sum_{i=1}^{n_b} M_i \langle {}^i\boldsymbol{r}_{G_i}, g{}^i\hat{\boldsymbol{e}}_{0_3}\rangle = g\sum_{i=1}^{n_b} M_i {}^i\boldsymbol{r}_{G_i}^T {}^i\hat{\boldsymbol{e}}_{0_3}. \tag{5.213}$$

So,

$$\frac{\partial V}{\partial \boldsymbol{q}} = g\sum_{i=1}^{n_b} M_i {}^i\boldsymbol{\Gamma}_{G_i}^T {}^i\hat{\boldsymbol{e}}_{0_3}. \tag{5.214}$$

For a system of rigid bodies, we had previously defined

$$g(q) \triangleq g\sum_{i=1}^{n_b} M_i {}^i\Gamma_{G_i}^{T}\, {}^i\hat{e}_{0_3}. \tag{5.215}$$

Thus,

$$\tau = \frac{d}{dt}\frac{\partial T}{\partial \dot{q}} - \frac{\partial T}{\partial q} + \frac{\partial V}{\partial q} = M\ddot{q} + \dot{M}\dot{q} - \frac{1}{2}\begin{pmatrix} \dot{q}^T \frac{\partial M}{\partial q_1}\dot{q} \\ \vdots \\ \dot{q}^T \frac{\partial M}{\partial q_n}\dot{q} \end{pmatrix} + g(q). \tag{5.216}$$

Defining $b(q, \dot{q})$ in an alternate, but consistent, manner as previously defined:

$$b(q, \dot{q}) \triangleq \dot{M}\dot{q} - \frac{1}{2}\begin{pmatrix} \dot{q}^T \frac{\partial M}{\partial q_1}\dot{q} \\ \vdots \\ \dot{q}^T \frac{\partial M}{\partial q_n}\dot{q} \end{pmatrix}: \tag{5.217}$$

we have

$$\tau = M(q)\ddot{q} + b(q, \dot{q}) + g(q). \tag{5.218}$$

Example: A gimballed gyroscope is depicted in Figure 5.11. The generalized coordinates of the 2-axis gimbal are q_1 and q_2, while the spin angle of the gyroscope is q_3. A

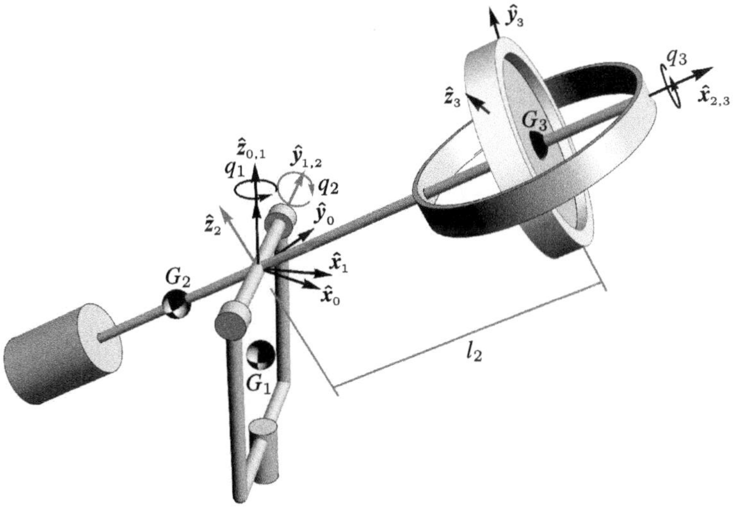

Figure 5.11 A gyroscope supported by a two-axis gimballed frame. The generalized coordinates, q_1 and q_2, parameterize the gimbal frame, and q_3 is the rotor angle. The rotor diameter is r.

zyx Euler sequence will be used to represent the orientation of the system. So,

$$ {}^0_1\boldsymbol{Q} = \boldsymbol{Q}_z(q_1), \tag{5.219} $$

$$ {}^0_2\boldsymbol{Q} = \boldsymbol{Q}_z(q_1)\boldsymbol{Q}_y(q_2), \tag{5.220} $$

$$ {}^0_3\boldsymbol{Q} = \boldsymbol{Q}_z(q_1)\boldsymbol{Q}_y(q_2)\boldsymbol{Q}_x(q_3). \tag{5.221} $$

The positions of the proximal ends and centers of mass of the links are given by

$$ {}^0\boldsymbol{r}_1 = \boldsymbol{0}, \tag{5.222} $$

$$ {}^0\boldsymbol{r}_{G_1} = {}^0\boldsymbol{r}_1 - l_{G_1}\hat{\boldsymbol{e}}_3, \tag{5.223} $$

$$ {}^0\boldsymbol{r}_2 = {}^0\boldsymbol{r}_1 + \boldsymbol{0}, \tag{5.224} $$

$$ {}^0\boldsymbol{r}_{G_2} = {}^0\boldsymbol{r}_2 - l_{G_2}{}^0_2\boldsymbol{Q}\hat{\boldsymbol{e}}_1, \tag{5.225} $$

$$ {}^0\boldsymbol{r}_3 = {}^0\boldsymbol{r}_3 + l_3{}^0_2\boldsymbol{Q}\hat{\boldsymbol{e}}_1, \tag{5.226} $$

$$ {}^0\boldsymbol{r}_{G_3} = {}^0\boldsymbol{r}_3. \tag{5.227} $$

We can compute the angular velocity in the local frame by noting that

$$ {}^1\boldsymbol{\Omega}_1 = {}^0_1\boldsymbol{Q}^T\,{}^0_1\dot{\boldsymbol{Q}}, \tag{5.228} $$

$$ {}^2\boldsymbol{\Omega}_2 = {}^0_2\boldsymbol{Q}^T\,{}^0_2\dot{\boldsymbol{Q}}, \tag{5.229} $$

$$ {}^3\boldsymbol{\Omega}_3 = {}^0_3\boldsymbol{Q}^T\,{}^0_3\dot{\boldsymbol{Q}}. \tag{5.230} $$

Carrying this operation out, we determine

$$ {}^1\boldsymbol{\omega}_1 = \begin{pmatrix} 0 \\ 0 \\ \dot{q}_1 \end{pmatrix} \tag{5.231} $$

and

$$ {}^2\boldsymbol{\omega}_2 = \begin{pmatrix} -\sin(q_2)\dot{q}_1 \\ \dot{q}_2 \\ \cos(q_2)\dot{q}_1 \end{pmatrix} \tag{5.232} $$

and

$$ {}^3\boldsymbol{\omega}_3 = \begin{pmatrix} \dot{q}_3 - \sin(q_2)\dot{q}_1 \\ \cos(q_2)\sin(q_3)\dot{q}_1 + \cos(q_3)\dot{q}_2 \\ \cos(q_2)\cos(q_3)\dot{q}_1 - \sin(q_3)\dot{q}_2 \end{pmatrix} \tag{5.233} $$

Alternately, we could have arrived at the same results by propagating the rotation rates forward, where

$$ {}^1\boldsymbol{\omega}_1 = \dot{q}_1\hat{\boldsymbol{e}}_3, \tag{5.234} $$

$$ {}^2\boldsymbol{\omega}_2 = \boldsymbol{Q}_y^T(q_2){}^1\boldsymbol{\omega}_1 + \dot{q}_2\hat{\boldsymbol{e}}_2, \tag{5.235} $$

$$ {}^3\boldsymbol{\omega}_3 = \boldsymbol{Q}_x^T(q_3){}^2\boldsymbol{\omega}_2 + \dot{q}_3\hat{\boldsymbol{e}}_1. \tag{5.236} $$

The translational velocity Jacobians are given by

$$ {}^0\boldsymbol{\Gamma}_{G_1} = \frac{\partial {}^0\boldsymbol{r}_{G_1}}{\partial \boldsymbol{q}} = \boldsymbol{0}, \tag{5.237} $$

$$ {}^0\boldsymbol{\Gamma}_{G_2} = \frac{\partial {}^0\boldsymbol{r}_{G_2}}{\partial \boldsymbol{q}} = \begin{pmatrix} l_{G_2}\sin(q_1)\cos(q_2) & l_{G_2}\cos(q_1)\sin(q_2) & 0 \\ -l_{G_2}\cos(q_1)\cos(q_2) & l_{G_2}\sin(q_1)\sin(q_2) & 0 \\ 0 & l_{G_2}\cos(q_2) & 0 \end{pmatrix}, \tag{5.238} $$

$$ {}^0\boldsymbol{\Gamma}_{G_3} = \frac{\partial {}^0\boldsymbol{r}_{G_3}}{\partial \boldsymbol{q}} = \begin{pmatrix} -l_3\sin(q_1)\cos(q_2) & -l_3\cos(q_1)\sin(q_2) & 0 \\ l_3\cos(q_1)\cos(q_2) & -l_3\sin(q_1)\sin(q_2) & 0 \\ 0 & -l_3\cos(q_2) & 0 \end{pmatrix}, \tag{5.239} $$

and the angular velocity Jacobians are given by

$$ {}^1\boldsymbol{\Pi}_1 = \frac{\partial {}^1\boldsymbol{\omega}_1}{\partial \dot{\boldsymbol{q}}} = \begin{pmatrix} 0 & 0 & 0 \\ 0 & 0 & 0 \\ 1 & 0 & 0 \end{pmatrix}, \tag{5.240} $$

$$ {}^2\boldsymbol{\Pi}_2 = \frac{\partial {}^2\boldsymbol{\omega}_2}{\partial \dot{\boldsymbol{q}}} = \begin{pmatrix} -\sin(q_2) & 0 & 0 \\ 0 & 1 & 0 \\ \cos(q_2) & 0 & 0 \end{pmatrix}, \tag{5.241} $$

$$ {}^3\boldsymbol{\Pi}_3 = \frac{\partial {}^3\boldsymbol{\omega}_3}{\partial \dot{\boldsymbol{q}}} = \begin{pmatrix} -\sin(q_2) & 0 & 1 \\ \cos(q_2)\sin(q_3) & \cos(q_3) & 0 \\ \cos(q_2)\cos(q_3) & -\sin(q_3) & 0 \end{pmatrix}. \tag{5.242} $$

Our dynamical terms are then

$$ \boldsymbol{M}(\boldsymbol{q}) = \sum_{i=1}^{3} \left(M_i {}^0\boldsymbol{\Gamma}_{G_i}^T {}^0\boldsymbol{\Gamma}_{G_i} + {}^i\boldsymbol{\Pi}_i^T {}^i\boldsymbol{I}_i^{Gi} {}^i\boldsymbol{\Pi}_i \right), \tag{5.243} $$

$$ \boldsymbol{b}(\boldsymbol{q}, \dot{\boldsymbol{q}}) = \dot{\boldsymbol{M}}\dot{\boldsymbol{q}} - \frac{1}{2} \begin{pmatrix} \dot{\boldsymbol{q}}^T \frac{\partial \boldsymbol{M}}{\partial q_1} \dot{\boldsymbol{q}} \\ \vdots \\ \dot{\boldsymbol{q}}^T \frac{\partial \boldsymbol{M}}{\partial q_3} \dot{\boldsymbol{q}} \end{pmatrix}, \tag{5.244} $$

$$ \boldsymbol{g}(\boldsymbol{q}) = g \sum_{i=1}^{3} M_i {}^0\boldsymbol{\Gamma}_{G_i}^T \hat{\boldsymbol{e}}_3. \tag{5.245} $$

The inertia tensors will be taken as

$$ {}^1\boldsymbol{I}_1^{G_1} = \begin{pmatrix} \frac{1}{3}M_1 & 0 & 0 \\ 0 & \frac{1}{3}M_1 & 0 \\ 0 & 0 & \frac{1}{2}M_1 \end{pmatrix}, \tag{5.246} $$

$$ {}^2\boldsymbol{I}_2^{G_2} = \begin{pmatrix} 0.125M_2 & 0 & 0 \\ 0 & 3.0625M_2 & 0 \\ 0 & 0 & 3.083M_2 \end{pmatrix}, \tag{5.247} $$

$$ {}^3\boldsymbol{I}_3^{G_3} = \begin{pmatrix} \frac{1}{2}M_3 r^2 & 0 & 0 \\ 0 & \frac{1}{12}h^2 M_3 + \frac{1}{4}M_3 r^2 & 0 \\ 0 & 0 & \frac{1}{12}h^2 M_3 + \frac{1}{4}M_3 r^2 \end{pmatrix}. \tag{5.248} $$

We then have

$$M(q) = \begin{pmatrix} M_{11} & 0 & -\frac{1}{2}M_3r^2\sin(q_2) \\ 0 & M_{22} & -\frac{1}{2}M_3r^2\sin(q_2) \\ -\frac{1}{2}M_3r^2\sin(q_2) & 0 & \frac{1}{2}M_3r^2 \end{pmatrix},$$

$$\begin{aligned} M_{11} &= 0.5M_1 + 1.604M_2 + 0.5l_{G_2}^2M_2 + 0.042h^2M_3 + 0.5l_3^2M_3 + 0.375M_3r^2 \\ &\quad + [(1.479 + 0.5l_{G_2}^2)M_2 + M_3(0.042h^2 + 0.5l_3^2 - 0.125r^2)]\cos(2q_2), \\ M_{22} &= (3.0625 + l_{G_2}^2)M_2 + M_3(0.083h^2 + l_3^2 + 0.25r^2), \end{aligned} \tag{5.249}$$

$$b(q, \dot{q}) = \begin{pmatrix} b_1 \\ b_2 \\ -\frac{1}{2}M_3r^2\cos(q_2)\dot{q}_1\dot{q}_2 \end{pmatrix},$$

$$\begin{aligned} b_1 &= -2\cos(q_2)\dot{q}_2([(2.958 + l_{G_2}^2)M_2 \\ &\quad + M_3(0.083h^2 + l_3^2 - 0.25r^2)]\sin(q_2)\dot{q}_1 + 0.25M_3r^2\dot{q}_3), \\ b_2 &= \cos(q_2)\dot{q}_1([(2.958 + l_{G_2}^2)M_2 \\ &\quad + M_3(0.083h^2 + l_3^2 - 0.25r^2)]\sin(q_2)\dot{q}_1 + 0.5M_3r^2\dot{q}_3), \end{aligned} \tag{5.250}$$

$$g(q) = \begin{pmatrix} 0 \\ g(l_{G_2}M_2 - l_3M_3)\cos(q_2) \\ 0 \end{pmatrix}. \tag{5.251}$$

The simulation results are displayed in Figure 5.12. The values of the constants were $M_1 = M_2 = M_3 = 1, l_{G_1} = l_{G_2} = 1, l_3 = 3, r = 1$, and $h = 1$. The initial conditions used were

$$q_o = \begin{pmatrix} 0 & -\pi/6 & 0 \end{pmatrix}^T \tag{5.252}$$

$$\dot{q}_o = \begin{pmatrix} 0 & 0 & 2\pi/8 \end{pmatrix}^T. \tag{5.253}$$

The kinetic energy of the system is

$$\begin{aligned} T &= \frac{1}{2}\dot{q}^T\left[\sum_{i=1}^{3}\left(M_i\,{}^0\Gamma_{G_i}^T\,{}^0\Gamma_{G_i} + {}^i\Pi_i^T\,{}^iI^{G_i}\,{}^i\Pi_i\right)\right]\dot{q} = \frac{1}{2}\dot{q}^TM(\dot{q}), \\ &= \frac{1}{2}[0.5M_1 + 1.604M_2 + 0.5l_{G_2}^2M_2 + 0.042h^2M_3 + 0.5l_3^2M_3 + 0.375M_3r^2 \\ &\quad + (1.479M_2 + 0.5l_{G_2}^2M_2 + 0.042h^2M_3 + 0.5l_3^2M_3 - 0.125M_3r^2)\cos(2q_2)]\dot{q}_1^2 \\ &\quad + (3.063M_2 + l_{G_2}^2M_2 + 0.083h^2M_3 + l_3^2M_3 + 0.25M_3r^2)\dot{q}_2^2 \\ &\quad - M_3r^2\sin(q_2)\dot{q}_1\dot{q}_3 + 0.5M_3r^2\dot{q}_3^2. \end{aligned} \tag{5.254}$$

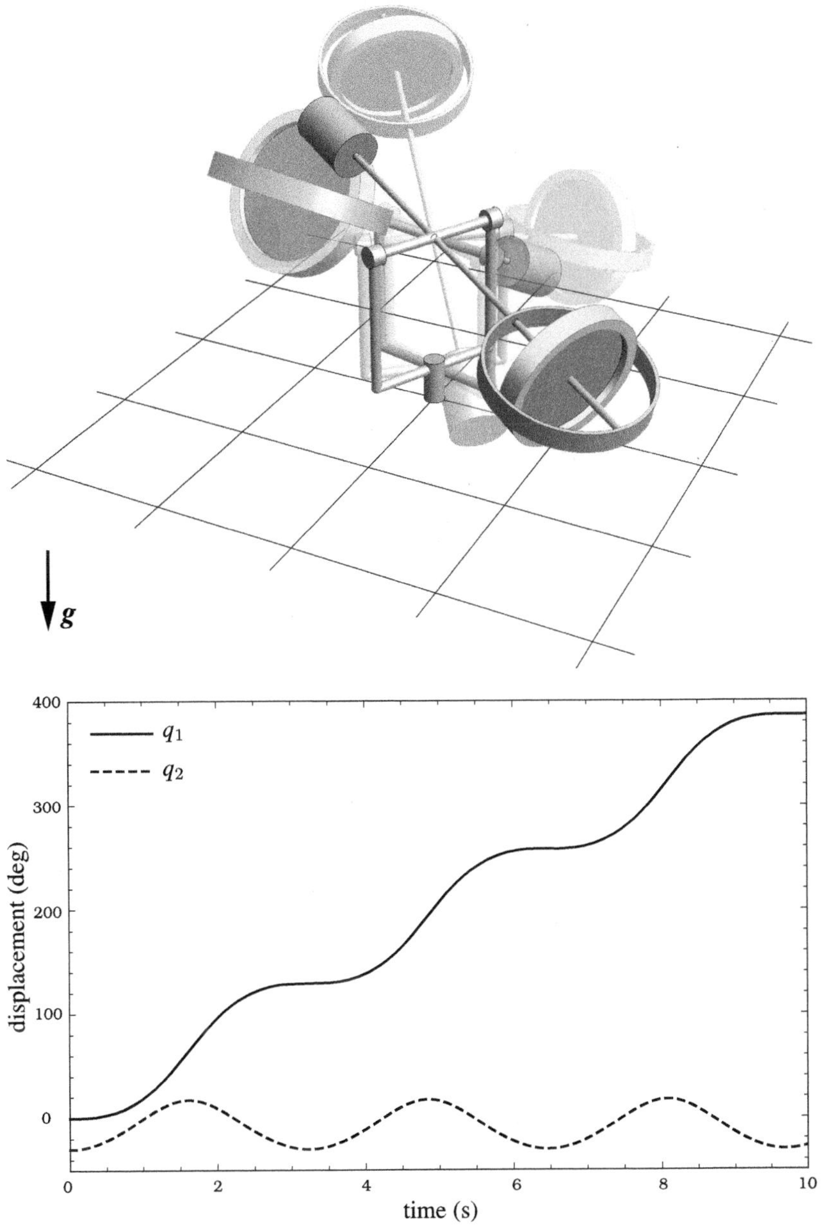

Figure 5.12 (Top) Animation frames from the simulation of the gimballed gyroscope. (Bottom) Time history of the gimbal angles, q_1 and q_2. The rotor of the gyroscope is spun up to $2\pi/8$ rad/s. Precession of the gyroscope can be observed.

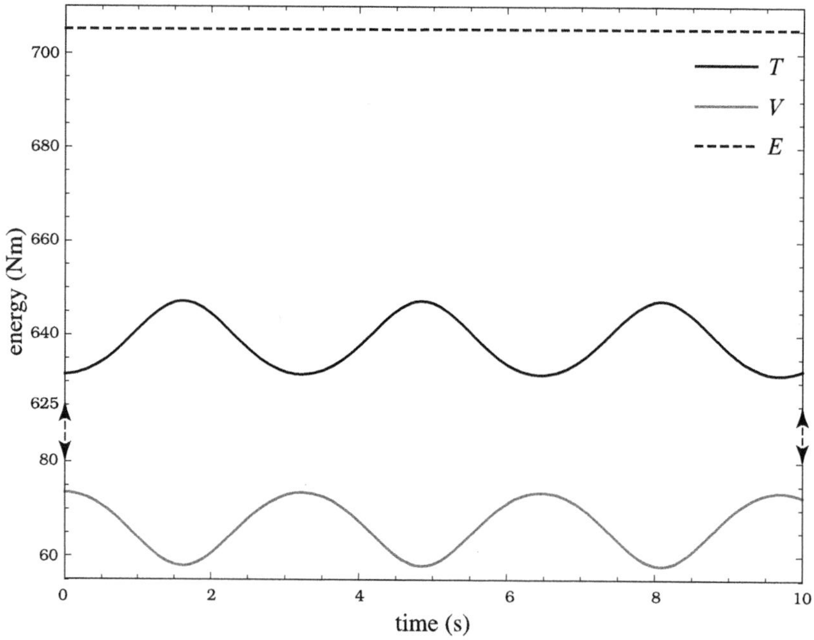

Figure 5.13 Fluctuations of kinetic and potential energy and conservation of total energy.

The potential energy (relative to $z = -2.5$) of the system is

$$V = g\sum_{i=1}^{n_b} M_i {}^0\boldsymbol{r}_{G_i}^T \hat{\boldsymbol{e}}_3 = g[(2.5 - l_{G_1})M_1 + 2.5M_2 + 2.5M_3 + (l_{G_2}M_2 - l_3M_3)\sin(q_2)], \tag{5.255}$$

and the total energy is

$$E = T + V. \tag{5.256}$$

The Lagrangian is

$$\mathcal{L} = T - V. \tag{5.257}$$

Figure 5.13 shows the fluctuations of kinetic and potential energy and the conservation of total energy for the gyroscope.

5.3.6 Constrained Least Action

Least action can be applied to multibody systems with auxiliary holonomic constraint equations. We introduce a set of m holonomic (and scleronomic) constraint equations, $\boldsymbol{\phi}(\boldsymbol{q}) = \boldsymbol{0}$. The zeroth-order variation of the constraint equations is $\delta\boldsymbol{\phi} = \boldsymbol{\Phi}\delta\boldsymbol{q} = \boldsymbol{0}$, where the matrix $\boldsymbol{\Phi}(\boldsymbol{q}) = \partial\boldsymbol{\phi}/\partial\boldsymbol{q} \in \mathbb{R}^{m_C \times n}$ is the constraint Jacobian. In this case, the Principle of Least Action can be stated as

$$\begin{gathered} \delta I = 0, \\ \forall\delta | \delta\boldsymbol{q}(t_o) = \delta\boldsymbol{q}(t_f) = \mathbf{0}, \quad \text{and} \quad \boldsymbol{\Phi}\delta\boldsymbol{q} = \mathbf{0}. \end{gathered} \tag{5.258}$$

Thus, least action seeks the path, $\boldsymbol{q}(t)$, in configuration space that results in a stationary value of action, I, under all path variations, $\delta\boldsymbol{q}$, that vanish at the endpoints and *satisfy the constraints*.

We recall the Euler-Lagrange equations for unconstrained systems:

$$\delta I = \int_{t_o}^{t_f} \left(\frac{\partial\mathcal{L}}{\partial\boldsymbol{q}} - \frac{d}{dt}\frac{\partial\mathcal{L}}{\partial\dot{\boldsymbol{q}}} + \boldsymbol{\tau} \right) \cdot \delta\boldsymbol{q}\, dt. \tag{5.259}$$

This makes use of the condition that the path variations vanish at the endpoints. The condition

$$\begin{gathered} \delta I = 0 \\ \forall\delta\boldsymbol{q} | \boldsymbol{\Phi}\delta\boldsymbol{q} = \mathbf{0} \end{gathered} \tag{5.260}$$

applied to (5.259) implies the following orthogonality relation at *any instant*:

$$\begin{gathered} \left(\frac{d}{dt}\frac{\partial\mathcal{L}}{\partial\dot{\boldsymbol{q}}} - \frac{\partial\mathcal{L}}{\partial\boldsymbol{q}} - \boldsymbol{\tau} \right) \cdot \delta\boldsymbol{q} = 0 \\ \forall\delta\boldsymbol{q} \in \ker(\boldsymbol{\Phi}). \end{gathered} \tag{5.261}$$

Thus,

$$\left(\frac{d}{dt}\frac{\partial\mathcal{L}}{\partial\dot{\boldsymbol{q}}} - \frac{\partial\mathcal{L}}{\partial\boldsymbol{q}} - \boldsymbol{\tau} \right) \in \ker(\boldsymbol{\Phi})^{\perp} = \operatorname{im}(\boldsymbol{\Phi}^T). \tag{5.262}$$

This implies the familiar constrained Euler-Lagrange equations:

$$\frac{d}{dt}\frac{\partial\mathcal{L}}{\partial\dot{\boldsymbol{q}}} - \frac{\partial\mathcal{L}}{\partial\boldsymbol{q}} = \boldsymbol{\tau} + \boldsymbol{\Phi}^T\boldsymbol{\lambda}. \tag{5.263}$$

Identical equations could have been obtained by embedding the constraints directly in the Lagrangian. In this case the Lagrangian in (5.259) would be replaced by the augmented Lagrangian with the constraints adjoined,

$$\mathcal{L}_{\text{aug}}(\boldsymbol{q}, \dot{\boldsymbol{q}}, \boldsymbol{\lambda}) \triangleq \mathcal{L}(\boldsymbol{q}, \dot{\boldsymbol{q}}) + \boldsymbol{\lambda}^T\boldsymbol{\phi}(\boldsymbol{q}), \tag{5.264}$$

and the stationary value of I would be sought for all variations that vanish at the endpoints. We note that

$$\delta\mathcal{L}_{\text{aug}} = \delta\mathcal{L} + \boldsymbol{\lambda}^T\delta\boldsymbol{\phi} = \delta\mathcal{L} + \boldsymbol{\lambda}^T\boldsymbol{\Phi}\delta\boldsymbol{q} \tag{5.265}$$

and

$$\frac{d}{dt}\frac{\partial\mathcal{L}_{\text{aug}}}{\partial\dot{\boldsymbol{q}}} - \frac{\partial\mathcal{L}_{\text{aug}}}{\partial\boldsymbol{q}} = \frac{d}{dt}\frac{\partial\mathcal{L}}{\partial\dot{\boldsymbol{q}}} - \frac{\partial\mathcal{L}}{\partial\boldsymbol{q}} - \boldsymbol{\Phi}^T\boldsymbol{\lambda} = \boldsymbol{\tau}. \tag{5.266}$$

In standard matrix form we have

$$\boldsymbol{M}(\boldsymbol{q})\ddot{\boldsymbol{q}} + \boldsymbol{b}(\boldsymbol{q}, \dot{\boldsymbol{q}}) + \boldsymbol{g}(\boldsymbol{q}) = \boldsymbol{\tau} + \boldsymbol{\Phi}^T \boldsymbol{\lambda}, \tag{5.267}$$

subject to $\boldsymbol{\phi}(\boldsymbol{q}) = \boldsymbol{0}$. For the stationary value of I to correspond to a minimum requires that the first-order variation of I be greater than or equal to *zero*. This condition is shown to be satisfied (Vujanovic and Atanackovic 2004) for sufficiently small time intervals, $[t_o, t_f]$, if the following is satisfied:

$$\int_{t_o}^{t_f} \delta\dot{\boldsymbol{q}}^T \boldsymbol{M} \delta\dot{\boldsymbol{q}}\, dt \geq 0, \tag{5.268}$$

which corresponds to $\boldsymbol{M}$ being positive definite over the actual path. For classical Lagrangian systems this condition is met.

Since (5.267) forms a set of second-order differential equations, it is appropriate to complement it with the second derivative of the constraint equations,

$$\ddot{\boldsymbol{\phi}} = \boldsymbol{\Phi}\ddot{\boldsymbol{q}} + \dot{\boldsymbol{\Phi}}\dot{\boldsymbol{q}} = \boldsymbol{0}. \tag{5.269}$$

5.4 Canonical Hamiltonian Formulation

5.4.1 Unconstrained Case

We can express a set of so-called canonical equations of motion by defining an additional set of states, the *generalized momenta*,

$$\boldsymbol{p}(\boldsymbol{q}, \dot{\boldsymbol{q}}) \triangleq \frac{\partial \mathcal{L}}{\partial \dot{\boldsymbol{q}}}. \tag{5.270}$$

In principle, we can invert this expression to represent the generalized velocities in terms of the generalized coordinates and the generalized momenta. With (5.270) we can express (5.169) as

$$\boldsymbol{\tau} = \dot{\boldsymbol{p}} - \frac{\partial \mathcal{L}}{\partial \boldsymbol{q}}. \tag{5.271}$$

Thus,

$$\frac{\partial \mathcal{L}}{\partial \boldsymbol{q}} = \dot{\boldsymbol{p}} - \boldsymbol{\tau}. \tag{5.272}$$

Given that $\mathcal{L} = \mathcal{L}(\boldsymbol{q}, \dot{\boldsymbol{q}})$, the total differential of the Lagrangian is

$$d\mathcal{L} = \frac{\partial \mathcal{L}}{\partial \boldsymbol{q}} \cdot d\boldsymbol{q} + \frac{\partial \mathcal{L}}{\partial \dot{\boldsymbol{q}}} \cdot d\dot{\boldsymbol{q}}. \tag{5.273}$$

Substituting (5.270) and (5.272) into (5.273), we have

$$d\mathcal{L} = (\dot{\boldsymbol{p}} - \boldsymbol{\tau}) \cdot d\boldsymbol{q} + \boldsymbol{p} \cdot d\dot{\boldsymbol{q}}. \tag{5.274}$$

Defining the *Hamiltonian*, $\mathcal{H}$, as

$$\mathcal{H} \triangleq \boldsymbol{p} \cdot \dot{\boldsymbol{q}} - \mathcal{L}, \tag{5.275}$$

we note that the total differential of the Hamiltonian is

$$d\mathcal{H} = d(\boldsymbol{p} \cdot \dot{\boldsymbol{q}} - \mathcal{L}) = d(\boldsymbol{p} \cdot \dot{\boldsymbol{q}}) - d\mathcal{L} = (\boldsymbol{\tau} - \dot{\boldsymbol{p}}) \cdot d\boldsymbol{q} + \dot{\boldsymbol{q}} \cdot d\boldsymbol{p}. \tag{5.276}$$

Given that we can express $\mathcal{H} = \mathcal{H}(\boldsymbol{q}, \boldsymbol{p})$ by replacing the generalized velocities in (5.275) with expressions in terms of the generalized momenta, we have

$$d\mathcal{H} = \frac{\partial \mathcal{H}}{\partial \boldsymbol{q}} \cdot d\boldsymbol{q} + \frac{\partial \mathcal{H}}{\partial \boldsymbol{p}} \cdot d\boldsymbol{p}. \tag{5.277}$$

Comparing this with (5.276), we note that

$$\frac{\partial \mathcal{H}}{\partial \boldsymbol{q}} = \boldsymbol{\tau} - \dot{\boldsymbol{p}} \quad \text{and} \quad \frac{\partial \mathcal{H}}{\partial \boldsymbol{p}} = \dot{\boldsymbol{q}}. \tag{5.278}$$

The canonical equations are thus the $2n$ first-order differential equations

$$\dot{\boldsymbol{q}} = \frac{\partial \mathcal{H}}{\partial \boldsymbol{p}} \tag{5.279}$$

$$\dot{\boldsymbol{p}} = \boldsymbol{\tau} - \frac{\partial \mathcal{H}}{\partial \boldsymbol{q}}. \tag{5.280}$$

where the state vector is $(\boldsymbol{q}\ \boldsymbol{p})^T$.

It is noted that, in practice, computing the Hamiltonian from the Lagrangian will result in an expression for the Hamiltonian in terms of the generalized coordinates, velocities, and momenta rather than just the generalized coordinates and momenta (canonical states). The procedure for expressing the Hamiltonian exclusively in terms of the generalized coordinates and momenta as follows:

1. Express the generalized momenta, $\boldsymbol{p}$, as a function of the generalized coordinates, $\boldsymbol{q}$, and the generalized velocities, $\dot{\boldsymbol{q}}$, using

$$\boldsymbol{p}(\boldsymbol{q}, \dot{\boldsymbol{q}}) = \frac{\partial \mathcal{L}}{\partial \dot{\boldsymbol{q}}}.$$

2. Invert this expression to represent the generalized velocities as a function of the generalized coordinates and the generalized momenta. That is, determine $\dot{\boldsymbol{q}} = \boldsymbol{f}(\boldsymbol{q}, \boldsymbol{p})$.
3. Compute the Hamiltonian, $\mathcal{H}$, using the Lagrangian

$$\mathcal{H}(\boldsymbol{q}, \dot{\boldsymbol{q}}, \boldsymbol{p}) = \boldsymbol{p} \cdot \dot{\boldsymbol{q}} - \mathcal{L}(\boldsymbol{q}, \dot{\boldsymbol{q}}).$$

4. Express the Hamiltonian as a function of the generalized coordinates and the generalized momenta by replacing the generalized velocities with expressions in terms of the generalized momenta using $\dot{\boldsymbol{q}} = \boldsymbol{f}(\boldsymbol{q}, \boldsymbol{p})$.

Numerical Integration

A first-order method for integrating Hamilton's equations can be summarized as shown in Algorithm 5.

Algorithm 5 First-order method for integrating Hamilton's equations

$\boldsymbol{q}_0 = \boldsymbol{q}_o$ {initialization}
$\boldsymbol{p}_0 = \boldsymbol{p}_o$ {initialization}
for $i = 0$ to $n_s - 1$ **do**
  $\dot{\boldsymbol{q}}_i = \frac{\partial \mathcal{H}}{\partial \boldsymbol{p}}|_i$
  $\dot{\boldsymbol{p}}_i = -\frac{\partial \mathcal{H}}{\partial \boldsymbol{q}}|_i$
  $\boldsymbol{p}_{i+1} = \boldsymbol{p}_i + \dot{\boldsymbol{p}}_i \Delta t$
  $\boldsymbol{q}_{i+1} = \boldsymbol{q}_i + \dot{\boldsymbol{q}}_i \Delta t$
end for

5.4.2 Auxiliary Constraints

As in the unconstrained case, we can express a set of canonical equations of motion. Beginning with the constrained Euler-Lagrange equations,

$$\boldsymbol{\tau} = \frac{d}{dt}\frac{\partial \mathcal{L}}{\partial \dot{\boldsymbol{q}}} - \frac{\partial \mathcal{L}}{\partial \boldsymbol{q}} - \boldsymbol{\Phi}^T \boldsymbol{\lambda}, \tag{5.281}$$

and following the same procedure as for the unconstrained case, we have the $2n$ first-order differential equations

$$\dot{\boldsymbol{q}} = \frac{\partial \mathcal{H}}{\partial \boldsymbol{p}} \tag{5.282}$$

$$\dot{\boldsymbol{p}} = \boldsymbol{\tau} + \boldsymbol{\Phi}^T \boldsymbol{\lambda} - \frac{\partial \mathcal{H}}{\partial \boldsymbol{q}}, \tag{5.283}$$

complemented by the m constraint equations $\boldsymbol{\Phi}\dot{\boldsymbol{q}} = \mathbf{0}$. This yields a set of $2n + m$ first-order differential equations. In practice the integration of the forward dynamics would also require constraint stabilization to mitigate drift in the constraints. Unlike second-order systems, where drift occurs at the position and velocity levels, we only need to be concerned with drift at the position level (Naudet et al. 2003). We can replace our original differential constraint equation with

$$\dot{\boldsymbol{\phi}} + \alpha \boldsymbol{\phi} = \mathbf{0}, \tag{5.284}$$

or

$$\boldsymbol{\Phi}\dot{\boldsymbol{q}} + \alpha \boldsymbol{\phi} = \mathbf{0}. \tag{5.285}$$

Thus, the constraint stabilized canonical equations of motion in compact form are

$$\begin{pmatrix} \mathbf{1} & 0 & 0 \\ 0 & \mathbf{1} & -\boldsymbol{\Phi}^T \\ \boldsymbol{\Phi} & 0 & 0 \end{pmatrix} \begin{pmatrix} \dot{\boldsymbol{q}} \\ \dot{\boldsymbol{p}} \\ \boldsymbol{\lambda} \end{pmatrix} = \begin{pmatrix} \frac{\partial \mathcal{H}}{\partial \boldsymbol{p}} \\ \boldsymbol{\tau} - \frac{\partial \mathcal{H}}{\partial \boldsymbol{q}} \\ -\alpha \boldsymbol{\phi} \end{pmatrix}. \tag{5.286}$$

Dirac generalized the handling of constraints in Hamiltonian dynamics (Dirac 1958). If we consider the constraints to be a function of the generalized momenta as well as the generalized coordinates, we have

$$\boldsymbol{\phi}(\boldsymbol{q}, \boldsymbol{p}) = \boldsymbol{0}. \tag{5.287}$$

We can define the augmented Hamiltonian as

$$\mathcal{H}_{\text{aug}}(\boldsymbol{q}, \boldsymbol{p}, \boldsymbol{\lambda}) \triangleq \mathcal{H}(\boldsymbol{q}, \boldsymbol{p}) - \boldsymbol{\lambda}^T \boldsymbol{\phi}(\boldsymbol{q}, \boldsymbol{p}). \tag{5.288}$$

The constrained canonical equations are then

$$\dot{\boldsymbol{q}} = \frac{\partial \mathcal{H}_{\text{aug}}}{\partial \boldsymbol{p}} \tag{5.289}$$

$$\dot{\boldsymbol{p}} = \boldsymbol{\tau} - \frac{\partial \mathcal{H}_{\text{aug}}}{\partial \boldsymbol{q}} \tag{5.290}$$

or

$$\dot{\boldsymbol{q}} = \frac{\partial \mathcal{H}}{\partial \boldsymbol{p}} - \frac{\partial \boldsymbol{\phi}}{\partial \boldsymbol{p}} \boldsymbol{\lambda} \tag{5.291}$$

$$\dot{\boldsymbol{p}} = \boldsymbol{\tau} - \frac{\partial \mathcal{H}}{\partial \boldsymbol{q}} + \frac{\partial \boldsymbol{\phi}}{\partial \boldsymbol{q}} \boldsymbol{\lambda}. \tag{5.292}$$

The stabilized differential constraint equation is

$$\dot{\boldsymbol{\phi}} + \alpha \boldsymbol{\phi} = \boldsymbol{0}, \tag{5.293}$$

or

$$\frac{\partial \boldsymbol{\phi}}{\partial \boldsymbol{q}} \dot{\boldsymbol{q}} + \frac{\partial \boldsymbol{\phi}}{\partial \boldsymbol{p}} \dot{\boldsymbol{p}} + \alpha \boldsymbol{\phi} = \boldsymbol{0}. \tag{5.294}$$

The generalized constraint stabilized canonical equations of motion in compact form are

$$\begin{pmatrix} \mathbf{1} & 0 & \frac{\partial \boldsymbol{\phi}}{\partial \boldsymbol{p}} \\ 0 & \mathbf{1} & -\frac{\partial \boldsymbol{\phi}}{\partial \boldsymbol{q}} \\ \frac{\partial \boldsymbol{\phi}}{\partial \boldsymbol{q}} & \frac{\partial \boldsymbol{\phi}}{\partial \boldsymbol{p}} & 0 \end{pmatrix} \begin{pmatrix} \dot{\boldsymbol{q}} \\ \dot{\boldsymbol{p}} \\ \boldsymbol{\lambda} \end{pmatrix} = \begin{pmatrix} \frac{\partial \mathcal{H}}{\partial \boldsymbol{p}} \\ \boldsymbol{\tau} - \frac{\partial \mathcal{H}}{\partial \boldsymbol{q}} \\ -\alpha \boldsymbol{\phi} \end{pmatrix}. \tag{5.295}$$

5.5 Elimination of Multipliers

The Lagrange multipliers can be eliminated from (5.86) by first expressing the zeroth-order variational equation:

$$\boldsymbol{\tau}_C \cdot \delta \boldsymbol{q} + (\boldsymbol{\tau} - \boldsymbol{M}\ddot{\boldsymbol{q}} - \boldsymbol{b} - \boldsymbol{g}) \cdot \delta \boldsymbol{q} = 0. \tag{5.296}$$

By restricting the variations to constraint-consistent virtual displacements, we have

$$\begin{gathered} \boldsymbol{\tau}_C \cdot \delta \boldsymbol{q} + (\boldsymbol{\tau} - \boldsymbol{M}\ddot{\boldsymbol{q}} - \boldsymbol{b} - \boldsymbol{g}) \cdot \delta \boldsymbol{q} = 0 \\ \forall \delta \boldsymbol{q} \in \ker(\boldsymbol{\Phi}). \end{gathered} \tag{5.297}$$

Recalling (5.83) we note that the generalized constraint forces produce no virtual work under virtual displacements that are consistent with the constraints. Thus, the term

$\boldsymbol{\tau}_C \cdot \delta \boldsymbol{q}$ vanishes from (5.297), and we have the orthogonality relation

$$\begin{gathered}(\boldsymbol{M}\ddot{\boldsymbol{q}} + \boldsymbol{b} + \boldsymbol{g} - \boldsymbol{\tau}) \cdot \delta\boldsymbol{q} = 0 \\ \forall \delta\boldsymbol{q} \in \ker(\boldsymbol{\Phi}).\end{gathered} \tag{5.298}$$

We now define a matrix, $\boldsymbol{W} \in \mathbb{R}^{n \times p}$, whose columns span the null space of $\boldsymbol{\Phi}$. This implies that $\operatorname{im}(\boldsymbol{W}) = \ker(\boldsymbol{\Phi})$. Thus, $\boldsymbol{\Phi W} = \boldsymbol{0}$ and $\boldsymbol{W}^T\boldsymbol{\Phi}^T = \boldsymbol{0}$. In this manner, $\boldsymbol{W}$ orthogonally complements $\boldsymbol{\Phi}$. That is,

$$\operatorname{im}(\boldsymbol{W}) = \ker(\boldsymbol{\Phi}) = \operatorname{im}(\boldsymbol{\Phi}^T)^{\perp}. \tag{5.299}$$

Geometrically, $\operatorname{im}(\boldsymbol{W})$ represents the tangent space of the constrained-motion manifold, Q^p (see Figure 5.5). These geometric properties are discussed in further detail in Blajer (1997) and Jungnickel (1994). While not required for the subsequent analysis, we specify that the columns of $\boldsymbol{W}$ be mutually orthogonal and thus form an orthogonal basis, $\mathcal{C}$, for the null space of $\boldsymbol{\Phi}$. The constraint-consistent virtual displacements, $\delta\boldsymbol{q} \in \ker(\boldsymbol{\Phi})$, can then be expressed in terms of the virtual displacements of a minimal set of p independent coordinates, $\boldsymbol{q}_p$:

$$\delta\boldsymbol{q} = \boldsymbol{W}\delta\boldsymbol{q}_p. \tag{5.300}$$

Using this relationship, we can express (5.298) over all possible variations of a minimal set of coordinates:

$$\begin{gathered}(\boldsymbol{W}^T\boldsymbol{M}\ddot{\boldsymbol{q}} + \boldsymbol{W}^T\boldsymbol{b} + \boldsymbol{W}^T\boldsymbol{g} - \boldsymbol{W}^T\boldsymbol{\tau}) \cdot \delta\boldsymbol{q}_p = 0, \\ \forall \delta\boldsymbol{q}_p \in \mathbb{R}^p, \\ \Downarrow \\ \boldsymbol{W}^T\boldsymbol{\tau} = \boldsymbol{W}^T\boldsymbol{M}\ddot{\boldsymbol{q}} + \boldsymbol{W}^T\boldsymbol{b} + \boldsymbol{W}^T\boldsymbol{g}.\end{gathered} \tag{5.301}$$

Noting that $\dot{\boldsymbol{q}} = \boldsymbol{W}\dot{\boldsymbol{q}}_p$ and $\ddot{\boldsymbol{q}} = \boldsymbol{W}\ddot{\boldsymbol{q}}_p + \dot{\boldsymbol{W}}\dot{\boldsymbol{q}}_p$, we can express (5.301) as

$$\boldsymbol{\tau}_p = \boldsymbol{M}_p(\boldsymbol{q})\ddot{\boldsymbol{q}}_p + \boldsymbol{b}_p(\boldsymbol{q}, \dot{\boldsymbol{q}}_p) + \boldsymbol{g}_p(\boldsymbol{q}), \tag{5.302}$$

where

$$\boldsymbol{M}_p(\boldsymbol{q}) = \boldsymbol{W}^T\boldsymbol{M}\boldsymbol{W}, \tag{5.303}$$

$$\boldsymbol{b}_p(\boldsymbol{q}, \dot{\boldsymbol{q}}_p) = \boldsymbol{W}^T\boldsymbol{b} + \boldsymbol{W}^T\boldsymbol{M}\dot{\boldsymbol{W}}\dot{\boldsymbol{q}}_p, \tag{5.304}$$

$$\boldsymbol{g}_p(\boldsymbol{q}) = \boldsymbol{W}^T\boldsymbol{g}, \tag{5.305}$$

$$\boldsymbol{\tau}_p = \boldsymbol{W}^T\boldsymbol{\tau}. \tag{5.306}$$

The approach outlined here is consistent with the projection method of Blajer (1997). This approach was also used by Russakow et al. for application to serial-to-parallel chain manipulators (Russakow, Khatib, and Rock 1995). We note that (5.302) includes a mix of our initial set of n generalized coordinates, $\boldsymbol{q}$, as well as the minimal set of p independent coordinates, $\boldsymbol{q}_p$. Since the constraints are holonomic, we would expect there to be a mapping, in principle, which could be derived from the constraints that would yield $\boldsymbol{q} = \boldsymbol{q}(\boldsymbol{q}_p)$. In this case $\boldsymbol{W}$ could be computed explicitly from the mapping rather than computing the null space of $\boldsymbol{\Phi}$; that is, $\boldsymbol{W} = \partial\boldsymbol{q}/\partial\boldsymbol{q}_p$. Additionally, the terms in (5.302) could be expressed as functions of $\boldsymbol{q}_p$ rather than $\boldsymbol{q}$. Since $\boldsymbol{q}_p$ are independent coordinates, the constraints would be implicitly addressed and the resulting system would be unconstrained with respect to configuration space. However, finding the mapping $\boldsymbol{q} = \boldsymbol{q}(\boldsymbol{q}_p)$ would be difficult in general. In such cases a null space method or a coordinate partitioning method (Wehage and Haug 1982) would need to be used to compute $\boldsymbol{W}$.

Additionally, the generalized coordinates, $\boldsymbol{q}_p$, and the generalized forces, $\boldsymbol{\tau}_p$, do not necessarily have a natural and physically intuitive meaning, making it difficult to standardize their use in a numerical algorithm. This is in contrast to the coordinates, $\boldsymbol{q}$, which are chosen specifically to describe the system in the most natural and physically intuitive manner. It is usually desirable to select $\boldsymbol{q}$ in a manner that preserves the physical meaning of the generalized forces as torques about individual joints. Often when using a minimal set of coordinates, this is not the case, since a single generalized coordinate may influence multiple joint displacements. Therefore, from an algorithmic perspective, it is often preferable to deal with a nonminimal but standardized set of generalized coordinates (like joint angles) that are amenable to numerical formulation and to compute the dynamical terms corresponding to that kinematic parametrization.

5.6 Exercises

1. Consider the serial chain robot from Section 4.2, Exercise 3 (shown in Figure 5.14). Recall that link lengths are l_1 and l_2. The link radii are r_1 and r_2, the link masses are M_1 and M_2, and the link inertia tensors are

$$
{}^{i}\boldsymbol{I}_{i}^{G_i} = \begin{pmatrix} \frac{1}{2}M_i r_i^2 & 0 & 0 \\ 0 & \frac{1}{12}l_i^2 M_i + \frac{1}{4}M_i r_i^2 & 0 \\ 0 & 0 & \frac{1}{12}l_i^2 M_i + \frac{1}{4}M_i r_i^2 \end{pmatrix}
$$

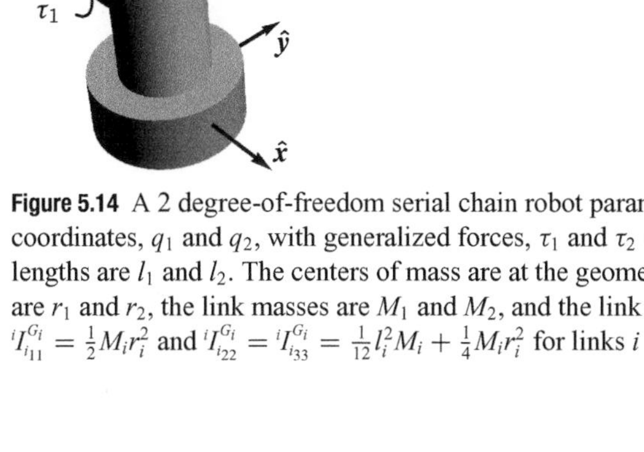

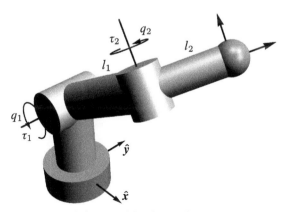

Figure 5.14 A 2 degree-of-freedom serial chain robot parameterized by the generalized coordinates, q_1 and q_2, with generalized forces, τ_1 and τ_2 (Exercises 1, 5, and 7). The link lengths are l_1 and l_2. The centers of mass are at the geometric centers of the links. The link radii are r_1 and r_2, the link masses are M_1 and M_2, and the link principal inertia components are ${}^{i}I_{i_{11}}^{G_i} = \frac{1}{2}M_i r_i^2$ and ${}^{i}I_{i_{22}}^{G_i} = {}^{i}I_{i_{33}}^{G_i} = \frac{1}{12}l_i^2 M_i + \frac{1}{4}M_i r_i^2$ for links $i = 1, 2$.

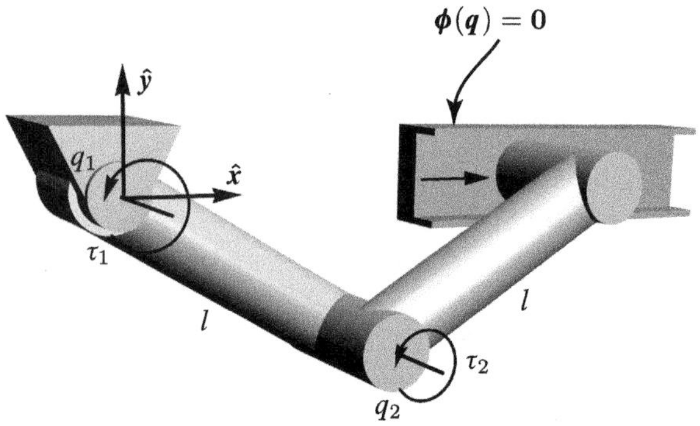

Figure 5.15 A two-link planar slider-crank mechanism (Exercises 2, 6, and 8). The unconstrained system is parameterized by the generalized coordinates, q_1 and q_2, with generalized forces, τ_1 and τ_2. The link lengths are each l and the link masses are each M. The centers of mass are at the geometric centers of the links. The link rotational inertias are taken to be *zero*. Link 2 is constrained from translating in the $\hat{y}$ direction.

for links $i = 1, 2$.

(a) Compute the mass matrix, $\boldsymbol{M}(\boldsymbol{q})$.

(b) Compute the vector of centrifugal and Coriolis forces, $\boldsymbol{b}(\boldsymbol{q}, \dot{\boldsymbol{q}})$.

(c) Compute the vector of gravity forces, $\boldsymbol{g}(\boldsymbol{q})$.

2. Consider the two-link planar slider-crank mechanism shown in Figure 5.15. The link lengths are each l and the link masses are each M. The link rotational inertias are taken to be *zero*.
   (a) How many degrees of freedom does the constrained system have?
   (b) Compute the mass matrix, $\boldsymbol{M}(\boldsymbol{q})$, for the unconstrained system.
   (c) Compute the vector of centrifugal and Coriolis forces, $\boldsymbol{b}(\boldsymbol{q}, \dot{\boldsymbol{q}})$, for the unconstrained system.
   (d) Compute the vector of gravity forces, $\boldsymbol{g}(\boldsymbol{q})$, for the unconstrained system.
   (e) Express the constraint equations, $\boldsymbol{\phi}(\boldsymbol{q})$, for the loop closure.
   (f) Compute the constraint matrix, $\boldsymbol{\Phi}(\boldsymbol{q})$.
3. Consider the planar four-bar linkage shown in Figure 5.16. The link lengths are each l and the link masses are each M. The link rotational inertias are taken to be *zero*.
   (a) How many degrees of freedom does the constrained system have?
   (b) Compute the mass matrix, $\boldsymbol{M}(\boldsymbol{q})$, for the unconstrained system.
   (c) Compute the vector of centrifugal and Coriolis forces, $\boldsymbol{b}(\boldsymbol{q}, \dot{\boldsymbol{q}})$, for the unconstrained system.
   (d) Compute the vector of gravity forces, $\boldsymbol{g}(\boldsymbol{q})$, for the unconstrained system.
   (e) Express the constraint equations, $\boldsymbol{\phi}(\boldsymbol{q})$ for the loop closure.
   (f) Compute the constraint matrix, $\boldsymbol{\Phi}(\boldsymbol{q})$.
4. Consider the three-link planar slider-crank mechanism shown in Figure 5.17. The unconstrained system is identical to the four-bar linkage of Exercise 3.
   (a) How many degrees of freedom does the constrained system have?

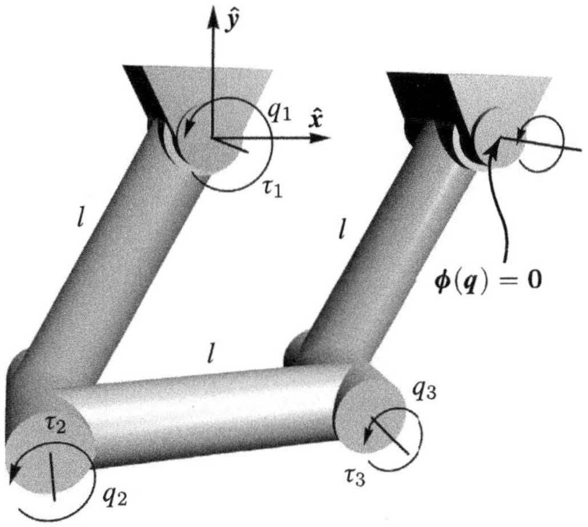

Figure 5.16 A planar four-bar linkage (Exercise 3). The unconstrained system is parameterized by the generalized coordinates, q_1, q_2, and q_3, with generalized forces, τ_1, τ_2, and τ_3. The link lengths are each l and the link masses are each M. The centers of mass are at the geometric centers of the links. The link rotational inertias are taken to be *zero*. The link 3 endpoint is constrained from translating.

(b) If not already completed, compute the mass matrix, $\boldsymbol{M}(\boldsymbol{q})$, the vector of centrifugal and Coriolis forces, $\boldsymbol{b}(\boldsymbol{q}, \dot{\boldsymbol{q}})$, and the gravity forces, $\boldsymbol{g}(\boldsymbol{q})$, for the unconstrained system.

(c) Express the constraint equations, $\boldsymbol{\phi}(\boldsymbol{q})$ for the loop closure.

(d) Compute the constraint matrix, $\boldsymbol{\Phi}(\boldsymbol{q})$.

5. Consider the serial chain robot from Exercise 1.

(a) Compute the kinetic energy, T, of the system.

(b) Compute the potential energy, V, of the system.

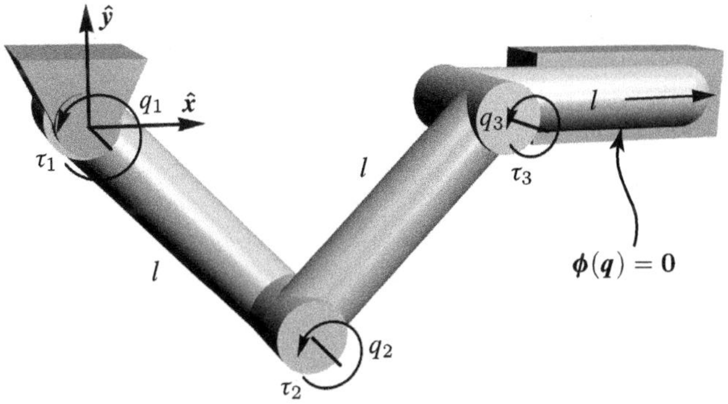

Figure 5.17 A three-link planar slider-crank mechanism (Exercise 4). The unconstrained system is parameterized by the generalized coordinates, q_1, q_2, and q_3, with generalized forces, τ_1, τ_2, and τ_3. The link lengths are each l and the link masses are each M. The centers of mass are at the geometric centers of the links. The link rotational inertias are taken to be *zero*. Link 3 is constrained from rotating as well as translating in the $\hat{y}$ direction.

(c) Compute the Lagrangian, $\mathcal{L}$, of the system.
(d) Generate the equations of motion, directly from the Lagrangian, using the Euler-Lagrange equations.

6. Consider the two-link planar slider-crank mechanism from Exercise 2.
   (a) Compute the kinetic energy, T, of the system.
   (b) Compute the potential energy, V, of the system.
   (c) Compute the Lagrangian, $\mathcal{L}$, of the system.
   (d) Generate the equations of motion, directly from the Lagrangian, using the Euler-Lagrange equations.
   (e) Express the constrained equations of motion.
7. Consider the serial chain robot from Exercise 1.
   (a) Express the generalized momenta, $\boldsymbol{p}$, as a function of the generalized coordinates, $\boldsymbol{q}$, and the generalized velocities, $\dot{\boldsymbol{q}}$. Invert this expression to represent the generalized velocities as a function of the generalized coordinates and the generalized momenta.
   (b) Compute the Hamiltonian, $\mathcal{H}$, using the Lagrangian. Express the Hamiltonian as a function of the generalized coordinates and the generalized momenta by replacing the generalized velocities with expressions in terms of the generalized momenta.
   (c) Generate the equations of motion, directly from the Hamiltonian, using Hamilton's canonical equations.
8. Consider the two-link planar slider-crank mechanism from Exercise 2.
   (a) Express the generalized momenta, $\boldsymbol{p}$, as a function of the generalized coordinates, $\boldsymbol{q}$, and the generalized velocities, $\dot{\boldsymbol{q}}$. Invert this expression to represent the generalized velocities as a function of the generalized coordinates and the generalized momenta.
   (b) Compute the Hamiltonian, $\mathcal{H}$, using the Lagrangian. Express the Hamiltonian as a function of the generalized coordinates and the generalized momenta by replacing the generalized velocities with expressions in terms of the generalized momenta.
   (c) Generate the equations of motion, directly from the Hamiltonian, using Hamilton's canonical equations.
   (d) Express the constrained equations of motion.

6 First-Order Variational Principles

In the previous chapter we addressed zeroth-order variational principles rooted in the fundamental principle of d'Alembert. In this chapter we will focus on the first-order variation of displacement. We will begin with Jourdain's Principle of Virtual Power (Jourdain 1909). The principle is based on the notion of virtual velocity. The derivation of the equations of motion for particles and rigid bodies using Jourdain's Principle closely parallels the derivation of the equations of motion using d'Alembert's Principle in the previous chapter. While Jourdain's Principle is analogous to d'Alembert's Principle, it can be seen as an independent principle of analytical dynamics. Therefore, the material in the previous chapter is not a prerequisite to this chapter, and this chapter will be treated in a stand-alone manner.

6.1 Virtual Velocities

Virtual velocities refer to all velocities of a system that satisfy the scleronomic constraints of the system. In the case of virtual velocities, time and position are frozen or stationary.

6.2 Jourdain's Principle of Virtual Power

PRINCIPLE 6.1 *The virtual power of a system is stationary. That is,*

$$\delta P = 0. \tag{6.1}$$

Additionally, the constraints of the system generate no virtual power,

$$\delta P_c = 0. \tag{6.2}$$

This is known as ***Jourdain's Principle.***

As with d'Alembert's Principle, it is noted that while Jourdain's Principle can be seen as providing an alternate statement of Newton's second law, for interacting bodies, a law of action and reaction (Newton's third law) is still needed. Therefore, when we use Jourdain's Principle to derive the equations of motion for systems of particles/bodies, we will invoke the law of action and reaction.

6.2.1 A Single Particle

Jourdain's Principle for a single point mass with a discrete set of n_f external forces, $\{\boldsymbol{f}_1, \ldots, \boldsymbol{f}_{n_f}\}$, acting on it is expressed as

$$\delta P = \sum_{i=1}^{n_f} \boldsymbol{f}_i \cdot \delta \boldsymbol{v} - Mg\hat{\boldsymbol{e}}_3 \cdot \delta \boldsymbol{v} - M\boldsymbol{a} \cdot \delta \boldsymbol{v} = 0 \quad \forall \delta \boldsymbol{v} \in \mathbb{R}^3, \tag{6.3}$$

where $\delta \boldsymbol{v}$ represents the velocity variations. During these variations, time and position are stationary. That is, $\delta t = 0$ and $\delta \boldsymbol{r} = 0$. More concisely, we can express

$$\delta P = \left(\sum_{i=1}^{n_f} \boldsymbol{f}_i - Mg\hat{\boldsymbol{e}}_3 - M\boldsymbol{a} \right) \cdot \delta \boldsymbol{v} = 0 \quad \forall \delta \boldsymbol{v} \in \mathbb{R}^3, \tag{6.4}$$

which implies

$$\sum_{i=1}^{n_f} \boldsymbol{f}_i - Mg\hat{\boldsymbol{e}}_3 - M\boldsymbol{a} = \boldsymbol{0}. \tag{6.5}$$

6.2.2 A Single Rigid Body

Jourdain's Principle for a single rigid body with a discrete set of external forces, $\boldsymbol{f}_i$, and moments, $\boldsymbol{\varphi}_j$, acting on it is expressed as

$$\delta P = \sum_{i=1}^{n_f} \boldsymbol{f}_i \cdot \delta \boldsymbol{v}_{P_i} + \sum_{j=1}^{n_\varphi} \boldsymbol{\varphi}_j \cdot \delta \boldsymbol{\omega} - Mg\hat{\boldsymbol{e}}_3 \cdot \delta \boldsymbol{v}_G - M\boldsymbol{a}_G \cdot \delta \boldsymbol{v}_G - (\boldsymbol{I}^G \boldsymbol{\alpha} + \boldsymbol{\omega} \times \boldsymbol{I}^G \boldsymbol{\omega}) \cdot \delta \boldsymbol{\omega} = 0, \quad \forall \delta \boldsymbol{v}_G \in \mathbb{R}^3, \quad \text{and} \quad \forall \delta \boldsymbol{\omega} \in \mathbb{R}^3, \tag{6.6}$$

where the $\delta \boldsymbol{v}$ and $\delta \boldsymbol{\omega}$ terms represent all velocity variations consistent with the rigid-body constraint. This is depicted in Figure 6.1. During these variations, time, position, and orientation are stationary. That is, $\delta t = 0$, $\delta \boldsymbol{r} = \boldsymbol{0}$, and $\delta \boldsymbol{\theta} = \boldsymbol{0}$. Jourdain's Principle states that the virtual power associated with all internal forces and moments consistent with the rigid-body constraint is *zero* ($P_c = 0$). We further note that

$$\delta \boldsymbol{v}_{P_i} = (\boldsymbol{v}_G + \delta \boldsymbol{v}_G) + (\boldsymbol{\omega} + \delta \boldsymbol{\omega}) \times \boldsymbol{d}_{\overrightarrow{GP_i}} - \boldsymbol{v}_G - \boldsymbol{\omega} \times \boldsymbol{d}_{\overrightarrow{GP_i}} = \delta \boldsymbol{v}_G + \delta \boldsymbol{\omega} \times \boldsymbol{d}_{\overrightarrow{GP_i}}, \tag{6.7}$$

Therefore,

$$\sum_{i=1}^{n_f} \boldsymbol{f}_i \cdot \delta \boldsymbol{v}_{P_i} = \sum_{i=1}^{n_f} \boldsymbol{f}_i \cdot (\delta \boldsymbol{v}_G + \delta \boldsymbol{\omega} \times \boldsymbol{d}_{\overrightarrow{GP_i}}) = \left(\sum_{i=1}^{n_f} \boldsymbol{f}_i \right) \cdot \delta \boldsymbol{v}_G + \sum_{i=1}^{n_f} \boldsymbol{f}_i \cdot (\delta \boldsymbol{\omega} \times \boldsymbol{d}_{\overrightarrow{GP_i}}). \tag{6.8}$$

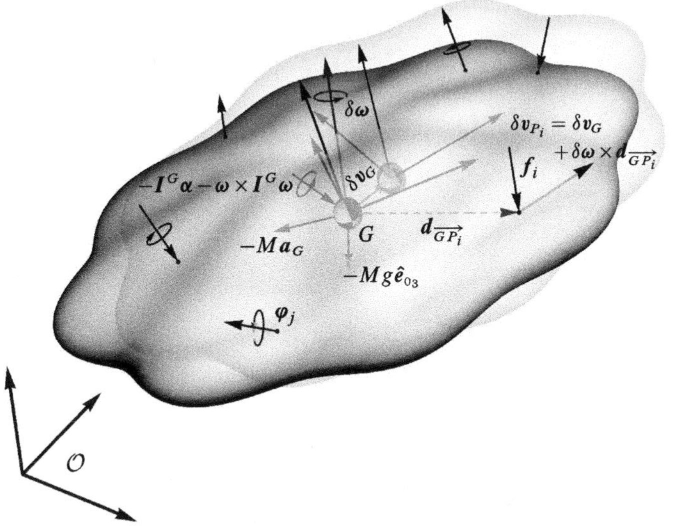

Figure 6.1 A virtual velocity of a rigid body consisting of a translational, δv_G, and rotational, $\delta\omega$, velocity variation. We are concerned with all such velocity variations consistent with the rigid-body constraint. During these variations, time, position, and orientation are stationary.

Since $\boldsymbol{a} \cdot (\boldsymbol{b} \times \boldsymbol{c}) = (\boldsymbol{c} \times \boldsymbol{a}) \cdot \boldsymbol{b}$,

$$\sum_{i=1}^{n_f} \boldsymbol{f}_i \cdot (\delta\boldsymbol{\omega} \times \boldsymbol{d}_{\overrightarrow{GP_i}}) = \sum_{i=1}^{n_f} (\boldsymbol{d}_{\overrightarrow{GP_i}} \times \boldsymbol{f}_i) \cdot \delta\boldsymbol{\omega} = \left(\sum_{i=1}^{n_f} \boldsymbol{d}_{\overrightarrow{GP_i}} \times \boldsymbol{f}_i \right) \cdot \delta\boldsymbol{\omega}. \tag{6.9}$$

So,

$$\sum_{i=1}^{n_f} \boldsymbol{f}_i \cdot \delta\boldsymbol{v}_{P_i} = \left(\sum_{i=1}^{n_f} \boldsymbol{f}_i \right) \cdot \delta\boldsymbol{v}_G + \left(\sum_{i=1}^{n_f} \boldsymbol{d}_{\overrightarrow{GP_i}} \times \boldsymbol{f}_i \right) \cdot \delta\boldsymbol{\omega}. \tag{6.10}$$

Substituting (6.10) into (6.6), we get

$$\begin{aligned} \delta P = {} & \left(\sum_{i=1}^{n_f} \boldsymbol{f}_i - Mg\hat{\boldsymbol{e}}_3 - M\boldsymbol{a}_G \right) \cdot \delta\boldsymbol{v}_G \\ & + \left(\sum_{i=1}^{n_f} \boldsymbol{d}_{\overrightarrow{GP_i}} \times \boldsymbol{f}_i + \sum_{j=1}^{n_\varphi} \boldsymbol{\varphi}_j - \boldsymbol{I}^G\boldsymbol{\alpha} - \boldsymbol{\omega} \times \boldsymbol{I}^G\boldsymbol{\omega} \right) \cdot \delta\boldsymbol{\omega} = 0, \end{aligned} \tag{6.11}$$

or

$$\begin{pmatrix} \sum_{i=1}^{n_f} \boldsymbol{f}_i - Mg\hat{\boldsymbol{e}}_3 - M\boldsymbol{a}_G \\ \sum_{i=1}^{n_f} \boldsymbol{d}_{\overrightarrow{GP_i}} \times \boldsymbol{f}_i + \sum_{j=1}^{n_\varphi} \boldsymbol{\varphi}_j - \boldsymbol{I}^G\boldsymbol{\alpha} - \boldsymbol{\omega} \times \boldsymbol{I}^G\boldsymbol{\omega} \end{pmatrix} \cdot \begin{pmatrix} \delta\boldsymbol{v}_G \\ \delta\boldsymbol{\omega} \end{pmatrix} = 0, \tag{6.12}$$
$$\forall \delta\boldsymbol{v}_G \in \mathbb{R}^3, \quad \text{and} \quad \forall \delta\boldsymbol{\omega} \in \mathbb{R}^3,$$

which implies

$$\sum_{i=1}^{n_f} f_i - Mg\hat{e}_3 - M\boldsymbol{a}_G = \mathbf{0} \tag{6.13}$$

$$\sum_{i=1}^{n_f} \boldsymbol{d}_{\overrightarrow{GP_i}} \times f_i + \sum_{j=1}^{n_\varphi} \varphi_j - \boldsymbol{I}^G\boldsymbol{\alpha} - \boldsymbol{\omega} \times \boldsymbol{I}^G\boldsymbol{\omega} = \mathbf{0}. \tag{6.14}$$

6.2.3 A System of Particles

We now apply Jourdain's Principle to a system of n_p particles with a discrete set of forces acting on them. The virtual power associated with a given particle i is given by

$$\delta P_i = \left(\sum_{j=0}^{n_p} \boldsymbol{f}_i^j\right) \cdot \delta\boldsymbol{v}_i - M_i(\boldsymbol{a}_i + g\hat{\boldsymbol{e}}_3) \cdot \delta\boldsymbol{v}_i = 0 \qquad \forall \delta\boldsymbol{v}_i \in \mathbb{R}^3 \tag{6.15}$$

for $i = 1, \dots, n_p$. The term $\boldsymbol{f}_i^j$ is the force that particle j exerts on particle i, where $\boldsymbol{f}_i^0$ is the force exerted by the ground (inertial reference frame) on the ith particle. Equation (6.15) is true for any particle i in the system, under all independent variations, $\delta\boldsymbol{v}_i$.

If we sum (6.15) over all particles, we obtain

$$\delta P = \sum_{i=1}^{n_p} \delta P_i = \sum_{i=1}^{n_p} \left(\sum_{j=0}^{n_p} \boldsymbol{f}_i^j\right) \cdot \delta\boldsymbol{v}_i - \sum_{i=1}^{n_p} M_i(\boldsymbol{a}_i + g\hat{\boldsymbol{e}}_3) \cdot \delta\boldsymbol{v}_i = 0. \tag{6.16}$$

The first summation is associated with the virtual power performed by the interparticle reaction forces. It will be useful to rearrange the summation so as to pair up the equal and opposite reaction forces, $\boldsymbol{f}_i^j$ and $\boldsymbol{f}_j^i$. Doing this, we have

$$\sum_{i=1}^{n_p} \left(\sum_{j=0}^{n_p} \boldsymbol{f}_i^j\right) \cdot \delta\boldsymbol{v}_i = \sum_{i=0}^{n_p-1} \sum_{j=i+1}^{n_p} (\boldsymbol{f}_i^j \cdot \delta\boldsymbol{v}_i + \boldsymbol{f}_j^i \cdot \delta\boldsymbol{v}_j) + \sum_{i=1}^{n_p} \boldsymbol{f}_i^i \cdot \delta\boldsymbol{v}_i - \sum_{j=1}^{n_p} \boldsymbol{f}_0^j \cdot \delta\boldsymbol{v}_0. \tag{6.17}$$

Since $\boldsymbol{f}_i^i = \mathbf{0}$ and $\delta\boldsymbol{v}_0 = \mathbf{0}$, we have

$$\sum_{i=1}^{n_p} \left(\sum_{j=0}^{n_p} \boldsymbol{f}_i^j\right) \cdot \delta\boldsymbol{v}_i = \sum_{i=0}^{n_p-1} \sum_{j=i+1}^{n_p} (\boldsymbol{f}_i^j \cdot \delta\boldsymbol{v}_i + \boldsymbol{f}_j^i \cdot \delta\boldsymbol{v}_j). \tag{6.18}$$

We note that $\boldsymbol{f}_j^i = -\boldsymbol{f}_i^j$. Thus,

$$\sum_{i=1}^{n_p} \left(\sum_{j=0}^{n_p} \boldsymbol{f}_i^j\right) \cdot \delta\boldsymbol{v}_i = \sum_{i=0}^{n_p-1} \sum_{j=i+1}^{n_p} \boldsymbol{f}_i^j \cdot (\delta\boldsymbol{v}_i - \delta\boldsymbol{v}_j). \tag{6.19}$$

This reflects the virtual power done by interparticle reaction forces, summed over all particles.

If we consider that the particles are subject to a set of holonomic constraints the positions, $\boldsymbol{r}_i$, and in turn velocities, $\boldsymbol{v}_i$, are not independent. In this case we will assume that the velocities can be expressed in terms of a set of n_q independent generalized velocities, $\dot{\boldsymbol{q}}$. If we now consider only the variations, $\delta \boldsymbol{v}_i$, that are consistent with the kinematic constraints, then (6.19) reflects a projection of the interparticle forces on the direction of the interparticle motion. Jourdain's Principle states that the virtual power associated with all forces orthogonal to the interparticle motion (reaction forces) is *zero* ($P_c = 0$), and only the generalized force acting in the direction of interparticle motion produces virtual power. Thus,

$$\sum_{i=0}^{n_p-1} \sum_{j=i+1}^{n_p} \boldsymbol{f}_i^j \cdot (\delta \boldsymbol{v}_i - \delta \boldsymbol{v}_j) = \sum_{i=1}^{n_q} \tau_i \cdot \delta \dot{q}_i, \tag{6.20}$$

and (6.16) can be expressed as

$$\sum_{i=1}^{n_q} \tau_i \cdot \delta \dot{q}_i - \sum_{i=0}^{n_p} M_i(\boldsymbol{a}_i + g\hat{\boldsymbol{e}}_3) \cdot \delta \boldsymbol{v}_i = 0 \tag{6.21}$$

for all variations, $\delta \boldsymbol{v}_i$, that are consistent with the kinematic constraints.

Expressing the variations in terms of variations in the generalized velocities (with stationary displacement and time), we have

$$\delta \boldsymbol{v}_i = \boldsymbol{\Gamma}_i \delta \dot{\boldsymbol{q}}. \tag{6.22}$$

So, (6.21) can be expressed as

$$\begin{gathered} \boldsymbol{\tau} \cdot \delta \dot{\boldsymbol{q}} - \sum_{i=1}^{n_p} M_i(\boldsymbol{a}_i + g\hat{\boldsymbol{e}}_3) \cdot (\boldsymbol{\Gamma}_i \delta \dot{\boldsymbol{q}}) = 0 \\ \forall \delta \dot{\boldsymbol{q}} \in \mathbb{R}^n, \end{gathered} \tag{6.23}$$

or

$$\begin{gathered} \left[\boldsymbol{\tau} - \sum_{i=1}^{n_p} \boldsymbol{\Gamma}_i^T (M_i \boldsymbol{a}_i + M_i g\hat{\boldsymbol{e}}_3) \right] \cdot \delta \dot{\boldsymbol{q}} = 0 \\ \forall \delta \dot{\boldsymbol{q}} \in \mathbb{R}^n, \end{gathered} \tag{6.24}$$

which implies

$$\boldsymbol{\tau} = \sum_{i=1}^{n_p} \boldsymbol{\Gamma}_i^T (M_i \boldsymbol{a}_i + M_i g\hat{\boldsymbol{e}}_3). \tag{6.25}$$

Noting that

$$\boldsymbol{a}_i = \boldsymbol{\Gamma}_i \ddot{\boldsymbol{q}}, \tag{6.26}$$

we have

$$\boldsymbol{\tau} = \left(\sum_{i=1}^{n_p} M_i \boldsymbol{\Gamma}_i^T \boldsymbol{\Gamma}_i \right) \ddot{\boldsymbol{q}} + g \sum_{i=1}^{n_q} M_i \boldsymbol{\Gamma}_i^T \hat{\boldsymbol{e}}_3. \tag{6.27}$$

For a system of particles, we had previously defined

$$M(q) \triangleq \sum_{i=1}^{n_p} M_i \Gamma_i^T \Gamma_i \tag{6.28}$$

$$g(q) \triangleq g \sum_{i=1}^{n_p} M_i \Gamma_i^T \hat{e}_3. \tag{6.29}$$

So,

$$\tau = M(q)\ddot{q} + g(q). \tag{6.30}$$

6.2.4 A System of Rigid Bodies

We now apply Jourdain's Principle to a system of n_b rigid bodies forming a serial chain, with a discrete set of forces acting on them. The virtual power associated with a given body i is given by

$$\begin{aligned} \delta P_i = f_i^{i-1} \cdot \delta v_i^{i-1} + f_i^{i+1} \cdot \delta v_i^{i+1} - M_i(a_{G_i} + g\hat{e}_{0_3}) \cdot \delta v_{G_i} + \varphi_i^{i-1} \cdot \delta \omega_i \\ + \varphi_i^{i+1} \cdot \delta \omega_i - (I_i^{G_i} \alpha_i + \omega_i \times I_i^{G_i} \omega_i) \cdot \delta \omega_i = 0, \\ \forall \delta v_{G_i} \in \mathbb{R}^3, \quad \text{and} \quad \forall \delta \omega_i \in \mathbb{R}^3, \end{aligned} \tag{6.31}$$

for $i = 1, \ldots, n_b$. The term f_i^{i-1} is the force that body $i-1$ exerts on body i. Likewise, φ_i^{i-1} is the moment that body $i-1$ exerts on body i. The term v_i^{i-1} is the velocity of the point on body i to which body $i-1$ attaches. Equation (6.31) is true for any rigid body i in the system, under all independent variations, δv_{G_i} and $\delta \omega_i$ (we note that δv_i^{i-1} and δv_i^{i+1} are functions of δv_{G_i} and $\delta \omega_i$ due to the rigid-body constraint). If we sum (6.31) over all rigid bodies, we obtain

$$\begin{aligned} \delta P = \sum_{i=1}^{n_b} \delta P_i = \sum_{i=1}^{n_b} (f_i^{i-1} \cdot \delta v_i^{i-1} + f_i^{i+1} \cdot \delta v_i^{i+1}) \\ - \sum_{i=1}^{n_b} M_i(a_{G_i} + g\hat{e}_{0_3}) \cdot \delta v_{G_i} + \sum_{i=1}^{n_b} (\varphi_i^{i-1} \cdot \delta \omega_i + \varphi_i^{i+1} \cdot \delta \omega_i) \\ - \sum_{i=1}^{n_b} (I_i^{G_i} \alpha_i + \omega_i \times I_i^{G_i} \omega_i) \cdot \delta \omega_i = 0. \end{aligned} \tag{6.32}$$

The first summation is associated with the virtual power performed by the interlink reaction forces. It will be useful to rearrange the summation so as to pair up the equal and opposite reaction forces, f_i^{i-1} and f_{i-1}^{i}, acting through each joint i. Doing this, we have

$$\sum_{i=1}^{n_b} (f_i^{i-1} \cdot \delta v_i^{i-1} + f_i^{i+1} \cdot \delta v_i^{i+1}) = \sum_{i=1}^{n_b} (f_i^{i-1} \cdot \delta v_i^{i-1} + f_{i-1}^{i} \cdot \delta v_{i-1}^{i}) + f_{n_b}^{n_b+1} \cdot \delta v_{n_b}^{n_b+1} - f_0^1 \cdot \delta v_0^1. \tag{6.33}$$

Since $f_{n_b}^{n_b+1} = \mathbf{0}$ and $\delta v_0^1 = \mathbf{0}$, we have

$$\sum_{i=1}^{n_b} (f_i^{i-1} \cdot \delta v_i^{i-1} + f_i^{i+1} \cdot \delta v_i^{i+1}) = \sum_{i=1}^{n_b} (f_i^{i-1} \cdot \delta v_i^{i-1} + f_{i-1}^{i} \cdot \delta v_{i-1}^{i}). \tag{6.34}$$

We note that $f_{i-1}^{i} = -f_i^{i-1}$. Thus,

$$\sum_{i=1}^{n_b} (f_i^{i-1} \cdot \delta v_i^{i-1} + f_i^{i+1} \cdot \delta v_i^{i+1}) = \sum_{i=1}^{n_b} f_i^{i-1} \cdot (\delta v_i^{i-1} - \delta v_{i-1}^{i}). \tag{6.35}$$

This reflects the virtual power done by reaction forces at each joint, summed over all joints.

As with the reaction forces, it will be useful to rearrange the summation associated with the virtual power performed by the interlink reaction moments so as to pair up the equal and opposite reaction moments, φ_i^{i-1} and φ_{i-1}^{i}, acting about each joint. Using the same procedure as with the reaction forces, we have

$$\begin{aligned}\sum_{i=1}^{n_b} (\varphi_i^{i-1} \cdot \delta\omega_i + \varphi_i^{i+1} \cdot \delta\omega_i) &= \sum_{i=1}^{n_b} (\varphi_i^{i-1} \cdot \delta\omega_i + \varphi_{i-1}^{i} \cdot \delta\omega_{i-1}) \\ &= \sum_{i=1}^{n_b} \varphi_i^{i-1} \cdot (\delta\omega_i - \delta\omega_{i-1}).\end{aligned} \tag{6.36}$$

This reflects the virtual power done by reaction moments at each joint, summed over all joints. So, the total virtual power performed by the interlink reaction forces and moments can be expressed compactly as

$$\begin{aligned}&\sum_{i=1}^{n_b} (f_i^{i-1} \cdot \delta v_i^{i-1} + f_i^{i+1} \cdot \delta v_i^{i+1}) + \sum_{i=1}^{n_b} (\varphi_i^{i-1} \cdot \delta\omega_i + \varphi_i^{i+1} \cdot \delta\omega_i) \\ &= \sum_{i=1}^{n_b} \begin{pmatrix} f_i^{i-1} \\ \varphi_i^{i-1} \end{pmatrix} \cdot \begin{pmatrix} \delta v_i^{i-1} - \delta v_{i-1}^{i} \\ \delta\omega_i - \delta\omega_{i-1} \end{pmatrix}.\end{aligned} \tag{6.37}$$

If we now consider only the variations δv_i^{i-1}, δv_{i-1}^{i}, $\delta\omega_i$, and $\delta\omega_{i-1}$ that are consistent with the kinematic constraints, then (6.37) reflects a projection of the interlink forces and moments on the direction of the joint motion. Jourdain's Principle states that the virtual power associated with all forces and moments orthogonal to the joint motion (reaction forces/moments) is *zero* ($P_c = 0$) and only the generalized force acting in the direction of joint motion produces virtual power. Thus,

$$\sum_{i=1}^{n_b} \begin{pmatrix} f_i^{i-1} \\ \varphi_i^{i-1} \end{pmatrix} \cdot \begin{pmatrix} \delta v_i^{i-1} - \delta v_{i-1}^{i} \\ \delta\omega_i - \delta\omega_{i-1} \end{pmatrix} = \sum_{i=1}^{n_q} \tau_i \cdot \delta\dot{q}_i, \tag{6.38}$$

and (6.32) can be expressed as

$$\sum_{i=1}^{n_q} \tau_i \cdot \delta\dot{q}_i - \sum_{i=1}^{n_b} M_i(a_{G_i} + g\hat{e}_{0_3}) \cdot \delta v_{G_i} - \sum_{i=1}^{n_b} (I_i^{G_i}\alpha_i + \omega_i \times I_i^{G_i}\omega_i) \cdot \delta\omega_i = 0 \tag{6.39}$$

for all variations, δv_{G_i} and $\delta\omega_i$, that are consistent with the kinematic constraints.

Using Jourdain's Principle, we can address a system of n_b rigid bodies forming a branching chain in a similar manner. With a serial chain the parent/child structure is implicit to the numbering scheme. Every link i has a single parent link, $i - 1$, at its proximal end and a single child link, $i + 1$, at its distal end (except for the nth link). The ith joint is at the proximal end of the ith link. With a branching chain the numbering of links is more arbitrary, without a parent/child structure implicit to the numbering scheme. We can explicitly capture the parent/child structure, however, by defining three additional parameters, λ_i, c_i, μ_{ij}. The term λ_i is the parent link number of the ith link, c_i is the number of child links for the ith link, and $\mu_{i1} \cdots \mu_{ic_i}$ are the child link numbers of the ith link. Given these parameters, the virtual power associated with a given body i is given by

$$\begin{aligned}\delta P_i = \boldsymbol{f}_i^{\lambda_i} \cdot \delta \boldsymbol{v}_i^{\lambda_i} + \sum_{j=1}^{c_i} \boldsymbol{f}_i^{\mu_{ij}} \cdot \delta \boldsymbol{v}_i^{\mu_{ij}} - M_i(\boldsymbol{a}_{G_i} + g\hat{\boldsymbol{e}}_{0_3}) \cdot \delta \boldsymbol{v}_{G_i} + \boldsymbol{\varphi}_i^{\lambda_i} \cdot \delta \boldsymbol{\omega}_i \\ + \sum_{j=1}^{c_i} \boldsymbol{\varphi}_i^{\mu_{ij}} \cdot \delta \boldsymbol{\omega}_i - (\boldsymbol{I}_i^{G_i} \boldsymbol{\alpha}_i + \boldsymbol{\omega}_i \times \boldsymbol{I}_i^{G_i} \boldsymbol{\omega}_i) \cdot \delta \boldsymbol{\omega}_i = 0, \\ \forall \delta \boldsymbol{v}_{G_i} \in \mathbb{R}^3, \quad \text{and} \quad \forall \delta \boldsymbol{\omega}_i \in \mathbb{R}^3,\end{aligned} \tag{6.40}$$

for $i = 1, \ldots, n_b$. The term $\boldsymbol{v}_i^{\lambda_i}$ is the velocity of the point on body i to which body (parent) λ_i attaches, and likewise $\boldsymbol{v}_i^{\mu_{i1}} \ldots \boldsymbol{v}_i^{\mu_{ic_i}}$ are the velocities of points on body i to which bodies (children) $\mu_{i1} \ldots \mu_{ic_i}$ attach. We note that $\delta \boldsymbol{v}_i^{\lambda_i}$ and $\delta \boldsymbol{v}_i^{\mu_{i1}} \ldots \delta \boldsymbol{v}_i^{\mu_{ic_i}}$ are functions of $\delta \boldsymbol{v}_{G_i}$ and $\delta \boldsymbol{\omega}_i$ due to the rigid-body constraint. If we sum (6.40) over all rigid bodies, we obtain

$$\begin{aligned}\delta P = \sum_{i=1}^{n_b} \delta P_i = \sum_{i=1}^{n_b} \left(\boldsymbol{f}_i^{\lambda_i} \cdot \delta \boldsymbol{v}_i^{\lambda_i} + \sum_{j=1}^{c_i} \boldsymbol{f}_i^{\mu_{ij}} \cdot \delta \boldsymbol{v}_i^{\mu_{ij}} \right) \\ - \sum_{i=1}^{n_b} M_i(\boldsymbol{a}_{G_i} + g\hat{\boldsymbol{e}}_{0_3}) \cdot \delta \boldsymbol{v}_{G_i} + \sum_{i=1}^{n_b} \left(\boldsymbol{\varphi}_i^{\lambda_i} \cdot \delta \boldsymbol{\omega}_i + \sum_{j=1}^{c_i} \boldsymbol{\varphi}_i^{\mu_{ij}} \cdot \delta \boldsymbol{\omega}_i \right) \\ - \sum_{i=1}^{n_b} (\boldsymbol{I}_i^{G_i} \boldsymbol{\alpha}_i + \boldsymbol{\omega}_i \times \boldsymbol{I}_i^{G_i} \boldsymbol{\omega}_i) \cdot \delta \boldsymbol{\omega}_i = 0.\end{aligned} \tag{6.41}$$

The term associated with the virtual power performed by the interlink reaction forces is

$$\sum_{i=1}^{n_b} \left(\boldsymbol{f}_i^{\lambda_i} \cdot \delta \boldsymbol{v}_i^{\lambda_i} + \sum_{j=1}^{c_i} \boldsymbol{f}_i^{\mu_{ij}} \cdot \delta \boldsymbol{v}_i^{\mu_{ij}} \right). \tag{6.42}$$

It will be useful to rearrange the summation so as to pair up the equal and opposite reaction forces, $\boldsymbol{f}_i^{\lambda_i}$ and $\boldsymbol{f}_{\lambda_i}^{i}$, acting through each joint i. We can rewrite the summation of (6.42) based on considering the virtual power at each joint. The sum over all joints then gives us

$$\sum_{i=1}^{n_b} (\boldsymbol{f}_i^{\lambda_i} \cdot \delta \boldsymbol{v}_i^{\lambda_i} + \boldsymbol{f}_{\lambda_i}^{i} \cdot \delta \boldsymbol{v}_{\lambda_i}^{i}) = \sum_{i=1}^{n_b} \boldsymbol{f}_i^{\lambda_i} \cdot (\delta \boldsymbol{v}_i^{\lambda_i} - \delta \boldsymbol{v}_{\lambda_i}^{i}). \tag{6.43}$$

The term associated with the virtual power performed by the interlink reaction moments is

$$\sum_{i=1}^{n_b}\left(\boldsymbol{\varphi}_i^{\lambda_i}\cdot\delta\boldsymbol{\omega}_i+\sum_{j=1}^{c_i}\boldsymbol{\varphi}_i^{\mu_{ij}}\cdot\delta\boldsymbol{\omega}_i\right). \tag{6.44}$$

In a similar manner as before, we can rewrite this summation based on considering the virtual power at each joint. The sum over all joints then gives us

$$\sum_{i=1}^{n_b}(\boldsymbol{\varphi}_i^{\lambda_i}\cdot\delta\boldsymbol{\omega}_i+\boldsymbol{\varphi}_{\lambda_i}^{i}\cdot\delta\boldsymbol{\omega}_{\lambda_i})=\sum_{i=1}^{n_b}\boldsymbol{\varphi}_i^{\lambda_i}\cdot(\delta\boldsymbol{\omega}_i-\delta\boldsymbol{\omega}_{\lambda_i}). \tag{6.45}$$

So, the total virtual power performed by the interlink reaction forces and moments can be expressed compactly as

$$\begin{aligned}\sum_{i=1}^{n_b}&\left(\boldsymbol{f}_i^{\lambda_i}\cdot\delta\boldsymbol{v}_i^{\lambda_i}+\sum_{j=1}^{c_i}\boldsymbol{f}_i^{\mu_{ij}}\cdot\delta\boldsymbol{v}_i^{\mu_{ij}}\right),+\sum_{i=1}^{n_b}\left(\boldsymbol{\varphi}_i^{\lambda_i}\cdot\delta\boldsymbol{\omega}_i+\sum_{j=1}^{c_i}\boldsymbol{\varphi}_i^{\mu_{ij}}\cdot\delta\boldsymbol{\omega}_i\right)\\&=\sum_{i=1}^{n_b}\begin{pmatrix}\boldsymbol{f}_i^{\lambda_i}\\\boldsymbol{\varphi}_i^{\lambda_i}\end{pmatrix}\cdot\begin{pmatrix}\delta\boldsymbol{v}_i^{\lambda_i}-\delta\boldsymbol{v}_{\lambda_i}^{i}\\\delta\boldsymbol{\omega}_i-\delta\boldsymbol{\omega}_{\lambda_i}\end{pmatrix}.\end{aligned} \tag{6.46}$$

If we now consider only the variations $\delta\boldsymbol{v}_i^{\lambda_i}$, $\delta\boldsymbol{v}_{\lambda_i}^{i}$, $\delta\boldsymbol{\omega}_i$, and $\delta\boldsymbol{\omega}_{\lambda_i}$ that are consistent with the kinematic constraints, then (6.46) reflects a projection of the interlink forces and moments on the direction of the joint motion. Jourdain's Principle states that the virtual power associated with all forces and moments orthogonal to the joint motion (reaction forces/moments) is *zero* ($P_c = 0$) and only the generalized force acting in the direction of joint motion produces virtual power. Thus,

$$\sum_{i=1}^{n_b}\begin{pmatrix}\boldsymbol{f}_i^{\lambda_i}\\\boldsymbol{\varphi}_i^{\lambda_i}\end{pmatrix}\cdot\begin{pmatrix}\delta\boldsymbol{v}_i^{\lambda_i}-\delta\boldsymbol{v}_{\lambda_i}^{i}\\\delta\boldsymbol{\omega}_i-\delta\boldsymbol{\omega}_{\lambda_i}\end{pmatrix}=\sum_{i=1}^{n_q}\tau_i\cdot\delta\dot{q}_i, \tag{6.47}$$

and (6.41) can be expressed as

$$\sum_{i=1}^{n_q}\tau_i\cdot\delta\dot{q}_i-\sum_{i=1}^{n_b}M_i(\boldsymbol{a}_{G_i}+g\hat{\boldsymbol{e}}_{0_3})\cdot\delta\boldsymbol{v}_{G_i}-\sum_{i=1}^{n_b}(\boldsymbol{I}_i^{G_i}\boldsymbol{\alpha}_i+\boldsymbol{\omega}_i\times\boldsymbol{I}_i^{G_i}\boldsymbol{\omega}_i)\cdot\delta\boldsymbol{\omega}_i=0 \tag{6.48}$$

for all variations, $\delta\boldsymbol{v}_{G_i}$ and $\delta\boldsymbol{\omega}_i$, that are consistent with the kinematic constraints. Expressing the velocity variations in terms of variations in the generalized velocities (with stationary displacement and time), we have

$${}^{i}\delta\boldsymbol{v}_{G_i}={}^{i}\boldsymbol{\Gamma}_{G_i}\delta\dot{\boldsymbol{q}}\quad\text{and}\quad{}^{i}\delta\boldsymbol{\omega}_i={}^{i}\boldsymbol{\Pi}_i\delta\dot{\boldsymbol{q}}, \tag{6.49}$$

where the terms are expressed in the local link frame i for convenience. So, (6.48) can be expressed as

$$\boldsymbol{\tau} \cdot \delta\dot{\boldsymbol{q}} - \sum_{i=1}^{n_b} M_i({}^i\boldsymbol{a}_{G_i} + g{}^i\hat{\boldsymbol{e}}_{0_3}) \cdot ({}^i\boldsymbol{\Gamma}_{G_i}\delta\dot{\boldsymbol{q}}) - \sum_{i=1}^{n_b} ({}^i\boldsymbol{I}_i^{G_i\,i}\boldsymbol{\alpha}_i + {}^i\boldsymbol{\omega}_i \times {}^i\boldsymbol{I}_i^{G_i\,i}\boldsymbol{\omega}_i) \cdot ({}^i\boldsymbol{\Pi}_i\delta\dot{\boldsymbol{q}}) = 0$$
$$\forall \delta\dot{\boldsymbol{q}} \in \mathbb{R}^n, \tag{6.50}$$

or

$$\boldsymbol{\tau} \cdot \delta\dot{\boldsymbol{q}} - \sum_{i=1}^{n_b} \left[M_i{}^i\boldsymbol{\Gamma}_{G_i}^{T\,i}\boldsymbol{a}_{G_i} + M_i g{}^i\boldsymbol{\Gamma}_{G_i}^{T\,i}\hat{\boldsymbol{e}}_{0_3} + {}^i\boldsymbol{\Pi}_i^T({}^i\boldsymbol{I}_i^{G_i\,i}\boldsymbol{\alpha}_i + {}^i\boldsymbol{\omega}_i \times {}^i\boldsymbol{I}_i^{G_i\,i}\boldsymbol{\omega}_i) \right] \cdot \delta\dot{\boldsymbol{q}} = 0$$
$$\forall \delta\dot{\boldsymbol{q}} \in \mathbb{R}^n. \tag{6.51}$$

In matrix form we have

$$\left[\boldsymbol{\tau} - \sum_{i=1}^{n_b} \begin{pmatrix} {}^i\boldsymbol{\Gamma}_{G_i}^T & {}^i\boldsymbol{\Pi}_i^T \end{pmatrix} \begin{pmatrix} M_i{}^i\boldsymbol{a}_{G_i} + M_i g{}^i\hat{\boldsymbol{e}}_{0_3} \\ {}^i\boldsymbol{I}_i^{G_i\,i}\boldsymbol{\alpha}_i + {}^i\boldsymbol{\omega}_i \times {}^i\boldsymbol{I}_i^{G_i\,i}\boldsymbol{\omega}_i \end{pmatrix} \right] \cdot \delta\dot{\boldsymbol{q}} = 0 \tag{6.52}$$
$$\forall \delta\dot{\boldsymbol{q}} \in \mathbb{R}^n,$$

which implies

$$\boldsymbol{\tau} = \sum_{i=1}^{n_b} \begin{pmatrix} {}^i\boldsymbol{\Gamma}_{G_i}^T & {}^i\boldsymbol{\Pi}_i^T \end{pmatrix} \begin{pmatrix} M_i{}^i\boldsymbol{a}_{G_i} + M_i g{}^i\hat{\boldsymbol{e}}_{0_3} \\ {}^i\boldsymbol{I}_i^{G_i\,i}\boldsymbol{\alpha}_i + {}^i\boldsymbol{\Omega}_i{}^i\boldsymbol{I}_i^{G_i\,i}\boldsymbol{\omega}_i \end{pmatrix}. \tag{6.53}$$

Since this is identical to (5.53), we have

$$\boldsymbol{\tau} = \boldsymbol{M}(\boldsymbol{q})\ddot{\boldsymbol{q}} + \boldsymbol{b}(\boldsymbol{q}, \dot{\boldsymbol{q}}) + \boldsymbol{g}(\boldsymbol{q}). \tag{6.54}$$

6.2.5 Auxiliary Constraints

First-Order Nonholonomic Constraints

Nonholonomic constraints are constraints that are *nonintegrable* functions of the generalized coordinates and generalized velocities (and possibly the generalized accelerations). The condition of nonintegrability means that not all constraints that are functions of the generalized velocities are in fact nonholonomic. For example, all smooth holonomic constraints can be differentiated, yielding constraints that are functions of the generalized velocities. However, these differentiated constraints are clearly integrable back to the original holonomic constraints. First-order nonholonomic constraints are nonintegrable functions exclusively of the generalized coordinates and generalized velocities. We will address these constraints here.

We previously considered the number of degrees of freedom for holonomically constrained systems. In systems with purely nonholomic constraints, all of the n-dimensional configuration space is accessible; however, the system is restricted in the *manner* in which points in configuration space can be reached. That is, the nonholmomic constraints do not place a restriction on points in configuration space that are allowable, but they do place a restriction on the differential motion allowed at a point in configuration space. The number of degrees of freedom of a nonholonomically constrained system is computed in the same way as that of a holonomically constrained system. That is,

$$p = n - m, \tag{6.55}$$

where n is the number of generalized coordinates and m is the number of independent nonholonomic constraint equations. The number of degrees of freedom refers to the dimensionality of the allowable differential motion of the system (not the dimensionality of the accessible configuration space).

We now consider the general case of auxiliary first-order nonholonomic constraints. Given the multibody system

$$\boldsymbol{\tau} = \boldsymbol{M}(\boldsymbol{q})\ddot{\boldsymbol{q}} + \boldsymbol{b}(\boldsymbol{q}, \dot{\boldsymbol{q}}) + \boldsymbol{g}(\boldsymbol{q}), \tag{6.56}$$

subject to the first-order nonholonomic constraints

$$\boldsymbol{\psi}(\boldsymbol{q}, \dot{\boldsymbol{q}}) = \boldsymbol{0}, \tag{6.57}$$

we begin by first expressing the first-order variational equation associated with Jourdain's Principle,

$$\boldsymbol{\tau}_C \cdot \delta\dot{\boldsymbol{q}} + (\boldsymbol{\tau} - \boldsymbol{M}\dot{\boldsymbol{q}} - \boldsymbol{b} - \boldsymbol{g}) \cdot \delta\dot{\boldsymbol{q}} = 0. \tag{6.58}$$

The virtual velocities, $\delta\dot{\boldsymbol{q}}$, refer to all velocity variations that satisfy the constraints, while time and displacement are fixed. With $\delta t = 0$ and $\delta\boldsymbol{q} = \boldsymbol{0}$, the variation of the constraint equation yields

$$\delta\boldsymbol{\psi} = \frac{\partial\boldsymbol{\psi}}{\partial\dot{\boldsymbol{q}}}\delta\dot{\boldsymbol{q}} = \boldsymbol{\Psi}\delta\dot{\boldsymbol{q}} = \boldsymbol{0}, \tag{6.59}$$

which implies that $\delta\dot{\boldsymbol{q}} \in \ker(\boldsymbol{\Psi})$. Under this condition, (6.58) can be restricted to constraint-consistent virtual displacements:

$$\begin{gathered} \boldsymbol{\tau}_C \cdot \delta\dot{\boldsymbol{q}} + (\boldsymbol{\tau} - \boldsymbol{M}\boldsymbol{q} - \boldsymbol{b} - \boldsymbol{g}) \cdot \delta\dot{\boldsymbol{q}} = 0 \\ \forall \delta\dot{\boldsymbol{q}} \in \ker(\boldsymbol{\Psi}). \end{gathered} \tag{6.60}$$

We have

$$\begin{gathered} \boldsymbol{\tau}_C \perp \delta\dot{\boldsymbol{q}} \\ \forall \delta\dot{\boldsymbol{q}} \in \ker(\boldsymbol{\Psi}). \end{gathered} \tag{6.61}$$

Thus, the term $\boldsymbol{\tau}_C \cdot \delta\dot{\boldsymbol{q}}$ vanishes from (6.60), and we have the orthogonality relation

$$\begin{gathered} (\boldsymbol{M}\boldsymbol{q} + \boldsymbol{b} + \boldsymbol{g} - \boldsymbol{\tau}) \cdot \delta\dot{\boldsymbol{q}} = 0 \\ \forall \delta\dot{\boldsymbol{q}} \in \ker(\boldsymbol{\Psi}). \end{gathered} \tag{6.62}$$

The constrained multibody equations of motion, expressed in the familiar multiplier form, are thus

$$\boldsymbol{\tau} = \boldsymbol{M}\ddot{\boldsymbol{q}} + \boldsymbol{b} + \boldsymbol{g} - \boldsymbol{\Psi}^T \boldsymbol{\lambda}, \tag{6.63}$$

subject to

$$\boldsymbol{\psi}(\boldsymbol{q}, \dot{\boldsymbol{q}}) = \mathbf{0} \quad \Rightarrow \quad \dot{\boldsymbol{\psi}} = \mathbf{0}. \tag{6.64}$$

In the special case of nonholonomic constraints that are linear in $\dot{\boldsymbol{q}}$, we have

$$\boldsymbol{\psi}(\boldsymbol{q}, \dot{\boldsymbol{q}}) = \boldsymbol{C}(\boldsymbol{q})\dot{\boldsymbol{q}} + \boldsymbol{d}(\boldsymbol{q}) = \mathbf{0}, \tag{6.65}$$

and the first-order variation is given by

$$\delta\boldsymbol{\psi} = \boldsymbol{C}\delta\dot{\boldsymbol{q}} = \mathbf{0}. \tag{6.66}$$

The constrained multibody equations of motion express as

$$\boldsymbol{\tau} = \boldsymbol{M}\ddot{\boldsymbol{q}} + \boldsymbol{b} + \boldsymbol{g} - \boldsymbol{C}^T \boldsymbol{\lambda}, \tag{6.67}$$

subject to

$$\boldsymbol{C}\dot{\boldsymbol{q}} + \boldsymbol{d} = \mathbf{0} \quad \Rightarrow \quad \dot{\boldsymbol{C}}\dot{\boldsymbol{q}} + \boldsymbol{C}\ddot{\boldsymbol{q}} + \dot{\boldsymbol{d}} = \mathbf{0}. \tag{6.68}$$

In the special case of holonomic constraints,

$$\boldsymbol{\phi}(\boldsymbol{q}) = \mathbf{0}, \tag{6.69}$$

we can differentiate once with respect to time to yield

$$\boldsymbol{\psi}(\boldsymbol{q}, \dot{\boldsymbol{q}}) = \dot{\boldsymbol{\phi}} = \frac{\partial \boldsymbol{\phi}}{\partial \boldsymbol{q}}\dot{\boldsymbol{q}} = \boldsymbol{\Phi}\dot{\boldsymbol{q}} = \mathbf{0}. \tag{6.70}$$

The variation is given by

$$\delta\boldsymbol{\psi} = \boldsymbol{\Phi}\delta\dot{\boldsymbol{q}} = \mathbf{0}, \tag{6.71}$$

and the constrained multibody equations of motion express identically to the zeroth-order variational case,

$$\boldsymbol{\tau} = \boldsymbol{M}\ddot{\boldsymbol{q}} + \boldsymbol{b} + \boldsymbol{g} - \boldsymbol{\Phi}^T \boldsymbol{\lambda}, \tag{6.72}$$

subject to

$$\boldsymbol{\phi}(\boldsymbol{q}) = \mathbf{0} \quad \Rightarrow \quad \dot{\boldsymbol{\phi}} = \mathbf{0}, \ddot{\boldsymbol{\phi}} = \mathbf{0}. \tag{6.73}$$

Numerical Integration

Given the nonholonomic system

$$\boldsymbol{\tau} = \boldsymbol{M}\ddot{\boldsymbol{q}} + \boldsymbol{b} + \boldsymbol{g} - \boldsymbol{C}^T \boldsymbol{\lambda}, \tag{6.74}$$

subject to

$$\boldsymbol{C}\dot{\boldsymbol{q}} = \mathbf{0} \quad \Rightarrow \quad \boldsymbol{C}\ddot{\boldsymbol{q}} + \dot{\boldsymbol{C}}\dot{\boldsymbol{q}} = \mathbf{0}, \tag{6.75}$$

Algorithm 6 Second-order method for integrating the first-order nonholonomically constrained equations of motion

1: $\boldsymbol{q}_0 = \boldsymbol{q}_o$ {initialization}
2: $\dot{\boldsymbol{q}}_0 = \dot{\boldsymbol{q}}_o$ {initialization}
3: **for** $i = 0$ to $n_s - 1$ **do**
4: $\quad \ddot{\boldsymbol{q}}_i = -\bar{\boldsymbol{C}}_i\dot{\boldsymbol{C}}_i\dot{\boldsymbol{q}}_i + \boldsymbol{\Theta}_i\boldsymbol{M}_i^{-1}(\boldsymbol{\tau}_i - \boldsymbol{b}_i - \boldsymbol{g}_i) - \beta\bar{\boldsymbol{C}}_i\boldsymbol{C}_i\dot{\boldsymbol{q}}_i$
5: $\quad \boldsymbol{\lambda}_i = -\bar{\boldsymbol{C}}_i^T(\boldsymbol{\tau}_i - \boldsymbol{b}_i - \boldsymbol{g}_i) - \mathbf{H}_i\dot{\boldsymbol{C}}_i\dot{\boldsymbol{q}}_i - \beta\mathbf{H}_i\boldsymbol{C}_i\dot{\boldsymbol{q}}_i$
6: $\quad \dot{\boldsymbol{q}}_{i+1} = \dot{\boldsymbol{q}}_i + \ddot{\boldsymbol{q}}_i\Delta t$
7: $\quad \boldsymbol{q}_{i+1} = \boldsymbol{q}_i + \dot{\boldsymbol{q}}_i\Delta t + \frac{1}{2}\ddot{\boldsymbol{q}}_i\Delta t^2$
8: **end for**

we wish to stabilize the constraints by replacing our original acceleration constraint equation with a linear combination of acceleration and velocity constraint terms,

$$\boldsymbol{C}\ddot{\boldsymbol{q}} + \dot{\boldsymbol{C}}\dot{\boldsymbol{q}} + \beta\boldsymbol{C}\dot{\boldsymbol{q}} = \mathbf{0}. \tag{6.76}$$

Thus, the constraint stabilized equations of motion are

$$\begin{pmatrix} \boldsymbol{M} & -\boldsymbol{C}^T \\ -\boldsymbol{C} & \mathbf{0} \end{pmatrix}\begin{pmatrix} \ddot{\boldsymbol{q}} \\ \boldsymbol{\lambda} \end{pmatrix} = \begin{pmatrix} \boldsymbol{\tau} - \boldsymbol{b} - \boldsymbol{g} \\ \dot{\boldsymbol{C}}\dot{\boldsymbol{q}} + \beta\boldsymbol{C}\dot{\boldsymbol{q}} \end{pmatrix}, \tag{6.77}$$

and the solution of this system is

$$\ddot{\boldsymbol{q}} = -\bar{\boldsymbol{C}}\dot{\boldsymbol{C}}\dot{\boldsymbol{q}} + \boldsymbol{\Theta}\boldsymbol{M}^{-1}(\boldsymbol{\tau} - \boldsymbol{b} - \boldsymbol{g}) - \beta\bar{\boldsymbol{C}}\boldsymbol{C}\dot{\boldsymbol{q}} \tag{6.78}$$

$$\boldsymbol{\lambda} = -\bar{\boldsymbol{C}}^T(\boldsymbol{\tau} - \boldsymbol{b} - \boldsymbol{g}) - \mathbf{H}\dot{\boldsymbol{C}}\dot{\boldsymbol{q}} - \beta\mathbf{H}\boldsymbol{C}\dot{\boldsymbol{q}}, \tag{6.79}$$

where $\boldsymbol{\Theta} = \mathbf{1} - \bar{\boldsymbol{C}}\boldsymbol{C}$ and $\mathbf{H} = (\boldsymbol{C}\boldsymbol{M}^{-1}\boldsymbol{C}^T)^{-1}$. The term $-\beta\mathbf{H}\boldsymbol{C}\dot{\boldsymbol{q}}$ in the expression for $\boldsymbol{\lambda}$ can be physically interpreted as a corrective constraint force term used to compensate for any drift in the constraints.

A second-order method for integrating this system can be summarized as shown in Algorithm 6.

Example: We consider a rolling disk which illustrates nonholonomic constraints. We parameterize the system using six generalized coordinates as shown in Figure 6.2. The order of rotations will be specified as a *zyx* Euler sequence:

$${}^{\mathcal{O}}_{\mathcal{D}}\boldsymbol{Q} = \boldsymbol{Q}_z(q_4)\boldsymbol{Q}_y(q_5)\boldsymbol{Q}_x(q_6) = \boldsymbol{Q}_{zyx}(\alpha, \beta, \gamma). \tag{6.80}$$

We can compute the angular velocity in the local frame by noting that

$${}^{\mathcal{D}}\boldsymbol{\Omega} = {}^{\mathcal{O}}_{\mathcal{D}}\boldsymbol{Q}^T\,{}^{\mathcal{O}}_{\mathcal{D}}\dot{\boldsymbol{Q}}. \tag{6.81}$$

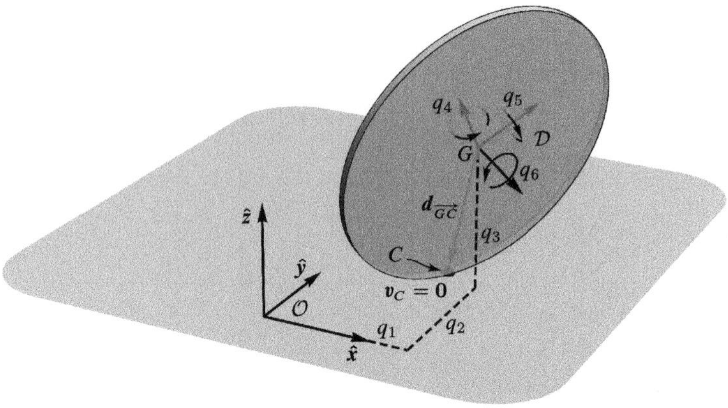

Figure 6.2 A rolling disk with nonholonomic constraints. Six generalized coordinates, $q_1, \ldots, q_6$, parameterize the position of the center of the disk and the disk orientation. Nonholonomic constraints arise out of the *zero* velocity (no-slip) condition imposed on the instantaneous contact point of the disk with the ground.

Carrying this operation out, we determine

$$^{\mathcal{D}}\boldsymbol{\omega} = \begin{pmatrix} -\sin(q_5) & 0 & 1 \\ \cos(q_5)\sin(q_6) & \cos(q_6) & 0 \\ \cos(q_5)\cos(q_6) & -\sin(q_6) & 0 \end{pmatrix} \begin{pmatrix} \dot{q}_4 \\ \dot{q}_5 \\ \dot{q}_6 \end{pmatrix} = \boldsymbol{E}(q_4, q_5, q_6) \begin{pmatrix} \dot{q}_4 \\ \dot{q}_5 \\ \dot{q}_6 \end{pmatrix}. \tag{6.82}$$

Alternately, we could have arrived at the same results by propagating the rotation rates forward, where

$$^{4}\boldsymbol{\omega}_4 = \dot{q}_4 \hat{\boldsymbol{e}}_3 = \begin{pmatrix} 0 \\ 0 \\ \dot{q}_4 \end{pmatrix} \tag{6.83}$$

and

$$^{5}\boldsymbol{\omega}_5 = \boldsymbol{Q}_y^T(q_5)\,^{4}\boldsymbol{\omega}_4 + \dot{q}_5 \hat{\boldsymbol{e}}_2 = \begin{pmatrix} -\sin(q_5)\dot{q}_4 \\ \dot{q}_5 \\ \cos(q_5)\dot{q}_4 \end{pmatrix}. \tag{6.84}$$

Finally,

$$^{\mathcal{D}}\boldsymbol{\omega} = \,^{6}\boldsymbol{\omega}_6 = \boldsymbol{Q}_x^T(q_6)\,^{5}\boldsymbol{\omega}_5 + \dot{q}_6 \hat{\boldsymbol{e}}_1 = \begin{pmatrix} -\sin(q_5)\dot{q}_4 + \dot{q}_6 \\ \cos(q_5)\sin(q_6)\dot{q}_4 + \cos(q_6)\dot{q}_5 \\ \cos(q_5)\cos(q_6)\dot{q}_4 - \sin(q_6)\dot{q}_5 \end{pmatrix}. \tag{6.85}$$

We now express the displacement vector from the center of the disk to the contact point of the disk with the plane. As the disk rotates about the x-axis, the displacement vector keeps the same orientation in the base frame. In the disk frame, the displacement vector rotates about the x-axis by $-q_6$. This can be expressed as

$$^{\mathcal{D}}\boldsymbol{d}_{\overrightarrow{GC}} = \boldsymbol{Q}_x(-q_6)[-r\hat{\boldsymbol{e}}_3], \tag{6.86}$$

or

$$ {}^{\mathcal{O}}\boldsymbol{d}_{\overrightarrow{GC}} = {}^{\mathcal{O}}_{\mathcal{D}}\boldsymbol{Q}\boldsymbol{Q}_x(-q_6)[-r\hat{\boldsymbol{e}}_3] = \boldsymbol{Q}_z(q_4)\boldsymbol{Q}_y(q_5)[-r\hat{\boldsymbol{e}}_3]. \tag{6.87} $$

The instantaneous velocity of the material point of the disk in contact with the plane is given by

$$ {}^{\mathcal{O}}\boldsymbol{v}_C = \begin{pmatrix} \dot{q}_1 \\ \dot{q}_2 \\ \dot{q}_3 \end{pmatrix} + {}^{\mathcal{O}}_{\mathcal{D}}\boldsymbol{Q}({}^{\mathcal{D}}\boldsymbol{\omega} \times {}^{\mathcal{D}}\boldsymbol{d}_{\overrightarrow{GC}}). \tag{6.88} $$

Under the condition that the disk does not slip relative to the plane, we have

$$ {}^{\mathcal{O}}\boldsymbol{v}_C = \boldsymbol{0}. \tag{6.89} $$

Evaluating this expression, we have

$$ {}^{\mathcal{O}}\boldsymbol{v}_C = \begin{pmatrix} \dot{q}_1 + r[\sin(q_4)\sin(q_5)\dot{q}_4 - \cos(q_4)\cos(q_5)\dot{q}_5 - \sin(q_4)\dot{q}_6] \\ \dot{q}_2 - r[\cos(q_4)\sin(q_5)\dot{q}_4 + \sin(q_4)\cos(q_5)\dot{q}_5 - \cos(q_4)\dot{q}_6] \\ \dot{q}_3 + r\sin(q_5)\dot{q}_5 \end{pmatrix} = \boldsymbol{0} \tag{6.90} $$

or

$$ \underbrace{\begin{pmatrix} 1 & 0 & 0 & r\sin(q_4)\sin(q_5) & -r\cos(q_4)\cos(q_5) & -r\sin(q_4) \\ 0 & 1 & 0 & -r\cos(q_4)\sin(q_5) & -r\sin(q_4)\cos(q_5) & r\cos(q_4) \\ 0 & 0 & 1 & 0 & r\sin(q_5) & 0 \end{pmatrix}}_{\boldsymbol{C}} \dot{\boldsymbol{q}} = \boldsymbol{C}\dot{\boldsymbol{q}} = \boldsymbol{0}, \tag{6.91} $$

which are our nonholonomic constraint equations.

We now note that

$$ {}^{\mathcal{O}}\boldsymbol{v}_G = {}^{\mathcal{O}}\boldsymbol{\Gamma}_G\dot{\boldsymbol{q}} \tag{6.92} $$

and

$$ {}^{\mathcal{D}}\boldsymbol{\omega} = {}^{\mathcal{D}}\boldsymbol{\Pi}\dot{\boldsymbol{q}}. \tag{6.93} $$

Evaluating the Jacobians, we have

$$ {}^{\mathcal{O}}\boldsymbol{\Gamma}_G = \begin{pmatrix} 1 & 0 & 0 & 0 & 0 & 0 \\ 0 & 1 & 0 & 0 & 0 & 0 \\ 0 & 0 & 1 & 0 & 0 & 0 \end{pmatrix} \tag{6.94} $$

and

$$ {}^{\mathcal{D}}\boldsymbol{\Pi} = \begin{pmatrix} \boldsymbol{0} & \boldsymbol{E}(q_4, q_5, q_6) \end{pmatrix} = \begin{pmatrix} 0 & 0 & 0 & -\sin(q_5) & 0 & 1 \\ 0 & 0 & 0 & \cos(q_5)\sin(q_6) & \cos(q_6) & 0 \\ 0 & 0 & 0 & \cos(q_5)\cos(q_6) & -\sin(q_6) & 0 \end{pmatrix}, \tag{6.95} $$

where

$$\boldsymbol{M} = M^{\mathcal{O}}\boldsymbol{\Gamma}_G^{T\,\mathcal{O}}\boldsymbol{\Gamma}_G + {}^{\mathcal{D}}\boldsymbol{\Pi}^{T\,\mathcal{D}}\boldsymbol{I}^{G\,\mathcal{D}}\boldsymbol{\Pi}, \tag{6.96}$$

$$\boldsymbol{b} = {}^{\mathcal{D}}\boldsymbol{\Pi}^T({}^{\mathcal{D}}\boldsymbol{I}^{G\,\mathcal{D}}\dot{\boldsymbol{\Pi}} + {}^{\mathcal{D}}\boldsymbol{\Omega}^{\mathcal{D}}\boldsymbol{I}^{G\,\mathcal{D}}\boldsymbol{\Pi})\dot{\boldsymbol{q}}, \tag{6.97}$$

$$\boldsymbol{g} = gM^{\mathcal{O}}\boldsymbol{\Gamma}_G^T\hat{\boldsymbol{e}}_3. \tag{6.98}$$

We will take the inertia tensor to be

$${}^{\mathcal{D}}\boldsymbol{I}^G = \begin{pmatrix} \frac{1}{2}Mr^2 & 0 & 0 \\ 0 & \frac{1}{4}Mr^2 & 0 \\ 0 & 0 & \frac{1}{4}Mr^2 \end{pmatrix}. \tag{6.99}$$

Evaluating the dynamical terms, we have

$$\boldsymbol{M} = \begin{pmatrix} M & 0 & 0 & 0 & 0 & 0 \\ 0 & M & 0 & 0 & 0 & 0 \\ 0 & 0 & M & 0 & 0 & 0 \\ 0 & 0 & 0 & \frac{1}{8}Mr^2[3-\cos(2q_5)] & 0 & -\frac{1}{2}Mr^2\sin(q_5) \\ 0 & 0 & 0 & 0 & \frac{1}{4}Mr^2 & 0 \\ 0 & 0 & 0 & -\frac{1}{2}Mr^2\sin(q_5) & 0 & \frac{1}{2}Mr^2 \end{pmatrix} \tag{6.100}$$

and

$$\boldsymbol{b} = \begin{pmatrix} 0 \\ 0 \\ 0 \\ \frac{1}{2}Mr^2\cos(q_5)\dot{q}_5[\sin(q_5)\dot{q}_4 - \dot{q}_6] \\ -\frac{1}{4}Mr^2\cos(q_5)\dot{q}_4[\sin(q_5)\dot{q}_4 - 2\dot{q}_6] \\ -\frac{1}{2}Mr^2\cos(q_5)\dot{q}_4\dot{q}_5 \end{pmatrix} \tag{6.101}$$

and

$$\boldsymbol{g} = \begin{pmatrix} 0 & 0 & Mg & 0 & 0 & 0 \end{pmatrix}^T. \tag{6.102}$$

The nonholonomically constrained equations of motion express as

$$\boldsymbol{\tau} = \boldsymbol{M}\ddot{\boldsymbol{q}} + \boldsymbol{b} + \boldsymbol{g} - \boldsymbol{C}^T\boldsymbol{\lambda}, \tag{6.103}$$

subject to

$$\boldsymbol{C}\dot{\boldsymbol{q}} = \boldsymbol{0} \quad \Rightarrow \quad \dot{\boldsymbol{C}}\dot{\boldsymbol{q}} + \boldsymbol{C}\ddot{\boldsymbol{q}} = \boldsymbol{0}. \tag{6.104}$$

We note that $n = 6$ and $m = 3$. The constrained system has $p = n - m = 3$ degrees of freedom.

The simulation results are displayed in Figures 6.3 and 6.4. All geometric and inertial constants were chosen to be 1. That is, $r = 1$ and $M = 1$. No generalized forces were

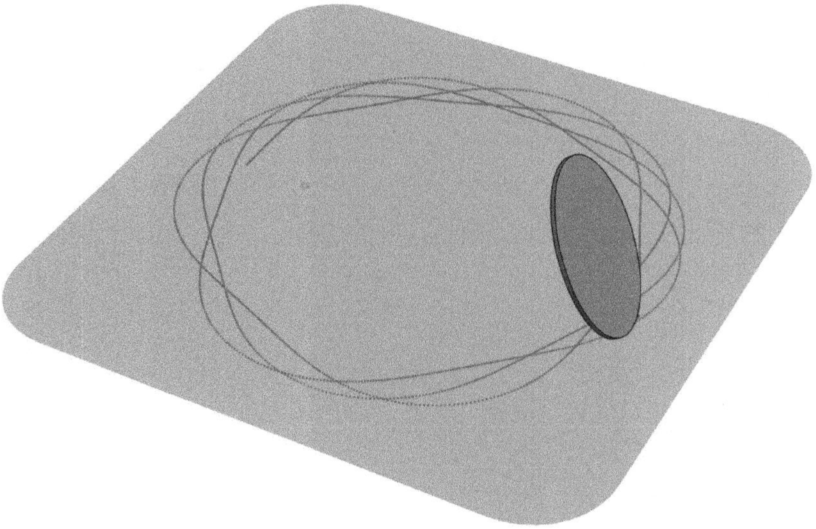

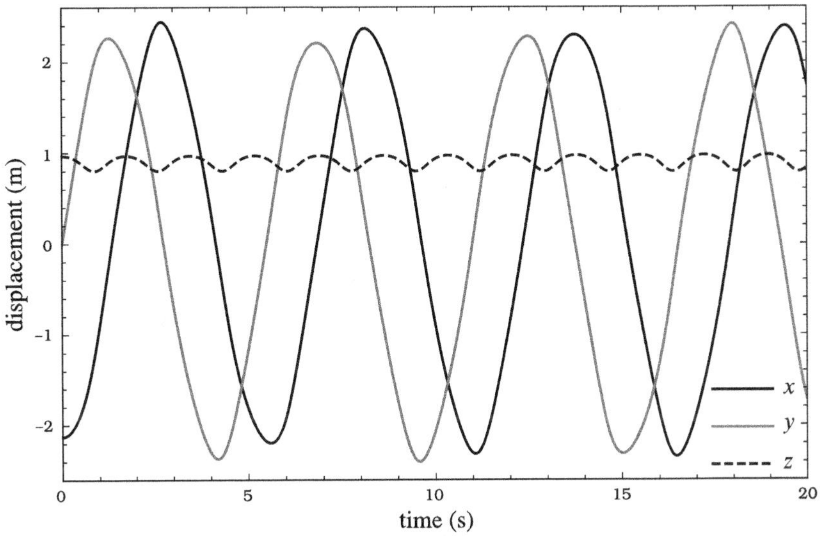

Figure 6.3 (Top) Animation frames from the simulation of a nonholonomic rolling disk. (Bottom) Time history of the position of the center of mass of the disk (x, y, z).

applied ($\boldsymbol{\tau} = \mathbf{0}$). The initial conditions used were

$$\boldsymbol{q}_o = \begin{pmatrix} -2.125 & 0 & \cos(\frac{\pi}{12}) & 0 & \frac{\pi}{12} & 0 \end{pmatrix}^T \tag{6.105}$$

$$\dot{\boldsymbol{q}}_o = \begin{pmatrix} 0 & 2.5 & 0 & 0 & 0 & -2.5 \end{pmatrix}^T. \tag{6.106}$$

It can easily be confirmed that these initial conditions satisfy the nonholonomic constraints, $\boldsymbol{C}\dot{\boldsymbol{q}} = \mathbf{0}$.

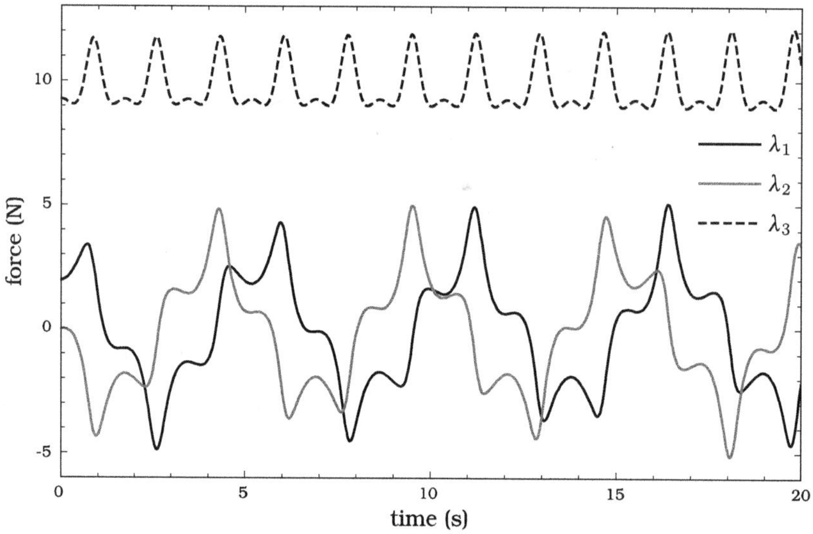

Figure 6.4 Time history of the constraint forces exerted on the disk at the contact point $(\lambda_1, \lambda_2, \lambda_3)$.

Example: A rolling ball can be modeled in a similar way as the rolling disk (see Figure 6.5). The order of rotations will be specified as a *zyx* Euler sequence, as with the disk,

$$ {}^{\mathcal{O}}_{\mathcal{B}}\boldsymbol{Q} = {}^{\mathcal{O}}_{6}\boldsymbol{Q} = \boldsymbol{Q}_z(q_4)\boldsymbol{Q}_y(q_5)\boldsymbol{Q}_x(q_6) = \boldsymbol{Q}_{zyx}(\alpha, \beta, \gamma). \tag{6.107} $$

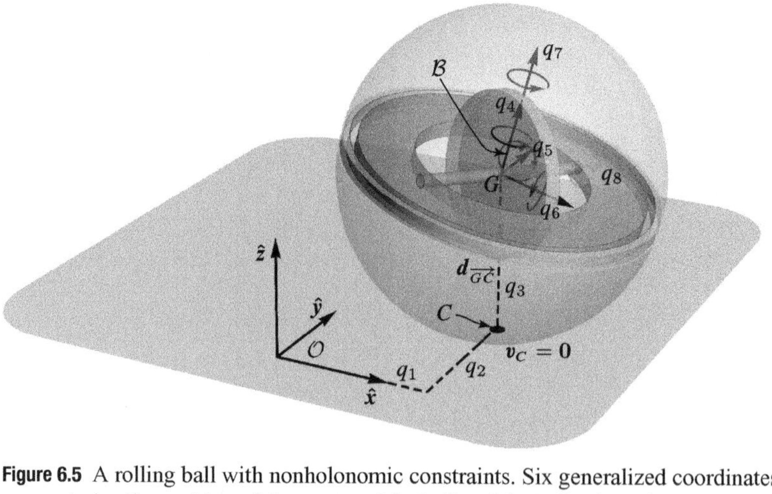

Figure 6.5 A rolling ball with nonholonomic constraints. Six generalized coordinates, $q_1, \ldots, q_6$, parameterize the position of the center of the ball and the ball orientation. Another two generalized coordinates, q_7 and q_8, parameterize the rotation of the two internal moment wheels. Nonholonomic constraints arise out of the *zero* velocity (no-slip) condition imposed on the instantaneous contact point of the ball with the ground.

We can compute the angular velocity of the ball in the local frame in the same manner as the disk:

$$ {}^{B}\boldsymbol{\omega}_{B} = {}^{6}\boldsymbol{\omega}_{6} = \begin{pmatrix} -\sin(q_5)\dot{q}_4 + \dot{q}_6 \\ \cos(q_5)\sin(q_6)\dot{q}_4 + \cos(q_6)\dot{q}_5 \\ \cos(q_5)\cos(q_6)\dot{q}_4 - \sin(q_6)\dot{q}_5 \end{pmatrix}. \tag{6.108} $$

The displacement vector from the center of the ball to the contact point of the ball with the plane is different than for the disk. The displacement vector remains fixed in the base frame. Thus,

$$ {}^{O}\boldsymbol{d}_{\overrightarrow{GC}} = -r\hat{\boldsymbol{e}}_3. \tag{6.109} $$

The instantaneous velocity of the material point of the disk in contact with the plane is given by

$$ {}^{O}\boldsymbol{v}_{C} = \begin{pmatrix} \dot{q}_1 \\ \dot{q}_2 \\ \dot{q}_3 \end{pmatrix} + \left({}^{O}_{B}\boldsymbol{Q}\,{}^{B}\boldsymbol{\omega}_{B}\right) \times {}^{O}\boldsymbol{d}_{\overrightarrow{GC}}. \tag{6.110} $$

Evaluating the no-slip condition yields

$$ {}^{O}\boldsymbol{v}_{C} = \begin{pmatrix} 1 & 0 & 0 & 0 & -r\cos(q_4) & -r\sin(q_4)\cos(q_5) \\ 0 & 1 & 0 & 0 & -r\sin(q_4) & r\cos(q_4)\cos(q_5) \\ 0 & 0 & 1 & 0 & 0 & 0 \end{pmatrix} \begin{pmatrix} q_1 \\ \vdots \\ q_6 \end{pmatrix} = \boldsymbol{0}. \tag{6.111} $$

We will consider that the ball has two internal actuators that generate torque at joints 7 and 8. These actuators behave as moment wheels. We will assume all masses to be the same (unity) and take the inertia tensors of the bodies associated with joints 6, 7, and 8 to be

$$ {}^{6}\boldsymbol{I}^{G}_{6} = \begin{pmatrix} 0.5 & 0 & 0 \\ 0 & 0.5 & 0 \\ 0 & 0 & 0.5 \end{pmatrix}, \tag{6.112} $$

$$ {}^{7}\boldsymbol{I}^{G}_{7} = \begin{pmatrix} 0.25 & 0 & 0 \\ 0 & 0.25 & 0 \\ 0 & 0 & 0.5 \end{pmatrix}, \tag{6.113} $$

$$ {}^{8}\boldsymbol{I}^{G}_{8} = \begin{pmatrix} 0.25 & 0 & 0 \\ 0 & 0.5 & 0 \\ 0 & 0 & 0.25 \end{pmatrix}. \tag{6.114} $$

The kinematics are given by

$$ {}^{O}_{7}\boldsymbol{Q} = {}^{O}_{6}\boldsymbol{Q}\boldsymbol{Q}_z(q_7) \tag{6.115} $$

$$ {}^{O}_{8}\boldsymbol{Q} = {}^{O}_{6}\boldsymbol{Q}\boldsymbol{Q}_z(q_7)\boldsymbol{Q}_y(q_8). \tag{6.116} $$

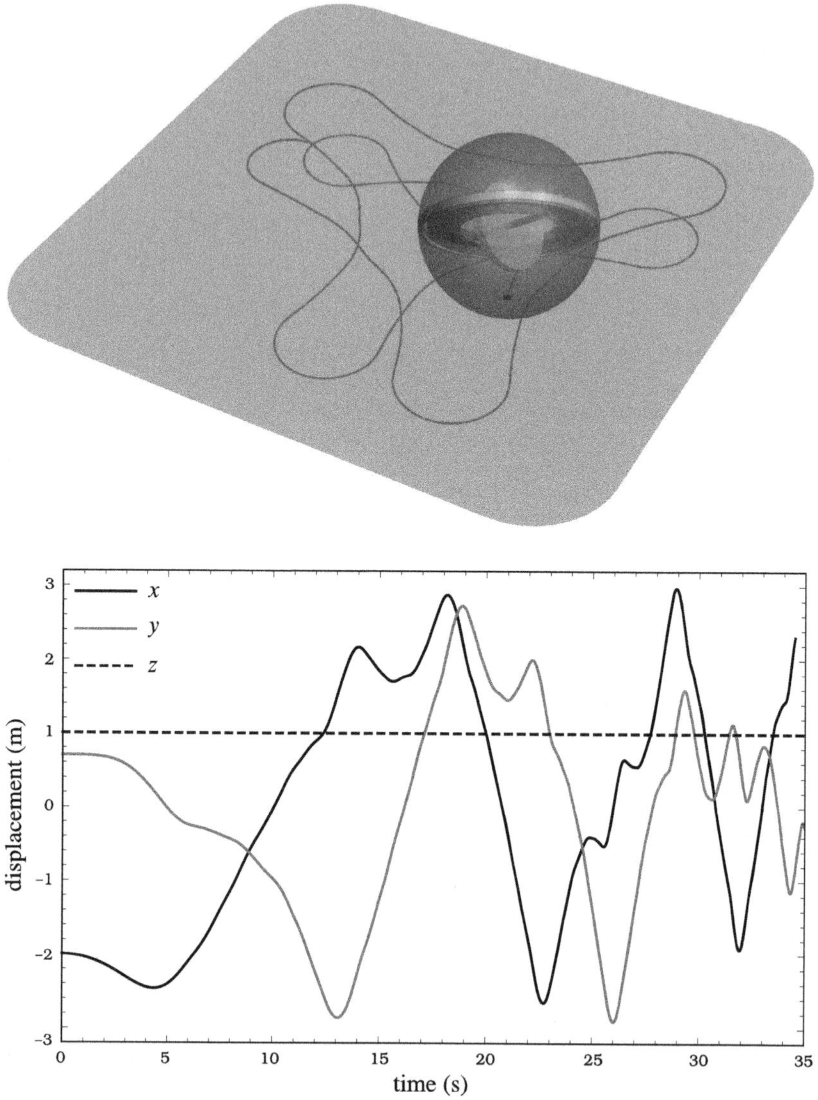

Figure 6.6 (Top) Animation frame from the simulation of the nonholonomic rolling ball. (Bottom) Time history of the position of the center of mass of the ball (x, y, z).

The angular velocities of the moment wheels (frames 7 and 8) can be computed from

$$ {}^7\boldsymbol{\Omega}_7 = {}^O_7\boldsymbol{Q}^T\,{}^O_7\dot{\boldsymbol{Q}} \tag{6.117}$$

$$ {}^8\boldsymbol{\Omega}_8 = {}^O_8\boldsymbol{Q}^T\,{}^O_8\dot{\boldsymbol{Q}}. \tag{6.118}$$

The angular velocity Jacobians, ${}^6\boldsymbol{\Pi}_6$, ${}^7\boldsymbol{\Pi}_7$, and ${}^8\boldsymbol{\Pi}_8$, follow from the angular velocities. For the translational velocity Jacobians, noting that all the mass centers are concentric,

we have

$$ {}^{O}\boldsymbol{\Gamma}_{G_6} = {}^{O}\boldsymbol{\Gamma}_{G_7} = {}^{O}\boldsymbol{\Gamma}_{G_8} = \begin{pmatrix} 1 & 0 & 0 & 0 & 0 & 0 & 0 & 0 \\ 0 & 1 & 0 & 0 & 0 & 0 & 0 & 0 \\ 0 & 0 & 1 & 0 & 0 & 0 & 0 & 0 \end{pmatrix} = {}^{O}\boldsymbol{\Gamma}_{G}. \tag{6.119} $$

Given

$$ \boldsymbol{M}(\boldsymbol{q}) = 3M{}^{O}\boldsymbol{\Gamma}_G^{T}{}^{O}\boldsymbol{\Gamma}_G + \sum_{i=6}^{8} {}^{i}\boldsymbol{\Pi}_i^{T}{}^{i}\boldsymbol{I}_i^{Gi}{}^{i}\boldsymbol{\Pi}_i, \tag{6.120} $$

$$ \boldsymbol{b}(\boldsymbol{q}, \dot{\boldsymbol{q}}) = \sum_{i=6}^{8} {}^{i}\boldsymbol{\Pi}_i^{T}({}^{i}\boldsymbol{I}_i^{Gi}{}^{i}\dot{\boldsymbol{\Pi}}_i + {}^{i}\boldsymbol{\Omega}_i{}^{i}\boldsymbol{I}_i^{Gi}{}^{i}\boldsymbol{\Pi}_i)\dot{\boldsymbol{q}}, \tag{6.121} $$

$$ \boldsymbol{g}(\boldsymbol{q}) = 3Mg{}^{O}\boldsymbol{\Gamma}_G^{T}\hat{\boldsymbol{e}}_3, \tag{6.122} $$

the nonholonomically constrained equations of motion express as

$$ \boldsymbol{\tau} = \boldsymbol{M}\ddot{\boldsymbol{q}} + \boldsymbol{b} + \boldsymbol{g} - \boldsymbol{C}^{T}\boldsymbol{\lambda}, \tag{6.123} $$

subject to

$$ \boldsymbol{C}\dot{\boldsymbol{q}} = \boldsymbol{0} \quad \Rightarrow \quad \dot{\boldsymbol{C}}\dot{\boldsymbol{q}} + \boldsymbol{C}\ddot{\boldsymbol{q}} = \boldsymbol{0}, \tag{6.124} $$

where

$$ \boldsymbol{C} = \begin{pmatrix} 1 & 0 & 0 & 0 & -r\cos(q_4) & -r\sin(q_4)\cos(q_5) & 0 & 0 \\ 0 & 1 & 0 & 0 & -r\sin(q_4) & r\cos(q_4)\cos(q_5) & 0 & 0 \\ 0 & 0 & 1 & 0 & 0 & 0 & 0 & 0 \end{pmatrix}. \tag{6.125} $$

We note that $n = 8$ and $m = 3$. The constrained system has $p = n - m = 5$ degrees of freedom.

The simulation results are displayed in Figures 6.6 and 6.7. All geometric and inertial constants were chosen to be 1. That is, the radius of the ball is 1 and all masses are 1. The initial conditions used were

$$ \boldsymbol{q}_o = (-2 \quad 0.7 \quad 1 \quad 0 \quad 0 \quad 0 \quad 0 \quad 0)^{T} \tag{6.126} $$

$$ \dot{\boldsymbol{q}}_o = (0 \quad 0 \quad 0 \quad 0 \quad 0 \quad 0 \quad 0 \quad 0)^{T}. \tag{6.127} $$

It can easily be confirmed that these initial conditions satisfy the nonholonomic constraints, $\boldsymbol{C}\dot{\boldsymbol{q}} = \boldsymbol{0}$. The applied torques used were

$$ \boldsymbol{\tau} = (0 \quad 0 \quad 0 \quad 0 \quad 0 \quad 0 \quad 0.1 \quad 0.275)^{T}. \tag{6.128} $$

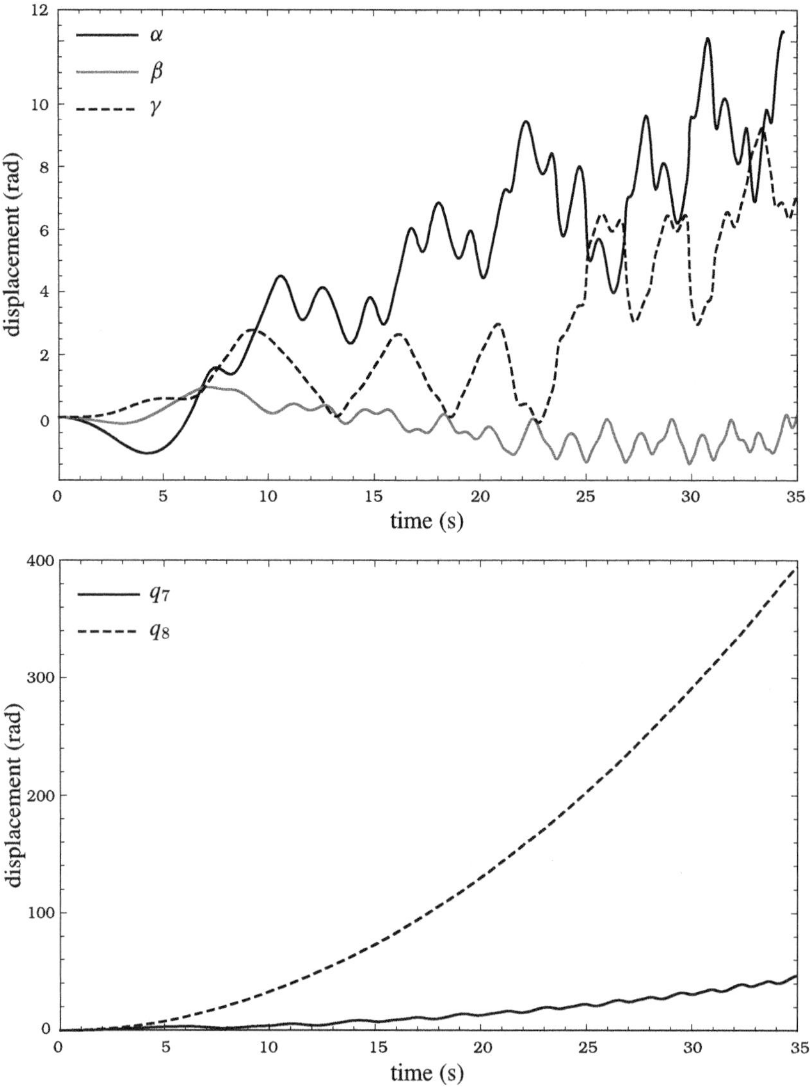

Figure 6.7 (Top) Time history of the orientation of the ball (α, β, γ). (Bottom) Time history of the moment wheel angles, q_7 and q_8.

Example: We now consider two wheels connected by an axle. We parameterize the system using eight generalized coordinates, as shown in Figure 6.8. The order of rotations of the axle will be specified as a *zyx* Euler sequence, as used previously:

$$ {}^{\mathcal{O}}_{\mathcal{A}}\boldsymbol{Q} = \boldsymbol{Q}_z(q_4)\boldsymbol{Q}_y(q_5)\boldsymbol{Q}_x(q_6). \tag{6.129}$$

The rotation matrices for the two wheels are

$$ {}^{\mathcal{O}}_{\mathcal{D}_1}\boldsymbol{Q} = \boldsymbol{Q}_z(q_4)\boldsymbol{Q}_y(q_5)\boldsymbol{Q}_x(q_6)\boldsymbol{Q}_x(q_7) \tag{6.130}$$

$$ {}^{\mathcal{O}}_{\mathcal{D}_2}\boldsymbol{Q} = \boldsymbol{Q}_z(q_4)\boldsymbol{Q}_y(q_5)\boldsymbol{Q}_x(q_6)\boldsymbol{Q}_x(q_8). \tag{6.131}$$

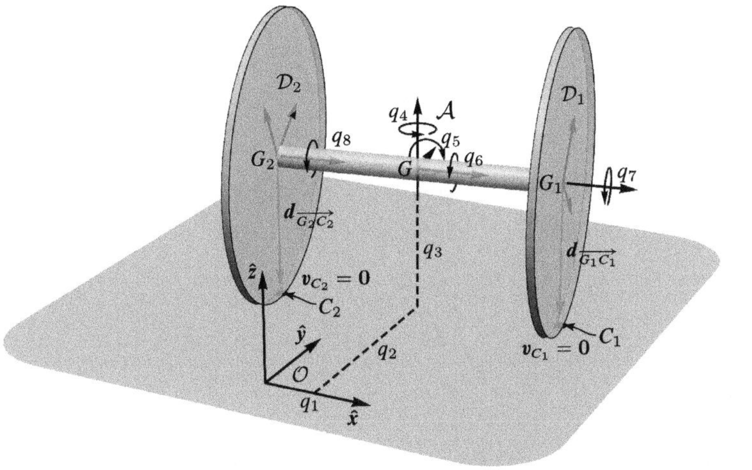

Figure 6.8 Two wheels connected by an axle with nonholonomic constraints. Six generalized coordinates, $q_1, \ldots, q_6$, parameterize the position of the center of the axle and the axle orientation. Another two generalized coordinates, q_7 and q_8, parameterize the rotation of the two wheels. The wheel radii are r, the axle radius is r_A, and the axle length is l.

The angular velocity for the axle in the local frame is identical to that of the disk and ball in the previous examples:

$$ {}^{\mathcal{A}}\boldsymbol{\omega}_{\mathcal{A}} = \begin{pmatrix} -\sin(q_5) & 0 & 1 \\ \cos(q_5)\sin(q_6) & \cos(q_6) & 0 \\ \cos(q_5)\cos(q_6) & -\sin(q_6) & 0 \end{pmatrix} \begin{pmatrix} \dot{q}_4 \\ \dot{q}_5 \\ \dot{q}_6 \end{pmatrix}. \tag{6.132} $$

Given

$$ {}^{\mathcal{D}_1}\boldsymbol{\Omega}_{\mathcal{D}_1} = {}^{\mathcal{O}}_{\mathcal{D}_1}\boldsymbol{Q}^{T}\, {}^{\mathcal{O}}_{\mathcal{D}_1}\dot{\boldsymbol{Q}}, \tag{6.133} $$

$$ {}^{\mathcal{D}_2}\boldsymbol{\Omega}_{\mathcal{D}_2} = {}^{\mathcal{O}}_{\mathcal{D}_2}\boldsymbol{Q}^{T}\, {}^{\mathcal{O}}_{\mathcal{D}_2}\dot{\boldsymbol{Q}}. \tag{6.134} $$

The angular velocities of the wheels in the local frames are

$$ {}^{\mathcal{D}_1}\boldsymbol{\omega}_{\mathcal{D}_1} = \begin{pmatrix} -\sin(q_5)\dot{q}_4 + \dot{q}_6 + \dot{q}_7 \\ \cos(q_5)\sin(q_6+q_7)\dot{q}_4 + \cos(q_6+q_7)\dot{q}_5 \\ \cos(q_5)\cos(q_6+q_7)\dot{q}_4 - \sin(q_6+q_7)\dot{q}_5 \end{pmatrix} \tag{6.135} $$

$$ {}^{\mathcal{D}_2}\boldsymbol{\omega}_{\mathcal{D}_2} = \begin{pmatrix} -\sin(q_5)\dot{q}_4 + \dot{q}_6 + \dot{q}_8 \\ \cos(q_5)\sin(q_6+q_8)\dot{q}_4 + \cos(q_6+q_8)\dot{q}_5 \\ \cos(q_5)\cos(q_6+q_8)\dot{q}_4 - \sin(q_6+q_8)\dot{q}_5 \end{pmatrix} \tag{6.136} $$

or, alternatively,

$$ {}^{\mathcal{D}_1}\boldsymbol{\omega}_{\mathcal{D}_1} = \boldsymbol{Q}_x(q_7)^{T}\, {}^{\mathcal{A}}\boldsymbol{\omega}_{\mathcal{A}} + \dot{q}_7\hat{\boldsymbol{e}}_1 \tag{6.137} $$

$$ {}^{\mathcal{D}_1}\boldsymbol{\omega}_{\mathcal{D}_2} = \boldsymbol{Q}_x(q_8)^{T}\, {}^{\mathcal{A}}\boldsymbol{\omega}_{\mathcal{A}} + \dot{q}_8\hat{\boldsymbol{e}}_1. \tag{6.138} $$

We now express the displacement vectors from the centers of the wheels to the contact points of the wheels with the plane. Since the wheels are always upright in the base frame, we have

$$ {}^{O}\boldsymbol{d}_{\overrightarrow{G_1C_1}} = {}^{O}\boldsymbol{d}_{\overrightarrow{G_2C_2}} = -r\hat{\boldsymbol{e}}_3. \tag{6.139}$$

The instantaneous velocity of the material point of each wheel in contact with the plane is given by

$$ {}^{O}\boldsymbol{v}_{C_1} = {}^{O}\boldsymbol{v}_{G_1} + ({}^{O}_{D_1}\boldsymbol{Q}\,{}^{D_1}\boldsymbol{\omega}_{D_1}) \times {}^{O}\boldsymbol{d}_{\overrightarrow{G_1C_1}} \tag{6.140}$$

$$ {}^{O}\boldsymbol{v}_{C_2} = {}^{O}\boldsymbol{v}_{G_2} + ({}^{O}_{D_2}\boldsymbol{Q}\,{}^{D_2}\boldsymbol{\omega}_{D_2}) \times {}^{O}\boldsymbol{d}_{\overrightarrow{G_2C_2}}, \tag{6.141}$$

where

$$ {}^{O}\boldsymbol{v}_{G_1} = {}^{O}\boldsymbol{v}_{G} + {}^{O}_{A}\boldsymbol{Q}\left({}^{A}\boldsymbol{\omega}_{A} \times \frac{l}{2}\hat{\boldsymbol{e}}_1\right), \tag{6.142}$$

$$ {}^{O}\boldsymbol{v}_{G_2} = {}^{O}\boldsymbol{v}_{G} + {}^{O}_{A}\boldsymbol{Q}\left({}^{A}\boldsymbol{\omega}_{A} \times \frac{-l}{2}\hat{\boldsymbol{e}}_1\right), \tag{6.143}$$

$$ {}^{O}\boldsymbol{v}_{G} = \begin{pmatrix} \dot{q}_1 \\ \dot{q}_2 \\ \dot{q}_3 \end{pmatrix}. \tag{6.144}$$

Under the condition that the wheels do not slip relative to the plane, we have

$$ \boldsymbol{v}_{C_1} = \boldsymbol{0} \tag{6.145}$$

$$ \boldsymbol{v}_{C_2} = \boldsymbol{0}. \tag{6.146}$$

We note that these two vector constraint equations contain a redundant constraint. Specifically, if one wheel is not allowed to move in the axial direction, the other wheel will automatically satisfy this constraint. Expressing the constraints in the axle frame and removing the first component (axial component) from ${}^{A}\boldsymbol{v}_{C_2}$ yields

$$ \begin{pmatrix} {}^{A}v_{C_{1x}} \\ {}^{A}v_{C_{1y}} \\ {}^{A}v_{C_{1z}} \\ {}^{A}v_{C_{2y}} \\ {}^{A}v_{C_{2z}} \end{pmatrix} = \boldsymbol{C}\dot{\boldsymbol{q}} = \boldsymbol{0}, \tag{6.147}$$

which are our nonholonomic constraint equations.

We now note that

$$ {}^{O}\boldsymbol{v}_{G} = {}^{O}\boldsymbol{\Gamma}_{G}\dot{\boldsymbol{q}}, \tag{6.148}$$

$$ {}^{O}\boldsymbol{v}_{G_1} = {}^{O}\boldsymbol{\Gamma}_{G_1}\dot{\boldsymbol{q}}, \tag{6.149}$$

$$ {}^{O}\boldsymbol{v}_{G_2} = {}^{O}\boldsymbol{\Gamma}_{G_2}\dot{\boldsymbol{q}}, \tag{6.150}$$

and that

$$ {}^{\mathcal{A}}\boldsymbol{\omega}_{\mathcal{A}} = {}^{\mathcal{A}}\boldsymbol{\Pi}_{\mathcal{A}}\dot{\boldsymbol{q}}, \tag{6.151}$$

$$ {}^{\mathcal{D}_1}\boldsymbol{\omega}_{\mathcal{D}_1} = {}^{\mathcal{D}_1}\boldsymbol{\Pi}_{\mathcal{D}_1}\dot{\boldsymbol{q}}, \tag{6.152}$$

$$ {}^{\mathcal{D}_2}\boldsymbol{\omega}_{\mathcal{D}_2} = {}^{\mathcal{D}_2}\boldsymbol{\Pi}_{\mathcal{D}_2}\dot{\boldsymbol{q}}. \tag{6.153}$$

Evaluating the Jacobians, we have

$$ {}^{\mathcal{O}}\boldsymbol{\Gamma}_{G} = \begin{pmatrix} 1 & 0 & 0 & 0 & 0 & 0 & 0 & 0 \\ 0 & 1 & 0 & 0 & 0 & 0 & 0 & 0 \\ 0 & 0 & 1 & 0 & 0 & 0 & 0 & 0 \end{pmatrix}, \tag{6.154}$$

$$ {}^{\mathcal{O}}\boldsymbol{\Gamma}_{G_1} = \begin{pmatrix} 1 & 0 & 0 & -\frac{1}{2}l\sin(q_4)\cos(q_5) & -\frac{1}{2}l\cos(q_4)\sin(q_5) & 0 & 0 & 0 \\ 0 & 1 & 0 & \frac{1}{2}l\cos(q_4)\cos(q_5) & -\frac{1}{2}l\sin(q_4)\sin(q_5) & 0 & 0 & 0 \\ 0 & 0 & 1 & 0 & -\frac{1}{2}l\cos(q_5) & 0 & 0 & 0 \end{pmatrix}, \tag{6.155}$$

$$ {}^{\mathcal{O}}\boldsymbol{\Gamma}_{G_2} = \begin{pmatrix} 1 & 0 & 0 & \frac{1}{2}l\sin(q_4)\cos(q_5) & \frac{1}{2}l\cos(q_4)\sin(q_5) & 0 & 0 & 0 \\ 0 & 1 & 0 & -\frac{1}{2}l\cos(q_4)\cos(q_5) & \frac{1}{2}l\sin(q_4)\sin(q_5) & 0 & 0 & 0 \\ 0 & 0 & 1 & 0 & \frac{1}{2}l\cos(q_5) & 0 & 0 & 0 \end{pmatrix}, \tag{6.156}$$

and

$$ {}^{\mathcal{A}}\boldsymbol{\Pi}_{\mathcal{A}} = \begin{pmatrix} 0 & 0 & 0 & -\sin(q_5) & 0 & 1 & 0 & 0 \\ 0 & 0 & 0 & \cos(q_5)\sin(q_6) & \cos(q_6) & 0 & 0 & 0 \\ 0 & 0 & 0 & \cos(q_5)\cos(q_6) & -\sin(q_6) & 0 & 0 & 0 \end{pmatrix}, \tag{6.157}$$

$$ {}^{\mathcal{D}_1}\boldsymbol{\Pi}_{\mathcal{D}_1} = \begin{pmatrix} 0 & 0 & 0 & -\sin(q_5) & 0 & 1 & 1 & 0 \\ 0 & 0 & 0 & \cos(q_5)\sin(q_6+q_7) & \cos(q_6+q_7) & 0 & 0 & 0 \\ 0 & 0 & 0 & \cos(q_5)\cos(q_6+q_7) & -\sin(q_6+q_7) & 0 & 0 & 0 \end{pmatrix}, \tag{6.158}$$

$$ {}^{\mathcal{D}_2}\boldsymbol{\Pi}_{\mathcal{D}_2} = \begin{pmatrix} 0 & 0 & 0 & -\sin(q_5) & 0 & 1 & 0 & 1 \\ 0 & 0 & 0 & \cos(q_5)\sin(q_6+q_8) & \cos(q_6+q_8) & 0 & 0 & 0 \\ 0 & 0 & 0 & \cos(q_5)\cos(q_6+q_8) & -\sin(q_6+q_8) & 0 & 0 & 0 \end{pmatrix}. \tag{6.159}$$

We will assume all masses to be the same and take the inertia tensors to be

$$
{}^{\mathcal{A}}\boldsymbol{I}^{G}_{\mathcal{A}} = \begin{pmatrix} \frac{1}{2}Mr_{\mathcal{A}}^2 & 0 & 0 \\ 0 & \frac{1}{2}Mr_{\mathcal{A}}^2 + \frac{1}{12}Ml^2 & 0 \\ 0 & 0 & \frac{1}{2}Mr_{\mathcal{A}}^2 + \frac{1}{12}Ml^2 \end{pmatrix} \tag{6.160}
$$

$$
{}^{\mathcal{D}_1}\boldsymbol{I}^{G_1}_{\mathcal{D}_1} = {}^{\mathcal{D}_2}\boldsymbol{I}^{G_2}_{\mathcal{D}_2} = \begin{pmatrix} \frac{1}{2}Mr^2 & 0 & 0 \\ 0 & \frac{1}{4}Mr^2 & 0 \\ 0 & 0 & \frac{1}{4}Mr^2 \end{pmatrix}, \tag{6.161}
$$

where

$$
\boldsymbol{M} = M{}^{\mathcal{O}}\boldsymbol{\Gamma}^T_{G}{}^{\mathcal{O}}\boldsymbol{\Gamma}_{G} + M\sum_{i=1}^{2}{}^{\mathcal{O}}\boldsymbol{\Gamma}^T_{G_i}{}^{\mathcal{O}}\boldsymbol{\Gamma}_{G_i} + {}^{\mathcal{A}}\boldsymbol{\Pi}^T_{\mathcal{A}}{}^{\mathcal{A}}\boldsymbol{I}^{G\mathcal{A}}_{\mathcal{A}}\boldsymbol{\Pi}_{\mathcal{A}} + \sum_{i=1}^{2}{}^{\mathcal{D}_i}\boldsymbol{\Pi}^T_{\mathcal{D}_i}{}^{\mathcal{D}_i}\boldsymbol{I}^{G_i\mathcal{D}_i}_{\mathcal{D}_i}\boldsymbol{\Pi}_{\mathcal{D}_i}, \tag{6.162}
$$

$$
\begin{aligned}
\boldsymbol{b} = {} & M\sum_{i=1}^{2}{}^{\mathcal{O}}\boldsymbol{\Gamma}^T_{G_i}{}^{\mathcal{O}}\dot{\boldsymbol{\Gamma}}_{G_i}\dot{\boldsymbol{q}} + {}^{\mathcal{A}}\boldsymbol{\Pi}^T_{\mathcal{A}}({}^{\mathcal{A}}\boldsymbol{I}^{G\mathcal{A}}_{\mathcal{A}}\dot{\boldsymbol{\Pi}}_{\mathcal{A}} + {}^{\mathcal{A}}\boldsymbol{\Omega}_{\mathcal{A}}{}^{\mathcal{A}}\boldsymbol{I}^{G\mathcal{A}}_{\mathcal{A}}\boldsymbol{\Pi}_{\mathcal{A}})\dot{\boldsymbol{q}} \\
& + \sum_{i=1}^{2}{}^{\mathcal{D}_i}\boldsymbol{\Pi}^T_{\mathcal{D}_i}({}^{\mathcal{D}_i}\boldsymbol{I}^{G_i\mathcal{D}_i}_{\mathcal{D}_i}\dot{\boldsymbol{\Pi}}_{\mathcal{D}_i} + {}^{\mathcal{D}_i}\boldsymbol{\Omega}_{\mathcal{D}_i}{}^{\mathcal{D}_i}\boldsymbol{I}^{G_i\mathcal{D}_i}_{\mathcal{D}_i}\boldsymbol{\Pi}_{\mathcal{D}_i})\dot{\boldsymbol{q}},
\end{aligned} \tag{6.163}
$$

$$
\boldsymbol{g} = gM{}^{\mathcal{O}}\boldsymbol{\Gamma}^T_{G}\hat{\boldsymbol{e}}_3 + gM\sum_{i=1}^{2}{}^{\mathcal{O}}\boldsymbol{\Gamma}^T_{G_i}\hat{\boldsymbol{e}}_3. \tag{6.164}
$$

Evaluating, we have

$$
\boldsymbol{M} = \left(\begin{matrix} 3M & 0 & 0 & 0 \\ 0 & 3M & 0 & 0 \\ 0 & 0 & 3M & 0 \\ 0 & 0 & 0 & \frac{1}{24}M[7l^2 + 9(r_{\mathcal{A}}^2 + 2r^2) + (7l^2 - 3(r_{\mathcal{A}}^2 + 2r^2))\cos(2q_5)] \\ 0 & 0 & 0 & 0 \\ 0 & 0 & 0 & -\frac{1}{2}M(r_{\mathcal{A}}^2 + 2r^2)\sin(q_5) \\ 0 & 0 & 0 & -\frac{1}{2}M(r^2)\sin(q_5) \\ 0 & 0 & 0 & -\frac{1}{2}M(r^2)\sin(q_5) \end{matrix}\right.
$$

$$
\left.\cdots \begin{matrix} 0 & 0 & 0 & 0 \\ 0 & 0 & 0 & 0 \\ 0 & 0 & 0 & 0 \\ 0 & 0 & 0 & 0 \\ \frac{1}{12}M(7l^2 + 3r_{\mathcal{A}}^2 + 6r^2) & 0 & 0 & 0 \\ 0 & \frac{1}{2}M(r_{\mathcal{A}}^2 + r^2) & \frac{1}{2}Mr^2 & \frac{1}{2}Mr^2 \\ 0 & \frac{1}{2}Mr^2 & \frac{1}{2}Mr^2 & 0 \\ 0 & \frac{1}{2}Mr^2 & 0 & \frac{1}{2}Mr^2 \end{matrix}\right), \tag{6.165}
$$

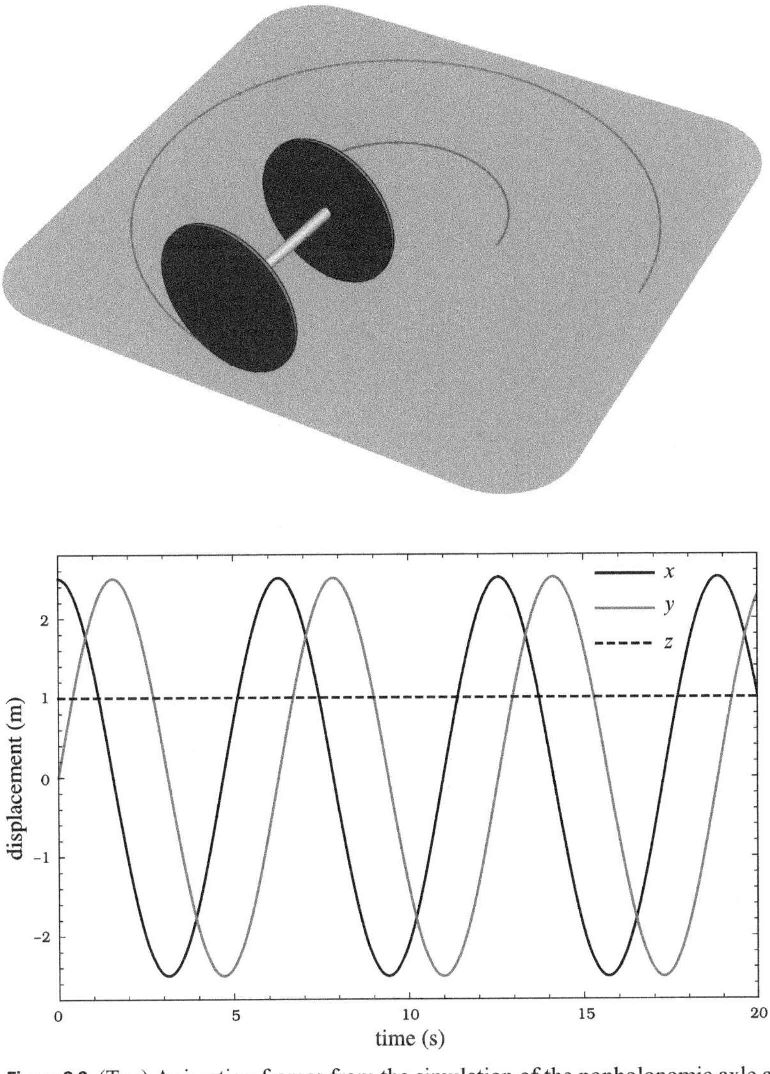

Figure 6.9 (Top) Animation frames from the simulation of the nonholonomic axle and rolling wheels. (Bottom) Time history of the position of the center of mass of the assembly (x, y, z).

and

$$
\boldsymbol{b} = \begin{pmatrix} 0 \\ 0 \\ 0 \\ -\frac{1}{6}M\cos(q_5)\dot{q}_5[(7l^2 - 3(r_A^2 + 2r^2))\sin(q_5)\dot{q}_4 + 3((r_A^2 + r^2)\dot{q}_6 + r^2(\dot{q}_7 + \dot{q}_8))] \\ -\frac{1}{12}M\cos(q_5)\dot{q}_4[(7l^2 - 3(r_A^2 + 2r^2))\sin(q_5)\dot{q}_4 + 6((r_A^2 + r^2)\dot{q}_6 + r^2(\dot{q}_7 + \dot{q}_8))] \\ -\frac{1}{2}M(r_A^2 + 2r^2)\cos(q_5)\dot{q}_4\dot{q}_5 \\ -\frac{1}{2}Mr^2\cos(q_5)\dot{q}_4\dot{q}_5 \\ -\frac{1}{2}Mr^2\cos(q_5)\dot{q}_4\dot{q}_5 \end{pmatrix}, \tag{6.166}
$$

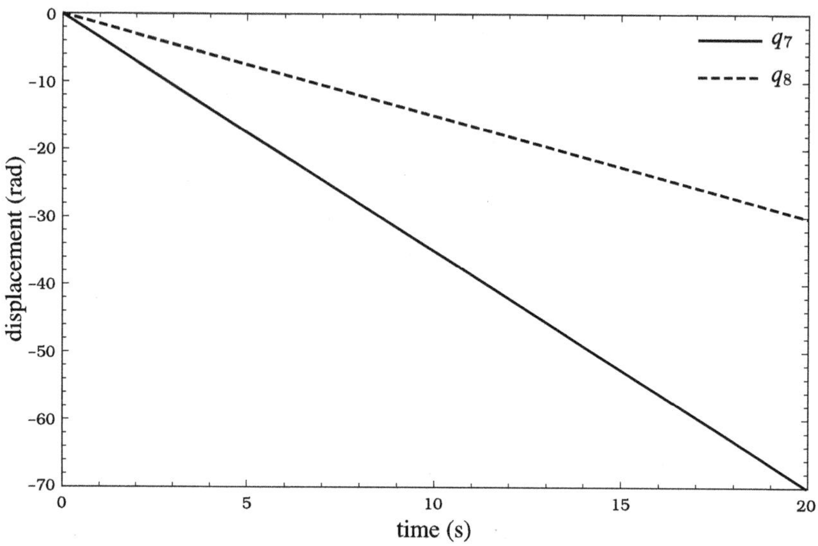

Figure 6.10 Time history of the wheel angles, q_7 and q_8, for the rolling wheel assembly.

and

$$\boldsymbol{g} = \begin{pmatrix} 0 & 0 & 3Mg & 0 & 0 & 0 & 0 & 0 \end{pmatrix}^T. \tag{6.167}$$

The nonholonomically constrained equations of motion express as

$$\boldsymbol{\tau} = \boldsymbol{M}\ddot{\boldsymbol{q}} + \boldsymbol{b} + \boldsymbol{g} - \boldsymbol{C}^T\boldsymbol{\lambda}, \tag{6.168}$$

subject to

$$\boldsymbol{C}\dot{\boldsymbol{q}} = \boldsymbol{0} \quad \Rightarrow \quad \dot{\boldsymbol{C}}\dot{\boldsymbol{q}} + \boldsymbol{C}\ddot{\boldsymbol{q}} = \boldsymbol{0}. \tag{6.169}$$

We note that $n = 8$ and $m = 5$. The constrained system has $p = n - m = 3$ degrees of freedom (this includes the internal rotation of the axle relative to the wheel hubs).

The simulation results are displayed in Figures 6.9 and 6.10. The geometric constants were chosen to be $r = 1$, $r_A = 0.075$, and $l = 2$. The masses of all bodies were chosen to be $M = 1$. No generalized forces were applied ($\boldsymbol{\tau} = \boldsymbol{0}$). The initial conditions used were

$$\boldsymbol{q}_o = (2.5 \quad 0 \quad 1 \quad 0 \quad 0 \quad 0 \quad 0 \quad 0)^T \tag{6.170}$$

$$\dot{\boldsymbol{q}}_o = (0 \quad 2.5 \quad 0 \quad 1 \quad 0 \quad 0 \quad -3.5 \quad -1.5)^T. \tag{6.171}$$

It can easily be confirmed that these initial conditions satisfy the nonholonomic constraints, $\boldsymbol{C}\dot{\boldsymbol{q}} = \boldsymbol{0}$.

6.3 Kane's Formulation

Kane's approach (Kane 1961) extended Jourdain's approach to rigid bodies and introduced quasi-velocities to implicitly handle nonholonomic constraints. We will present Kane's method here in a different way from the original presentation. This will be consistent with the idea of virtual velocities.

6.3.1 A System of Rigid Bodies

Given a set of m nonholonomic constraints that are linear in $\dot{\boldsymbol{q}}$,

$$\boldsymbol{C}(\boldsymbol{q})\dot{\boldsymbol{q}} = \boldsymbol{0}, \tag{6.172}$$

we can define a set of independent *quasi-velocities*, $\boldsymbol{s}$. The generalized velocities are related to these quasi-velocities by the null space of $\boldsymbol{C}$. That is,

$$\dot{\boldsymbol{q}} = \boldsymbol{W}(\boldsymbol{q})\boldsymbol{s}, \tag{6.173}$$

where

$$\operatorname{im}(\boldsymbol{W}) = \ker(\boldsymbol{C}). \tag{6.174}$$

We note that

$$\boldsymbol{v}(\boldsymbol{q}, \dot{\boldsymbol{q}}) = \boldsymbol{\Gamma}\dot{\boldsymbol{q}} \quad \text{and} \quad \boldsymbol{\omega}(\boldsymbol{q}, \dot{\boldsymbol{q}}) = \boldsymbol{\Pi}\dot{\boldsymbol{q}}. \tag{6.175}$$

Making use of (6.173), we can replace the $\dot{\boldsymbol{q}}$ terms in favor of the $\boldsymbol{s}$ terms in our expressions for $\boldsymbol{v}$ and $\boldsymbol{\omega}$. That is,

$$\tilde{\boldsymbol{v}}(\boldsymbol{q}, \boldsymbol{s}) = \boldsymbol{\Gamma}\boldsymbol{W}\boldsymbol{s} = \tilde{\boldsymbol{\Gamma}}\boldsymbol{s} \quad \text{and} \quad \tilde{\boldsymbol{\omega}}(\boldsymbol{q}, \boldsymbol{s}) = \boldsymbol{\Pi}\boldsymbol{W}\boldsymbol{s} = \tilde{\boldsymbol{\Pi}}\boldsymbol{s}. \tag{6.176}$$

Clearly

$$\tilde{\boldsymbol{v}}(\boldsymbol{q}, \boldsymbol{s}) = \boldsymbol{v}(\boldsymbol{q}, \dot{\boldsymbol{q}}) \quad \text{and} \quad \tilde{\boldsymbol{\omega}}(\boldsymbol{q}, \boldsymbol{s}) = \boldsymbol{\omega}(\boldsymbol{q}, \dot{\boldsymbol{q}}). \tag{6.177}$$

The terms $\tilde{\boldsymbol{\Gamma}}$ and $\tilde{\boldsymbol{\Pi}}$ are the *quasi-Jacobians*, given by

$$\tilde{\boldsymbol{\Gamma}} = \frac{\partial \tilde{\boldsymbol{v}}}{\partial \boldsymbol{s}} = \boldsymbol{\Gamma}\boldsymbol{W} \quad \text{and} \quad \tilde{\boldsymbol{\Pi}} = \frac{\partial \tilde{\boldsymbol{\omega}}}{\partial \boldsymbol{s}} = \boldsymbol{\Pi}\boldsymbol{W}. \tag{6.178}$$

The virtual velocity variations are given by

$$\delta\tilde{\boldsymbol{v}} = \tilde{\boldsymbol{\Gamma}}(\boldsymbol{s} + \delta\boldsymbol{s}) - \tilde{\boldsymbol{\Gamma}}\boldsymbol{s} = \tilde{\boldsymbol{\Gamma}}\delta\boldsymbol{s} \tag{6.179}$$

and

$$\delta\tilde{\boldsymbol{\omega}} = \tilde{\boldsymbol{\Pi}}(\boldsymbol{s} + \delta\boldsymbol{s}) - \tilde{\boldsymbol{\Pi}}\boldsymbol{s} = \tilde{\boldsymbol{\Pi}}\delta\boldsymbol{s}. \tag{6.180}$$

Since the nonholonomic constraint forces vanish under these variations, we have

$$\sum_{i=1}^{p} \tilde{\tau}_i \cdot \delta s_i - \sum_{i=1}^{n_b} M_i(\boldsymbol{a}_{G_i} + g\hat{\boldsymbol{e}}_{0_3}) \cdot \delta\tilde{\boldsymbol{v}}_{G_i} - \sum_{i=1}^{n_b} (\boldsymbol{I}_i^{G_i}\boldsymbol{\alpha}_i + \boldsymbol{\omega}_i \times \boldsymbol{I}_i^{G_i}\boldsymbol{\omega}_i) \cdot \delta\tilde{\boldsymbol{\omega}}_i = 0 \tag{6.181}$$

for all variations, $\delta\tilde{\boldsymbol{v}}_{G_i}$ and $\delta\tilde{\boldsymbol{\omega}}_i$, that are consistent with the kinematic constraints. The terms $\tilde{\tau}_i$ are the *quasi-forces*.

Expressing the velocity variations in terms of variations in the quasi-velocities (with stationary displacement and time), we have

$$^i\delta\tilde{\boldsymbol{v}}_{G_i} = {^i\tilde{\boldsymbol{\Gamma}}}_{G_i}\delta\boldsymbol{s} \quad \text{and} \quad {^i\delta\tilde{\boldsymbol{\omega}}_i} = {^i\tilde{\boldsymbol{\Pi}}_i}\delta\boldsymbol{s}, \tag{6.182}$$

where the velocities are expressed in the local link frame i for convenience. So, we have

$$\tilde{\boldsymbol{\tau}}\cdot\delta\boldsymbol{s} - \sum_{i=1}^{n_b} M_i({^i\boldsymbol{a}_{G_i}} + g\,{^i\hat{\boldsymbol{e}}_{0_3}})\cdot({^i\tilde{\boldsymbol{\Gamma}}_{G_i}}\delta\boldsymbol{s}) - \sum_{i=1}^{n_b}({^i\boldsymbol{I}_i^{G_i}}\,{^i\boldsymbol{\alpha}_i} + {^i\boldsymbol{\omega}_i}\times{^i\boldsymbol{I}_i^{G_i}}\,{^i\boldsymbol{\omega}_i})\cdot({^i\tilde{\boldsymbol{\Pi}}_i}\delta\boldsymbol{s}) = 0$$
$$\forall\delta\boldsymbol{s}\in\mathbb{R}^p, \tag{6.183}$$

or

$$\tilde{\boldsymbol{\tau}}\cdot\delta\boldsymbol{s} - \sum_{i=1}^{n_b}\left[M_i\,{^i\boldsymbol{\Gamma}_{G_i}^T}\,{^i\boldsymbol{a}_{G_i}} + M_i g\,{^i\tilde{\boldsymbol{\Gamma}}_{G_i}^T}\,{^i\hat{\boldsymbol{e}}_{0_3}} + {^i\tilde{\boldsymbol{\Pi}}_i^T}({^i\boldsymbol{I}_i^{G_i}}\,{^i\boldsymbol{\alpha}_i} + {^i\boldsymbol{\omega}_i}\times{^i\boldsymbol{I}_i^{G_i}}\,{^i\boldsymbol{\omega}_i})\right]\cdot\delta\boldsymbol{s} = 0$$
$$\forall\delta\boldsymbol{s}\in\mathbb{R}^p. \tag{6.184}$$

In matrix form we have

$$\left[\tilde{\boldsymbol{\tau}} - \sum_{i=1}^{n_b}\begin{pmatrix}{^i\tilde{\boldsymbol{\Gamma}}_{G_i}^T} & {^i\tilde{\boldsymbol{\Pi}}_i^T}\end{pmatrix}\begin{pmatrix}M_i\,{^i\boldsymbol{a}_{G_i}} + M_i g\,{^i\hat{\boldsymbol{e}}_{0_3}}\\ {^i\boldsymbol{I}_i^{G_i}}\,{^i\boldsymbol{\alpha}_i} + {^i\boldsymbol{\omega}_i}\times{^i\boldsymbol{I}_i^{G_i}}\,{^i\boldsymbol{\omega}_i}\end{pmatrix}\right]\cdot\delta\boldsymbol{s} = 0$$
$$\forall\delta\boldsymbol{s}\in\mathbb{R}^p, \tag{6.185}$$

which implies

$$\tilde{\boldsymbol{\tau}} = \sum_{i=1}^{n_b}\begin{pmatrix}{^i\tilde{\boldsymbol{\Gamma}}_{G_i}^T} & {^i\tilde{\boldsymbol{\Pi}}_i^T}\end{pmatrix}\begin{pmatrix}M_i\,{^i\boldsymbol{a}_{G_i}} + M_i g\,{^i\hat{\boldsymbol{e}}_{0_3}}\\ {^i\boldsymbol{I}_i^{G_i}}\,{^i\boldsymbol{\alpha}_i} + {^i\boldsymbol{\omega}_i}\times{^i\boldsymbol{I}_i^{G_i}}\,{^i\boldsymbol{\omega}_i}\end{pmatrix}. \tag{6.186}$$

Replacing the $\dot{\boldsymbol{q}}$ and $\ddot{\boldsymbol{q}}$ terms in favor of the $\boldsymbol{s}$ and $\dot{\boldsymbol{s}}$ terms, we have

$$\tilde{\boldsymbol{\tau}} = \sum_{i=1}^{n_b}\begin{pmatrix}{^i\tilde{\boldsymbol{\Gamma}}_{G_i}^T} & {^i\tilde{\boldsymbol{\Pi}}_i^T}\end{pmatrix}\begin{pmatrix}M_i\,{^i\tilde{\boldsymbol{a}}_{G_i}} + M_i g\,{^i\hat{\boldsymbol{e}}_{0_3}}\\ {^i\boldsymbol{I}_i^{G_i}}\,{^i\tilde{\boldsymbol{\alpha}}_i} + {^i\tilde{\boldsymbol{\omega}}_i}\times{^i\boldsymbol{I}_i^{G_i}}\,{^i\tilde{\boldsymbol{\omega}}_i}\end{pmatrix}, \tag{6.187}$$

where

$$\tilde{\boldsymbol{a}}_{G_i}(\boldsymbol{q},\boldsymbol{s},\dot{\boldsymbol{s}}) = \boldsymbol{a}_{G_i}(\boldsymbol{q},\dot{\boldsymbol{q}},\ddot{\boldsymbol{q}}) \quad \text{and} \quad \tilde{\boldsymbol{\alpha}}_i(\boldsymbol{q},\boldsymbol{s},\dot{\boldsymbol{s}}) = \boldsymbol{\alpha}_i(\boldsymbol{q},\dot{\boldsymbol{q}},\ddot{\boldsymbol{q}}). \tag{6.188}$$

Equation (6.187) is a system of p first-order differential equations. We need to complement this with the set of n equations

$$\dot{\boldsymbol{q}} = \boldsymbol{W}\boldsymbol{s}. \tag{6.189}$$

This yields a total system of $n+p$ first-order differential equations in the $n+p$ states, $\boldsymbol{q}$ and $\boldsymbol{s}$. Noting that

$${^i\tilde{\boldsymbol{\omega}}_i} = {^i\tilde{\boldsymbol{\Pi}}_i}\boldsymbol{s}, \quad {^i\tilde{\boldsymbol{a}}_{G_i}} = {^i\tilde{\boldsymbol{\Gamma}}_{G_i}}\dot{\boldsymbol{s}} + ({^i\dot{\tilde{\boldsymbol{\Gamma}}}_{G_i}} + {^i\tilde{\boldsymbol{\Omega}}_i}\,{^i\tilde{\boldsymbol{\Gamma}}_{G_i}})\boldsymbol{s} \quad \text{and} \quad {^i\tilde{\boldsymbol{\alpha}}_i} = {^i\tilde{\boldsymbol{\Pi}}_i}\dot{\boldsymbol{s}} + {^i\dot{\tilde{\boldsymbol{\Pi}}}_i}\boldsymbol{s}, \tag{6.190}$$

Algorithm 7 First-order method for integrating Kane's equations

$\boldsymbol{q}_0 = \boldsymbol{q}_o$ {initialization}
$\boldsymbol{s}_0 = \boldsymbol{s}_o$ {initialization}
for $i = 0$ to $n_s - 1$ **do**
  $\dot{\boldsymbol{q}}_i = \boldsymbol{W}_i \boldsymbol{s}_i$
  $\dot{\boldsymbol{s}}_i = \tilde{\boldsymbol{M}}_i^{-1}(\tilde{\boldsymbol{\tau}}_i - \tilde{\boldsymbol{b}}_i - \tilde{\boldsymbol{g}}_i)$
  $\boldsymbol{s}_{i+1} = \boldsymbol{s}_i + \dot{\boldsymbol{s}}_i \Delta t$
  $\boldsymbol{q}_{i+1} = \boldsymbol{q}_i + \dot{\boldsymbol{q}}_i \Delta t$
end for

we have

$$\begin{aligned}\tilde{\boldsymbol{\tau}} = &\left[\sum_{i=1}^{n_b} (M_i {}^i\tilde{\boldsymbol{\Gamma}}^T_{G_i} {}^i\tilde{\boldsymbol{\Gamma}}_{G_i} + {}^i\tilde{\boldsymbol{\Pi}}^T_i {}^i\boldsymbol{I}^{G_i}_i {}^i\tilde{\boldsymbol{\Pi}}_i)\right] \dot{\boldsymbol{s}} \\ &+ \left[\sum_{i=1}^{n_b} \left(M_i {}^i\tilde{\boldsymbol{\Gamma}}^T_{G_i} ({}^i\dot{\tilde{\boldsymbol{\Gamma}}}_{G_i} + {}^i\tilde{\boldsymbol{\Omega}}_i {}^i\tilde{\boldsymbol{\Gamma}}_{G_i}) + {}^i\tilde{\boldsymbol{\Pi}}^T_i ({}^i\boldsymbol{I}^{G_i}_i {}^i\dot{\tilde{\boldsymbol{\Pi}}}_i + {}^i\tilde{\boldsymbol{\Omega}}_i {}^i\boldsymbol{I}^{G_i}_i {}^i\tilde{\boldsymbol{\Pi}}_i)\right)\right] \boldsymbol{s} \\ &+ g \sum_{i=1}^{n_b} M_i {}^i\tilde{\boldsymbol{\Gamma}}^T_{G_i} {}^i\hat{\boldsymbol{e}}_{0_3}.\end{aligned} \tag{6.191}$$

Defining

$$\tilde{\boldsymbol{M}}(\boldsymbol{q}) \triangleq \sum_{i=1}^{n_b} (M_i {}^i\tilde{\boldsymbol{\Gamma}}^T_{G_i} {}^i\tilde{\boldsymbol{\Gamma}}_{G_i} + {}^i\boldsymbol{\Pi}^T_i {}^i\boldsymbol{I}^{G_i}_i {}^i\tilde{\boldsymbol{\Pi}}_i), \tag{6.192}$$

$$\tilde{\boldsymbol{b}}(\boldsymbol{q}, \boldsymbol{s}) \triangleq \left[\sum_{i=1}^{n_b} \left(M_i {}^i\tilde{\boldsymbol{\Gamma}}^T_{G_i} ({}^i\dot{\tilde{\boldsymbol{\Gamma}}}_{G_i} + {}^i\tilde{\boldsymbol{\Omega}}_i {}^i\tilde{\boldsymbol{\Gamma}}_{G_i}) + {}^i\tilde{\boldsymbol{\Pi}}^T_i ({}^i\boldsymbol{I}^{G_i}_i {}^i\dot{\tilde{\boldsymbol{\Pi}}}_i + {}^i\tilde{\boldsymbol{\Omega}}_i {}^i\boldsymbol{I}^{G_i}_i {}^i\tilde{\boldsymbol{\Pi}}_i)\right)\right] \boldsymbol{s}, \tag{6.193}$$

$$\tilde{\boldsymbol{g}}(\boldsymbol{q}) \triangleq g \sum_{i=1}^{n_b} M_i {}^i\tilde{\boldsymbol{\Gamma}}^T_{G_i} {}^i\hat{\boldsymbol{e}}_{0_3}, \tag{6.194}$$

we have

$$\tilde{\boldsymbol{\tau}} = \tilde{\boldsymbol{M}}(\boldsymbol{q})\dot{\boldsymbol{s}} + \tilde{\boldsymbol{b}}(\boldsymbol{q}, \boldsymbol{s}) + \tilde{\boldsymbol{g}}(\boldsymbol{q}). \tag{6.195}$$

Numerical Integration

The state equation for our system is

$$\begin{pmatrix} \dot{\boldsymbol{q}} \\ \dot{\boldsymbol{s}} \end{pmatrix} = \begin{pmatrix} \boldsymbol{W}\boldsymbol{s} \\ \tilde{\boldsymbol{M}}(\boldsymbol{q})^{-1}[\tilde{\boldsymbol{\tau}} - \tilde{\boldsymbol{b}}(\boldsymbol{q}, \boldsymbol{s}) - \tilde{\boldsymbol{g}}(\boldsymbol{q})] \end{pmatrix}. \tag{6.196}$$

A first-order method for integrating this system can be summarized as shown in Algorithm 7.

Example: We revisit the rolling disk. As before, we parameterize the system using six generalized coordinates as shown in Figure 6.11. We recall the nonholonomic constraint equations,

$$
{}^{\mathcal{O}}\boldsymbol{v}_C = \begin{pmatrix} \dot{q}_1 + r[\sin(q_4)\sin(q_5)\dot{q}_4 - \cos(q_4)\cos(q_5)\dot{q}_5 - \sin(q_4)\dot{q}_6] \\ \dot{q}_2 - r[\cos(q_4)\sin(q_5)\dot{q}_4 + \sin(q_4)\cos(q_5)\dot{q}_5 - \cos(q_4)\dot{q}_6] \\ \dot{q}_3 + r\sin(q_5)\dot{q}_5 \end{pmatrix} = \mathbf{0}. \quad (6.197)
$$

We can introduce a set of quasi-velocities

$$
s_1 = \dot{q}_4, \qquad s_2 = \dot{q}_5 \qquad \text{and} \qquad s_3 = \dot{q}_6. \quad (6.198)
$$

Given the constraint equations, the first three generalized velocities can then be expressed as

$$
\dot{q}_1 = -r[\sin(q_4)\sin(q_5)s_1 + \cos(q_4)\cos(q_5)s_2 + \sin(q_4)s_3], \quad (6.199)
$$
$$
\dot{q}_2 = r[\cos(q_4)\sin(q_5)s_1 - \sin(q_4)\cos(q_5)s_2 + \cos(q_4)s_3], \quad (6.200)
$$
$$
\dot{q}_3 = -r\sin(q_5)s_2. \quad (6.201)
$$

So, we have

$$
\dot{\boldsymbol{q}} = \begin{pmatrix} -r\sin(q_4)\sin(q_5) & r\cos(q_4)\cos(q_5) & r\sin(q_4) \\ r\cos(q_4)\sin(q_5) & -r\sin(q_4)\cos(q_5) & r\cos(q_4) \\ 0 & -r\sin(q_5) & 0 \\ 1 & 0 & 0 \\ 0 & 1 & 0 \\ 0 & 0 & 1 \end{pmatrix} \boldsymbol{s} = \boldsymbol{W}\boldsymbol{s}. \quad (6.202)
$$

We can confirm that $\text{im}(\boldsymbol{W}) = \ker(\boldsymbol{C})$ since $\boldsymbol{CW} = \mathbf{0}$.

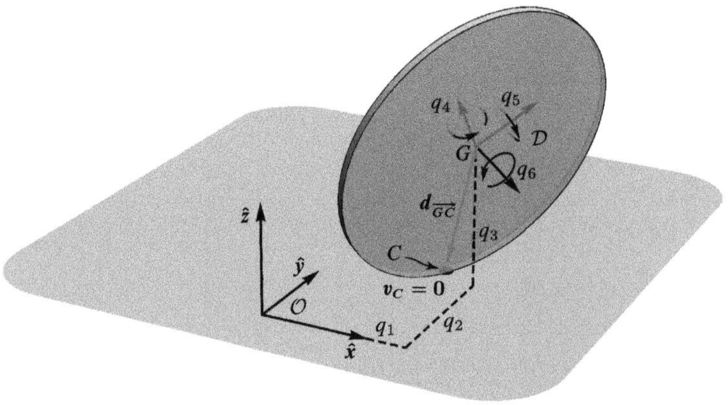

Figure 6.11 A rolling disk with nonholonomic constraints. Six generalized coordinates, $q_1, \ldots, q_6$, parameterize the position of the center of the disk and the disk orientation. Nonholonomic constraints arise out of the *zero* velocity (no-slip) condition imposed on the instantaneous contact point of the disk with the ground.

Evaluating the Jacobians, we have

$$ {}^{\mathcal{O}}\tilde{\boldsymbol{\Gamma}}_{G} = {}^{\mathcal{O}}\boldsymbol{\Gamma}_{G}\boldsymbol{W} = \begin{pmatrix} -r\sin(q_4)\sin(q_5) & r\cos(q_4)\cos(q_5) & r\sin(q_4) \\ r\cos(q_4)\sin(q_5) & -r\sin(q_4)\cos(q_5) & r\cos(q_4) \\ 0 & -r\sin(q_5) & 0 \end{pmatrix} \quad (6.203) $$

and

$$ {}^{\mathcal{D}}\tilde{\boldsymbol{\Pi}} = {}^{\mathcal{D}}\boldsymbol{\Pi}\boldsymbol{W} = \begin{pmatrix} -\sin(q_5) & 0 & 1 \\ \cos(q_5)\sin(q_6) & \cos(q_6) & 0 \\ \cos(q_5)\cos(q_6) & -\sin(q_6) & 0 \end{pmatrix}. \quad (6.204) $$

The dynamical terms are

$$ \tilde{\boldsymbol{M}} = M\,{}^{\mathcal{O}}\tilde{\boldsymbol{\Gamma}}_{G}^{T}\,{}^{\mathcal{O}}\tilde{\boldsymbol{\Gamma}}_{G} + {}^{\mathcal{D}}\tilde{\boldsymbol{\Pi}}^{T}\,{}^{\mathcal{D}}\boldsymbol{I}^{G\mathcal{D}}\tilde{\boldsymbol{\Pi}}, \quad (6.205) $$

$$ \tilde{\boldsymbol{b}} = [M\,{}^{\mathcal{O}}\tilde{\boldsymbol{\Gamma}}_{G}^{T}\,{}^{\mathcal{O}}\dot{\tilde{\boldsymbol{\Gamma}}}_{G} + {}^{\mathcal{D}}\tilde{\boldsymbol{\Pi}}^{T}({}^{\mathcal{D}}\boldsymbol{I}^{G\mathcal{D}}\dot{\tilde{\boldsymbol{\Pi}}} + {}^{\mathcal{D}}\tilde{\boldsymbol{\Omega}}^{\mathcal{D}}\boldsymbol{I}^{G\mathcal{D}}\tilde{\boldsymbol{\Pi}})]\boldsymbol{s}, \quad (6.206) $$

$$ \tilde{\boldsymbol{g}} = gM\,{}^{\mathcal{O}}\tilde{\boldsymbol{\Gamma}}_{G}^{T}\hat{\boldsymbol{e}}_3. \quad (6.207) $$

As before, we will take the inertia tensor to be

$$ {}^{\mathcal{D}}\boldsymbol{I}^{G} = \begin{pmatrix} \frac{1}{2}Mr^2 & 0 & 0 \\ 0 & \frac{1}{4}Mr^2 & 0 \\ 0 & 0 & \frac{1}{4}Mr^2 \end{pmatrix}. \quad (6.208) $$

Evaluating, we have

$$ \tilde{\boldsymbol{M}} = \begin{pmatrix} \frac{1}{8}Mr^2[7-5\cos(2q_5)] & 0 & -\frac{3}{2}Mr^2\sin(q_5) \\ 0 & \frac{5}{4}Mr^2 & 0 \\ -\frac{3}{2}Mr^2\sin(q_5) & 0 & \frac{3}{2}Mr^2 \end{pmatrix} \quad (6.209) $$

and

$$ \tilde{\boldsymbol{b}} = \begin{pmatrix} -\frac{1}{2}Mr^2\cos(q_5)s_2[s_3 - 5\sin(q_5)s_1] \\ -\frac{1}{4}Mr^2\cos(q_5)s_1[-6s_3 + 5\sin(q_5)s_1] \\ -\frac{5}{2}Mr^2\cos(q_5)s_1s_2 \end{pmatrix} \quad (6.210) $$

and

$$ \tilde{\boldsymbol{g}} = \begin{pmatrix} 0 & Mgr\sin(q_5) & 0 \end{pmatrix}^T. \quad (6.211) $$

The equations of motion express as

$$ \tilde{\boldsymbol{\tau}} = \tilde{\boldsymbol{M}}\dot{\boldsymbol{s}} + \tilde{\boldsymbol{b}} + \tilde{\boldsymbol{g}}. \quad (6.212) $$

It can be confirmed that, for the same initial conditions, (6.212) yield the same results as (6.103) and (6.104). In this example, we explore different initial conditions

$$ \boldsymbol{q}_o = \begin{pmatrix} 1 & -0.25 & \cos\left(\frac{\pi}{8}\right) & 0 & -\frac{\pi}{8} & 0 \end{pmatrix}^T \quad (6.213) $$

$$ \boldsymbol{s}_o = \begin{pmatrix} 0.5 & 0 & -2 \end{pmatrix}^T, \quad (6.214) $$

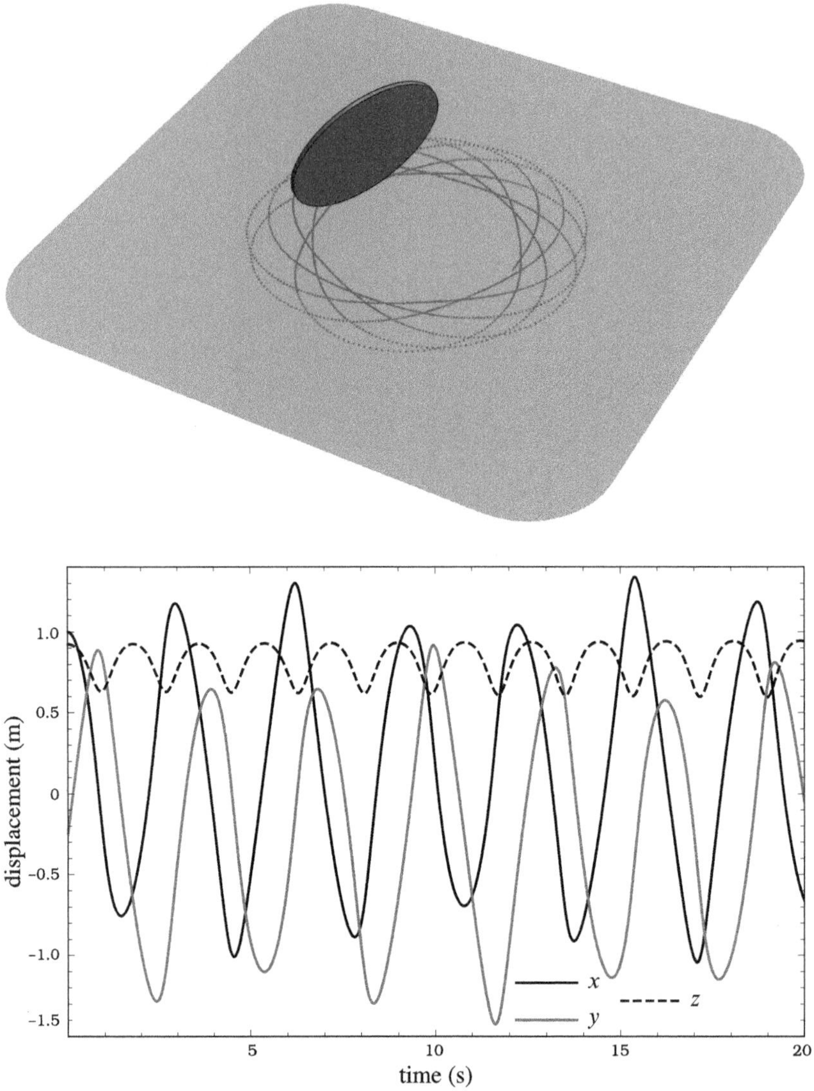

Figure 6.12 (Top) Animation frames from the simulation of the nonholonomic rolling disk using Kane's method. (Bottom) Time history of the position of the center of mass of the disk (x, y, z).

which correspond to the following conditions on the generalized velocities:

$$\dot{\boldsymbol{q}}_o = \boldsymbol{W}(\boldsymbol{q}_o)\boldsymbol{s}_o = \begin{pmatrix} 0 & 1.809 & 0 & 0.5 & 0 & -2 \end{pmatrix}^T. \tag{6.215}$$

The simulation results are displayed in Figure 6.12. As before, all geometric and inertial constants were chosen to be 1. That is, $r = 1$ and $M = 1$.

6.4 Exercises

1. Are rolling constraints necessarily nonholonomic? Demonstrate this.
2. Consider the rolling disk shown in Figure 6.13. This is a simplified version of the rolling disk from Section 6.2.5 in that the disk always remains upright (it cannot pitch from side to side). Use a $\boldsymbol{Q}_z(q_4)\boldsymbol{Q}_x(q_5)$ Euler sequence to represent the orientation of the disk. The disk radius is r and the mass is M. The principal inertia components are ${}^{\mathcal{D}}I^G_{11} = \frac{1}{2}Mr^2$ and ${}^{\mathcal{D}}I^G_{22} = {}^{\mathcal{D}}I^G_{33} = \frac{1}{4}Mr^2$. Using Jourdain's Principle and Lagrange multipliers, answer the following.
   (a) How many degrees of freedom does the constrained system have?
   (b) Compute the mass matrix, $\boldsymbol{M}(\boldsymbol{q})$, for the unconstrained system.
   (c) Compute the vector of centrifugal and Coriolis forces, $\boldsymbol{b}(\boldsymbol{q}, \dot{\boldsymbol{q}})$, for the unconstrained system.
   (d) Compute the vector of gravity forces, $\boldsymbol{g}(\boldsymbol{q})$, for the unconstrained system.
   (e) Express the velocity condition, ${}^{\mathcal{O}}\boldsymbol{v}_C$, and the constraint matrix, $\boldsymbol{C}(\boldsymbol{q})$, for the rolling contact.

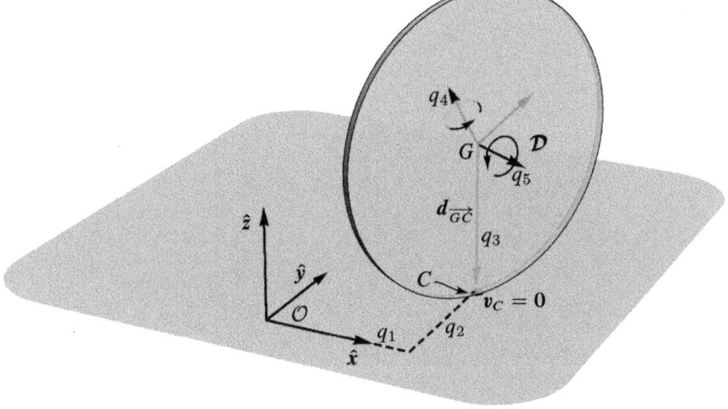

Figure 6.13 A simplified version of the rolling disk that always remains upright (it cannot pitch from side to side, Exercises 2 and 3). A $\boldsymbol{Q}_z(q_4)\boldsymbol{Q}_x(q_5)$ Euler sequence is used to represent the orientation of the disk. The disk radius is r and the mass is M. The principal inertia components are ${}^{\mathcal{D}}I^G_{11} = \frac{1}{2}Mr^2$ and ${}^{\mathcal{D}}I^G_{22} = {}^{\mathcal{D}}I^G_{33} = \frac{1}{4}Mr^2$.

3. Consider the rolling disk from Exercise 2 shown in Figure 6.13. Using Kane's method and defining independent quasi-velocities, $s_1 = \dot{q}_4$ and $s_2 = \dot{q}_5$, answer the following.
   (a) From the rolling contact constraint equations, determine the relationship, $\dot{\boldsymbol{q}} = \boldsymbol{W}\boldsymbol{s}$, between the generalized velocities and the quasi-velocities.
   (b) Compute the mass matrix, $\tilde{\boldsymbol{M}}(\boldsymbol{q})$.
   (c) Compute the vector of centrifugal and Coriolis forces, $\tilde{\boldsymbol{b}}(\boldsymbol{q}, \boldsymbol{s})$.
   (d) Compute the vector of gravity forces, $\tilde{\boldsymbol{g}}(\boldsymbol{q})$.

4. Consider the two-wheeled apparatus shown in Figure 6.14. The apparatus always remains upright (the wheels cannot pitch from side to side). The radius and mass of each wheel are r and M, respectively. The principal inertia components of each wheel are ${}^{\mathcal{D}}I^G_{22} = \frac{1}{2}Mr^2$ and ${}^{\mathcal{D}}I^G_{11} = {}^{\mathcal{D}}I^G_{33} = \frac{1}{4}Mr^2$. Each link of the connecting linkage has length l, mass M, and center of mass located at the geometric center. Assume the rotational inertia of the connecting links to be *zero*. Use (q_1, q_2, q_3) to locate the center of the forward link of the connecting linkage, and use a $\boldsymbol{Q}_z(q_4)\boldsymbol{Q}_y(q_5)$ Euler sequence to represent its orientation, as shown. Use q_6 for the elbow joint of the connecting linkage, and use q_7 and q_8 for the wheel angles.

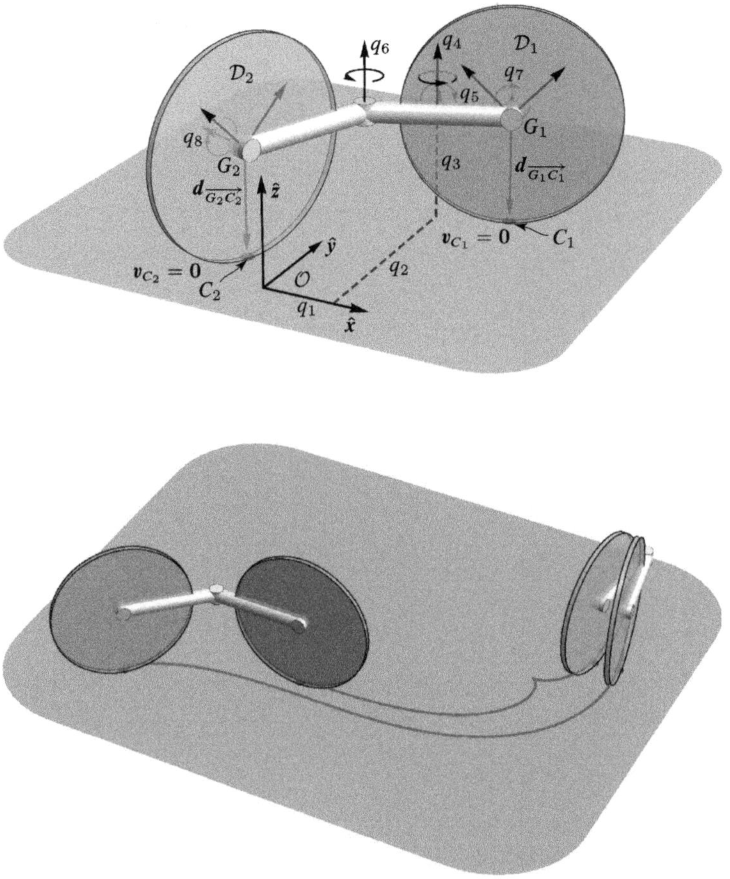

Figure 6.14 (Top) A two-wheeled apparatus (Exercise 4). The apparatus always remains upright (the wheels cannot pitch from side to side). The radius and mass of each wheel are r and M, respectively. The principal inertia components of each wheel are ${}^{\mathcal{D}}I^G_{22} = \frac{1}{2}Mr^2$ and ${}^{\mathcal{D}}I^G_{11} = {}^{\mathcal{D}}I^G_{33} = \frac{1}{4}Mr^2$. Each link of the connecting linkage has length l, mass M, and center of mass located at the geometric center. The rotational inertia of the connecting links are *zero*. The center of the forward link of the connecting linkage is located by (q_1, q_2, q_3), and a $\boldsymbol{Q}_z(q_4)\boldsymbol{Q}_y(q_5)$ Euler sequence is used to represent its orientation, as shown. The elbow joint of the connecting linkage is q_6, and the wheel angles are q_7 and q_8. (Bottom) Animation showing example operation of the two-wheeled apparatus.

Consider the following conditions,

$$q_0 = (0 \quad 0 \quad 0.75 \quad 0 \quad 0 \quad \pi/4 \quad 0 \quad 0)^T,$$
$$\dot{q}_0 = (0.75 \quad 0 \quad 0 \quad 0 \quad 0 \quad -0.75\sin(\pi/4) \quad 1 \quad \sqrt{2}/2)^T.$$

Using Jourdain's Principle and Lagrange multipliers, answer the following.

(a) How many degrees of freedom does the constrained system have?

(b) Compute the mass matrix, $\boldsymbol{M}(\boldsymbol{q})$, for the unconstrained system at $\boldsymbol{q}_0$.

(c) Compute the vector of centrifugal and Coriolis forces, $\boldsymbol{b}(\boldsymbol{q}, \dot{\boldsymbol{q}})$, for the unconstrained system at $\boldsymbol{q}_0$ and $\dot{\boldsymbol{q}}_0$.

(d) Compute the vector of gravity forces, $\boldsymbol{g}(\boldsymbol{q})$, for the unconstrained system at $\boldsymbol{q}_0$.

(e) Express the velocity condition, ${}^{O}\boldsymbol{v}_C$, and the constraint matrix, $\boldsymbol{C}(\boldsymbol{q})$, for the rolling contact at $\boldsymbol{q}_0$.

7 Second-Order Variational Principles

In the previous two chapters we addressed zeroth- and first-order variational principles. In this chapter we will focus on the second-order variation of displacement. We will begin with Gauss's Principle (Gauss 1829). The principle is based on the notion of virtual acceleration. The derivation of the equations of motion for particles and rigid bodies using Gauss's Principle closely parallels the derivation of the equations of motion using d'Alembert's Principle and Jourdain's Principle in the previous chapters. As with d'Alembert's and Jourdain's Principles, Gauss's Principle can be seen as an independent principle of analytical dynamics. Therefore, as with the previous chapters, this chapter is presented in a stand-alone manner.

7.1 Virtual Accelerations

Virtual accelerations refer to all accelerations of a system that satisfy the scleronomic constraints. In the case of virtual accelerations, time, position, and velocity are frozen or stationary.

7.2 Gauss's Principle

PRINCIPLE 7.1 *The virtual acceleration work of a system is stationary. That is,*

$$\delta A = 0. \tag{7.1}$$

Additionally, the constraints of the system generate no virtual acceleration work,

$$\delta A_c = 0. \tag{7.2}$$

This is known as ***Gauss's Principle****.*

As with d'Alembert's Principle and Jourdain's Principle, it is noted that while Gauss's Principle can be seen as providing an alternate statement of Newton's second law, for interacting bodies, a law of action and reaction (Newton's third law) is still needed. Therefore, when we use Gauss's Principle to derive the equations of motion for systems of particles/bodies, we will invoke the law of action and reaction.

7.2.1 A Single Particle

Gauss's Principle for a single point mass with a discrete set of n_f external forces, $\{\boldsymbol{f}_1, \ldots, \boldsymbol{f}_{n_f}\}$, acting on it is expressed as

$$\delta A = \sum_{i=1}^{n_f} \boldsymbol{f}_i \cdot \delta \boldsymbol{a} - Mg\hat{\boldsymbol{e}}_3 \cdot \delta \boldsymbol{a} - M\boldsymbol{a} \cdot \delta \boldsymbol{a} = 0 \qquad \forall \delta \boldsymbol{a} \in \mathbb{R}^3, \tag{7.3}$$

where $\delta \boldsymbol{a}$ represents the acceleration variations. During these variations, time, position, and velocity are stationary. That is, $\delta t = 0$, $\delta \boldsymbol{r} = 0$, and $\delta \boldsymbol{v} = 0$. More concisely, we can express

$$\delta A = \left(\sum_{i=1}^{n_f} \boldsymbol{f}_i - Mg\hat{\boldsymbol{e}}_3 - M\boldsymbol{a} \right) \cdot \delta \boldsymbol{a} = 0 \qquad \forall \delta \boldsymbol{a} \in \mathbb{R}^3, \tag{7.4}$$

which implies

$$\sum_{i=1}^{n_f} \boldsymbol{f}_i - Mg\hat{\boldsymbol{e}}_3 - M\boldsymbol{a} = \boldsymbol{0}. \tag{7.5}$$

7.2.2 A Single Rigid Body

Gauss's Principle for a single rigid body with a discrete set of external forces, $\boldsymbol{f}_i$, and moments, $\boldsymbol{\varphi}_j$, acting on it is expressed as

$$\delta A = \sum_{i=1}^{n_f} \boldsymbol{f}_i \cdot \delta \boldsymbol{a}_{P_i} + \sum_{j=1}^{n_\varphi} \boldsymbol{\varphi}_j \cdot \delta \boldsymbol{\alpha} - Mg\hat{\boldsymbol{e}}_3 \cdot \delta \boldsymbol{a}_G - M\boldsymbol{a} \cdot \delta \boldsymbol{a}_G - (\boldsymbol{I}^G \boldsymbol{\alpha} + \boldsymbol{\omega} \times \boldsymbol{I}^G \boldsymbol{\omega}) \cdot \delta \boldsymbol{\alpha} = 0,$$

$$\forall \delta \boldsymbol{a}_G \in \mathbb{R}^3, \quad \text{and} \quad \forall \delta \boldsymbol{\alpha} \in \mathbb{R}^3, \tag{7.6}$$

where the $\delta \boldsymbol{a}$ and $\delta \boldsymbol{\alpha}$ terms represent all acceleration variations consistent with the rigid-body constraint. This is depicted in Figure 7.1. During these variations, time, position, orientation, and translational and angular velocities are stationary. That is, $\delta t = 0$, $\delta \boldsymbol{r} = 0$, $\delta \boldsymbol{\theta} = 0$, $\delta \boldsymbol{v} = 0$, and $\delta \boldsymbol{\omega} = 0$. We further note that

$$\begin{aligned} \delta \boldsymbol{a}_{P_i} &= (\boldsymbol{a}_G + \delta \boldsymbol{a}_G) + (\boldsymbol{\alpha} + \delta \boldsymbol{\alpha}) \times \boldsymbol{d}_{\overrightarrow{GP_i}} + \boldsymbol{\omega} \times (\boldsymbol{\omega} \times \boldsymbol{d}_{\overrightarrow{GP_i}}) \\ &\quad - \boldsymbol{a}_G - \boldsymbol{\alpha} \times \boldsymbol{d}_{\overrightarrow{GP_i}} - \boldsymbol{\omega} \times (\boldsymbol{\omega} \times \boldsymbol{d}_{\overrightarrow{GP_i}}) \\ &= \delta \boldsymbol{a}_G + \delta \boldsymbol{\alpha} \times \boldsymbol{d}_{\overrightarrow{GP_i}}; \end{aligned} \tag{7.7}$$

then,

$$\sum_{i=1}^{n_f} \boldsymbol{f}_i \cdot \delta \boldsymbol{a}_{P_i} = \sum_{i=1}^{n_f} \boldsymbol{f}_i \cdot (\delta \boldsymbol{a}_G + \delta \boldsymbol{\alpha} \times \boldsymbol{d}_{\overrightarrow{GP_i}}) = \left(\sum_{i=1}^{n_f} \boldsymbol{f}_i \right) \cdot \delta \boldsymbol{a}_G + \sum_{i=1}^{n_f} \boldsymbol{f}_i \cdot (\delta \boldsymbol{\alpha} \times \boldsymbol{d}_{\overrightarrow{GP_i}}). \tag{7.8}$$

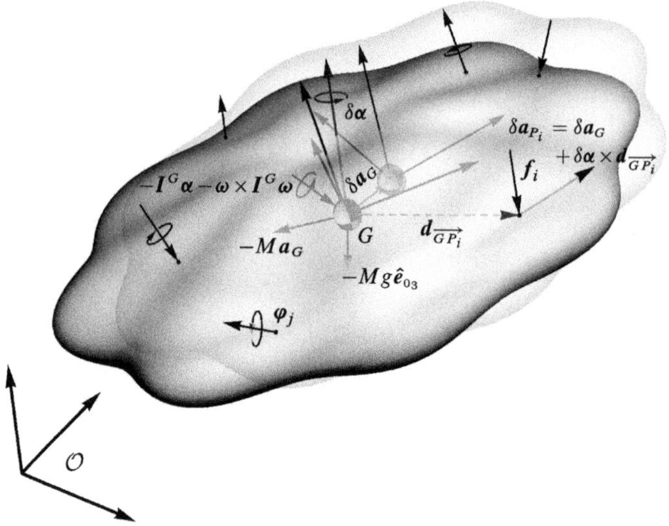

Figure 7.1 A virtual acceleration of a rigid body consisting of a translational, $\delta\boldsymbol{a}_G$, and rotational, $\delta\boldsymbol{\alpha}$, acceleration variation. We are concerned with all such acceleration variations consistent with the rigid-body constraint. During these variations, time, position, orientation, and translational and angular velocities are stationary.

Since $\boldsymbol{a} \cdot (\boldsymbol{b} \times \boldsymbol{c}) = (\boldsymbol{c} \times \boldsymbol{a}) \cdot \boldsymbol{b}$,

$$\sum_{i=1}^{n_f} \boldsymbol{f}_i \cdot (\delta\boldsymbol{\alpha} \times \boldsymbol{d}_{\overrightarrow{GP_i}}) = \sum_{i=1}^{n_f} (\boldsymbol{d}_{\overrightarrow{GP_i}} \times \boldsymbol{f}_i) \cdot \delta\boldsymbol{\alpha} = \left(\sum_{i=1}^{n_f} \boldsymbol{d}_{\overrightarrow{GP_i}} \times \boldsymbol{f}_i \right) \cdot \delta\boldsymbol{\alpha}. \tag{7.9}$$

So,

$$\sum_{i=1}^{n_f} \boldsymbol{f}_i \cdot \delta\boldsymbol{a}_{P_i} = \left(\sum_{i=1}^{n_f} \boldsymbol{f}_i \right) \cdot \delta\boldsymbol{a}_G + \left(\sum_{i=1}^{n_f} \boldsymbol{d}_{\overrightarrow{GP_i}} \times \boldsymbol{f}_i \right) \cdot \delta\boldsymbol{\alpha}. \tag{7.10}$$

Substituting (7.10) into (7.6), we get

$$\begin{aligned} \delta A = & \left(\sum_{i=1}^{n_f} \boldsymbol{f}_i - Mg\hat{\boldsymbol{e}}_3 - M\boldsymbol{a} \right) \cdot \delta\boldsymbol{a}_G \\ & + \left(\sum_{i=1}^{n_f} \boldsymbol{d}_{\overrightarrow{GP_i}} \times \boldsymbol{f}_i + \sum_{j=1}^{n_\varphi} \boldsymbol{\varphi}_j - \boldsymbol{I}^G\boldsymbol{\alpha} - \boldsymbol{\omega} \times \boldsymbol{I}^G\boldsymbol{\omega} \right) \cdot \delta\boldsymbol{\alpha} = 0, \end{aligned} \tag{7.11}$$

or

$$\begin{pmatrix} \sum_{i=1}^{n_f} \boldsymbol{f}_i - Mg\hat{\boldsymbol{e}}_3 - M\boldsymbol{a} \\ \sum_{i=1}^{n_f} \boldsymbol{d}_{\overrightarrow{GP_i}} \times \boldsymbol{f}_i + \sum_{j=1}^{n_\varphi} \boldsymbol{\varphi}_j - \boldsymbol{I}^G\boldsymbol{\alpha} - \boldsymbol{\omega} \times \boldsymbol{I}^G\boldsymbol{\omega} \end{pmatrix} \cdot \begin{pmatrix} \delta\boldsymbol{a}_G \\ \delta\boldsymbol{\alpha} \end{pmatrix} = 0, \tag{7.12}$$

$$\forall \delta\boldsymbol{a}_G \in \mathbb{R}^3, \quad \text{and} \quad \forall \delta\boldsymbol{\alpha} \in \mathbb{R}^3,$$

which implies

$$\sum_{i=1}^{n_f} \boldsymbol{f}_i - Mg\hat{\boldsymbol{e}}_3 - M\boldsymbol{a} = \boldsymbol{0} \tag{7.13}$$

$$\sum_{i=1}^{n_f} \boldsymbol{d}_{\overrightarrow{GP_i}} \times \boldsymbol{f}_i + \sum_{j=1}^{n_\varphi} \boldsymbol{\varphi}_j - \boldsymbol{I}^G\boldsymbol{\alpha} - \boldsymbol{\omega} \times \boldsymbol{I}^G\boldsymbol{\omega} = \boldsymbol{0}. \tag{7.14}$$

7.2.3 A System of Particles

We now apply Gauss's Principle to a system of n_p particles with a discrete set of forces acting on them. The virtual acceleration work associated with a given particle i is given by

$$\delta A_i = \left(\sum_{j=0}^{n_p} \boldsymbol{f}_i^j\right) \cdot \delta\boldsymbol{a}_i - M_i(\boldsymbol{a}_i + g\hat{\boldsymbol{e}}_3) \cdot \delta\boldsymbol{a}_i = 0 \qquad \forall \delta\boldsymbol{a}_i \in \mathbb{R}^3, \tag{7.15}$$

for $i = 1, \ldots, n_p$. The term $\boldsymbol{f}_i^j$ is the force that particle j exerts on particle i, where $\boldsymbol{f}_i^0$ is the force exerted by the ground (inertial reference frame) on the ith particle. Equation (7.15) is true for any particle i in the system, under all independent variations, $\delta\boldsymbol{a}_i$.

If we sum (7.15) over all particles, we obtain

$$\delta A = \sum_{i=1}^{n_p} \delta A_i = \sum_{i=1}^{n_p} \left(\sum_{j=0}^{n_p} \boldsymbol{f}_i^j\right) \cdot \delta\boldsymbol{a}_i - \sum_{i=1}^{n_p} M_i(\boldsymbol{a}_i + g\hat{\boldsymbol{e}}_3) \cdot \delta\boldsymbol{a}_i = 0. \tag{7.16}$$

The first summation is associated with the virtual acceleration work performed by the interparticle reaction forces. It will be useful to rearrange the summation so as to pair up the equal and opposite reaction forces, $\boldsymbol{f}_i^j$ and $\boldsymbol{f}_j^i$. Doing this, we have

$$\sum_{i=1}^{n_p} \left(\sum_{j=0}^{n_p} \boldsymbol{f}_i^j\right) \cdot \delta\boldsymbol{a}_i = \sum_{i=0}^{n_p-1} \sum_{j=i+1}^{n_p} (\boldsymbol{f}_i^j \cdot \delta\boldsymbol{a}_i + \boldsymbol{f}_j^i \cdot \delta\boldsymbol{a}_j) + \sum_{i=1}^{n_p} \boldsymbol{f}_i^i \cdot \delta\boldsymbol{a}_i - \sum_{j=1}^{n_p} \boldsymbol{f}_0^j \cdot \delta\boldsymbol{a}_0. \tag{7.17}$$

Since $\boldsymbol{f}_i^i = \boldsymbol{0}$ and $\delta\boldsymbol{a}_0 = \boldsymbol{0}$, we have

$$\sum_{i=1}^{n_p} \left(\sum_{j=0}^{n_p} \boldsymbol{f}_i^j\right) \cdot \delta\boldsymbol{a}_i = \sum_{i=0}^{n_p-1} \sum_{j=i+1}^{n_p} (\boldsymbol{f}_i^j \cdot \delta\boldsymbol{a}_i + \boldsymbol{f}_j^i \cdot \delta\boldsymbol{a}_j). \tag{7.18}$$

We note that $\boldsymbol{f}_j^i = -\boldsymbol{f}_i^j$. Thus,

$$\sum_{i=1}^{n_p} \left(\sum_{j=0}^{n_p} \boldsymbol{f}_i^j\right) \cdot \delta\boldsymbol{a}_i = \sum_{i=0}^{n_p-1} \sum_{j=i+1}^{n_p} \boldsymbol{f}_i^j \cdot (\delta\boldsymbol{a}_i - \delta\boldsymbol{a}_j). \tag{7.19}$$

This reflects the virtual acceleration work done by interparticle reaction forces, summed over all particles.

If we consider that the particles are subject to a set of holonomic constraints, the positions, $\boldsymbol{r}_i$, and in turn velocities, $\boldsymbol{v}_i$, and accelerations, $\boldsymbol{a}_i$, are not independent. In this case we will assume that the accelerations can be expressed in terms of a set of n_q independent generalized accelerations, $\ddot{\boldsymbol{q}}$. If we now consider only the variations, $\delta\boldsymbol{a}_i$, that are consistent with the kinematic constraints, then (7.19) reflects a projection of the interparticle forces on the direction of the interparticle motion. Gauss's Principle states that the virtual acceleration work associated with all forces orthogonal to the interparticle motion (reaction forces) is *zero* ($A_c = 0$), and only the generalized force acting in the direction of interparticle motion produces virtual acceleration work. Thus,

$$\sum_{i=0}^{n_p-1}\sum_{j=i+1}^{n_p} \boldsymbol{f}_i^j \cdot (\delta\boldsymbol{a}_i - \delta\boldsymbol{a}_j) = \sum_{i=1}^{n_q} \tau_i \cdot \delta\ddot{q}_i, \tag{7.20}$$

and (7.16) can be expressed as

$$\sum_{i=1}^{n_q} \tau_i \cdot \delta\ddot{q}_i - \sum_{i=0}^{n_p} M_i(\boldsymbol{a}_i + g\hat{\boldsymbol{e}}_3) \cdot \delta\boldsymbol{a}_i = 0 \tag{7.21}$$

for all variations, $\delta\boldsymbol{v}_i$, that are consistent with the kinematic constraints.

Expressing the variations in terms of variations in the generalized velocities (with stationary velocity, displacement, and time), we have

$$\delta\boldsymbol{a}_i = \boldsymbol{\Gamma}_i \delta\ddot{\boldsymbol{q}}. \tag{7.22}$$

So, (7.21) can be expressed as

$$\boldsymbol{\tau} \cdot \delta\ddot{\boldsymbol{q}} - \sum_{i=1}^{n_p} M_i(\boldsymbol{a}_i + g\hat{\boldsymbol{e}}_3) \cdot (\boldsymbol{\Gamma}_i \delta\ddot{\boldsymbol{q}}) = 0 \qquad \forall \delta\ddot{\boldsymbol{q}} \in \mathbb{R}^n, \tag{7.23}$$

or

$$\left[\boldsymbol{\tau} - \sum_{i=1}^{n_p} \boldsymbol{\Gamma}_i^T (M_i\boldsymbol{a}_i + M_i g\hat{\boldsymbol{e}}_3)\right] \cdot \delta\ddot{\boldsymbol{q}} = 0 \qquad \forall \delta\ddot{\boldsymbol{q}} \in \mathbb{R}^n, \tag{7.24}$$

which implies

$$\boldsymbol{\tau} = \sum_{i=1}^{n_p} \boldsymbol{\Gamma}_i^T (M_i\boldsymbol{a}_i + M_i g\hat{\boldsymbol{e}}_3). \tag{7.25}$$

Noting that

$$\boldsymbol{a}_i = \boldsymbol{\Gamma}_i \ddot{\boldsymbol{q}}, \tag{7.26}$$

we have

$$\boldsymbol{\tau} = \left(\sum_{i=1}^{n_p} M_i \boldsymbol{\Gamma}_i^T \boldsymbol{\Gamma}_i \right) \ddot{\boldsymbol{q}} + g \sum_{i=1}^{n_q} M_i \boldsymbol{\Gamma}_i^T \hat{\boldsymbol{e}}_3. \tag{7.27}$$

For a system of particles, we had previously defined

$$\boldsymbol{M}(\boldsymbol{q}) \triangleq \sum_{i=1}^{n_p} M_i \boldsymbol{\Gamma}_i^T \boldsymbol{\Gamma}_i \tag{7.28}$$

$$\boldsymbol{g}(\boldsymbol{q}) \triangleq g \sum_{i=1}^{n_p} M_i \boldsymbol{\Gamma}_i^T \hat{\boldsymbol{e}}_3. \tag{7.29}$$

So,

$$\boldsymbol{\tau} = \boldsymbol{M}(\boldsymbol{q}) \ddot{\boldsymbol{q}} + \boldsymbol{g}(\boldsymbol{q}). \tag{7.30}$$

7.2.4 A System of Rigid Bodies

We now apply Gauss's Principle to a system of n_b rigid bodies forming a serial chain, with a discrete set of forces acting on them. The virtual acceleration work associated with a given body i is given by

$$\begin{aligned} \delta A_i = {} & \boldsymbol{f}_i^{i-1} \cdot \delta \boldsymbol{a}_i^{i-1} + \boldsymbol{f}_i^{i+1} \cdot \delta \boldsymbol{a}_i^{i+1} - M_i(\boldsymbol{a}_{G_i} + g\hat{\boldsymbol{e}}_{0_3}) \cdot \delta \boldsymbol{a}_{G_i} + \boldsymbol{\varphi}_i^{i-1} \cdot \delta \boldsymbol{\alpha}_i \\ & + \boldsymbol{\varphi}_i^{i+1} \cdot \delta \boldsymbol{\alpha}_i - (\boldsymbol{I}_i^{G_i} \boldsymbol{\alpha}_i + \boldsymbol{\omega}_i \times \boldsymbol{I}_i^{G_i} \boldsymbol{\omega}_i) \cdot \delta \boldsymbol{\alpha}_i = 0, \\ & \forall \delta \boldsymbol{a}_{G_i} \in \mathbb{R}^3, \quad \text{and} \quad \forall \delta \boldsymbol{\alpha}_i \in \mathbb{R}^3, \end{aligned} \tag{7.31}$$

for $i = 1, \ldots, n_b$. The term $\boldsymbol{f}_i^{i-1}$ is the force that body $i-1$ exerts on body i. Likewise, $\boldsymbol{\varphi}_i^{i-1}$ is the moment that body $i-1$ exerts on body i. The term $\boldsymbol{a}_i^{i-1}$ is the acceleration of the point on body i to which body $i-1$ attaches. Equation (7.31) is true for any rigid-body i in the system, under all independent variations, $\delta \boldsymbol{a}_{G_i}$ and $\delta \boldsymbol{\alpha}_i$ (we note that $\delta \boldsymbol{a}_i^{i-1}$ and $\delta \boldsymbol{a}_i^{i+1}$ are functions of $\delta \boldsymbol{a}_{G_i}$ and $\delta \boldsymbol{\alpha}_i$ due to the rigid body constraint). If we sum (7.31) over all rigid bodies, we obtain

$$\begin{aligned} \delta A = \sum_{i=1}^{n_b} \delta A_i = {} & \sum_{i=1}^{n_b} (\boldsymbol{f}_i^{i-1} \cdot \delta \boldsymbol{a}_i^{i-1} + \boldsymbol{f}_i^{i+1} \cdot \delta \boldsymbol{a}_i^{i+1}) \\ & - \sum_{i=1}^{n_b} M_i(\boldsymbol{a}_{G_i} + g\hat{\boldsymbol{e}}_{0_3}) \cdot \delta \boldsymbol{a}_{G_i} + \sum_{i=1}^{n_b} (\boldsymbol{\varphi}_i^{i-1} \cdot \delta \boldsymbol{\alpha}_i + \boldsymbol{\varphi}_i^{i+1} \cdot \delta \boldsymbol{\alpha}_i) \\ & - \sum_{i=1}^{n_b} (\boldsymbol{I}_i^{G_i} \boldsymbol{\alpha}_i + \boldsymbol{\omega}_i \times \boldsymbol{I}_i^{G_i} \boldsymbol{\omega}_i) \cdot \delta \boldsymbol{\alpha}_i = 0. \end{aligned} \tag{7.32}$$

The first summation is associated with the virtual acceleration work performed by the interlink reaction forces. It will be useful to rearrange the summation so as to pair up the equal and opposite reaction forces, $\boldsymbol{f}_i^{i-1}$ and $\boldsymbol{f}_{i-1}^i$, acting through each joint i. Doing this, we have

$$\sum_{i=1}^{n_b}(\boldsymbol{f}_i^{i-1}\cdot\delta\boldsymbol{a}_i^{i-1}+\boldsymbol{f}_i^{i+1}\cdot\delta\boldsymbol{a}_i^{i+1})=\sum_{i=1}^{n_b}(\boldsymbol{f}_i^{i-1}\cdot\delta\boldsymbol{a}_i^{i-1}+\boldsymbol{f}_{i-1}^{i}\cdot\delta\boldsymbol{a}_{i-1}^{i})+\boldsymbol{f}_{n_b}^{n_b+1}\cdot\delta\boldsymbol{a}_{n_b}^{n_b+1}-\boldsymbol{f}_0^1\cdot\delta\boldsymbol{a}_0^1. \tag{7.33}$$

Since $\boldsymbol{f}_{n_b}^{n_b+1}=\mathbf{0}$ and $\delta\boldsymbol{a}_0^1=\mathbf{0}$, we have

$$\sum_{i=1}^{n_b}(\boldsymbol{f}_i^{i-1}\cdot\delta\boldsymbol{a}_i^{i-1}+\boldsymbol{f}_i^{i+1}\cdot\delta\boldsymbol{a}_i^{i+1})=\sum_{i=1}^{n_b}(\boldsymbol{f}_i^{i-1}\cdot\delta\boldsymbol{a}_i^{i-1}+\boldsymbol{f}_{i-1}^{i}\cdot\delta\boldsymbol{a}_{i-1}^{i}). \tag{7.34}$$

We note that $\boldsymbol{f}_{i-1}^{i}=-\boldsymbol{f}_i^{i-1}$. Thus,

$$\sum_{i=1}^{n_b}(\boldsymbol{f}_i^{i-1}\cdot\delta\boldsymbol{a}_i^{i-1}+\boldsymbol{f}_i^{i+1}\cdot\delta\boldsymbol{a}_i^{i+1})=\sum_{i=1}^{n_b}\boldsymbol{f}_i^{i-1}\cdot(\delta\boldsymbol{a}_i^{i-1}-\delta\boldsymbol{a}_{i-1}^{i}). \tag{7.35}$$

This reflects the virtual acceleration work done by reaction forces at each joint, summed over all joints.

As with the reaction forces, it will be useful to rearrange the summation associated with the virtual acceleration work performed by the interlink reaction moments so as to pair up the equal and opposite reaction moments, $\boldsymbol{\varphi}_i^{i-1}$ and $\boldsymbol{\varphi}_{i-1}^{i}$, acting about each joint. Using the same procedure as with the reaction forces, we have

$$\begin{aligned}\sum_{i=1}^{n_b}(\boldsymbol{\varphi}_i^{i-1}\cdot\delta\boldsymbol{\alpha}_i+\boldsymbol{\varphi}_i^{i+1}\cdot\delta\boldsymbol{\alpha}_i)&=\sum_{i=1}^{n_b}(\boldsymbol{\varphi}_i^{i-1}\cdot\delta\boldsymbol{\alpha}_i+\boldsymbol{\varphi}_{i-1}^{i}\cdot\delta\boldsymbol{\alpha}_{i-1})\\&=\sum_{i=1}^{n_b}\boldsymbol{\varphi}_i^{i-1}\cdot(\delta\boldsymbol{\alpha}_i-\delta\boldsymbol{\alpha}_{i-1}).\end{aligned} \tag{7.36}$$

This reflects the virtual acceleration work done by reaction moments at each joint, summed over all joints.

So, the total virtual acceleration work performed by the interlink reaction forces and moments can be expressed compactly as

$$\begin{aligned}&\sum_{i=1}^{n_b}(\boldsymbol{f}_i^{i-1}\cdot\delta\boldsymbol{a}_i^{i-1}+\boldsymbol{f}_i^{i+1}\cdot\delta\boldsymbol{a}_i^{i+1})+\sum_{i=1}^{n_b}(\boldsymbol{\varphi}_i^{i-1}\cdot\delta\boldsymbol{\alpha}_i+\boldsymbol{\varphi}_i^{i+1}\cdot\delta\boldsymbol{\alpha}_i)\\&\quad=\sum_{i=1}^{n_b}\begin{pmatrix}\boldsymbol{f}_i^{i-1}\\ \boldsymbol{\varphi}_i^{i-1}\end{pmatrix}\cdot\begin{pmatrix}\delta\boldsymbol{a}_i^{i-1}-\delta\boldsymbol{a}_{i-1}^{i}\\ \delta\boldsymbol{\alpha}_i-\delta\boldsymbol{\alpha}_{i-1}\end{pmatrix}.\end{aligned} \tag{7.37}$$

If we now consider only the variations $\delta\boldsymbol{a}_i^{i-1}$, $\delta\boldsymbol{a}_{i-1}^{i}$, $\delta\boldsymbol{\alpha}_i$, and $\delta\boldsymbol{\alpha}_{i-1}$ that are consistent with the kinematic constraints, then (7.37) reflects a projection of the interlink forces and moments on the direction of the joint motion. Gauss's Principle states that the virtual acceleration work associated with all forces and moments orthogonal to the joint motion (reaction forces/moments) is *zero* ($A_c=0$) and only the generalized force acting in the direction of joint motion produces virtual acceleration work. Thus,

$$\sum_{i=1}^{n_b} \begin{pmatrix} \boldsymbol{f}_i^{i-1} \\ \boldsymbol{\varphi}_i^{i-1} \end{pmatrix} \cdot \begin{pmatrix} \delta \boldsymbol{a}_i^{i-1} - \delta \boldsymbol{a}_{i-1}^{i} \\ \delta \boldsymbol{\alpha}_i - \delta \boldsymbol{\alpha}_{i-1} \end{pmatrix} = \sum_{i=1}^{n_q} \tau_i \cdot \delta \ddot{q}_i, \tag{7.38}$$

and (7.32) can be expressed as

$$\sum_{i=1}^{n_q} \tau_i \cdot \delta \ddot{q}_i - \sum_{i=1}^{n_b} M_i(\boldsymbol{a}_{G_i} + g\hat{\boldsymbol{e}}_{0_3}) \cdot \delta \boldsymbol{a}_{G_i} - \sum_{i=1}^{n_b} (\boldsymbol{I}_i^{G_i} \boldsymbol{\alpha}_i + \boldsymbol{\omega}_i \times \boldsymbol{I}_i^{G_i} \boldsymbol{\omega}_i) \cdot \delta \boldsymbol{\alpha}_i = 0 \tag{7.39}$$

for all variations, $\delta \boldsymbol{a}_{G_i}$ and $\delta \boldsymbol{\alpha}_i$, that are consistent with the kinematic constraints.

Using Gauss's Principle, we can address a system of n_b rigid bodies forming a branching chain in a similar manner. With a serial chain the parent/child structure is implicit to the numbering scheme. Every link i has a single parent link, $i-1$, at its proximal end and a single child link, $i+1$, at its distal end (except for the nth link). The ith joint is at the proximal end of the ith link. With a branching chain, the numbering of links is more arbitrary, without a parent/child structure implicit to the numbering scheme. We can explicitly capture the parent/child structure, however, by defining three additional parameters, λ_i, c_i, μ_{ij}. The term λ_i is the parent link number of the ith link, c_i is the number of child links for the ith link, and $\mu_{i1} \ldots \mu_{ic_i}$ are the child link numbers of the ith link. Given these parameters, the virtual acceleration work associated with a given body i is given by

$$\begin{aligned} \delta A_i = {} & \boldsymbol{f}_i^{\lambda_i} \cdot \delta \boldsymbol{a}_i^{\lambda_i} + \sum_{j=1}^{c_i} \boldsymbol{f}_i^{\mu_{ij}} \cdot \delta \boldsymbol{a}_i^{\mu_{ij}} - M_i(\boldsymbol{a}_{G_i} + g\hat{\boldsymbol{e}}_{0_3}) \cdot \delta \boldsymbol{a}_{G_i} + \boldsymbol{\varphi}_i^{\lambda_i} \cdot \delta \boldsymbol{\alpha}_i \\ & + \sum_{j=1}^{c_i} \boldsymbol{\varphi}_i^{\mu_{ij}} \cdot \delta \boldsymbol{\alpha}_i - (\boldsymbol{I}_i^{G_i} \boldsymbol{\alpha}_i + \boldsymbol{\omega}_i \times \boldsymbol{I}_i^{G_i} \boldsymbol{\omega}_i) \cdot \delta \boldsymbol{\alpha}_i = 0, \\ & \forall \delta \boldsymbol{a}_{G_i} \in \mathbb{R}^3, \quad \text{and} \quad \forall \delta \boldsymbol{\alpha}_i \in \mathbb{R}^3, \end{aligned} \tag{7.40}$$

for $i = 1, \ldots, n_b$. The term $\boldsymbol{a}_i^{\lambda_i}$ is the acceleration of the point on body i to which body (parent) λ_i attaches, and likewise $\boldsymbol{a}_i^{\mu_{i1}} \ldots \boldsymbol{a}_i^{\mu_{ic_i}}$ are the accelerations of points on body i to which bodies (children) $\mu_{i1} \ldots \mu_{ic_i}$ attach. We note that $\delta \boldsymbol{a}_i^{\lambda_i}$ and $\delta \boldsymbol{a}_i^{\mu_{i1}} \ldots \delta \boldsymbol{a}_i^{\mu_{ic_i}}$ are functions of $\delta \boldsymbol{a}_{G_i}$ and $\delta \boldsymbol{\alpha}_i$ due to the rigid-body constraint. If we sum (7.40) over all rigid bodies, we obtain

$$\begin{aligned} \delta A = \sum_{i=1}^{n_b} \delta A_i = {} & \sum_{i=1}^{n_b} \left(\boldsymbol{f}_i^{\lambda_i} \cdot \delta \boldsymbol{a}_i^{\lambda_i} + \sum_{j=1}^{c_i} \boldsymbol{f}_i^{\mu_{ij}} \cdot \delta \boldsymbol{a}_i^{\mu_{ij}} \right) \\ & - \sum_{i=1}^{n_b} M_i(\boldsymbol{a}_{G_i} + g\hat{\boldsymbol{e}}_{0_3}) \cdot \delta \boldsymbol{a}_{G_i} + \sum_{i=1}^{n_b} \left(\boldsymbol{\varphi}_i^{\lambda_i} \cdot \delta \boldsymbol{\alpha}_i + \sum_{j=1}^{c_i} \boldsymbol{\varphi}_i^{\mu_{ij}} \cdot \delta \boldsymbol{\alpha}_i \right) \\ & - \sum_{i=1}^{n_b} (\boldsymbol{I}_i^{G_i} \boldsymbol{\alpha}_i + \boldsymbol{\omega}_i \times \boldsymbol{I}_i^{G_i} \boldsymbol{\omega}_i) \cdot \delta \boldsymbol{\alpha}_i = 0. \end{aligned} \tag{7.41}$$

The term associated with the virtual acceleration work performed by the interlink reaction forces is

$$\sum_{i=1}^{n_b}\left(\boldsymbol{f}_i^{\lambda_i}\cdot\delta\boldsymbol{a}_i^{\lambda_i}+\sum_{j=1}^{c_i}\boldsymbol{f}_i^{\mu_{ij}}\cdot\delta\boldsymbol{a}_i^{\mu_{ij}}\right). \tag{7.42}$$

It will be useful to rearrange the summation so as to pair up the equal and opposite reaction forces, $\boldsymbol{f}_i^{\lambda_i}$ and $\boldsymbol{f}_{\lambda_i}^{i}$, acting through each joint i. We can rewrite the summation of (7.42) based on considering the virtual acceleration work at each joint. The sum over all joints then gives us

$$\sum_{i=1}^{n_b}(\boldsymbol{f}_i^{\lambda_i}\cdot\delta\boldsymbol{a}_i^{\lambda_i}+\boldsymbol{f}_{\lambda_i}^{i}\cdot\delta\boldsymbol{a}_{\lambda_i}^{i})=\sum_{i=1}^{n_b}\boldsymbol{f}_i^{\lambda_i}\cdot(\delta\boldsymbol{a}_i^{\lambda_i}-\delta\boldsymbol{a}_{\lambda_i}^{i}). \tag{7.43}$$

The term associated with the virtual acceleration work performed by the interlink reaction moments is

$$\sum_{i=1}^{n_b}\left(\boldsymbol{\varphi}_i^{\lambda_i}\cdot\delta\boldsymbol{\alpha}_i+\sum_{j=1}^{c_i}\boldsymbol{\varphi}_i^{\mu_{ij}}\cdot\delta\boldsymbol{\alpha}_i\right). \tag{7.44}$$

In a similar manner as before, we can rewrite this summation based on considering the virtual acceleration work at each joint. The sum over all joints then gives us

$$\sum_{i=1}^{n_b}(\boldsymbol{\varphi}_i^{\lambda_i}\cdot\delta\boldsymbol{\alpha}_i+\boldsymbol{\varphi}_{\lambda_i}^{i}\cdot\delta\boldsymbol{\alpha}_{\lambda_i})=\sum_{i=1}^{n_b}\boldsymbol{\varphi}_i^{\lambda_i}\cdot(\delta\boldsymbol{\alpha}_i-\delta\boldsymbol{\alpha}_{\lambda_i}). \tag{7.45}$$

So, the total virtual acceleration work performed by the interlink reaction forces and moments can be expressed compactly as

$$\begin{aligned}\sum_{i=1}^{n_b}\left(\boldsymbol{f}_i^{\lambda_i}\cdot\delta\boldsymbol{a}_i^{\lambda_i}+\sum_{j=1}^{c_i}\boldsymbol{f}_i^{\mu_{ij}}\cdot\delta\boldsymbol{a}_i^{\mu_{ij}}\right)+\sum_{i=1}^{n_b}\left(\boldsymbol{\varphi}_i^{\lambda_i}\cdot\delta\boldsymbol{\alpha}_i+\sum_{j=1}^{c_i}\boldsymbol{\varphi}_i^{\mu_{ij}}\cdot\delta\boldsymbol{\alpha}_i\right)\\=\sum_{i=1}^{n_b}\begin{pmatrix}\boldsymbol{f}_i^{\lambda_i}\\ \boldsymbol{\varphi}_i^{\lambda_i}\end{pmatrix}\cdot\begin{pmatrix}\delta\boldsymbol{a}_i^{\lambda_i}-\delta\boldsymbol{a}_{\lambda_i}^{i}\\ \delta\boldsymbol{\alpha}_i-\delta\boldsymbol{\alpha}_{\lambda_i}\end{pmatrix}.\end{aligned} \tag{7.46}$$

If we now consider only the variations $\delta\boldsymbol{a}_i^{\lambda_i}$, $\delta\boldsymbol{a}_{\lambda_i}^{i}$, $\delta\boldsymbol{\alpha}_i$, and $\delta\boldsymbol{\alpha}_{\lambda_i}$ that are consistent with the kinematic constraints, then (7.46) reflects a projection of the interlink forces and moments on the direction of the joint motion. Gauss's Principle states that the virtual acceleration work associated with all forces and moments orthogonal to the joint motion (reaction forces/moments) is *zero* ($A_c = 0$) and only the generalized force acting in the direction of joint motion produces virtual acceleration work. Thus,

$$\sum_{i=1}^{n_b}\begin{pmatrix}\boldsymbol{f}_i^{\lambda_i}\\ \boldsymbol{\varphi}_i^{\lambda_i}\end{pmatrix}\cdot\begin{pmatrix}\delta\boldsymbol{a}_i^{\lambda_i}-\delta\boldsymbol{a}_{\lambda_i}^{i}\\ \delta\boldsymbol{\alpha}_i-\delta\boldsymbol{\alpha}_{\lambda_i}\end{pmatrix}=\sum_{i=1}^{n_q}\tau_i\cdot\delta\ddot{q}_i, \tag{7.47}$$

and (7.41) can be expressed as

$$\sum_{i=1}^{n_q} \tau_i \cdot \delta\ddot{q}_i - \sum_{i=1}^{n_b} M_i(\boldsymbol{a}_{G_i} + g\hat{\boldsymbol{e}}_{0_3}) \cdot \delta\boldsymbol{a}_{G_i} - \sum_{i=1}^{n_b} (\boldsymbol{I}_i^{G_i}\boldsymbol{\alpha}_i + \boldsymbol{\omega}_i \times \boldsymbol{I}_i^{G_i}\boldsymbol{\omega}_i) \cdot \delta\boldsymbol{\alpha}_i = 0 \tag{7.48}$$

for all variations, $\delta\boldsymbol{a}_{G_i}$ and $\delta\boldsymbol{\alpha}_i$, that are consistent with the kinematic constraints. Expressing the variations in terms of variations in the generalized accelerations (with stationary velocity, displacement, and time), we have

$${}^i\delta\boldsymbol{a}_{G_i} = {}^i\boldsymbol{\Gamma}_{G_i}\delta\ddot{\boldsymbol{q}} \quad \text{and} \quad {}^i\delta\boldsymbol{\alpha}_i = {}^i\boldsymbol{\Pi}_i\delta\ddot{\boldsymbol{q}}, \tag{7.49}$$

where the accelerations are expressed in the local link frame i for convenience. So, (7.48) can be expressed as

$$\boldsymbol{\tau} \cdot \delta\ddot{\boldsymbol{q}} - \sum_{i=1}^{n_b} M_i({}^i\boldsymbol{a}_{G_i} + g{}^i\hat{\boldsymbol{e}}_{0_3}) \cdot ({}^i\boldsymbol{\Gamma}_{G_i}\delta\ddot{\boldsymbol{q}}) - \sum_{i=1}^{n_b} ({}^i\boldsymbol{I}_i^{G_i}{}^i\boldsymbol{\alpha}_i + {}^i\boldsymbol{\omega}_i \times {}^i\boldsymbol{I}_i^{G_i}{}^i\boldsymbol{\omega}_i) \cdot ({}^i\boldsymbol{\Pi}_i\delta\ddot{\boldsymbol{q}}) = 0$$
$$\forall \delta\ddot{\boldsymbol{q}} \in \mathbb{R}^n, \tag{7.50}$$

or

$$\boldsymbol{\tau} \cdot \delta\ddot{\boldsymbol{q}} - \sum_{i=1}^{n_b} \left[M_i{}^i\boldsymbol{\Gamma}_{G_i}^{T}{}^i\boldsymbol{a}_{G_i} + M_i g{}^i\boldsymbol{\Gamma}_{G_i}^{T}{}^i\hat{\boldsymbol{e}}_{0_3} + {}^i\boldsymbol{\Pi}_i^T({}^i\boldsymbol{I}_i^{G_i}{}^i\boldsymbol{\alpha}_i + {}^i\boldsymbol{\omega}_i \times {}^i\boldsymbol{I}_i^{G_i}{}^i\boldsymbol{\omega}_i)\right] \cdot \delta\ddot{\boldsymbol{q}} = 0$$
$$\forall \delta\ddot{\boldsymbol{q}} \in \mathbb{R}^n. \tag{7.51}$$

In matrix form we have

$$\left[\boldsymbol{\tau} - \sum_{i=1}^{n_b} \begin{pmatrix} {}^i\boldsymbol{\Gamma}_{G_i}^T & {}^i\boldsymbol{\Pi}_i^T \end{pmatrix} \begin{pmatrix} M_i{}^i\boldsymbol{a}_{G_i} + M_i g{}^i\hat{\boldsymbol{e}}_{0_3} \\ {}^i\boldsymbol{I}_i^{G_i}{}^i\boldsymbol{\alpha}_i + {}^i\boldsymbol{\omega}_i \times {}^i\boldsymbol{I}_i^{G_i}{}^i\boldsymbol{\omega}_i \end{pmatrix}\right] \cdot \delta\ddot{\boldsymbol{q}} = 0$$
$$\forall \delta\ddot{\boldsymbol{q}} \in \mathbb{R}^n, \tag{7.52}$$

which implies

$$\boldsymbol{\tau} = \sum_{i=1}^{n_b} \begin{pmatrix} {}^i\boldsymbol{\Gamma}_{G_i}^T & {}^i\boldsymbol{\Pi}_i^T \end{pmatrix} \begin{pmatrix} M_i{}^i\boldsymbol{a}_{G_i} + M_i g{}^i\hat{\boldsymbol{e}}_{0_3} \\ {}^i\boldsymbol{I}_i^{G_i}{}^i\boldsymbol{\alpha}_i + {}^i\boldsymbol{\Omega}_i{}^i\boldsymbol{I}_i^{G_i}{}^i\boldsymbol{\omega}_i \end{pmatrix}. \tag{7.53}$$

Since this is identical to (5.53), we have

$$\boldsymbol{\tau} = \boldsymbol{M}(\boldsymbol{q})\ddot{\boldsymbol{q}} + \boldsymbol{b}(\boldsymbol{q}, \dot{\boldsymbol{q}}) + \boldsymbol{g}(\boldsymbol{q}). \tag{7.54}$$

7.2.5 Auxiliary Constraints

Second-Order Nonholonomic Constraints

Second-order nonholonomic constraints are nonintegrable functions of the generalized coordinates, generalized velocities, and generalized accelerations (Udwadia and Kalaba 1992). We will address these constraints here.

We now consider the general case of auxiliary second-order nonholonomic constraints. Given the multibody system,

$$\boldsymbol{\tau} = \boldsymbol{M}(\boldsymbol{q})\ddot{\boldsymbol{q}} + \boldsymbol{b}(\boldsymbol{q}, \dot{\boldsymbol{q}}) + \boldsymbol{g}(\boldsymbol{q}), \tag{7.55}$$

subject to the second-order nonholonomic constraints

$$\boldsymbol{A}(\boldsymbol{q}, \dot{\boldsymbol{q}})\ddot{\boldsymbol{q}} + \boldsymbol{\psi}(\boldsymbol{q}, \dot{\boldsymbol{q}}) = \boldsymbol{0}. \tag{7.56}$$

We begin by first expressing the second-order variational equation associated with Gauss's Principle,

$$\boldsymbol{\tau}_C \cdot \delta\ddot{\boldsymbol{q}} + (\boldsymbol{\tau} - \boldsymbol{M}\ddot{\boldsymbol{q}} - \boldsymbol{b} - \boldsymbol{g}) \cdot \delta\ddot{\boldsymbol{q}} = 0. \tag{7.57}$$

The virtual accelerations, $\delta\ddot{\boldsymbol{q}}$, refer to all acceleration variations that satisfy the constraints, while time, displacement, and velocity are fixed. With $\delta t = 0$, $\delta\boldsymbol{q} = \boldsymbol{0}$, and $\delta\dot{\boldsymbol{q}} = \boldsymbol{0}$, the second-order variation of the constraint equation yields

$$\boldsymbol{A}\delta\ddot{\boldsymbol{q}} = \boldsymbol{0}, \tag{7.58}$$

which implies that $\delta\ddot{\boldsymbol{q}} \in \ker(\boldsymbol{A})$. Under this condition, (7.57) can be restricted to constraint-consistent virtual displacements,

$$\begin{gathered} \boldsymbol{\tau}_C \cdot \delta\ddot{\boldsymbol{q}} + (\boldsymbol{\tau} - \boldsymbol{M}\boldsymbol{q} - \boldsymbol{b} - \boldsymbol{g}) \cdot \delta\ddot{\boldsymbol{q}} = 0 \\ \forall \delta\ddot{\boldsymbol{q}} \in \ker(\boldsymbol{A}). \end{gathered} \tag{7.59}$$

We have

$$\begin{gathered} \boldsymbol{\tau}_C \perp \delta\ddot{\boldsymbol{q}} \\ \forall \delta\ddot{\boldsymbol{q}} \in \ker(\boldsymbol{A}). \end{gathered} \tag{7.60}$$

Thus the term $\boldsymbol{\tau}_C \cdot \delta\ddot{\boldsymbol{q}}$ vanishes from (7.59), and we have the orthogonality relation

$$\begin{gathered} (\boldsymbol{M}\boldsymbol{q} + \boldsymbol{b} + \boldsymbol{g} - \boldsymbol{\tau}) \cdot \delta\ddot{\boldsymbol{q}} = 0 \\ \forall \delta\ddot{\boldsymbol{q}} \in \ker(\boldsymbol{A}). \end{gathered} \tag{7.61}$$

The constrained multibody equations of motion, expressed in the familiar multiplier form, are thus

$$\boldsymbol{\tau} = \boldsymbol{M}\ddot{\boldsymbol{q}} + \boldsymbol{b} + \boldsymbol{g} - \boldsymbol{A}^T\boldsymbol{\lambda}, \tag{7.62}$$

subject to

$$\boldsymbol{A}\ddot{\boldsymbol{q}} + \boldsymbol{\psi} = \boldsymbol{0}. \tag{7.63}$$

In the special case of first-order nonholonomic constraints,

$$\boldsymbol{\psi}(\boldsymbol{q}, \dot{\boldsymbol{q}}) = \boldsymbol{0}, \tag{7.64}$$

we can differentiate once with respect to time to yield

$$\dot{\boldsymbol{\psi}} = \frac{\partial \boldsymbol{\psi}}{\partial \boldsymbol{q}}\dot{\boldsymbol{q}} + \frac{\partial \boldsymbol{\psi}}{\partial \dot{\boldsymbol{q}}}\ddot{\boldsymbol{q}} = \boldsymbol{0}. \tag{7.65}$$

The second-order variation is given by

$$\delta\dot{\boldsymbol{\psi}} = \frac{\partial \boldsymbol{\psi}}{\partial \dot{\boldsymbol{q}}}\delta\ddot{\boldsymbol{q}} = \boldsymbol{\Psi}\delta\ddot{\boldsymbol{q}} = \boldsymbol{0}, \tag{7.66}$$

and the constrained multibody equations of motion express identically to the first-order case,

$$\boldsymbol{\tau} = \boldsymbol{M}\ddot{\boldsymbol{q}} + \boldsymbol{b} + \boldsymbol{g} - \boldsymbol{\Psi}^T\boldsymbol{\lambda}, \tag{7.67}$$

subject to

$$\boldsymbol{\psi}(\boldsymbol{q}, \dot{\boldsymbol{q}}) = \boldsymbol{0} \quad \Rightarrow \quad \dot{\boldsymbol{\psi}} = \boldsymbol{0}. \tag{7.68}$$

In the special case of first-order nonholonomic constraints that are linear in $\dot{\boldsymbol{q}}$,

$$\boldsymbol{\psi}(\boldsymbol{q}, \dot{\boldsymbol{q}}) = \boldsymbol{C}(\boldsymbol{q})\dot{\boldsymbol{q}} + \boldsymbol{d}(\boldsymbol{q}) = \boldsymbol{0}, \tag{7.69}$$

we can differentiate once with respect to time to yield

$$\dot{\boldsymbol{\psi}} = \boldsymbol{C}\ddot{\boldsymbol{q}} + \dot{\boldsymbol{C}}\dot{\boldsymbol{q}} + \dot{\boldsymbol{d}} = \boldsymbol{0}. \tag{7.70}$$

The second-order variation is given by

$$\delta\dot{\boldsymbol{\psi}} = \boldsymbol{C}\delta\ddot{\boldsymbol{q}} = \boldsymbol{0}, \tag{7.71}$$

and the constrained multibody equations of motion express identically to the first-order case,

$$\boldsymbol{\tau} = \boldsymbol{M}\ddot{\boldsymbol{q}} + \boldsymbol{b} + \boldsymbol{g} - \boldsymbol{C}^T\boldsymbol{\lambda}, \tag{7.72}$$

subject to

$$\boldsymbol{C}\dot{\boldsymbol{q}} + \boldsymbol{d} = \boldsymbol{0} \quad \Rightarrow \quad \boldsymbol{C}\ddot{\boldsymbol{q}} + \dot{\boldsymbol{C}}\dot{\boldsymbol{q}} + \dot{\boldsymbol{d}} = \boldsymbol{0}. \tag{7.73}$$

In the special case of zeroth-order holonomic constraints,

$$\boldsymbol{\phi}(\boldsymbol{q}) = \boldsymbol{0}, \tag{7.74}$$

we can differentiate twice with respect to time to yield

$$\ddot{\boldsymbol{\phi}} = \boldsymbol{\Phi}\ddot{\boldsymbol{q}} + \dot{\boldsymbol{\Phi}}\dot{\boldsymbol{q}} = \boldsymbol{0}. \tag{7.75}$$

Algorithm 8 Second-order method for integrating the second-order nonholonomically constrained equations of motion

1: $\boldsymbol{q}_0 = \boldsymbol{q}_o$ {initialization}
2: $\dot{\boldsymbol{q}}_0 = \dot{\boldsymbol{q}}_o$ {initialization}
3: **for** $i = 0$ to $n_s - 1$ **do**
4: $\quad \ddot{\boldsymbol{q}}_i = -\bar{\boldsymbol{A}}_i\boldsymbol{\psi}_i + \boldsymbol{\Theta}_i\boldsymbol{M}_i^{-1}(\boldsymbol{\tau}_i - \boldsymbol{b}_i - \boldsymbol{g}_i)$
5: $\quad \boldsymbol{\lambda}_i = -\bar{\boldsymbol{A}}_i^T(\boldsymbol{\tau}_i - \boldsymbol{b}_i - \boldsymbol{g}_i) - \mathbf{H}_i\boldsymbol{\psi}_i$
6: $\quad \dot{\boldsymbol{q}}_{i+1} = \dot{\boldsymbol{q}}_i + \ddot{\boldsymbol{q}}_i\Delta t$
7: $\quad \boldsymbol{q}_{i+1} = \boldsymbol{q}_i + \dot{\boldsymbol{q}}_i\Delta t + \frac{1}{2}\ddot{\boldsymbol{q}}_i\Delta t^2$
8: **end for**

The second-order variation is given by

$$\delta\ddot{\boldsymbol{\phi}} = \boldsymbol{\Phi}\delta\ddot{\boldsymbol{q}} = \boldsymbol{0}, \tag{7.76}$$

and the constrained multibody equations of motion express identically to the zeroth-order case,

$$\boldsymbol{\tau} = \boldsymbol{M}\ddot{\boldsymbol{q}} + \boldsymbol{b} + \boldsymbol{g} - \boldsymbol{\Phi}^T\boldsymbol{\lambda}, \tag{7.77}$$

subject to

$$\boldsymbol{\phi}(\boldsymbol{q}) = \boldsymbol{0} \quad \Rightarrow \quad \dot{\boldsymbol{\phi}} = \boldsymbol{0}, \ddot{\boldsymbol{\phi}} = \boldsymbol{0}. \tag{7.78}$$

Numerical Integration

Given the nonholonomic system

$$\boldsymbol{\tau} = \boldsymbol{M}\ddot{\boldsymbol{q}} + \boldsymbol{b} + \boldsymbol{g} - \boldsymbol{A}^T\boldsymbol{\lambda}, \tag{7.79}$$

subject to

$$\boldsymbol{A}\ddot{\boldsymbol{q}} + \boldsymbol{\psi} = \boldsymbol{0}, \tag{7.80}$$

the equations of motion are

$$\begin{pmatrix} \boldsymbol{M} & -\boldsymbol{A}^T \\ -\boldsymbol{A} & \boldsymbol{0} \end{pmatrix} \begin{pmatrix} \ddot{\boldsymbol{q}} \\ \boldsymbol{\lambda} \end{pmatrix} = \begin{pmatrix} \boldsymbol{\tau} - \boldsymbol{b} - \boldsymbol{g} \\ \boldsymbol{\psi} \end{pmatrix}, \tag{7.81}$$

and the solution of this system is

$$\ddot{\boldsymbol{q}} = -\bar{\boldsymbol{A}}\boldsymbol{\psi} + \boldsymbol{\Theta}\boldsymbol{M}^{-1}(\boldsymbol{\tau} - \boldsymbol{b} - \boldsymbol{g}) \tag{7.82}$$

$$\boldsymbol{\lambda} = -\bar{\boldsymbol{A}}^T(\boldsymbol{\tau} - \boldsymbol{b} - \boldsymbol{g}) - \mathbf{H}\boldsymbol{\psi}, \tag{7.83}$$

where $\boldsymbol{\Theta} = \mathbf{1} - \bar{\boldsymbol{A}}\boldsymbol{A}$ and $\mathbf{H} = (\boldsymbol{A}\boldsymbol{M}^{-1}\boldsymbol{A}^T)^{-1}$.

A second-order method for integrating this system can be summarized as shown in Algorithm 8.

7.3 Gauss's Principle of Least Constraint

Gauss's Principle of Least Constraint uses the second variation to express the dynamics of a system as a minimization problem. The Principle of Least Constraint is a differential local minimum principle, as opposed to Hamilton's Principle, which is an integral extremal principle. In the next section we will apply Gauss's Principle of Least Constraint to rigid-body systems.

7.3.1 A System of Rigid Bodies

We begin by noting the second-order orthogonality relation for a multibody system,

$$\begin{gathered}(\boldsymbol{M}\boldsymbol{q}+\boldsymbol{b}+\boldsymbol{g}-\boldsymbol{\tau})\cdot\delta\ddot{\boldsymbol{q}}=0\\ \forall\delta\ddot{\boldsymbol{q}}\in\ker(\boldsymbol{A}),\end{gathered}\tag{7.84}$$

subject to the constraints

$$\boldsymbol{A}(\boldsymbol{q},\dot{\boldsymbol{q}})\ddot{\boldsymbol{q}}+\boldsymbol{\psi}(\boldsymbol{q},\dot{\boldsymbol{q}})=\boldsymbol{0}.\tag{7.85}$$

If we define the quadratic *Gauss* function, $\mathcal{G}$, as

$$\mathcal{G}(\ddot{\boldsymbol{q}})=\frac{1}{2}(\boldsymbol{M}\ddot{\boldsymbol{q}}+\boldsymbol{b}+\boldsymbol{g}-\boldsymbol{\tau})^T\boldsymbol{M}^{-1}(\boldsymbol{M}\ddot{\boldsymbol{q}}+\boldsymbol{b}+\boldsymbol{g}-\boldsymbol{\tau}),\tag{7.86}$$

the variation, $\delta\mathcal{G}$, with respect to the virtual accelerations is

$$\delta\mathcal{G}=\frac{\partial\mathcal{G}}{\partial\ddot{\boldsymbol{q}}}\cdot\delta\ddot{\boldsymbol{q}}=(\boldsymbol{M}\ddot{\boldsymbol{q}}+\boldsymbol{b}+\boldsymbol{g}-\boldsymbol{\tau})\cdot\delta\ddot{\boldsymbol{q}}.\tag{7.87}$$

Thus, we can write (7.84) as

$$\begin{gathered}\delta\mathcal{G}=0\\ \forall\delta\ddot{\boldsymbol{q}}\in\ker(\boldsymbol{A}).\end{gathered}\tag{7.88}$$

This implies a constrained minimization of $\mathcal{G}(\ddot{\boldsymbol{q}})$, subject to

$$\boldsymbol{A}(\boldsymbol{q},\dot{\boldsymbol{q}})\ddot{\boldsymbol{q}}+\boldsymbol{\psi}(\boldsymbol{q},\dot{\boldsymbol{q}})=\boldsymbol{0}.\tag{7.89}$$

Since we can express $\mathcal{G}$ as

$$\mathcal{G}=\frac{1}{2}\boldsymbol{\tau_C}^T\boldsymbol{M}^{-1}\boldsymbol{\tau_C}=\frac{1}{2}\left\|\boldsymbol{\tau_C}\right\|_{M^{-1}}^2,\tag{7.90}$$

this constrained minimization seeks a constraint-consistent generalized acceleration that minimizes the $\boldsymbol{M}^{-1}$-weighted norm of the generalized constraint forces, $\boldsymbol{\tau_C}$. Alternatively, we can write $\mathcal{G}$ as

$$\mathcal{G}=\frac{1}{2}\left[\ddot{\boldsymbol{q}}+\boldsymbol{M}^{-1}(\boldsymbol{b}+\boldsymbol{g}-\boldsymbol{\tau})\right]^T\boldsymbol{M}\left[\ddot{\boldsymbol{q}}+\boldsymbol{M}^{-1}(\boldsymbol{b}+\boldsymbol{g}-\boldsymbol{\tau})\right].\tag{7.91}$$

Noting that the unconstrained generalized acceleration, $\ddot{\boldsymbol{q}}_\star$, is

$$\ddot{\boldsymbol{q}}_\star=\boldsymbol{M}^{-1}(\boldsymbol{\tau}-\boldsymbol{b}-\boldsymbol{g}),\tag{7.92}$$

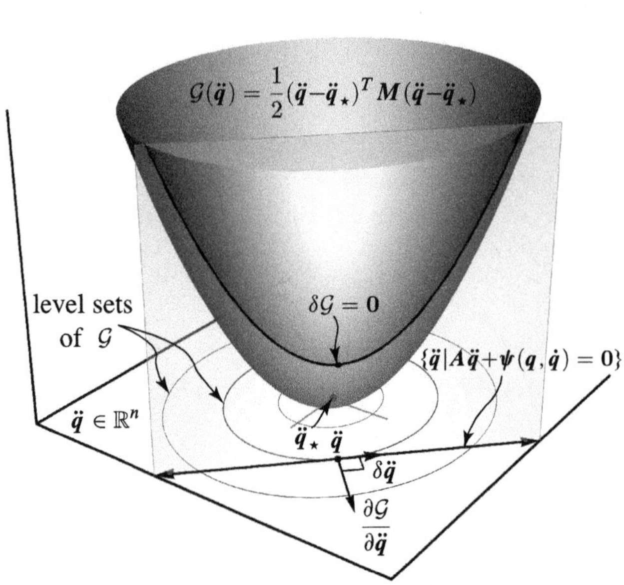

Figure 7.2 The Gauss function, $\mathcal{G}$, is minimized subject to the constraints. At the solution point, $\ddot{q}$, the gradient of $\mathcal{G}$ is orthogonal to the space of virtual accelerations. Geometrically, the solution minimizes the distance (mass weighted) to the unconstrained acceleration, $\ddot{q}_\star$.

we can write $\mathcal{G}$ as

$$\mathcal{G}(\ddot{q}) = \frac{1}{2}(\ddot{q} - \ddot{q}_\star)^T M(\ddot{q} - \ddot{q}_\star) = \frac{1}{2}\,\|\ddot{q} - \ddot{q}_\star\|_M^2\,. \tag{7.93}$$

Thus, the constrained minimization of $\mathcal{G}$ also seeks a constraint-consistent generalized acceleration that minimizes the M-weighted norm of the difference between the actual constrained generalized acceleration and the unconstrained generalized acceleration. This is illustrated in Figure 7.2.

7.3.2 Hertz's Principle of Least Curvature

Using Gauss's Principle, we can examine the special case of a particle moving in n dimensions, $r(t) = q(t) \in \mathbb{R}^n$, under holonomic constraints but no applied force. That is, $M\ddot{r} = \Phi^T\lambda$ subject to $\phi(r) = 0$. For this system we have the Gauss function

$$\mathcal{G} = \frac{1}{2}M(\ddot{r} - \ddot{r}_\star)^T(\ddot{r} - \ddot{r}_\star). \tag{7.94}$$

In the absence of applied forces, $\ddot{r}_\star = 0$, so

$$\mathcal{G} = \frac{1}{2}M\ddot{r}^T\ddot{r} = \frac{1}{2}M\,\langle\ddot{r}, \ddot{r}\rangle\,. \tag{7.95}$$

The vector r can be parameterized in terms of arc length, s, in which case we have the following relationships:

$$\dot{r} = \frac{dr}{dt} = \frac{dr}{ds}\frac{ds}{dt}, \tag{7.96}$$

and

$$\ddot{\boldsymbol{r}} = \frac{d}{dt}\frac{d\boldsymbol{r}}{dt} = \frac{d^2\boldsymbol{r}}{ds^2}\left(\frac{ds}{dt}\right)^2 + \frac{d\boldsymbol{r}}{ds}\frac{d^2 s}{dt^2}. \tag{7.97}$$

The Gauss function can then be expressed as

$$\mathcal{G} = \frac{1}{2}M\left[\left\|\frac{d^2\boldsymbol{r}}{ds^2}\right\|^2\left(\frac{ds}{dt}\right)^4 + \left\|\frac{d\boldsymbol{r}}{ds}\right\|^2\left(\frac{d^2s}{dt^2}\right)^2 + 2\left\langle\frac{d^2\boldsymbol{r}}{ds^2},\frac{d\boldsymbol{r}}{ds}\right\rangle\left(\frac{ds}{dt}\right)^2\frac{d^2s}{dt^2}\right]. \tag{7.98}$$

We note that

$$\left\|\frac{d\boldsymbol{r}}{ds}\right\|^2 = \left\langle\frac{d\boldsymbol{r}}{ds},\frac{d\boldsymbol{r}}{ds}\right\rangle = 1 \tag{7.99}$$

and

$$\frac{d}{ds}\left\langle\frac{d\boldsymbol{r}}{ds},\frac{d\boldsymbol{r}}{ds}\right\rangle = 0 = 2\left\langle\frac{d^2\boldsymbol{r}}{ds^2},\frac{d\boldsymbol{r}}{ds}\right\rangle. \tag{7.100}$$

Thus,

$$\mathcal{G} = \frac{1}{2}M\left[\left\|\frac{d^2\boldsymbol{r}}{ds^2}\right\|^2\left(\frac{ds}{dt}\right)^4 + \left(\frac{d^2s}{dt^2}\right)^2\right]. \tag{7.101}$$

Noting that the extrinsic path curvature is $k = \left\|d^2\boldsymbol{r}/ds^2\right\|$, we have

$$\mathcal{G} = \frac{1}{2}M\left[k^2\left(\frac{ds}{dt}\right)^4 + \left(\frac{d^2s}{dt^2}\right)^2\right], \tag{7.102}$$

subject to the constraints. The term ds/dt is the particle speed, determined from the system state, and d^2s/dt^2 is the tangential acceleration. Both k^2 and $(d^2s/dt^2)^2$ are positive numbers. Additionally, they can be selected (and minimized) independently of each other. Therefore, minimizing $\mathcal{G}$ subject to the constraints implies that k^2 (and k) is minimized subject to the constraints and that $d^2s/dt^2 = 0$.

The implication that curvature is minimized reflects *Hertz's Principle of Least Curvature* (Hertz 1894; Lanczos 1986; Papastavridis 2002), which states that under force-free constrained motion, a system will follow the path of least extrinsic curvature, k, on the constrained-motion manifold, Q^p. Furthermore, this constrained minimization implies $\nabla k^2 \perp T_r(Q^p)$, where $T_r(Q^p)$ denotes the tangent space of Q^p at the point $\boldsymbol{r}$. We note that

$$\nabla k^2 = \nabla\left\|\frac{d^2\boldsymbol{r}}{ds^2}\right\|^2 = 2\frac{d^2\boldsymbol{r}}{ds^2}, \tag{7.103}$$

so

$$\frac{d^2\boldsymbol{r}}{ds^2}\perp T_r(Q^p). \tag{7.104}$$

This implies that the covariant derivative, D/ds, of the path tangent vanishes,

$$\frac{D}{ds}\frac{d\boldsymbol{r}}{ds} = \mathrm{proj}_T\left(\frac{d}{ds}\frac{d\boldsymbol{r}}{ds}\right) = \mathrm{proj}_T\left(\frac{d^2\boldsymbol{r}}{ds^2}\right) = \boldsymbol{0}, \tag{7.105}$$

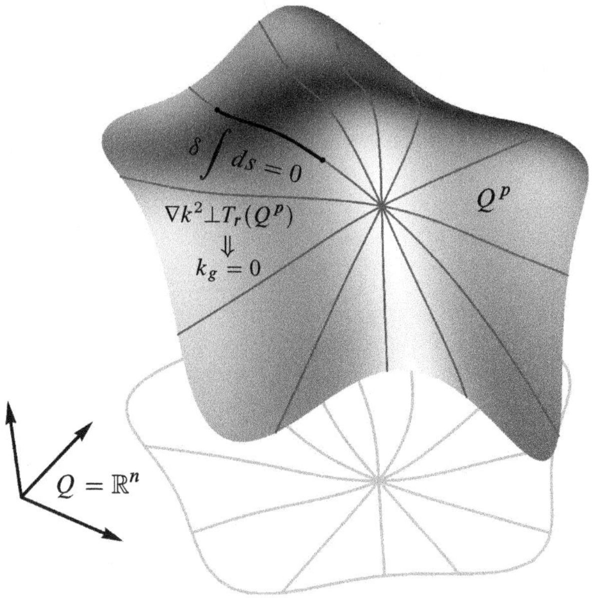

Figure 7.3 Geodesic force-free paths on a constrained-motion manifold. Force-free motion minimizes the extrinsic path curvature, k, subject to the constraints, yielding zero geodesic (intrinsic) curvature, k_g.

where $\text{proj}_T()$ denotes the projection of a vector onto the tangent space of Q^p. The intrinsic geodesic curvature of the path,

$$k_g = \left\| \frac{D}{ds}\frac{d\boldsymbol{r}}{ds} \right\|, \tag{7.106}$$

is thus *zero*. This implies that under force-free constrained motion, a system will follow geodesics (paths for which $k_g = 0$) on the constrained-motion manifold. This is illustrated in Figure 7.3.

Since a geodesic path minimizes arc length the condition of *zero* geodesic curvature is equivalent to finding a path which minimizes the action defined in terms of arc length. That is,

$$\delta I = \delta \int ds = 0, \tag{7.107}$$

subject to the constraints. Equivalently, for a system with no external forces this fact can be concluded from Jacobi's form of least action (Goldstein et al. 2002), which states,

$$\delta I = \delta \int_{t_o}^{t_f} T\, dt = 0, \tag{7.108}$$

subject to the constraints. In this case the Lagrangian has been replaced by the kinetic energy alone (no potential energy). Since

$$dt = \sqrt{M/2T}\, ds, \tag{7.109}$$

we have

$$\delta I = \delta \int \sqrt{MT/2}\, ds = 0. \tag{7.110}$$

Because T is constant for this system, (7.110) implies that arc length is minimized on the constrained-motion manifold.

One application of the least curvature principle involves computing the geodesics of a surface by solving an analogous mechanical system. By the Principle of Least Curvature, solutions of this system will trace out geodesics on the constrained-motion surface $Q^p \subset \mathbb{R}^3$.

In three dimensions, geodesics can be computed by solving the set of second-order nonlinear differential equations (2.153), given in *Chapter 2*. Alternatively, we can solve the analogous mechanical system, which, by the Principle of Least Curvature, will trace out geodesics on the constrained-motion surface $Q^p \subset \mathbb{R}^3$. This system is

$$M\ddot{\boldsymbol{r}} = \boldsymbol{\Phi}^T \lambda \qquad \text{subject to,} \quad \phi(\boldsymbol{r}) = 0, \tag{7.111}$$

where $\phi(\boldsymbol{r}) = 0$ is an implicit representation of the surface represented parametrically by $\boldsymbol{r}(u, v)$. Using the acceleration form of the constraint equations, the system of (7.111) can be solved to yield the differential equations

$$\ddot{\boldsymbol{r}} = -\boldsymbol{\Phi}^T(\boldsymbol{\Phi}\boldsymbol{\Phi}^T)^{-1}\dot{\boldsymbol{\Phi}}\dot{\boldsymbol{r}}, \tag{7.112}$$

or, incorporating constraint stabilization,

$$\ddot{\boldsymbol{r}} = -\boldsymbol{\Phi}^T(\boldsymbol{\Phi}\boldsymbol{\Phi}^T)^{-1}(\dot{\boldsymbol{\Phi}}\dot{\boldsymbol{r}} + \beta\boldsymbol{\Phi}\dot{\boldsymbol{r}} + \alpha\boldsymbol{\phi}). \tag{7.113}$$

Thus, (7.113) represents a mechanically derived approach for computing geodesics. Equation (7.113) is not limited to $\mathbb{R}^3$ and can be used to compute geodesics in $\mathbb{R}^n$.

Example: We can apply (7.113) to the problem of computing geodesics for the surface:

$$\phi(x, y, z) = -\frac{x^2 + y^2}{2} + \frac{4\sin(4x)\sin(4y) + 7}{10} - z = 0. \tag{7.114}$$

The constraint Jacobian is computed directly from ϕ as

$$\boldsymbol{\Phi} = \begin{pmatrix} \partial\phi/\partial x & \partial\phi/\partial y & \partial\phi/\partial z \end{pmatrix}, \tag{7.115}$$

and (7.113) yields a system of three second-order nonlinear differential equations in x, y, and z:

$$(\ddot{x}\,\ddot{y}\,\ddot{z})^T = -\boldsymbol{\Phi}^T(\boldsymbol{\Phi}\boldsymbol{\Phi}^T)^{-1}(\dot{\boldsymbol{\Phi}}\dot{\boldsymbol{r}} + \beta\boldsymbol{\Phi}\dot{\boldsymbol{r}} + \alpha\boldsymbol{\phi}). \tag{7.116}$$

Specifying the point $(x_o, y_o, z_o) = (0.175, -0.175, 0.503)$ as one initial condition, we can solve (7.116) using different departure directions, $(\dot{x}_o, \dot{y}_o, \dot{z}_o)$. The resulting geodesics are shown in Figure 7.4. It is noted that the specific time parameterization used does not affect the shape of the paths, only the speed at which they are traversed.

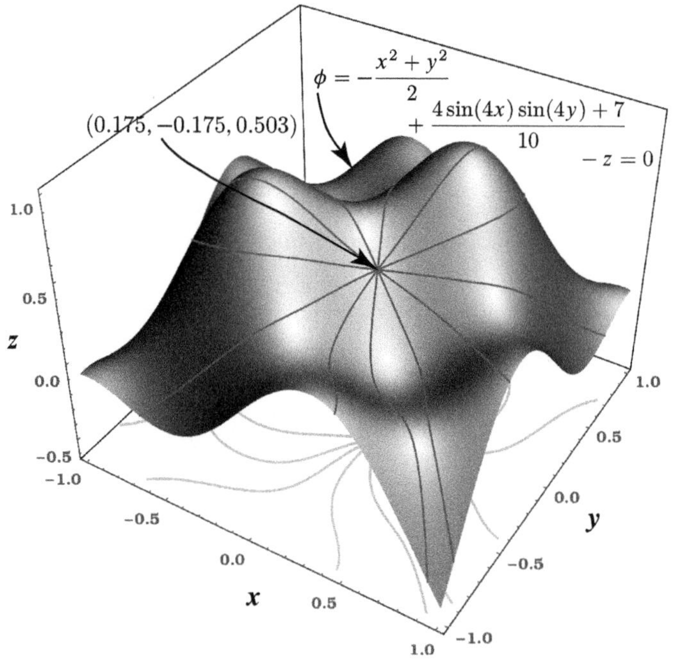

Figure 7.4 Solving the system of (7.113), geodesic force-free paths were computed for the surface $\phi = -\left(x^2+y^2\right)/2 + [\sin(4x)\sin(4y)+7]/10 - z = 0$. All paths were chosen to emanate from the point (0.175, −0.175, 0.503).

Therefore, only the direction (not the magnitude) of the initial velocity dictates the path.

7.4 Gibbs-Appell Formulation

The Gibbs-Appell formulation is a second-order variational method that introduces quasi-velocities and quasi-accelerations to implicitly handle nonholonomic constraints in much the same way that Kane's method introduces quasi-velocities. The formulation proceeds using virtual variations on the accelerations rather than virtual variations on velocities. As we shall see, since the Gibbs-Appell equations are derived from a second-order variational principle, they can be stated as the gradient of a scalar function with respect to the quasi-accelerations. In the next section we will formulate the Gibbs-Appell equations for rigid-body systems.

7.4.1 A System of Rigid Bodies

Given a set of m nonholonomic constraints described by

$$\boldsymbol{C}(\boldsymbol{q})\dot{\boldsymbol{q}} = \boldsymbol{0}, \tag{7.117}$$

we can define a set of independent quasi-velocities, $\boldsymbol{s}$. The generalized velocities are related to these quasi-velocities by the null space of $\boldsymbol{C}$. That is,

$$\dot{\boldsymbol{q}} = \boldsymbol{W}(\boldsymbol{q})\boldsymbol{s}, \tag{7.118}$$

where

$$\text{im}(\boldsymbol{W}) = \ker(\boldsymbol{C}). \tag{7.119}$$

We note that

$$\boldsymbol{v}(\boldsymbol{q}, \dot{\boldsymbol{q}}) = \boldsymbol{\Gamma}\dot{\boldsymbol{q}} \quad \text{and} \quad \boldsymbol{\omega}(\boldsymbol{q}, \dot{\boldsymbol{q}}) = \boldsymbol{\Pi}\dot{\boldsymbol{q}}. \tag{7.120}$$

Making use of (7.118), we can replace the $\dot{\boldsymbol{q}}$ terms in favor of the $\boldsymbol{s}$ terms in our expressions for $\boldsymbol{v}$ and $\boldsymbol{\omega}$. That is,

$$\tilde{\boldsymbol{v}}(\boldsymbol{q}, \boldsymbol{s}) = \boldsymbol{\Gamma}\boldsymbol{W}\boldsymbol{s} = \tilde{\boldsymbol{\Gamma}}\boldsymbol{s} \quad \text{and} \quad \tilde{\boldsymbol{\omega}}(\boldsymbol{q}, \boldsymbol{s}) = \boldsymbol{\Pi}\boldsymbol{W}\boldsymbol{s} = \tilde{\boldsymbol{\Pi}}\boldsymbol{s}. \tag{7.121}$$

Differentiating (7.118), we have

$$\ddot{\boldsymbol{q}} = \boldsymbol{W}\dot{\boldsymbol{s}} + \dot{\boldsymbol{W}}\boldsymbol{s}. \tag{7.122}$$

The term, $\dot{\boldsymbol{s}}$, is the vector of quasi-accelerations. We note that

$$\boldsymbol{a}(\boldsymbol{q}, \dot{\boldsymbol{q}}, \ddot{\boldsymbol{q}}) = \boldsymbol{\Gamma}\ddot{\boldsymbol{q}} + (\dot{\boldsymbol{\Gamma}} + \boldsymbol{\Omega}\boldsymbol{\Gamma})\dot{\boldsymbol{q}} \quad \text{and} \quad \boldsymbol{\alpha}(\boldsymbol{q}, \dot{\boldsymbol{q}}, \ddot{\boldsymbol{q}}) = \boldsymbol{\Pi}\ddot{\boldsymbol{q}} + \dot{\boldsymbol{\Pi}}\dot{\boldsymbol{q}}. \tag{7.123}$$

Making use of (7.118) and (7.122), we can replace the $\dot{\boldsymbol{q}}$ and $\ddot{\boldsymbol{q}}$ terms in favor of the $\boldsymbol{s}$ and $\dot{\boldsymbol{s}}$ terms in our expressions for $\boldsymbol{a}$ and $\boldsymbol{\alpha}$. That is,

$$\tilde{\boldsymbol{a}}(\boldsymbol{q}, \boldsymbol{s}, \dot{\boldsymbol{s}}) = \boldsymbol{\Gamma}\boldsymbol{W}\dot{\boldsymbol{s}} + [(\boldsymbol{\Gamma}\dot{\boldsymbol{W}} + \dot{\boldsymbol{\Gamma}}\boldsymbol{W}) + \boldsymbol{\Omega}\boldsymbol{\Gamma}\boldsymbol{W}]\boldsymbol{s} = \tilde{\boldsymbol{\Gamma}}\dot{\boldsymbol{s}} + (\dot{\tilde{\boldsymbol{\Gamma}}} + \boldsymbol{\Omega}\tilde{\boldsymbol{\Gamma}})\boldsymbol{s} \tag{7.124}$$

and

$$\tilde{\boldsymbol{\alpha}}(\boldsymbol{q}, \boldsymbol{s}, \dot{\boldsymbol{s}}) = \boldsymbol{\Pi}\boldsymbol{W}\dot{\boldsymbol{s}} + (\boldsymbol{\Pi}\dot{\boldsymbol{W}} + \dot{\boldsymbol{\Pi}}\boldsymbol{W})\boldsymbol{s} = \tilde{\boldsymbol{\Pi}}\dot{\boldsymbol{s}} + \dot{\tilde{\boldsymbol{\Pi}}}\boldsymbol{s}, \tag{7.125}$$

where $p = n - m$. Clearly

$$\tilde{\boldsymbol{v}}(\boldsymbol{q}, \boldsymbol{s}) = \boldsymbol{v}(\boldsymbol{q}, \dot{\boldsymbol{q}}) \quad \text{and} \quad \tilde{\boldsymbol{\omega}}(\boldsymbol{q}, \boldsymbol{s}) = \boldsymbol{\omega}(\boldsymbol{q}, \dot{\boldsymbol{q}}), \tag{7.126}$$

and

$$\tilde{\boldsymbol{a}}(\boldsymbol{q}, \boldsymbol{s}, \dot{\boldsymbol{s}}) = \boldsymbol{a}(\boldsymbol{q}, \dot{\boldsymbol{q}}, \dot{\boldsymbol{q}}) \quad \text{and} \quad \tilde{\boldsymbol{\alpha}}(\boldsymbol{q}, \boldsymbol{s}, \dot{\boldsymbol{s}}) = \boldsymbol{\alpha}(\boldsymbol{q}, \dot{\boldsymbol{q}}, \ddot{\boldsymbol{q}}). \tag{7.127}$$

The terms $\tilde{\boldsymbol{\Gamma}}$ and $\tilde{\boldsymbol{\Pi}}$ are the quasi-Jacobians given by

$$\tilde{\boldsymbol{\Gamma}} = \frac{\partial \tilde{\boldsymbol{a}}}{\partial \dot{\boldsymbol{s}}} = \boldsymbol{\Gamma}\boldsymbol{W} \quad \text{and} \quad \tilde{\boldsymbol{\Pi}} = \frac{\partial \tilde{\boldsymbol{\alpha}}}{\partial \dot{\boldsymbol{s}}} = \boldsymbol{\Pi}\boldsymbol{W}. \tag{7.128}$$

The virtual acceleration variations are then given by

$$\delta\tilde{\boldsymbol{a}} = \tilde{\boldsymbol{\Gamma}}(\dot{\boldsymbol{s}} + \delta\dot{\boldsymbol{s}}) + (\dot{\tilde{\boldsymbol{\Gamma}}} + \boldsymbol{\Omega}\tilde{\boldsymbol{\Gamma}})\boldsymbol{s} - \tilde{\boldsymbol{\Gamma}}\dot{\boldsymbol{s}} - (\dot{\tilde{\boldsymbol{\Gamma}}} + \boldsymbol{\Omega}\tilde{\boldsymbol{\Gamma}})\boldsymbol{s} = \tilde{\boldsymbol{\Gamma}}\delta\dot{\boldsymbol{s}} \tag{7.129}$$

and

$$\delta\tilde{\boldsymbol{\alpha}} = \tilde{\boldsymbol{\Pi}}(\dot{\boldsymbol{s}} + \delta\dot{\boldsymbol{s}}) + \dot{\tilde{\boldsymbol{\Pi}}}\boldsymbol{s} - \tilde{\boldsymbol{\Pi}}\dot{\boldsymbol{s}} - \dot{\tilde{\boldsymbol{\Pi}}}\boldsymbol{s} = \tilde{\boldsymbol{\Pi}}\delta\dot{\boldsymbol{s}}. \tag{7.130}$$

Since the nonholonomic constraint forces vanish under these variations, we have

$$\sum_{i=1}^{p} \tilde{\tau}_i \cdot \delta \dot{s}_i - \sum_{i=1}^{n_b} M_i(\boldsymbol{a}_{G_i} + g\hat{\boldsymbol{e}}_{0_3}) \cdot \delta\tilde{\boldsymbol{a}}_{G_i} - \sum_{i=1}^{n_b} (\boldsymbol{I}_i^{G_i}\boldsymbol{\alpha}_i + \boldsymbol{\omega}_i \times \boldsymbol{I}_i^{G_i}\boldsymbol{\omega}_i) \cdot \delta\tilde{\boldsymbol{\alpha}}_i = 0 \quad (7.131)$$

for all variations, $\delta\tilde{\boldsymbol{a}}_{G_i}$ and $\delta\tilde{\boldsymbol{\alpha}}_i$, that are consistent with the kinematic constraints. The terms $\tilde{\tau}_i$ are the quasi-forces.

Expressing the acceleration variations in terms of variations in the quasi-accelerations (with stationary displacement and time), we have

$$^i\delta\tilde{\boldsymbol{a}}_{G_i} = {^i\tilde{\boldsymbol{\Gamma}}}_{G_i}\delta\dot{\boldsymbol{s}} \quad \text{and} \quad {^i\delta\tilde{\boldsymbol{\alpha}}_i} = {^i\tilde{\boldsymbol{\Pi}}_i}\delta\dot{\boldsymbol{q}}, \quad (7.132)$$

where the accelerations are expressed in the local link frame i for convenience. So, we have

$$\tilde{\boldsymbol{\tau}} \cdot \delta\dot{\boldsymbol{s}} - \sum_{i=1}^{n_b} M_i({^i\boldsymbol{a}}_{G_i} + g{^i\hat{\boldsymbol{e}}}_{0_3}) \cdot ({^i\tilde{\boldsymbol{\Gamma}}}_{G_i}\delta\dot{\boldsymbol{s}}) - \sum_{i=1}^{n_b} ({^i\boldsymbol{I}_i^{G_i}}{^i\boldsymbol{\alpha}_i} + {^i\boldsymbol{\omega}_i} \times {^i\boldsymbol{I}_i^{G_i}}{^i\boldsymbol{\omega}_i}) \cdot ({^i\tilde{\boldsymbol{\Pi}}_i}\delta\dot{\boldsymbol{s}}) = 0$$
$$\forall \delta\dot{\boldsymbol{s}} \in \mathbb{R}^p, \quad (7.133)$$

or

$$\tilde{\boldsymbol{\tau}} \cdot \delta\dot{\boldsymbol{s}} - \sum_{i=1}^{n_b} \left[M_i {^i\boldsymbol{\Gamma}}_{G_i}^{T}{^i\boldsymbol{a}}_{G_i} + M_i g {^i\tilde{\boldsymbol{\Gamma}}}_{G_i}^{T}{^i\hat{\boldsymbol{e}}}_{0_3} + {^i\tilde{\boldsymbol{\Pi}}_i^T}({^i\boldsymbol{I}_i^{G_i}}{^i\boldsymbol{\alpha}_i} + {^i\boldsymbol{\omega}_i} \times {^i\boldsymbol{I}_i^{G_i}}{^i\boldsymbol{\omega}_i}) \right] \cdot \delta\dot{\boldsymbol{s}} = 0 \quad (7.134)$$
$$\forall \delta\dot{\boldsymbol{s}} \in \mathbb{R}^p.$$

In matrix form we have

$$\left[\tilde{\boldsymbol{\tau}} - \sum_{i=1}^{n_b} \begin{pmatrix} {^i\tilde{\boldsymbol{\Gamma}}}_{G_i}^T & {^i\tilde{\boldsymbol{\Pi}}_i^T} \end{pmatrix} \begin{pmatrix} M_i{^i\boldsymbol{a}}_{G_i} + M_i g {^i\hat{\boldsymbol{e}}}_{0_3} \\ {^i\boldsymbol{I}_i^{G_i}}{^i\boldsymbol{\alpha}_i} + {^i\boldsymbol{\omega}_i} \times {^i\boldsymbol{I}_i^{G_i}}{^i\boldsymbol{\omega}_i}, \end{pmatrix} \right] \cdot \delta\dot{\boldsymbol{s}} = 0, \quad (7.135)$$
$$\forall \delta\dot{\boldsymbol{s}} \in \mathbb{R}^p$$

which implies

$$\tilde{\boldsymbol{\tau}} = \sum_{i=1}^{n_b} \begin{pmatrix} {^i\tilde{\boldsymbol{\Gamma}}}_{G_i}^T & {^i\tilde{\boldsymbol{\Pi}}_i^T} \end{pmatrix} \begin{pmatrix} M_i{^i\boldsymbol{a}}_{G_i} + M_i g {^i\hat{\boldsymbol{e}}}_{0_3} \\ {^i\boldsymbol{I}_i^{G_i}}{^i\boldsymbol{\alpha}_i} + {^i\boldsymbol{\omega}_i} \times {^i\boldsymbol{I}_i^{G_i}}{^i\boldsymbol{\omega}_i} \end{pmatrix}. \quad (7.136)$$

Replacing the $\dot{\boldsymbol{q}}$ and $\ddot{\boldsymbol{q}}$ terms in favor of the $\boldsymbol{s}$ and $\dot{\boldsymbol{s}}$ terms, we have

$$\tilde{\boldsymbol{\tau}} = \sum_{i=1}^{n_b} \begin{pmatrix} {^i\tilde{\boldsymbol{\Gamma}}}_{G_i}^T & {^i\tilde{\boldsymbol{\Pi}}_i^T} \end{pmatrix} \begin{pmatrix} M_i{^i\tilde{\boldsymbol{a}}}_{G_i} + M_i g {^i\hat{\boldsymbol{e}}}_{0_3} \\ {^i\boldsymbol{I}_i^{G_i}}{^i\tilde{\boldsymbol{\alpha}}_i} + {^i\tilde{\boldsymbol{\omega}}_i} \times {^i\boldsymbol{I}_i^{G_i}}{^i\tilde{\boldsymbol{\omega}}_i} \end{pmatrix}. \quad (7.137)$$

Equation (7.137) is a system of p first-order differential equations. We need to complement this with the set of n equations

$$\dot{\boldsymbol{q}} = \boldsymbol{W}\boldsymbol{s}. \quad (7.138)$$

This yields a total system of $n + p$ first-order differential equations in the $n + p$ states, $\boldsymbol{q}$ and $\boldsymbol{s}$. Noting that

$${^i\tilde{\boldsymbol{\omega}}_i} = {^i\tilde{\boldsymbol{\Pi}}_i}\boldsymbol{s}, \quad {^i\tilde{\boldsymbol{a}}}_{G_i} = {^i\tilde{\boldsymbol{\Gamma}}}_{G_i}\dot{\boldsymbol{s}} + ({^i\dot{\tilde{\boldsymbol{\Gamma}}}}_{G_i} + {^i\tilde{\boldsymbol{\Omega}}_i}{^i\tilde{\boldsymbol{\Gamma}}}_{G_i})\boldsymbol{s} \quad \text{and} \quad {^i\tilde{\boldsymbol{\alpha}}_i} = {^i\tilde{\boldsymbol{\Pi}}_i}\dot{\boldsymbol{s}} + {^i\dot{\tilde{\boldsymbol{\Pi}}}_i}\boldsymbol{s}, \quad (7.139)$$

we have

$$\tilde{\tau} = \left[\sum_{i=1}^{n_b} (M_i {}^i\tilde{\Gamma}_{G_i}^T {}^i\tilde{\Gamma}_{G_i} + {}^i\tilde{\Pi}_i^T {}^iI_i^{G_i} {}^i\tilde{\Pi}_i)\right] \dot{s}$$
$$+ \left[\sum_{i=1}^{n_b} \left(M_i {}^i\tilde{\Gamma}_{G_i}^T ({}^i\dot{\tilde{\Gamma}}_{G_i} + {}^i\tilde{\Omega}_i {}^i\tilde{\Gamma}_{G_i}) + {}^i\tilde{\Pi}_i^T ({}^iI_i^{G_i} {}^i\dot{\tilde{\Pi}}_i + {}^i\tilde{\Omega}_i {}^iI_i^{G_i} {}^i\Pi_i)\right)\right] s$$
$$+ g \sum_{i=1}^{n_b} M_i {}^i\tilde{\Gamma}_{G_i}^T {}^i\hat{e}_{0_3}. \tag{7.140}$$

As with Kane's method, we have

$$\tilde{\tau} = \tilde{M}(q)\dot{s} + \tilde{b}(q, s) + \tilde{g}(q). \tag{7.141}$$

Now let us define the *Gibbs* function, $\mathcal{S}$,

$$\mathcal{S}(\dot{s}) = \frac{1}{2} \sum_{i=1}^{n_b} [M_i {}^i\tilde{a}_{G_i}^T {}^i\tilde{a}_{G_i} + 2 {}^i\tilde{a}_{G_i}^T M_i g {}^i\hat{e}_{0_3} + {}^i\tilde{\alpha}_i^T {}^iI_i^{G_i} {}^i\tilde{\alpha}_i + 2 {}^i\tilde{\alpha}_i^T ({}^i\tilde{\omega}_i \times {}^iI_i^{G_i} {}^i\tilde{\omega}_i)]. \tag{7.142}$$

Recognizing that

$$\frac{\partial \tilde{a}_{G_i}}{\partial \dot{s}} = \tilde{\Gamma}_{G_i} \quad \text{and} \quad \frac{\partial \tilde{\alpha}_i}{\partial \dot{s}} = \tilde{\Pi}_i, \tag{7.143}$$

we observe that the gradient of $\mathcal{S}$ is

$$\frac{\partial \mathcal{S}}{\partial \dot{s}} = \sum_{i=1}^{n_b} \begin{pmatrix} {}^i\tilde{\Gamma}_{G_i}^T & {}^i\tilde{\Pi}_i^T \end{pmatrix} \begin{pmatrix} M_i \tilde{a}_{G_i} + M_i g {}^i\hat{e}_{0_3} \\ {}^iI_i^{G_i} {}^i\tilde{\alpha}_i + {}^i\tilde{\omega}_i \times {}^iI_i^{G_i} {}^i\tilde{\omega}_i \end{pmatrix} = \tilde{\tau}. \tag{7.144}$$

So,

$$\tilde{\tau} = \frac{\partial \mathcal{S}}{\partial \dot{s}} \tag{7.145}$$

and

$$\dot{q} = W s \tag{7.146}$$

form the *Gibbs-Appell* dynamical equations.

We can relate the Gibbs function to the Gauss function by first noting that the Gauss function can be written as

$$\delta\mathcal{G} = \left[\sum_{i=1}^{n_b} \begin{pmatrix} {}^i\Gamma_{G_i}^T & {}^i\Pi_i^T \end{pmatrix} \begin{pmatrix} M_i {}^ia_{G_i} + M_i g {}^i\hat{e}_{0_3} \\ {}^iI_i^{G_i} {}^i\alpha_i + {}^i\omega_i \times {}^iI_i^{G_i} {}^i\omega_i \end{pmatrix} - \tau\right] \cdot \delta\ddot{q}. \tag{7.147}$$

Replacing the $\dot{q}$ and $\ddot{q}$ terms in favor of the s and $\dot{s}$ terms, we have

$$\delta\tilde{\mathcal{G}} = \left[\sum_{i=1}^{n_b} \begin{pmatrix} {}^i\Gamma_{G_i}^T & {}^i\Pi_i^T \end{pmatrix} \begin{pmatrix} M_i {}^i\tilde{a}_{G_i} + M_i g {}^i\hat{e}_{0_3} \\ {}^iI_i^{G_i} {}^i\tilde{\alpha}_i + {}^i\tilde{\omega}_i \times {}^iI_i^{G_i} {}^i\tilde{\omega}_i \end{pmatrix} - \tilde{\tau}\right] \cdot W\delta\dot{s}$$
$$= \left[\sum_{i=1}^{n_b} W^T \begin{pmatrix} {}^i\Gamma_{G_i}^T & {}^i\Pi_i^T \end{pmatrix} \begin{pmatrix} M_i {}^i\tilde{a}_{G_i} + M_i g {}^i\hat{e}_{0_3} \\ {}^iI_i^{G_i} {}^i\tilde{\alpha}_i + {}^i\tilde{\omega}_i \times {}^iI_i^{G_i} {}^i\tilde{\omega}_i \end{pmatrix} - \tau\right] \cdot \delta\dot{s}, \tag{7.148}$$

and finally,

$$\delta\tilde{\mathcal{G}} = \left[\sum_{i=1}^{n_b} \begin{pmatrix} {}^i\tilde{\boldsymbol{\Gamma}}_{G_i}^T & {}^i\tilde{\boldsymbol{\Pi}}_i^T \end{pmatrix} \begin{pmatrix} M_i{}^i\tilde{\boldsymbol{a}}_{G_i} + M_i g{}^i\hat{\boldsymbol{e}}_{0_3} \\ {}^i\boldsymbol{I}_i^{G_i}{}^i\tilde{\boldsymbol{\alpha}}_i + {}^i\tilde{\boldsymbol{\omega}}_i \times {}^i\boldsymbol{I}_i^{G_i}{}^i\tilde{\boldsymbol{\omega}}_i \end{pmatrix}\right] \cdot \delta\dot{\boldsymbol{s}} - \tilde{\boldsymbol{\tau}} \cdot \delta\dot{\boldsymbol{s}}. \tag{7.149}$$

Noting that

$$\delta\mathcal{S} = \frac{\partial\mathcal{S}}{\partial\dot{\boldsymbol{s}}} \cdot \delta\dot{\boldsymbol{s}} = \left[\sum_{i=1}^{n_b} \begin{pmatrix} {}^i\tilde{\boldsymbol{\Gamma}}_{G_i}^T & {}^i\tilde{\boldsymbol{\Pi}}_i^T \end{pmatrix} \begin{pmatrix} M_i{}^i\tilde{\boldsymbol{a}}_{G_i} + M_i g{}^i\hat{\boldsymbol{e}}_{0_3} \\ {}^i\boldsymbol{I}_i^{G_i}{}^i\tilde{\boldsymbol{\alpha}}_i + {}^i\tilde{\boldsymbol{\omega}}_i \times {}^i\boldsymbol{I}_i^{G_i}{}^i\tilde{\boldsymbol{\omega}}_i \end{pmatrix}\right] \cdot \delta\dot{\boldsymbol{s}}, \tag{7.150}$$

we have

$$\delta\mathcal{S} = \delta\tilde{\mathcal{G}} + \tilde{\boldsymbol{\tau}} \cdot \delta\dot{\boldsymbol{s}}. \tag{7.151}$$

Example: Recalling the rolling disk from Chapter 6 (see Figure 7.5), the Gibbs function can be formulated as

$$\mathcal{S} = \frac{1}{2}[M{}^{\mathcal{O}}\tilde{\boldsymbol{a}}_G^{T\,\mathcal{O}}\tilde{\boldsymbol{a}}_G + 2{}^{\mathcal{O}}\tilde{\boldsymbol{a}}_G^T M g\hat{\boldsymbol{e}}_3 + {}^{\mathcal{D}}\tilde{\boldsymbol{\alpha}}^{T\,\mathcal{D}}\boldsymbol{I}^{G\mathcal{D}}\tilde{\boldsymbol{\alpha}}_{\mathcal{D}} + 2{}^{\mathcal{D}}\tilde{\boldsymbol{\alpha}}^T({}^{\mathcal{D}}\tilde{\boldsymbol{\omega}} \times {}^{\mathcal{D}}\boldsymbol{I}^{G\mathcal{D}}\tilde{\boldsymbol{\omega}})]. \tag{7.152}$$

The translational and angular velocities can be expressed in terms of the quasi-velocities, $\boldsymbol{s}$:

$${}^{\mathcal{O}}\tilde{\boldsymbol{v}}_G = {}^{\mathcal{O}}\tilde{\boldsymbol{\Gamma}}_G\boldsymbol{s} = \begin{pmatrix} r[\cos(q_4)\cos(q_5)s_2 + \sin(q_4)(s_3 - s_1\sin(q_5))] \\ r[-\cos(q_4)s_3 + \cos(q_5)s_2\sin(q_4) + \cos(q_4)s_1\sin(q_5)] \\ -rs_2\sin(q_5) \end{pmatrix} \tag{7.153}$$

$${}^{\mathcal{D}}\tilde{\boldsymbol{\omega}} = {}^{\mathcal{D}}\tilde{\boldsymbol{\Pi}}\boldsymbol{s} = \begin{pmatrix} s_3 - s_1\sin(q_5) \\ \cos(q_6)s_2 + \cos(q_5)s_1\sin(q_6) \\ \cos(q_5)\cos(q_6)s_1 - s_2\sin(q_6) \end{pmatrix}, \tag{7.154}$$

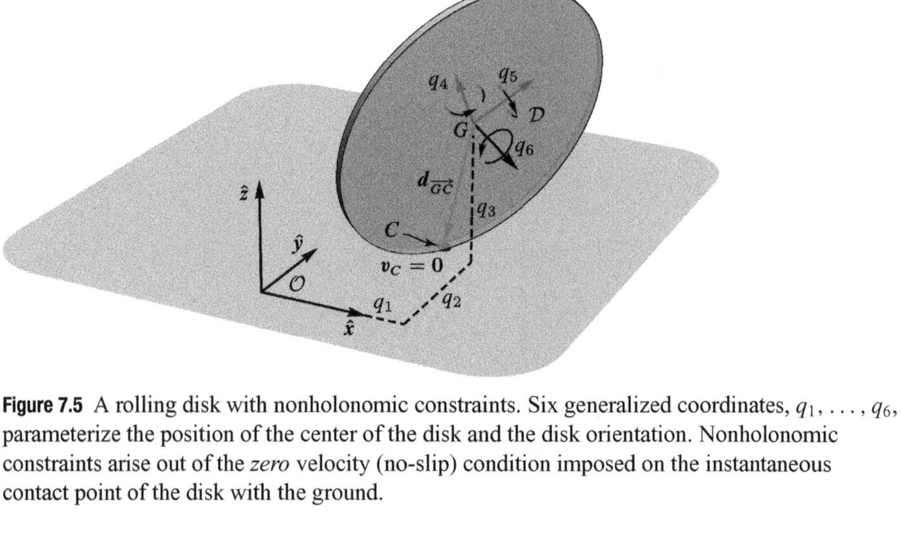

Figure 7.5 A rolling disk with nonholonomic constraints. Six generalized coordinates, $q_1, \ldots, q_6$, parameterize the position of the center of the disk and the disk orientation. Nonholonomic constraints arise out of the *zero* velocity (no-slip) condition imposed on the instantaneous contact point of the disk with the ground.

and the translational and angular accelerations can be expressed in terms of the quasi-accelerations, $\dot{s}$, as

$$^{O}\tilde{\boldsymbol{a}}_G = {}^{O}\dot{\boldsymbol{v}}_G|_{\dot{q}\to s} \tag{7.155}$$

$$^{D}\tilde{\boldsymbol{\alpha}} = {}^{D}\dot{\boldsymbol{\omega}}|_{\dot{q}\to s}. \tag{7.156}$$

We can now derive the Gibbs function:

$$\begin{aligned}\mathcal{S} = \frac{1}{16}Mr\big(&8rs_2^4 + 8rs_1^4\sin^2(q5) - rs_1^3 s_3[17\sin(q_5) + \sin(3q_5)] + 7r\dot{s}_1^2 \\ &+ s_2^2[-16g\cos(q_5) + r(25 + 11\cos(2q_5))s_1^2 + 6rs_3^2 - 12rs_1s_3\sin(q_5)] \\ &- 5r\cos(2q_5)\dot{s}_1^2 + 24r\cos(q_5)s_1s_3\dot{s}_2 - 16g\sin(q_5)\dot{s}_2 + 10r\dot{s}_2^2 \\ &+ rs_1^2[(11 + 3\cos(2q_5))s_3^2 - 10\sin(2q_5)\dot{s}_2] - 24r\sin(q_5)\dot{s}_1\dot{s}_3 + 12r\dot{s}_3^2 \\ &- 8r\cos(q_5)s_2[s_3\dot{s}_1 + 5s_1(-\sin(q_5)\dot{s}_1 + \dot{s}_3)]\big).\end{aligned} \tag{7.157}$$

The gradient of $\mathcal{S}$ is then

$$\frac{\partial\mathcal{S}}{\partial\dot{s}} = \begin{pmatrix} \frac{1}{8}Mr^2[-4\cos(q_5)s_2(s_3 - 5s_1\sin(q_5)) + (7 - 5\cos(2q_5))\dot{s}_1 - 12\sin(q_5)\dot{s}_3] \\ \frac{1}{8}Mr[12r\cos(q_5)s_1s_3 - 8g\sin(q_5) - 5Rs_1^2\sin(2q_5) + 10r\dot{s}_2] \\ -\frac{1}{2}Mr^2[5\cos(q_5)s_1s_2 + 3\sin(q_5)\dot{s}_1 - 3\dot{s}_3] \end{pmatrix}. \tag{7.158}$$

It can be verified that this is equal to the expression

$$\tilde{\boldsymbol{M}}(\boldsymbol{q})\dot{\boldsymbol{s}} + \tilde{\boldsymbol{b}}(\boldsymbol{q}, \boldsymbol{s}) + \tilde{\boldsymbol{g}}(\boldsymbol{q}), \tag{7.159}$$

generated using Kane's method in the example of Section 6.3.1.

7.5 Exercises

1. Consider the planar four-bar linkage from Section 5.6, Exercise 3 (shown in Figure 7.6). The link lengths are each $l = 1$ and the link masses are each $M = 1$. The link rotational inertias are taken to be *zero*: Let $\boldsymbol{\tau} = \boldsymbol{0}$ and,

$$\boldsymbol{q}_o = \begin{pmatrix} -2\pi/3 & 2\pi/3 & \pi/3 \end{pmatrix}^T \quad \text{and} \quad \dot{\boldsymbol{q}}_o = \begin{pmatrix} 1 & -1 & 1 \end{pmatrix}^T.$$

   (a) If not already completed, compute the configuration space mass matrix, $\boldsymbol{M}(\boldsymbol{q})$, the vector of centrifugal and Coriolis forces, $\boldsymbol{b}(\boldsymbol{q}, \dot{\boldsymbol{q}})$, and the vector of gravity forces, $\boldsymbol{g}(\boldsymbol{q})$, for the unconstrained system.
   (b) If not already completed, compute the constraint matrix, $\boldsymbol{\Phi}(\boldsymbol{q})$, for the loop closure.
   (c) Compute the Gauss function, $\mathcal{G}(\ddot{\boldsymbol{q}})$, at $\boldsymbol{q}_o$ and $\dot{\boldsymbol{q}}_o$.
   (d) Compute the gradient, $\partial\mathcal{G}/\partial\ddot{\boldsymbol{q}}$.
   (e) Determine a basis for $\ker(\boldsymbol{\Phi})$ at $\boldsymbol{q}_o$ and $\dot{\boldsymbol{q}}_o$.

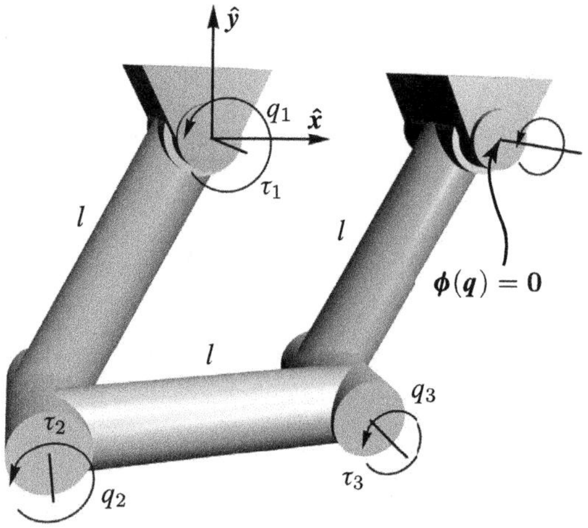

Figure 7.6 A planar four-bar linkage (Exercise 1). The unconstrained system is parameterized by the generalized coordinates, q_1, q_2, and q_3, with generalized forces, τ_1, τ_2, and τ_3. The link lengths are each l and the link masses are each M. The centers of mass are at the geometric centers of the links. The link rotational inertias are taken to be *zero*. The link 3 endpoint is constrained from translating.

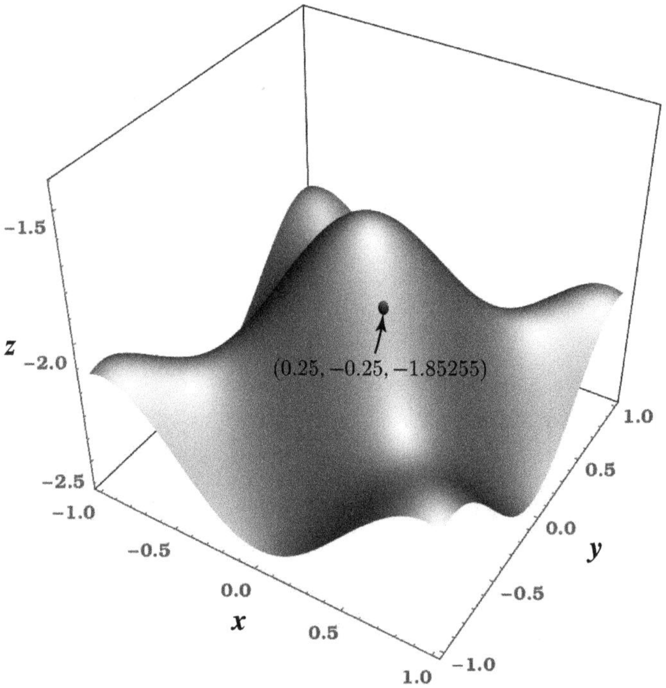

Figure 7.7 Implicit surface $\phi = 4 + x^2 + y^2 - \left[z - \frac{1}{3}\cos(3x)\cos(3y)\right]^2 = 0$ (Exercise 2).

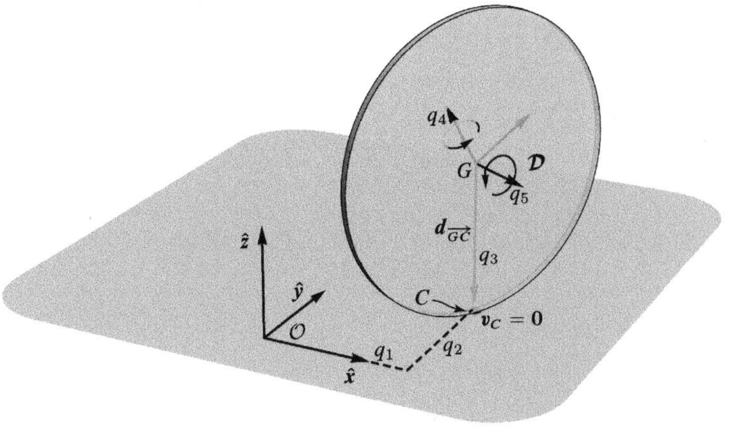

Figure 7.8 A simplified version of the rolling disk (Gibbs-Appell method) that always remains upright (it cannot pitch from side to side, Exercise 3). A $\boldsymbol{Q}_z(q_4)\boldsymbol{Q}_x(q_5)$ Euler sequence is used to represent the orientation of the disk. The disk radius is r and the mass is M. The principal inertia components are ${}^{\mathcal{D}}I_{11}^G = \frac{1}{2}Mr^2$ and ${}^{\mathcal{D}}I_{22}^G = {}^{\mathcal{D}}I_{33}^G = \frac{1}{4}Mr^2$.

(f) Use Gauss's Principle of Least Constraint to solve for the constrained accelerations, $\ddot{\boldsymbol{q}}$, at $\boldsymbol{q}_o$ and $\dot{\boldsymbol{q}}_o$. You will have two scalar equations in $\ddot{\boldsymbol{q}}$ by taking the second derivative of $\boldsymbol{\phi}(\boldsymbol{q}) = \boldsymbol{0}$ and one scalar equation from the orthogonality relation,

$$\delta\mathcal{G} = \frac{\partial \mathcal{G}}{\partial \ddot{\boldsymbol{q}}} = 0.$$

2. Numerically generate geodesics for the surface (see Figure 7.7)

$$\phi = 4 + x^2 + y^2 - \left[z - \frac{1}{3}\cos(3x)\cos(3y)\right]^2 = 0$$

in the region $x, y = [-1, 1]$ by interpreting the problem in terms of constrained dynamics. Choose several departure directions emanating from the point $(0.25, -0.25, -1.85255)$ to generate the geodesics.

3. Consider the rolling disk from Section 6.4, Exercise 3 shown in Figure 7.8. Use the Gibbs-Appell method and define independent quasi-velocities, $s_1 = \dot{q}_4$ and $s_2 = \dot{q}_5$.
   (a) If not already completed, from the rolling contact constraint equations, determine the relationship, $\dot{\boldsymbol{q}} = \boldsymbol{W}\boldsymbol{s}$, between the generalized velocities and the quasi-velocities.
   (b) Derive the Gibbs function, $\mathcal{S}$.
   (c) Compute the gradient of the Gibbs function, $\partial\mathcal{S}/\partial\dot{\boldsymbol{s}}$, to obtain the equations of motion.

8 Dynamics in Task Space

In the previous chapters we considered configuration space descriptions of the dynamics of constrained multibody systems. In this chapter our objective is to reformulate these descriptions in the context of task space. As a starting point, we begin with a review of the basic operational space framework developed by Khatib (1987, 1995). Operational space is synonymous with task space. It refers to a motion space that is a mapping from configuration space to a convenient coordinate space with which to formulate the system dynamics.

8.1 Task Space Framework

The basic task space framework addresses the dynamics of branching chain mechanical systems. Given a branching chain system, the initial step involves defining a set of m_T task, or task space, coordinates, $\mathbf{x}$. The function $\mathbf{x}(\boldsymbol{q})$ represents a kinematic mapping from the set of generalized coordinates to the set of task space coordinates. The task space coordinates can represent any function of the generalized coordinates but typically are chosen to describe the set of coordinates associated with a motion task. Figure 8.1 illustrates simple branching chain systems where the task space coordinates are chosen to be the coordinates associated with positioning the terminal point(s) of the chain. Furthermore, by taking the gradient of $\mathbf{x}$, we have the relationship

$$\dot{\mathbf{x}} = \frac{\partial \mathbf{x}}{\partial \boldsymbol{q}} \dot{\boldsymbol{q}} = \boldsymbol{J}(\boldsymbol{q}) \dot{\boldsymbol{q}}, \tag{8.1}$$

where $\boldsymbol{J}(\boldsymbol{q})$ is the $m_T \times n$ task Jacobian matrix. This relationship applies to both kinematically nonredundant and redundant systems. Kinematically redundant systems are those where the number of degrees of freedom exceed, the number of task coordinates ($m_T < n$), resulting in an infinite number of configurations for a given task. In the case of nonredundant systems the inverse of this relationship is well defined outside of singularities. For such cases we have

$$\dot{\boldsymbol{q}} = \boldsymbol{J}^{-1} \dot{\mathbf{x}}. \tag{8.2}$$

In the redundant case we can define the right inverse of $\boldsymbol{J}$ as

$$\{\boldsymbol{J}^{\#} | \boldsymbol{J}\boldsymbol{J}^{\#} = \mathbf{1}\}. \tag{8.3}$$

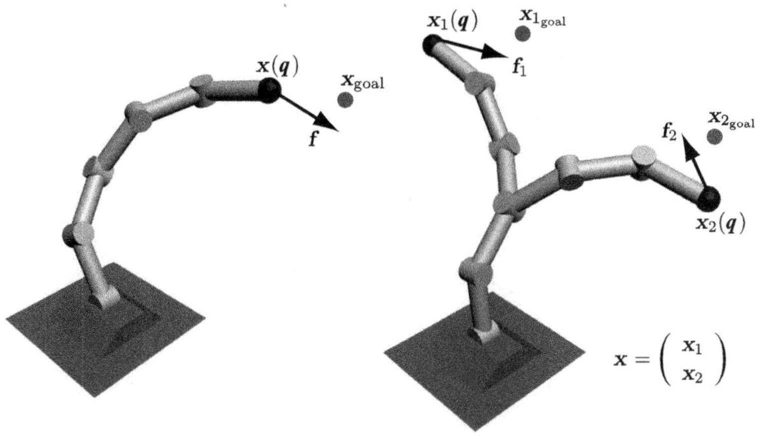

Figure 8.1 Task descriptions for serial and branching chains. (Left) A simple task description for a serial chain where the task space coordinates describe the Cartesian position of the terminal point of the chain. (Right) A branching chain where the task space coordinates describe the positions of both terminal points.

The solutions to the underdetermined equation, $\boldsymbol{J}\dot{\boldsymbol{q}} = \dot{\mathbf{x}}$, are thus given by

$$\dot{\boldsymbol{q}} = \boldsymbol{J}^{\#}\dot{\mathbf{x}} + \boldsymbol{N}\dot{\boldsymbol{q}}_o, \tag{8.4}$$

where $\boldsymbol{N} \triangleq \mathbf{1} - \boldsymbol{J}^{\#}\boldsymbol{J}$ and $\dot{\boldsymbol{q}}_o$ is an arbitrary vector in $\mathbb{R}^n$.

We now define a matrix, $\boldsymbol{W}$, whose columns span the null space of $\boldsymbol{J}$. This implies that $\mathrm{im}(\boldsymbol{W}) = \ker(\boldsymbol{J})$. Thus, $\boldsymbol{J}\boldsymbol{W} = \mathbf{0}$ and $\boldsymbol{W}^T\boldsymbol{J}^T = \mathbf{0}$. In this manner, $\boldsymbol{W}$ orthogonally complements $\boldsymbol{J}^T$. That is,

$$\mathrm{im}(\boldsymbol{W}) = \ker(\boldsymbol{J}) = \mathrm{im}(\boldsymbol{J}^T)^{\perp}. \tag{8.5}$$

If we form the projection matrix, $\boldsymbol{P} = \boldsymbol{P}^T$, that projects any vector in $\mathbb{R}^n$ onto the null space of $\boldsymbol{J}$, we have

$$\boldsymbol{P} = \boldsymbol{W}\boldsymbol{W}^T = \mathbf{1} - \boldsymbol{J}^+\boldsymbol{J} = \mathbf{1} - \boldsymbol{J}^T(\boldsymbol{J}\boldsymbol{J}^T)^{-1}\boldsymbol{J}, \tag{8.6}$$

and a specific solution of the inverse kinematics problem, using the pseudoinverse $\boldsymbol{J}^+$, is

$$\dot{\boldsymbol{q}} = \boldsymbol{J}^+\dot{\mathbf{x}} + \boldsymbol{P}\dot{\boldsymbol{q}}_o = \boldsymbol{J}^+\dot{\mathbf{x}} + (\mathbf{1} - \boldsymbol{J}^+\boldsymbol{J})\dot{\boldsymbol{q}}_o. \tag{8.7}$$

This can be used for kinematic path generation given a task trajectory, $\mathbf{x}(t)$. If some minimization criterion, $U(\boldsymbol{q})$, is specified we have,

$$\dot{\boldsymbol{q}} = \boldsymbol{J}^+\dot{\mathbf{x}} - \boldsymbol{P}\frac{\partial U}{\partial \boldsymbol{q}}. \tag{8.8}$$

At this point we can address task space kinetics. In the nonredundant case, any generalized force can be produced by a task space force, $\boldsymbol{f}$, acting at the task point along the task coordinates. Figure 8.1 illustrates the action of the task space force for the intuitive case of Cartesian positioning tasks. The generalized force is then composed as $\boldsymbol{J}^T\boldsymbol{f}$, by the Principle of Virtual Work (3.184). In the redundant case an additional term needs to complement the task term to realize any arbitrary generalized force. We will refer to this term as the null space term, and it can be composed as $\boldsymbol{N}^T\boldsymbol{\tau}_o$, where $\boldsymbol{N}^T$ is the null space projection matrix. An arbitrary generalized force, $\boldsymbol{\tau}$, can be expressed as

$$\boldsymbol{\tau} = \boldsymbol{J}^T\boldsymbol{f} + \boldsymbol{N}^T\boldsymbol{\tau}_o = \boldsymbol{M}\ddot{\boldsymbol{q}} + \boldsymbol{b} + \boldsymbol{g}. \tag{8.9}$$

We can premultiply (8.9) by $\boldsymbol{J}\boldsymbol{M}^{-1}$ and rearrange to get

$$\ddot{\boldsymbol{x}} = \boldsymbol{J}\boldsymbol{M}^{-1}\boldsymbol{J}^T\boldsymbol{f} + \boldsymbol{J}\boldsymbol{M}^{-1}\boldsymbol{N}^T\boldsymbol{\tau}_o - \boldsymbol{J}\boldsymbol{M}^{-1}\boldsymbol{b} - \boldsymbol{J}\boldsymbol{M}^{-1}\boldsymbol{g} + \dot{\boldsymbol{J}}\dot{\boldsymbol{q}}, \tag{8.10}$$

where we note that $\ddot{\boldsymbol{x}} = \boldsymbol{J}\ddot{\boldsymbol{q}} + \dot{\boldsymbol{J}}\dot{\boldsymbol{q}}$. We can now impose the condition that the term associated with the null space, $\boldsymbol{N}^T\boldsymbol{\tau}_o$, does not contribute to the task space acceleration. This is referred to as *dynamic consistency* (Khatib 1995) and is expressed as

$$\begin{gathered} \boldsymbol{J}\boldsymbol{M}^{-1}\boldsymbol{N}^T\boldsymbol{\tau}_o = \boldsymbol{J}\boldsymbol{M}^{-1}(\boldsymbol{1} - \boldsymbol{J}^T\boldsymbol{J}^{T\#})\boldsymbol{\tau}_o = \boldsymbol{0} \\ \forall\boldsymbol{\tau}_o \in \mathbb{R}^n. \end{gathered} \tag{8.11}$$

We can solve for $\boldsymbol{J}^{\#}$ under this condition and denote this solution as $\bar{\boldsymbol{J}}$, the dynamically consistent inverse of $\boldsymbol{J}$ (Khatib 1995):

$$\bar{\boldsymbol{J}} = \boldsymbol{M}^{-1}\boldsymbol{J}^T(\boldsymbol{J}\boldsymbol{M}^{-1}\boldsymbol{J}^T)^{-1}. \tag{8.12}$$

This represents a unique right inverse of $\boldsymbol{J}$ where, by construction, the null space projection matrix, $\boldsymbol{N}^T = \boldsymbol{1} - \boldsymbol{J}^T\bar{\boldsymbol{J}}^T$, will not influence the task acceleration. The matrix $\boldsymbol{N}^T$ can be regarded as a *mass-weighted* null space projection matrix, compared to $\boldsymbol{P}^T = \boldsymbol{W}\boldsymbol{W}^T$, which can be regarded as a *kinematic* null space projection matrix. We can manipulate (8.10) to arrive at

$$\boldsymbol{f} = (\boldsymbol{J}\boldsymbol{M}^{-1}\boldsymbol{J}^T)^{-1}\ddot{\boldsymbol{x}} + (\boldsymbol{J}\boldsymbol{M}^{-1}\boldsymbol{J}^T)^{-1}(\boldsymbol{J}\boldsymbol{M}^{-1}\boldsymbol{b} - \dot{\boldsymbol{J}}\dot{\boldsymbol{q}}) + (\boldsymbol{J}\boldsymbol{M}^{-1}\boldsymbol{J}^T)^{-1}\boldsymbol{J}\boldsymbol{M}^{-1}\boldsymbol{g}. \tag{8.13}$$

This expresses the task space dynamical equation

$$\boldsymbol{f} = \boldsymbol{\Lambda}(\boldsymbol{q})\ddot{\boldsymbol{x}} + \boldsymbol{\mu}(\boldsymbol{q}, \dot{\boldsymbol{q}}) + \boldsymbol{p}(\boldsymbol{q}), \tag{8.14}$$

where $\boldsymbol{\Lambda}(\boldsymbol{q})$ is the $m_T \times m_T$ symmetric positive definite task space mass matrix, $\boldsymbol{\mu}(\boldsymbol{q}, \dot{\boldsymbol{q}})$ is the $m_T \times 1$ task space centrifugal and Coriolis force vector, and $\boldsymbol{p}(\boldsymbol{q})$ is the $m_T \times 1$ task space gravity force vector:

$$\boldsymbol{\Lambda}(\boldsymbol{q}) = (\boldsymbol{J}\boldsymbol{M}^{-1}\boldsymbol{J}^T)^{-1}, \tag{8.15}$$

$$\boldsymbol{\mu}(\boldsymbol{q}, \dot{\boldsymbol{q}}) = \bar{\boldsymbol{J}}^T\boldsymbol{b}(\boldsymbol{q}, \dot{\boldsymbol{q}}) - \boldsymbol{\Lambda}\dot{\boldsymbol{J}}\dot{\boldsymbol{q}}, \tag{8.16}$$

$$\boldsymbol{p}(\boldsymbol{q}) = \bar{\boldsymbol{J}}^T\boldsymbol{g}(\boldsymbol{q}), \tag{8.17}$$

$$\bar{\boldsymbol{J}}^T = \boldsymbol{\Lambda}\boldsymbol{J}\boldsymbol{M}^{-1}. \tag{8.18}$$

Thus, the overall dynamics of our multibody system can be mapped into task space using $\bar{\boldsymbol{J}}^T$:

$$\boldsymbol{\tau} = \boldsymbol{M}\ddot{\boldsymbol{q}} + \boldsymbol{b} + \boldsymbol{g} \ \overset{\bar{\boldsymbol{J}}^T}{\rightarrow} \ \boldsymbol{f} = \boldsymbol{\Lambda}\ddot{\boldsymbol{x}} + \boldsymbol{\mu} + \boldsymbol{p}. \tag{8.19}$$

In a complementary manner the overall dynamics can be mapped into the task-consistent null space (or self-motion space) using $\boldsymbol{N}^T$. The null space term is guaranteed not to interfere with the task dynamics of (8.14) due to the condition of dynamic consistency. Finally, the overall torque applied to the system is composed as in (8.9).

Example: We consider the serial chain robot of Section 5.2.4 (see Figure 8.2). We can use the configuration space properties already computed to formulate the task space dynamics. We will specify the task to be the position of the distal end of link 4 (the terminal point in the chain denoted by $\boldsymbol{r}_E$).

$$\mathbf{x} = {}^0\boldsymbol{r}_E = \begin{pmatrix} l_1\cos(q_1) + l_2\cos(q_1+q_2) + l_3\cos(q_1+q_2+q_3) \\ l_1\sin(q_1) + l_2\sin(q_1+q_2) + l_3\sin(q_1+q_2+q_3) \end{pmatrix.$$
$$\cdots \begin{matrix} +l_4\cos(q_1+q_2+q_3+q_4) \\ +l_4\sin(q_1+q_2+q_3+q_4) \end{pmatrix}. \tag{8.20}$$

The task Jacobian is,

$$\boldsymbol{J}(\boldsymbol{q}) = \frac{\partial \mathbf{x}}{\partial \boldsymbol{q}}. \tag{8.21}$$

Using $\boldsymbol{J}(\boldsymbol{q})$, as well as $\boldsymbol{M}(\boldsymbol{q})$, $\boldsymbol{b}(\boldsymbol{q}, \dot{\boldsymbol{q}})$, and $\mathbf{g}(\boldsymbol{q})$, which were computed in the example of Section 5.2.4, we can compute the task space terms,

$$\boldsymbol{\Lambda}(\boldsymbol{q}) = (\boldsymbol{J}\boldsymbol{M}^{-1}\boldsymbol{J}^T)^{-1}, \tag{8.22}$$

$$\boldsymbol{\mu}(\boldsymbol{q}, \dot{\boldsymbol{q}}) = \boldsymbol{\Lambda}\boldsymbol{J}\boldsymbol{M}^{-1}\boldsymbol{b}(\boldsymbol{q}, \dot{\boldsymbol{q}}) - \boldsymbol{\Lambda}\dot{\boldsymbol{J}}\dot{\boldsymbol{q}}, \tag{8.23}$$

$$\boldsymbol{p}(\boldsymbol{q}) = \boldsymbol{\Lambda}\boldsymbol{J}\boldsymbol{M}^{-1}\mathbf{g}(\boldsymbol{q}). \tag{8.24}$$

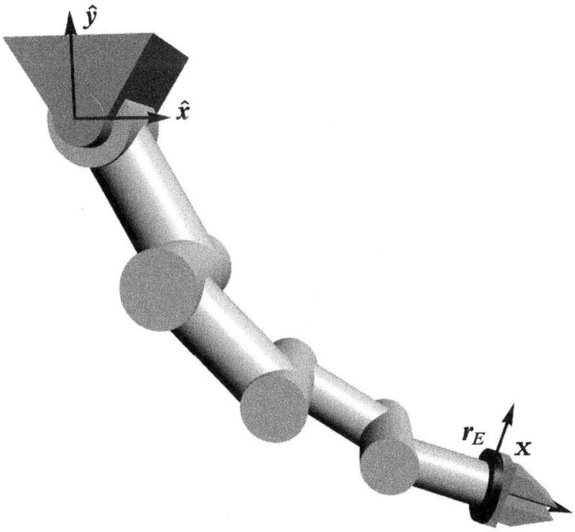

Figure 8.2 A 4 degree-of-freedom serial chain robot parameterized using four generalized coordinates. The task point, $\mathbf{x}$, is shown.

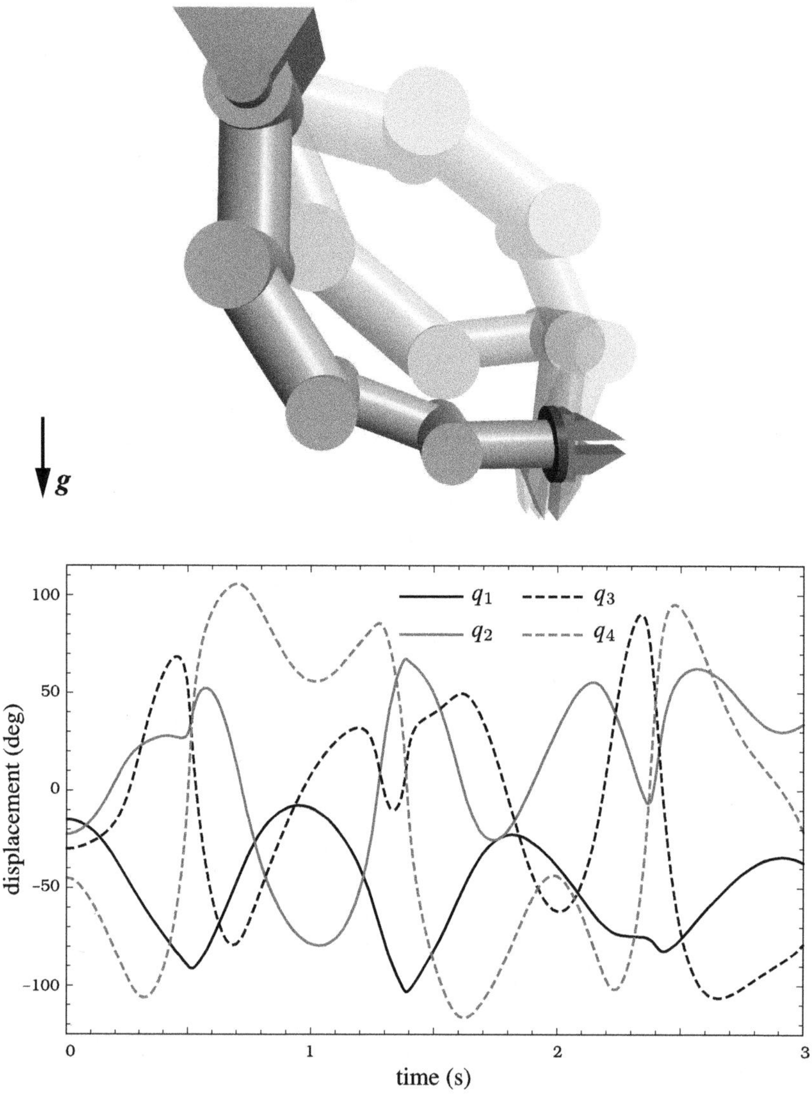

Figure 8.3 (Top) Animation frames from the simulation of a serial chain robot. (Bottom) Time history of the robot generalized coordinates (q_1, q_2, q_3, q_4).

Since these terms involve matrix inverses it is preferable to compute them numerically at each time step within the numerical integration loop, rather then to compute them symbolically. Our task space equation of motion for the dynamics at the task point is,

$$\boldsymbol{f} = \boldsymbol{\Lambda}(\boldsymbol{q})\ddot{\boldsymbol{x}} + \boldsymbol{\mu}(\boldsymbol{q}, \dot{\boldsymbol{q}}) + \boldsymbol{p}(\boldsymbol{q}). \tag{8.25}$$

We will specify a task space force at the task point to cancel out the task space centrifugal and Coriolis force, and the task space gravity force. That is,

$$\boldsymbol{f} = \boldsymbol{\mu}(\boldsymbol{q}, \dot{\boldsymbol{q}}) + \boldsymbol{p}(\boldsymbol{q}). \tag{8.26}$$

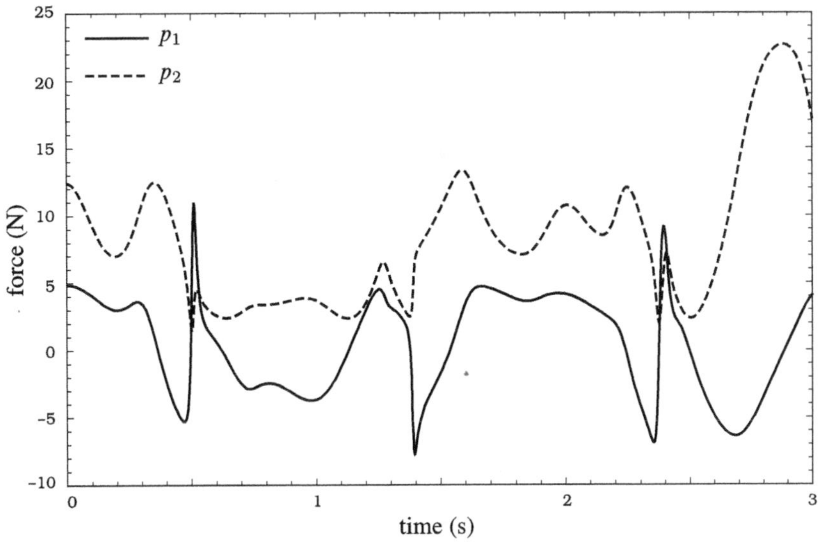

Figure 8.4 Time history of the robot task space gravity vector.

This will result in the following dynamics,

$$\Lambda(q)\ddot{x} = \mathbf{0}. \tag{8.27}$$

Therefore, the task point will have *zero* acceleration. Given *zero* velocity initial conditions, the task point will not move. The redundant chain will however move in the null space.

The simulation results are displayed in Figures 8.3 and 8.4. The following values were used for the initial conditions,

$$q_o = \begin{pmatrix} -\frac{\pi}{12} & -\frac{\pi}{8} & -\frac{\pi}{6} & -\frac{\pi}{4} \end{pmatrix}^T, \tag{8.28}$$

$$\dot{q}_o = \begin{pmatrix} 0 & 0 & 0 & 0 \end{pmatrix}^T. \tag{8.29}$$

8.1.1 Inertial Properties

The task space kinetic energy is given by,

$$T = \frac{1}{2}\dot{x}^T \Lambda \dot{x}. \tag{8.30}$$

We will assume this expression is represented in the base frame for convenience. Since Λ is symmetric positive definite it has positive eigenvalues and an orthogonal eigenbasis (principal axes), $\mathcal{E}$. Therefore,

$$T = \frac{1}{2}{}^{O}\dot{x}^{T\,O}\Lambda\,{}^{O}\dot{x} = \frac{1}{2}{}^{\mathcal{E}}\dot{x}^{T\,O}_{\mathcal{E}}Q^{T\,O}\Lambda\,{}^{O}_{\mathcal{E}}Q\,{}^{\mathcal{E}}\dot{x}, \tag{8.31}$$

where the columns of ${}^{\mathcal{O}}_{\mathcal{E}}\boldsymbol{Q}$ are the eigenvectors, $\{\boldsymbol{v}_1, \boldsymbol{v}_2, \boldsymbol{v}_3\}$, of ${}^{\mathcal{O}}\boldsymbol{\Lambda}$,

$$ {}^{\mathcal{O}}_{\mathcal{E}}\boldsymbol{Q} = \begin{pmatrix} \uparrow & \uparrow & \uparrow \\ \boldsymbol{v}_1 & \boldsymbol{v}_2 & \boldsymbol{v}_3 \\ \downarrow & \downarrow & \downarrow \end{pmatrix}. \tag{8.32} $$

We then have,

$$ {}^{\mathcal{O}}_{\mathcal{E}}\boldsymbol{Q}^{T}\, {}^{\mathcal{O}}\boldsymbol{\Lambda}\, {}^{\mathcal{O}}_{\mathcal{E}}\boldsymbol{Q} = {}^{\mathcal{E}}\boldsymbol{\Lambda}, \tag{8.33} $$

where ${}^{\mathcal{E}}\boldsymbol{\Lambda}$ is a diagonal matrix of eigenvalues, $\{\lambda_1, \lambda_2, \lambda_3\}$, of ${}^{\mathcal{O}}\boldsymbol{\Lambda}$,

$$ {}^{\mathcal{E}}\boldsymbol{\Lambda} = \begin{pmatrix} \lambda_1 & 0 & 0 \\ 0 & \lambda_2 & 0 \\ 0 & 0 & \lambda_3 \end{pmatrix}. \tag{8.34} $$

In terms of the eigenbasis,

$$ T = \frac{1}{2} {}^{\mathcal{E}}\dot{\mathbf{x}}^{T}\, {}^{\mathcal{E}}\boldsymbol{\Lambda}\, {}^{\mathcal{E}}\dot{\mathbf{x}}. \tag{8.35} $$

For constant values of the kinetic energy T, this represents an ellipsoid expressed with respect to the principal axes. In scalar form,

$$ T = \frac{1}{2}\left(\lambda_1 {}^{\mathcal{E}}\dot{x}_1^2 + \lambda_2 {}^{\mathcal{E}}\dot{x}_2^2 + \lambda_3 {}^{\mathcal{E}}\dot{x}_3^2\right). \tag{8.36} $$

The semiaxes of the ellipsoid in frame $\mathcal{E}$ are then,

$$ a_i = \sqrt{2T/\lambda_i}. \tag{8.37} $$

Khatib introduced the concept of the belted ellipsoid as another means of visualizing task space inertial properties (Khatib 1995). The belted ellipsoid is given by the following equation,

$$ \frac{\mathbf{v}^T \boldsymbol{\Lambda}^{-1} \mathbf{v}}{\|\mathbf{v}\|} = 1. \tag{8.38} $$

This representation inverts the task space mass matrix in the quadratic form and normalizes by $\|\mathbf{v}\|$. The points on the belted ellipsoid correspond to the effective masses/inertias in different unit directions.

Example: We consider the serial chain robot of Section 4.1.4 (see Figure 8.5). We can use the equations of motion already computed in Section 4.1.4 to form $\boldsymbol{M}(\boldsymbol{q})$, $\boldsymbol{b}(\boldsymbol{q}, \dot{\boldsymbol{q}})$, and $\boldsymbol{g}(\boldsymbol{q})$. This will allow us to formulate the task space dynamics. We will specify the task to be the position of the distal end of link 3 (the terminal point in the chain denoted by $\boldsymbol{r}_E$).

$$ \mathbf{x} = {}^{0}\boldsymbol{r}_E = \begin{pmatrix} \cos(q_1)[l_1 + l_2\cos(q_2) + l_3\cos(q_2+q_3)] \\ [l_1 + l_2\cos(q_2) + l_3\cos(q_2+q_3)]\sin(q_1) \\ l_2\sin(q_2) + l_3\sin(q_2+q_3) \end{pmatrix}. \tag{8.39} $$

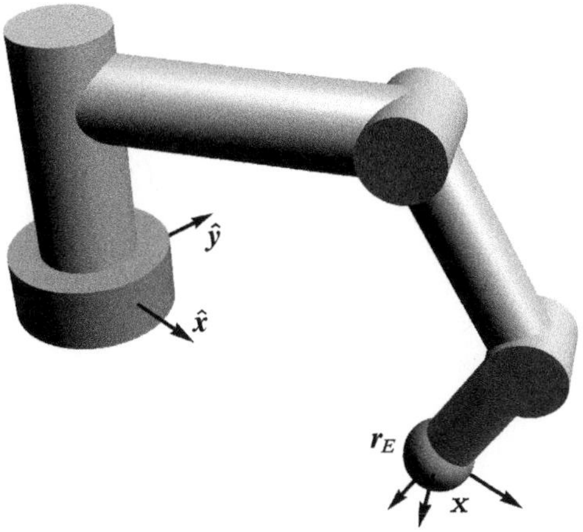

Figure 8.5 A 3 degree-of-freedom serial chain robot parameterized by the generalized coordinates, q_1, q_2, and q_3. The link lengths are l_1, l_2, and l_3, the radii are r_1, r_2, and r_3, and the masses of the links are M_1, M_2, and M_3. The task point, $\mathbf{x}$, is shown.

The task Jacobian is,

$$\boldsymbol{J}(\boldsymbol{q}) = \left(\begin{matrix} -[l_1 + l_2\cos(q_2) + l_3\cos(q_2+q_3)]\sin(q_1) \\ [l_1 + l_2\cos(q_2) + l_3\cos(q_2+q_3)]\cos(q_1) \\ 0 \end{matrix} \right.$$

$$\cdots \begin{matrix} -\cos(q_1)[l_2\sin(q_2) + l_3\sin(q_2+q_3)] \\ -\sin(q_1)[l_2\sin(q_2) + l_3\sin(q_2+q_3)] \\ l_2\cos(q_2) + l_3\cos(q_2+q_3) \end{matrix}$$

$$\cdots \left. \begin{matrix} -l_3\cos(q_1)\sin(q_2+q_3)] \\ -l_3\sin(q_1)\sin(q_2+q_3)] \\ l_3\cos(q_2+q_3) \end{matrix} \right).$$

Using $\boldsymbol{J}(\boldsymbol{q})$ and $\boldsymbol{M}(\boldsymbol{q})$, we can compute the task space mass matrix, $\boldsymbol{\Lambda}(\boldsymbol{q}) = (\boldsymbol{J}\boldsymbol{M}^{-1}\boldsymbol{J}^T)^{-1}$. As noted in the previous example, since this term involves matrix inverses it is preferable to compute it numerically rather then symbolically.

Using the following values for the constants,

$$\begin{aligned} r_1 &= 0.125, \quad r_2 = 0.1, \quad r_3 = 0.09, \\ l_1 &= 1, \quad l_2 = 0.75, \quad l_3 = 0.5, \\ M_1 &= 1, \quad M_2 = 0.75, \quad M_3 = 0.5, \end{aligned} \tag{8.40}$$

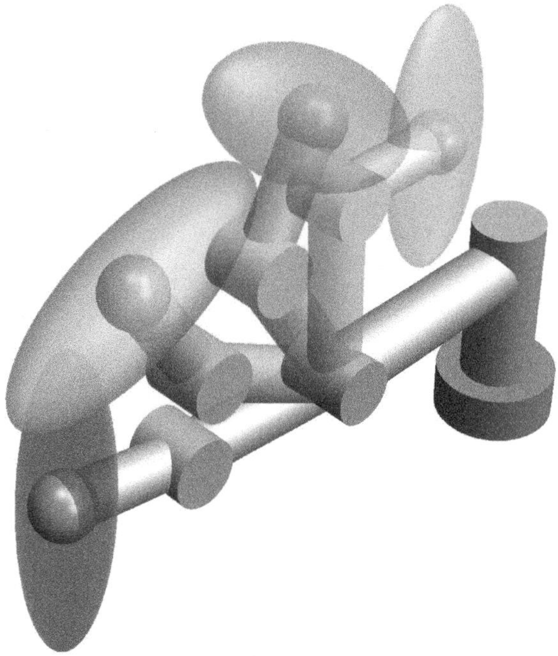

Figure 8.6 Kinetic energy ellipsoids for a sequence of configurations.

we can evaluate $\mathbf{\Lambda}(\boldsymbol{q})$ for an arbitrary configuration. At $q_1 = -\pi/2$, $q_2 = \pi/3$, $q_3 = \pi/3$ we have,

$$\mathbf{\Lambda}(-\pi/2, \pi/3, \pi/3) = \begin{pmatrix} 1.732 & 0 & 0 \\ 0 & 0.391 & 0.336 \\ 0 & 0.336 & 0.674 \end{pmatrix}, \tag{8.41}$$

and, solving the eigenvalue problem yields,

$$^{\mathcal{O}}_{\mathcal{E}}\boldsymbol{Q} = \begin{pmatrix} \uparrow & \uparrow & \uparrow \\ \boldsymbol{v}_1 & \boldsymbol{v}_2 & \boldsymbol{v}_3 \\ \downarrow & \downarrow & \downarrow \end{pmatrix} = \begin{pmatrix} 1 & 0 & 0 \\ 0 & 0.553 & 0.833 \\ 0 & 0.833 & -0.553 \end{pmatrix}, \tag{8.42}$$

and,

$$^{\mathcal{E}}\mathbf{\Lambda} = \begin{pmatrix} \lambda_1 & 0 & 0 \\ 0 & \lambda_2 & 0 \\ 0 & 0 & \lambda_3 \end{pmatrix} = \begin{pmatrix} 1.732 & 0 & 0 \\ 0 & 0.898 & 0 \\ 0 & 0 & 0.168 \end{pmatrix}. \tag{8.43}$$

Figure 8.6 displays kinetic energy ellipsoids for a sequence of configurations. The belted ellipsoids, given by,

$$\frac{\mathbf{v}^T \mathbf{\Lambda}^{-1} \mathbf{v}}{\|\mathbf{v}\|} = 1, \tag{8.44}$$

for a sequence of configurations are shown in Figure 8.7.

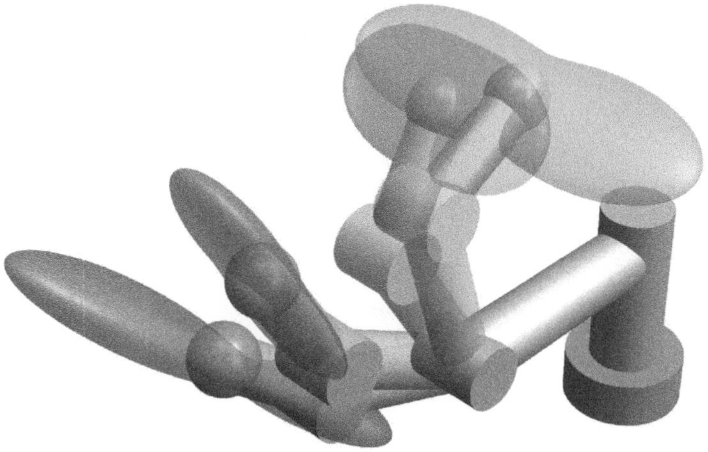

Figure 8.7 Belted ellipsoids for a sequence of configurations.

8.1.2 Energy Minimization

Minimization of Gravitational Potential Energy

Given the self-motion manifold associated with a fixed task point, $\mathbf{x}_O$,

$$M(\mathbf{x}_O) = \{\boldsymbol{q} | \mathbf{x}(\boldsymbol{q}) = \mathbf{x}_O\}, \tag{8.45}$$

there is a configuration, $\boldsymbol{q}_o$, which minimizes the potential energy, $V(\boldsymbol{q})$, on $M(\boldsymbol{x}_O)$. This configuration satisfies,

$$\nabla V|_{q_o} \in T_{q_o}(M)^{\perp} = \ker(\boldsymbol{J})^{\perp}|_{q_o} = \mathrm{im}(\boldsymbol{J}^T)|_{q_o}. \tag{8.46}$$

This implies,

$$\boldsymbol{g}(\boldsymbol{q}_o) = \nabla V|_{q_o} = \boldsymbol{J}^T, (\boldsymbol{q}_o)\boldsymbol{c} \tag{8.47}$$

for some vector, $\boldsymbol{c}$. Thus,

$$\boldsymbol{c} = \bar{\boldsymbol{J}}^T(\boldsymbol{q}_o)\boldsymbol{g}(\boldsymbol{q}_o) = \boldsymbol{p}(\boldsymbol{q}_o), \tag{8.48}$$

so,

$$\boldsymbol{g}(\boldsymbol{q}_o) = \boldsymbol{J}^T(\boldsymbol{q}_o)\boldsymbol{p}(\boldsymbol{q}_o). \tag{8.49}$$

That is, under static equilibrium the configuration associated with the minimum potential energy on the self-motion manifold for a fixed task point corresponds to a force of $\boldsymbol{p}(\boldsymbol{q}_o)$ acting at the task point.

Minimization of Kinetic Energy

We can seek a solution to the kinematic relationship, $\boldsymbol{J}\dot{\boldsymbol{q}} = \dot{\mathbf{x}}$, which minimizes the kinetic energy,

$$T = \frac{1}{2}\dot{\boldsymbol{q}}^T \boldsymbol{M}\dot{\boldsymbol{q}}. \tag{8.50}$$

The solution to this constrained minimization problem is straightforward and yields,

$$\dot{\boldsymbol{q}} = \boldsymbol{M}^{-1}\boldsymbol{J}^T(\boldsymbol{J}\boldsymbol{M}^{-1}\boldsymbol{J}^T)^{-1}\dot{\mathbf{x}}. \tag{8.51}$$

Noting that the dynamically consistent inverse of $\boldsymbol{J}$ is given by,

$$\bar{\boldsymbol{J}} = \boldsymbol{M}^{-1}\boldsymbol{J}^T(\boldsymbol{J}\boldsymbol{M}^{-1}\boldsymbol{J}^T)^{-1}, \tag{8.52}$$

we have that $\dot{\boldsymbol{q}} = \bar{\boldsymbol{J}}\dot{\mathbf{x}}$ yields the kinetic energy minimizing solution of (8.1). Similarly, we can seek a solution of the acceleration expression,

$$\boldsymbol{J}\ddot{\boldsymbol{q}} = \ddot{\mathbf{x}} - \dot{\boldsymbol{J}}\dot{\boldsymbol{q}}, \tag{8.53}$$

which minimizes the *acceleration energy*, defined as the following mass weighted quadratic form,

$$\frac{1}{2}\ddot{\boldsymbol{q}}^T \boldsymbol{M}\ddot{\boldsymbol{q}}. \tag{8.54}$$

This yields,

$$\ddot{\boldsymbol{q}} = \boldsymbol{M}^{-1}\boldsymbol{J}^T(\boldsymbol{J}\boldsymbol{M}^{-1}\boldsymbol{J}^T)^{-1}(\ddot{\mathbf{x}} - \dot{\boldsymbol{J}}\dot{\boldsymbol{q}}), \tag{8.55}$$

and we have that $\ddot{\boldsymbol{q}} = \bar{\boldsymbol{J}}(\ddot{\mathbf{x}} - \dot{\boldsymbol{J}}\dot{\boldsymbol{q}})$ yields the acceleration energy minimizing solution of (8.53).

8.1.3 Task and Constraint Symmetry

There are parallels between the structure of the constrained multibody dynamics problem, in both the multiplier and minimization (Gauss) forms, and the task space formulation. These parallels are derived from the common mathematical description used for tasks and constraints. Both utilize a Jacobian representation (constraint matrix, $\boldsymbol{\Phi}$, or task Jacobian, $\boldsymbol{J}$) and tasks can be viewed as rheonomic constraints of the form $\mathbf{x}(\boldsymbol{q}) = \mathbf{x}_d(t)$.

Despite the common form used in specifying them, the mechanism by which constraints and tasks are satisfied differs. Constraints are *imposed* by the physical structure of the multibody system, whereas tasks are *achieved* by means of an applied control input. Nevertheless, due to their common mathematical form there are similarities between the structure of task dynamics and constrained dynamics.

Symmetry Between Projection Matrices

The generalized constrained equation of motion (5.154) provides a unique perspective into constrained dynamics. The projection matrix, $\boldsymbol{\Theta}^T$, filters out the component of the generalized force which acts in the direction of the constraint force. That is,

$$\boldsymbol{\Theta}^T\boldsymbol{\tau} = \boldsymbol{\tau} - \boldsymbol{\Phi}^T\bar{\boldsymbol{\Phi}}^T\boldsymbol{\tau}. \tag{8.56}$$

Consequently, only the component of the generalized force which influences the motion of the system is preserved. Equivalent motion (Huston, Liu, and Li 2003) is produced by all choices of $\boldsymbol{\tau}$ which differ by a vector lying in the im($\boldsymbol{\Phi}^T$). We also note that the complementary spaces defined by im($\boldsymbol{\Phi}^T$) and im($\boldsymbol{\Theta}^T$) are orthogonal in a mass weighted sense,

$$\left\langle\boldsymbol{\Phi}^T, \boldsymbol{\Theta}^T\right\rangle_{M^{-1}} = \boldsymbol{\Phi}\boldsymbol{M}^{-1}\boldsymbol{\Theta}^T = \mathbf{0}. \tag{8.57}$$

Similarly, in task space, the projection matrix, $\boldsymbol{N}^T$, filters out the component of the generalized force which produces acceleration in the task direction. That is,

$$\boldsymbol{N}^T\boldsymbol{\tau} = \boldsymbol{\tau} - \boldsymbol{J}^T\bar{\boldsymbol{J}}^T\boldsymbol{\tau}. \tag{8.58}$$

Consequently, only the component of the generalized force which influences the internal self-motion of the system is preserved. We also note that the complementary spaces defined by im($\boldsymbol{J}^T$) and im($\boldsymbol{N}^T$) are orthogonal in a mass weighted sense,

$$\left\langle\boldsymbol{J}^T, \boldsymbol{N}^T\right\rangle_{M^{-1}} = \boldsymbol{J}\boldsymbol{M}^{-1}\boldsymbol{N}^T = \mathbf{0}. \tag{8.59}$$

To summarize, in the case of constrained motion, a projection matrix, $\boldsymbol{\Theta}^T$, is used to project the overall system dynamics into the constraint null space. This preserves only the dynamics which influences the constrained motion of the system. In the case of task space dynamics a projection matrix, $\boldsymbol{N}^T$, is used to project the overall system dynamics into the task null space. This preserves only the dynamics which influences the task-consistent self-motion of the system.

Energy Minimization

We observed in Section 8.1.2 that the dynamically consistent inverse, $\bar{\boldsymbol{J}}$, of the task Jacobian provides the unique task-consistent solution of $\ddot{\boldsymbol{q}}$ which minimizes both the kinetic energy and acceleration energy of the system. This is analogous to the manner in which the mass weighted inverse, $\bar{\boldsymbol{\Phi}}$, of the constraint matrix yields a constraint-consistent solution that minimizes the Gauss function. That is,

$$\ddot{\boldsymbol{q}} = \bar{\boldsymbol{J}}(\ddot{\boldsymbol{x}} - \dot{\boldsymbol{J}}\dot{\boldsymbol{q}}) \tag{8.60}$$

is the solution of

$$\boldsymbol{J}\ddot{\boldsymbol{q}} = \ddot{\boldsymbol{x}} - \dot{\boldsymbol{J}}\dot{\boldsymbol{q}}, \tag{8.61}$$

which minimizes

$$\frac{1}{2}\,\|\ddot{\boldsymbol{q}}\|^2_M, \tag{8.62}$$

and

$$\ddot{\boldsymbol{q}} = -\bar{\boldsymbol{\Phi}}\dot{\boldsymbol{\Phi}}\dot{\boldsymbol{q}} + (\mathbf{1} - \bar{\boldsymbol{\Phi}}\boldsymbol{\Phi})\ddot{\boldsymbol{q}}_\star \tag{8.63}$$

is the solution of

$$\boldsymbol{\Phi}\ddot{\boldsymbol{q}} = -\dot{\boldsymbol{\Phi}}\dot{\boldsymbol{q}}, \tag{8.64}$$

which minimizes

$$\mathcal{G} = \frac{1}{2}\left\|\ddot{\boldsymbol{q}} - \ddot{\boldsymbol{q}}_{\star}\right\|_{M}^{2}. \tag{8.65}$$

Least Action in Task Space

It was noted earlier that tasks can be viewed as rheonomic constraints. We will now interpret the task space representation as a rheonomically constrained least action problem. Noting that $\delta\mathbf{x}_d(t) = \mathbf{0}$, since time is fixed under all virtual variations, we have $\delta\mathbf{x} = \boldsymbol{J}\delta\boldsymbol{q} = \mathbf{0}$. Application of the Principle of Least Action then yields

$$\begin{gathered} \delta I = 0, \\ \forall \delta \,|\, \delta\boldsymbol{q}(t_o) = \delta\boldsymbol{q}(t_f) = \mathbf{0}, \quad \text{and} \quad \boldsymbol{J}\delta\boldsymbol{q} = \mathbf{0}. \end{gathered} \tag{8.66}$$

Thus,

$$\begin{gathered} \delta I = \int_{t_o}^{t_f} \left(\frac{\partial \mathcal{L}}{\partial \boldsymbol{q}} - \frac{d}{dt}\frac{\partial \mathcal{L}}{\partial \dot{\boldsymbol{q}}}\right) \cdot \delta\boldsymbol{q}dt = 0 \\ \forall \delta\boldsymbol{q} \,|\, \boldsymbol{J}\delta\boldsymbol{q} = \mathbf{0}. \end{gathered} \tag{8.67}$$

This implies the following Euler-Lagrange equations:

$$\frac{d}{dt}\frac{\partial \mathcal{L}}{\partial \dot{\boldsymbol{q}}} - \frac{\partial \mathcal{L}}{\partial \boldsymbol{q}} = \boldsymbol{J}^T\boldsymbol{f}. \tag{8.68}$$

Identical equations could have been obtained by embedding the rheonomic constraints directly in the Lagrangian. Adjoining the constraints yields

$$\mathcal{L}_{\text{aug}}(\boldsymbol{q}, \dot{\boldsymbol{q}}, \boldsymbol{\lambda}) \triangleq \mathcal{L}(\boldsymbol{q}, \dot{\boldsymbol{q}}) + \boldsymbol{f}^T[\mathbf{x}(\boldsymbol{q}) - \mathbf{x}_d(t)], \tag{8.69}$$

and the stationary value of I would be sought for all variations that vanish at the endpoints. Recalling that $\delta\mathbf{x}_d = \mathbf{0}$,

$$\delta\mathcal{L}_{\text{aug}} = \delta\mathcal{L} + \boldsymbol{f}^T\delta\mathbf{x} = \delta\mathcal{L} + \boldsymbol{f}^T\boldsymbol{J}\delta\boldsymbol{q}. \tag{8.70}$$

In either case, (8.68) can be expressed as

$$\boldsymbol{J}^T\boldsymbol{f} = \boldsymbol{M}\ddot{\boldsymbol{q}} + \boldsymbol{b} + \boldsymbol{g}, \tag{8.71}$$

subject to $\ddot{\mathbf{x}}(\boldsymbol{q}) = \ddot{\mathbf{x}}_d(t)$. This is similar to (8.9), except that no null space term is present. That is, $\boldsymbol{\tau} = \boldsymbol{J}^T\boldsymbol{f}$ (absent any null space torque) achieves the task space

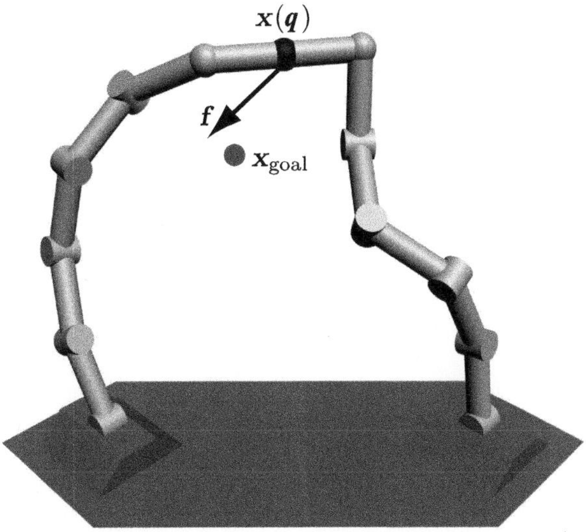

Figure 8.8 Task description for a multibody system with loop constraints. The task coordinates, $\mathbf{x}$, can be assigned to a point on one of the links. More generally, the task can be defined as any function of the generalized coordinates that does not violate the constraints.

objective, consistent with least action. Projecting (8.71) into task space yields the task space equations

$$\boldsymbol{f} = \boldsymbol{\Lambda}\ddot{\mathbf{x}} + \boldsymbol{\mu} + \boldsymbol{p}. \tag{8.72}$$

8.2 Constrained Dynamics in Task Space

We will now present task space methodologies for addressing constrained systems in the task space framework. As an example, Figure 8.8 illustrates a task description for a multibody system with loop constraints. The task coordinates, $\mathbf{x}$, are assigned to a point on one of the links in a manner that does not violate the constraints. In the following sections we will formalize approaches for representing constrained system dynamics in a task space form.

8.2.1 Direct Task Space Mapping

This formulation involves directly mapping the generalized constrained equation of motion (5.154) into task space coordinates using the dynamically consistent inverse of the task Jacobian. Alternatively, we will make use of the constrained equation of motion in which the Lagrange multipliers have been eliminated through the introduction of a minimal set of independent coordinates (5.302). As with (5.154), this equation can be directly mapped into task space coordinates using the dynamically consistent inverse of the task Jacobian.

We begin by recalling the generalized constrained equation of motion (5.154)

$$\boldsymbol{\Theta}^T \boldsymbol{\tau} = \boldsymbol{M}\ddot{\boldsymbol{q}} + \boldsymbol{b} + \boldsymbol{g} - \boldsymbol{\Phi}^T(\boldsymbol{\rho} + \boldsymbol{\alpha}). \tag{8.73}$$

We can relate a set of task coordinates, $\mathbf{x}$, to the set of generalized coordinates, $\boldsymbol{q}$, by

$$\dot{\mathbf{x}} = \boldsymbol{J}\dot{\boldsymbol{q}}. \tag{8.74}$$

Mapping (8.73) into any appropriate task space via the dynamically consistent inverse of $\boldsymbol{J}$ yields

$$\bar{\boldsymbol{J}}^T \boldsymbol{\Theta}^T \boldsymbol{\tau} = \boldsymbol{\Lambda}(\boldsymbol{q})\ddot{\mathbf{x}} + \boldsymbol{\mu}(\boldsymbol{q}, \dot{\boldsymbol{q}}) + \boldsymbol{p}(\boldsymbol{q}) + \boldsymbol{\gamma}(\boldsymbol{q}), \tag{8.75}$$

where

$$\boldsymbol{\Lambda}(\boldsymbol{q}) = (\boldsymbol{J}\boldsymbol{M}^{-1}\boldsymbol{J}^T)^{-1}, \tag{8.76}$$

$$\boldsymbol{\mu}(\boldsymbol{q}, \dot{\boldsymbol{q}}) = \bar{\boldsymbol{J}}^T \boldsymbol{b} - \boldsymbol{\Lambda}\dot{\boldsymbol{J}}\dot{\boldsymbol{q}}, \tag{8.77}$$

$$\boldsymbol{p}(\boldsymbol{q}) = \bar{\boldsymbol{J}}^T \boldsymbol{g}, \tag{8.78}$$

$$\boldsymbol{\gamma} = -\bar{\boldsymbol{J}}^T \boldsymbol{\Phi}^T(\boldsymbol{p} + \boldsymbol{\alpha}), \tag{8.79}$$

$$\bar{\boldsymbol{J}}^T = \boldsymbol{\Lambda}\boldsymbol{J}\boldsymbol{M}^{-1}. \tag{8.80}$$

The procedure applied to (5.154) can also be applied to (5.302). In this case, however, the task space mapping is associated with the minimal set of generalized coordinates, $\boldsymbol{q}_p$. We begin by recalling (5.302):

$$\boldsymbol{\tau}_p = \boldsymbol{M}_p(\boldsymbol{q})\ddot{\boldsymbol{q}}_p + \boldsymbol{b}_p(\boldsymbol{q}, \dot{\boldsymbol{q}}, \dot{\boldsymbol{q}}_p) + \boldsymbol{g}_p(\boldsymbol{q}). \tag{8.81}$$

We can relate a set of task coordinates, $\mathbf{x}$, to the coordinates, $\boldsymbol{q}_p$, by

$$\dot{\mathbf{x}} = \boldsymbol{J}\dot{\boldsymbol{q}} = \boldsymbol{J}_p(\boldsymbol{q})\dot{\boldsymbol{q}}_p, \tag{8.82}$$

where

$$\boldsymbol{J}_p(\boldsymbol{q}) = \boldsymbol{J}\boldsymbol{W} \tag{8.83}$$

is the task Jacobian with respect to the minimal set of coordinates. We can map (8.81) into any appropriate task space via the dynamically consistent inverse of $\boldsymbol{J}_p$. This yields (Russakow et al. 1995)

$$\boldsymbol{f} = \boldsymbol{\Lambda}(\boldsymbol{q})\ddot{\mathbf{x}} + \boldsymbol{\mu}(\boldsymbol{q}, \dot{\boldsymbol{q}}, \dot{\boldsymbol{q}}_p) + \boldsymbol{p}(\boldsymbol{q}), \tag{8.84}$$

where

$$\boldsymbol{\Lambda}(\boldsymbol{q}) = (\boldsymbol{J}_p\boldsymbol{M}_p^{-1}\boldsymbol{J}_p^T)^{-1}, \tag{8.85}$$

$$\boldsymbol{\mu}(\boldsymbol{q}, \dot{\boldsymbol{q}}, \dot{\boldsymbol{q}}_p) = \bar{\boldsymbol{J}}_p^T \boldsymbol{b}_p - \boldsymbol{\Lambda}\dot{\boldsymbol{J}}_p\dot{\boldsymbol{q}}_p, \tag{8.86}$$

$$\boldsymbol{p}(\boldsymbol{q}) = \bar{\boldsymbol{J}}_p^T \boldsymbol{g}_p, \tag{8.87}$$

$$\bar{\boldsymbol{J}}_p^T = \boldsymbol{\Lambda}\boldsymbol{J}_p\boldsymbol{M}_p^{-1}. \tag{8.88}$$

We can express (8.84) in terms of the generalized forces:

$$\bar{J}_p^T \tau_p = \bar{J}_p^T W^T \tau = \Lambda \ddot{x} + \mu + p. \tag{8.89}$$

8.2.2 Task/Constraint Partitioning of Dynamics

In this formulation the multiplier form of the constrained equation of motion (5.86) is mapped into task space using the dynamically consistent inverse of a Jacobian that characterizes both task and constraints (De Sapio and Khatib 2005; De Sapio, Khatib, and Delp 2006). The resulting task space equation is then partitioned into an equation corresponding to task motion and an equation corresponding to constraint forces.

We begin by recalling the multiplier form of the constrained equation of motion (5.86):

$$\tau = M\ddot{q} + b + g - \Phi^T \lambda. \tag{8.90}$$

Again, we can relate a set of task coordinates, x, to the set of generalized coordinates, q, by

$$\dot{x} = J\dot{q}, \tag{8.91}$$

in addition to the constraint condition

$$\dot{\phi} = \Phi \dot{q} = 0. \tag{8.92}$$

We can concatenate (8.91) and (8.92) into a single vector:

$$\dot{\check{x}} = \begin{pmatrix} \dot{x} \\ \dot{\phi} \end{pmatrix} = \begin{pmatrix} J \\ \Phi \end{pmatrix} \dot{q} = \check{J}\dot{q}, \tag{8.93}$$

where we use the notation $\check{\square}$ to represent a quantity that is formed from the composition of task and constraint terms. The applied generalized force can be decomposed into a task space component and null space component as in (8.9):

$$\tau = \check{J}^T \check{f} + \check{N}^T \tau_o = \begin{pmatrix} J^T & \Phi^T \end{pmatrix} \begin{pmatrix} f \\ f_c \end{pmatrix} + \check{N}^T \tau_o, \tag{8.94}$$

where f_c is the component of the applied task space force acting along the constraint direction. Equation (8.90) can thus be written as

$$\check{J}^T \begin{pmatrix} f \\ f_c \end{pmatrix} + \check{N}^T \tau_o = M\ddot{q} + b + g - \check{J}^T \begin{pmatrix} 0 \\ \lambda \end{pmatrix}. \tag{8.95}$$

We can now map (8.95) into task space by premultiplying (8.95) by the dynamically consistent inverse, $\bar{\check{J}}^T$, of $\check{J}^T$. This yields

$$\begin{pmatrix} f \\ f_c \end{pmatrix} = \check{\Lambda}(q) \begin{pmatrix} \ddot{x} \\ 0 \end{pmatrix} + \check{\mu}(q, \dot{q}) + \check{p}(q) - \begin{pmatrix} 0 \\ \lambda \end{pmatrix}, \tag{8.96}$$

where the constraint condition, $\ddot{\boldsymbol{\phi}} = \mathbf{0}$, has been imposed. We have the following identities:

$$\check{\boldsymbol{\Lambda}}(\boldsymbol{q}) = \begin{pmatrix} \boldsymbol{J}\boldsymbol{M}^{-1}\boldsymbol{J}^T & \boldsymbol{J}\boldsymbol{M}^{-1}\boldsymbol{\Phi}^T \\ \boldsymbol{\Phi}\boldsymbol{M}^{-1}\boldsymbol{J}^T & \boldsymbol{\Phi}\boldsymbol{M}^{-1}\boldsymbol{\Phi}^T \end{pmatrix}^{-1}, \tag{8.97}$$

$$\check{\boldsymbol{\mu}}(\boldsymbol{q}, \dot{\boldsymbol{q}}) = \check{\boldsymbol{\Lambda}} \begin{pmatrix} \boldsymbol{J}\boldsymbol{M}^{-1}\boldsymbol{b} - \dot{\boldsymbol{J}}\dot{\boldsymbol{q}} \\ \boldsymbol{\Phi}\boldsymbol{M}^{-1}\boldsymbol{b} - \dot{\boldsymbol{\Phi}}\dot{\boldsymbol{q}} \end{pmatrix}, \tag{8.98}$$

$$\check{\boldsymbol{p}}(\boldsymbol{q}) = \check{\boldsymbol{\Lambda}} \begin{pmatrix} \boldsymbol{J}\boldsymbol{M}^{-1}\boldsymbol{g} \\ \boldsymbol{\Phi}\boldsymbol{M}^{-1}\boldsymbol{g} \end{pmatrix}. \tag{8.99}$$

While (8.96) expresses the combined task-constraint dynamics of our system, it is useful to partition the dynamics in the following manner:

$$\begin{pmatrix} \boldsymbol{f} \\ \boldsymbol{f}_C \end{pmatrix} = \begin{pmatrix} \check{\boldsymbol{\Lambda}}_{11} & \check{\boldsymbol{\Lambda}}_{12} \\ \check{\boldsymbol{\Lambda}}_{21} & \check{\boldsymbol{\Lambda}}_{22} \end{pmatrix} \begin{pmatrix} \ddot{\mathbf{x}} \\ \mathbf{0} \end{pmatrix} + \begin{pmatrix} \check{\boldsymbol{\mu}}_1 \\ \check{\boldsymbol{\mu}}_2 \end{pmatrix} + \begin{pmatrix} \check{\boldsymbol{p}}_1 \\ \check{\boldsymbol{p}}_2 \end{pmatrix} - \begin{pmatrix} \mathbf{0} \\ \boldsymbol{\lambda} \end{pmatrix}. \tag{8.100}$$

From this partitioning we have an equation corresponding to task motion,

$$\boldsymbol{f} = \check{\boldsymbol{\Lambda}}_{11}\ddot{\mathbf{x}} + \check{\boldsymbol{\mu}}_1 + \check{\boldsymbol{p}}_1, \tag{8.101}$$

and an equation corresponding to the constraint forces,

$$\boldsymbol{f}_C + \boldsymbol{\lambda} = \check{\boldsymbol{\Lambda}}_{21}\ddot{\mathbf{x}} + \check{\boldsymbol{\mu}}_2 + \check{\boldsymbol{p}}_2. \tag{8.102}$$

The constraint force vector, $\boldsymbol{\lambda}$, will always arise so as to satisfy (8.102), as dictated by constraint consistency. The component of applied task space force, $\boldsymbol{f}_C$, acting along the constraint direction has no impact on the motion of the task. Its only effect is on the constraint forces (values of $\boldsymbol{\lambda}$) that arise. Thus, all choices for $\boldsymbol{f}_C$ result in equivalent motion of the system (Huston et al. 2003).

Example: We consider the parallel mechanism of Section 5.2.5 (see Figure 8.9). We can use the configuration space properties already computed to formulate the task space dynamics. We will specify the task to be the center point, $\boldsymbol{r}_{G_p}$, of the platform

$$\mathbf{x} = {}^0\boldsymbol{r}_{G_p} = \begin{pmatrix} q_7 & q_8 \end{pmatrix}^T. \tag{8.103}$$

The task Jacobian is

$$\boldsymbol{J}(\boldsymbol{q}) = \frac{\partial \mathbf{x}}{\partial \boldsymbol{q}} = \begin{pmatrix} 0 & 0 & 0 & 0 & 0 & 0 & 1 & 0 & 0 \\ 0 & 0 & 0 & 0 & 0 & 0 & 0 & 1 & 0 \end{pmatrix}. \tag{8.104}$$

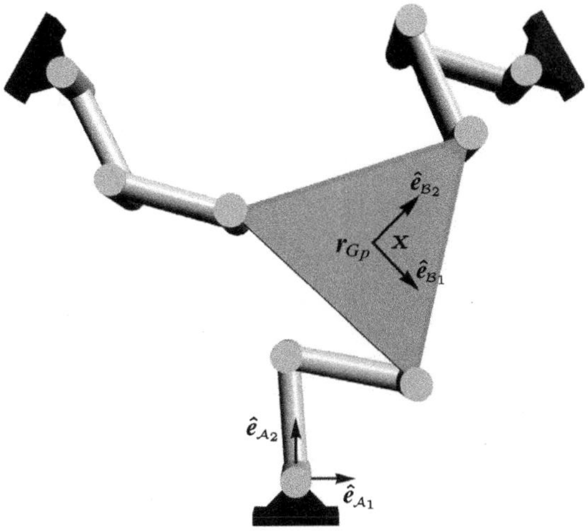

Figure 8.9 Parallel mechanism consisting of serial chains with loop closures. The three elbow joints are actuated while the remaining joints are passive. The task point, $\mathbf{x}$, is shown.

Using $\boldsymbol{J}(\boldsymbol{q})$, as well as $\boldsymbol{\Phi}(\boldsymbol{q})$, $\boldsymbol{M}(\boldsymbol{q})$, $\boldsymbol{b}(\boldsymbol{q}, \dot{\boldsymbol{q}})$, and $\boldsymbol{g}(\boldsymbol{q})$, which were computed in the example of Section 5.2.5, we can compute the task/constraint space terms:

$$\check{\boldsymbol{\Lambda}}(\boldsymbol{q}) = \begin{pmatrix} \boldsymbol{J}\boldsymbol{M}^{-1}\boldsymbol{J}^T & \boldsymbol{J}\boldsymbol{M}^{-1}\boldsymbol{\Phi}^T \\ \boldsymbol{\Phi}\boldsymbol{M}^{-1}\boldsymbol{J}^T & \boldsymbol{\Phi}\boldsymbol{M}^{-1}\boldsymbol{\Phi}^T \end{pmatrix}^{-1}, \tag{8.105}$$

$$\check{\boldsymbol{\mu}}(\boldsymbol{q}, \dot{\boldsymbol{q}}) = \check{\boldsymbol{\Lambda}} \begin{pmatrix} \boldsymbol{J}\boldsymbol{M}^{-1}\boldsymbol{b} - \dot{\boldsymbol{J}}\dot{\boldsymbol{q}} \\ \boldsymbol{\Phi}\boldsymbol{M}^{-1}\boldsymbol{b} - \dot{\boldsymbol{\Phi}}\dot{\boldsymbol{q}} \end{pmatrix}, \tag{8.106}$$

$$\check{\boldsymbol{p}}(\boldsymbol{q}) = \check{\boldsymbol{\Lambda}} \begin{pmatrix} \boldsymbol{J}\boldsymbol{M}^{-1}\boldsymbol{g} \\ \boldsymbol{\Phi}\boldsymbol{M}^{-1}\boldsymbol{g} \end{pmatrix}. \tag{8.107}$$

As in the previous examples, since these terms involve matrix inverses, it is preferable to compute them numerically at each time step within the numerical integration loop rather then to compute them symbolically.

Our task/constraint space equation of motion for the dynamics at the task point is

$$\boldsymbol{f} = \check{\boldsymbol{\Lambda}}_{11}\ddot{\mathbf{x}} + \check{\boldsymbol{\mu}}_1 + \check{\boldsymbol{p}}_1, \tag{8.108}$$

and the equation corresponding to constraint forces in the system is

$$\boldsymbol{f}_c + \boldsymbol{\lambda} = \check{\boldsymbol{\Lambda}}_{21}\ddot{\mathbf{x}} + \check{\boldsymbol{\mu}}_2 + \check{\boldsymbol{p}}_2. \tag{8.109}$$

We will specify a task space force at the task point to cancel out the task/constraint space centrifugal and Coriolis force and the task/constraint space gravity force. That is,

$$\boldsymbol{f} = \check{\boldsymbol{\mu}}_1 + \check{\boldsymbol{p}}_1. \tag{8.110}$$

This will result in the following dynamics:

$$\check{\boldsymbol{\Lambda}}_{11}\ddot{\mathbf{x}} = \mathbf{0}. \tag{8.111}$$

Therefore, the task point will have *zero* acceleration. Given *zero* velocity initial conditions, the task point will not move. The system will, however, move in the null space.

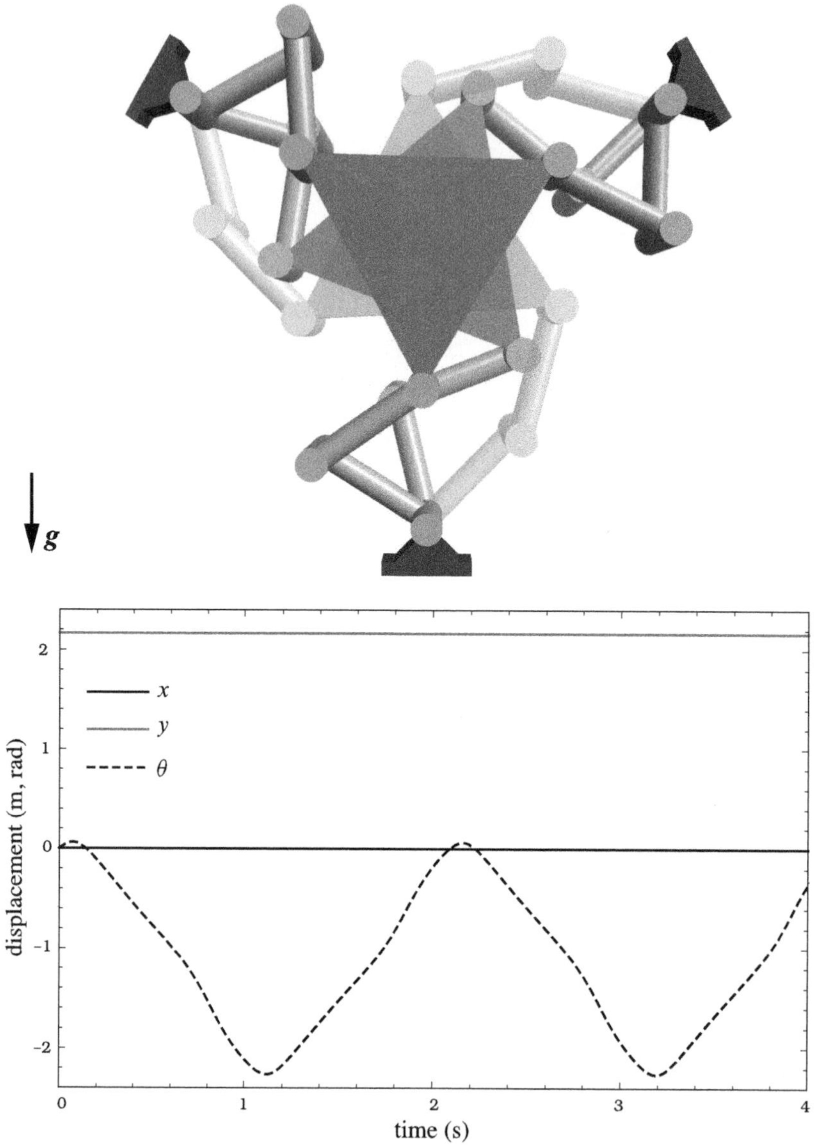

Figure 8.10 (Top) Animation frames from the simulation of a parallel mechanism. (Bottom) Time history of the platform motion.

Results from simulation of the system are shown in Figure 8.10. The initial conditions used were as follows:

$$\boldsymbol{q}_o = \begin{pmatrix} \pi/4 & \pi/6 & 11\pi/12 & \pi/6 & -5\pi/12 & \pi/6 & 0 & 2.165 & 0 \end{pmatrix}^T \tag{8.112}$$

$$\dot{\boldsymbol{q}}_o = \begin{pmatrix} 3.331 & -6.57 & 3.331 & -6.57 & 3.331 & -6.57 & 0 & 0 & \pi/2 \end{pmatrix}^T. \tag{8.113}$$

It can be verified that these satisfy the constraints.

8.3 Exercises

1. Consider the serial chain robot from Section 5.6, Exercise 1 (shown in Figure 8.11). The link lengths are $l_1 = 0.85$ and $l_2 = 0.5$. The link radii are $r_1 = 0.125$ and $r_2 = 0.0875$, the link masses are $M_1 = 1$ and $M_2 = 1.5$, and the link inertia tensors are

$$
{}^{i}\boldsymbol{I}_{i}^{G_i} = \begin{pmatrix} \frac{1}{2}M_i r_i^2 & 0 & 0 \\ 0 & \frac{1}{12}l_i^2 M_i + \frac{1}{4}M_i r_i^2 & 0 \\ 0 & 0 & \frac{1}{12}l_i^2 M_i + \frac{1}{4}M_i r_i^2 \end{pmatrix}
$$

for links $i = 1, 2$. The task, $\mathbf{x}$, is defined as the x and y components of the endpoint of link 2. Let

$$
\boldsymbol{q}_o = \begin{pmatrix} \pi/4 & \pi/3 \end{pmatrix}^T \quad \text{and} \quad \dot{\boldsymbol{q}}_o = \begin{pmatrix} 1 & 2 \end{pmatrix}^T.
$$

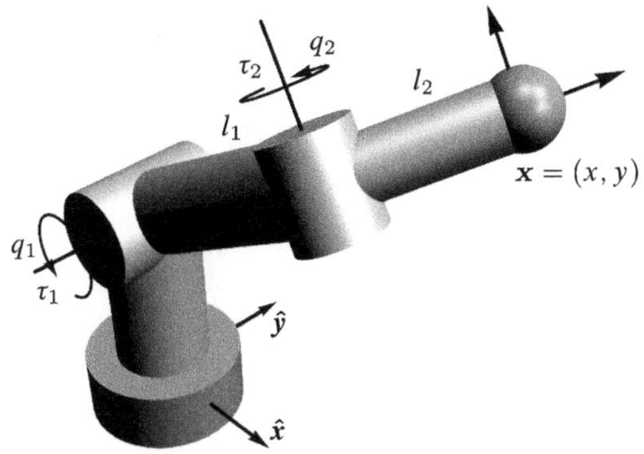

Figure 8.11 A 2 degree-of-freedom serial chain robot parameterized by the generalized coordinates, q_1 and q_2, with generalized forces, τ_1 and τ_2 (Exercise 1). The link lengths are $l_1 = 0.85$ and $l_2 = 0.5$. The centers of mass are at the geometric centers of the links. The link radii are $r_1 = 0.125$ and $r_2 = 0.0875$, the link masses are $M_1 = 1$ and $M_2 = 1.5$, and the link principal inertia components are ${}^{i}I_{i_{11}}^{G_i} = \frac{1}{2}M_i r_i^2$ and ${}^{i}I_{i_{22}}^{G_i} = {}^{i}I_{i_{33}}^{G_i} = \frac{1}{12}l_i^2 M_i + \frac{1}{4}M_i r_i^2$ for links $i = 1, 2$. The task, $\mathbf{x}$, is defined as the x and y components of the endpoint of link 2.

(a) If not already completed, compute the configuration space mass matrix, $\boldsymbol{M}(\boldsymbol{q})$, the vector of centrifugal and Coriolis forces, $\boldsymbol{b}(\boldsymbol{q}, \dot{\boldsymbol{q}})$, and the vector of gravity forces, $\boldsymbol{g}(\boldsymbol{q})$.

(b) Express the task vector, $\mathbf{x}(\boldsymbol{q})$, and the task Jacobian, $\boldsymbol{J}(\boldsymbol{q})$.

(c) Compute the task space mass matrix, $\boldsymbol{\Lambda}(\boldsymbol{q})$, at $\boldsymbol{q}_o$.

(d) Find the eigenvalues and eigenvectors of $\boldsymbol{\Lambda}(\boldsymbol{q})$ at $\boldsymbol{q}_o$ and specify the kinetic energy ellipsoid.

(e) Compute the task space vector of centrifugal and Coriolis forces, $\boldsymbol{\mu}(\boldsymbol{q}, \dot{\boldsymbol{q}})$, at $\boldsymbol{q}_o$ and $\dot{\boldsymbol{q}}_o$.

(f) Compute the task space vector of gravity forces, $\boldsymbol{p}(\boldsymbol{q})$, at $\boldsymbol{q}_o$.

2. Consider the two-link planar slider-crank mechanism from Section 5.6, Exercise 2 shown in Figure 8.12. The link lengths are each $l = 1$ and the link masses are each

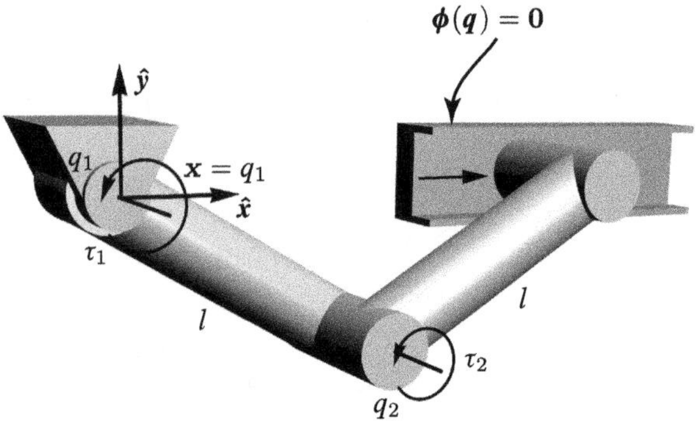

Figure 8.12 A two-link planar slider-crank mechanism (Exercise 2). The unconstrained system is parameterized by the generalized coordinates, q_1 and q_2, with generalized forces, τ_1 and τ_2. The link lengths are each $l = 1$ and the link masses are each $M = 1$. The centers of mass are at the geometric centers of the links. The link rotational inertias are taken to be *zero*. Link 2 is constrained from translating in the $\hat{y}$ direction. The task, $\mathbf{x}$, is defined as q_1.

$M = 1$. The link rotational inertias are taken to be *zero*. The task, $\mathbf{x}$, is defined as q_1. Let

$$\boldsymbol{q}_o = \begin{pmatrix} -\pi/6 & 2\pi/6 \end{pmatrix}^T \quad \text{and} \quad \dot{\boldsymbol{q}}_o = \begin{pmatrix} \sqrt{3}/2 & -\sqrt{3} \end{pmatrix}^T.$$

(a) If not already completed, compute the configuration space mass matrix, $\boldsymbol{M}(\boldsymbol{q})$, the vector of centrifugal and Coriolis forces, $\boldsymbol{b}(\boldsymbol{q}, \dot{\boldsymbol{q}})$, and the vector of gravity forces, $\boldsymbol{g}(\boldsymbol{q})$, for the unconstrained system.

(b) If not already completed, compute the constraint matrix, $\boldsymbol{\Phi}(\boldsymbol{q})$, for the loop closure.

(c) Express the task vector, $\mathbf{x}(\boldsymbol{q})$, and the task Jacobian, $\boldsymbol{J}(\boldsymbol{q})$.

(d) Compute the task space mass matrix, $\boldsymbol{\Lambda}(\boldsymbol{q})$, at $\boldsymbol{q}_o$ for the unconstrained system.

(e) Compute the task space vector of centrifugal and Coriolis forces, $\boldsymbol{\mu}(\boldsymbol{q}, \dot{\boldsymbol{q}})$, at $\boldsymbol{q}_o$ and $\dot{\boldsymbol{q}}_o$ for the unconstrained system.

(f) Compute the task space vector of gravity forces, $\boldsymbol{p}(\boldsymbol{q})$, at $\boldsymbol{q}_o$ for the the unconstrained system.

(g) Compute the task/constraint space mass matrix, $\check{\boldsymbol{\Lambda}}(\boldsymbol{q})$, at $\boldsymbol{q}_o$ for the constrained system.

(h) Compute the task/constraint space vector of centrifugal and Coriolis forces, $\check{\boldsymbol{\mu}}(\boldsymbol{q}, \dot{\boldsymbol{q}})$, at $\boldsymbol{q}_o$ and $\dot{\boldsymbol{q}}_o$ for the constrained system.

(i) Compute the task/constraint space vector of gravity forces, $\check{\boldsymbol{p}}(\boldsymbol{q})$, at $\boldsymbol{q}_o$ for the constrained system.

9 Applications to Biomechanical Systems

The biomechanics of motion provides a rich domain for the application of the principles of analytical dynamics. The skeletal systems of vertebrate animal species are well modeled as collections of interconnected rigid bodies (bones) with joints being parameterized by generalized coordinates. Additional complex coupling between the generalized coordinates/velocities can be accommodated with constraints – either holonomic constraints or nonholonomic rolling constraints. The musculoskeletal system comprises the skeletal system and a collection of musculotendon actuators that generate moments about the joints. In this chapter we will address dynamical models of the musculoskeletal and neuromuscular systems. The latter model describes how neural excitations result in contractile forces in the muscles, and the former model describes how musculotendon forces drive biomechanical motion.

9.1 Musculoskeletal and Neuromuscular Dynamics

The musculoskeletal model presented here involves a description of the skeletal system as a rigid multibody system spanned by a set of musculotendon actuators. The musculotendon actuator model consists of a standard two-state Hill-type model (Schutte 1992; Zajac 1993). A stiff tendon simplification of the two-state model will also be presented.

9.1.1 Musculoskeletal Dynamics

We begin by providing a description of the musculoskeletal system. A general branching chain multibody system is described by the standard system of n equations in configuration space

$$\boldsymbol{\tau} = \boldsymbol{M}(\boldsymbol{q})\ddot{\boldsymbol{q}} + \boldsymbol{b}(\boldsymbol{q}, \dot{\boldsymbol{q}}) + \boldsymbol{g}(\boldsymbol{q}). \tag{9.1}$$

Equation (9.1) can be used to model musculoskeletal dynamics, however, idealized torques at the joints, $\boldsymbol{\tau}$, do not represent the mechanism by which the musculoskeletal system is actuated.

We now consider a set of r musculotendon actuators spanning the skeletal system. We will assume that the vector of musculotendon lengths, $\boldsymbol{l} \in \mathbb{R}^r$, can be uniquely determined from the system configuration, $\boldsymbol{q}$. That is, $\boldsymbol{l} = \boldsymbol{l}(\boldsymbol{q})$. As a consequence of this assumption, differential variations in $\boldsymbol{l}$ are given by

$$\delta \boldsymbol{l} = \frac{\partial \boldsymbol{l}}{\partial \boldsymbol{q}} \delta \boldsymbol{q} = \boldsymbol{L}(\boldsymbol{q}) \delta \boldsymbol{q}, \tag{9.2}$$

where $\boldsymbol{L}(\boldsymbol{q}) \in \mathbb{R}^{r \times n}$ is the muscle Jacobian. From the principle of virtual work, we conclude that

$$\boldsymbol{\tau} = -\boldsymbol{L}^T \boldsymbol{f}_T = \boldsymbol{R}(\boldsymbol{q}) \boldsymbol{f}_T, \tag{9.3}$$

where $\boldsymbol{f}_T \in \mathbb{R}^r$ is the vector of tendon forces. The negative sign is due to the convention of taking contractile muscle forces as positive. The matrix of muscle moment arms is denoted $\boldsymbol{R}(\boldsymbol{q}) \in \mathbb{R}^{n \times r}$. From (9.1) and (9.3), the dynamics of the musculoskeletal system can be modeled as

$$\boldsymbol{R}(\boldsymbol{q}) \boldsymbol{f}_T = \boldsymbol{M}(\boldsymbol{q}) \ddot{\boldsymbol{q}} + \boldsymbol{b}(\boldsymbol{q}, \dot{\boldsymbol{q}}) + \boldsymbol{g}(\boldsymbol{q}). \tag{9.4}$$

9.1.2 Neuromuscular Dynamics

The behavior of the musculotendon actuators can be modeled as a set of Hill-type active state force-generating units (Zajac 1989, 1993). Activation dynamics refers to the process of muscle activation in response to neural excitation. This process can be modeled by the following equation of state written in terms of the vector of muscle activations, $\boldsymbol{a} \in \mathbb{R}^r$:

$$\dot{\boldsymbol{a}} = \begin{pmatrix} \tau(u_1, a_1) & 0 & 0 \\ 0 & \ddots & 0 \\ 0 & 0 & \tau(u_r, a_r) \end{pmatrix}^{-1} (\boldsymbol{u} - \boldsymbol{a}), \tag{9.5}$$

where $a_i \in [0, 1]$ and where $u_i \in [0, 1]$ is the neural input (excitation). The term $\tau(u_i, a_i)$ is a time constant given by

$$\tau(u_i, a_i) = \begin{cases} (\tau_a - \tau_d) u_i + \tau_d, & \text{if } u_i \geq a_i \\ \tau_d, & \text{if } u_i < a_i, \end{cases} \tag{9.6}$$

where τ_a and τ_d are the activation and deactivation time constants, respectively.

Contraction dynamics refers to the process of force generation in the muscle based on muscle contraction, rate of contraction, and activation. This process can be modeled as the lumped parameter system of Figure 9.1, which describes the configuration of forces. There is an active element, a passive viscoelastic element (in parallel), and an elastic tendon element (in series). The relative angle associated with the muscle fibers, α, is referred to as the pennation angle.

From Figure 9.1 we have the following relations:

$$\begin{aligned} \boldsymbol{l}(\boldsymbol{q}) &= \boldsymbol{l}_M \cos(\alpha) + \boldsymbol{l}_T \\ \dot{\boldsymbol{l}}(\boldsymbol{q}) &= \dot{\boldsymbol{l}}_M \cos(\alpha) + \dot{\boldsymbol{l}}_T, \end{aligned} \tag{9.7}$$

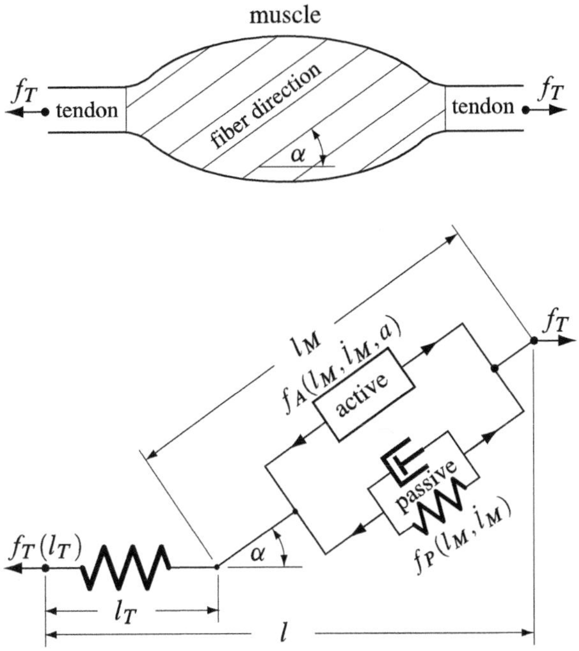

Figure 9.1 Active state musculotendon model. The active contractile element and passive viscoelastic element are in parallel. The passive elastic tendon element is in series.

where $\boldsymbol{l}_M$ is the vector of muscle lengths and $\boldsymbol{l}_T$ is the vector of tendon lengths. The following force equilibrium equation can be expressed:

$$\boldsymbol{f}_T = (\boldsymbol{f}_A + \boldsymbol{f}_P)\cos(\alpha) = \boldsymbol{f}_M \cos(\alpha), \tag{9.8}$$

where $\boldsymbol{f}_M$ and $\boldsymbol{f}_T$ are the vectors of forces in the muscles and tendons, respectively, and $\boldsymbol{f}_A$ and $\boldsymbol{f}_P$ are the vectors of active and passive forces in the muscles, respectively.

Using this force equilibrium equation as well as constitutive relationships describing muscle forces as a function of muscle length and contraction rate, $\boldsymbol{f}_A(\boldsymbol{l}_M, \dot{\boldsymbol{l}}_M, \boldsymbol{a})$ and $\boldsymbol{f}_P(\boldsymbol{l}_M, \dot{\boldsymbol{l}}_M)$, and tendon force as a function of tendon length, $\boldsymbol{f}_T(\boldsymbol{l}_T)$, we can express the following equation of state in functional form:

$$\dot{\boldsymbol{l}}_M = \dot{\boldsymbol{l}}_M(\boldsymbol{l}(\boldsymbol{q}), \dot{\boldsymbol{l}}(\boldsymbol{q}, \dot{\boldsymbol{q}}), \boldsymbol{l}_M, \boldsymbol{a}). \tag{9.9}$$

So, for a system of r musculotendon actuators, we can express the following system of $2r$ first-order state equations:

$$\dot{\boldsymbol{a}} = \begin{pmatrix} \tau(u_1, a_1) & 0 & 0 \\ 0 & \ddots & 0 \\ 0 & 0 & \tau(u_r, a_r) \end{pmatrix}^{-1} (\boldsymbol{u} - \boldsymbol{a}) \tag{9.10}$$

$$\dot{\boldsymbol{l}}_M = \dot{\boldsymbol{l}}_M(\boldsymbol{l}(\boldsymbol{q}), \dot{\boldsymbol{l}}(\boldsymbol{q}, \dot{\boldsymbol{q}}), \boldsymbol{l}_M, \boldsymbol{a}) \tag{9.11}$$

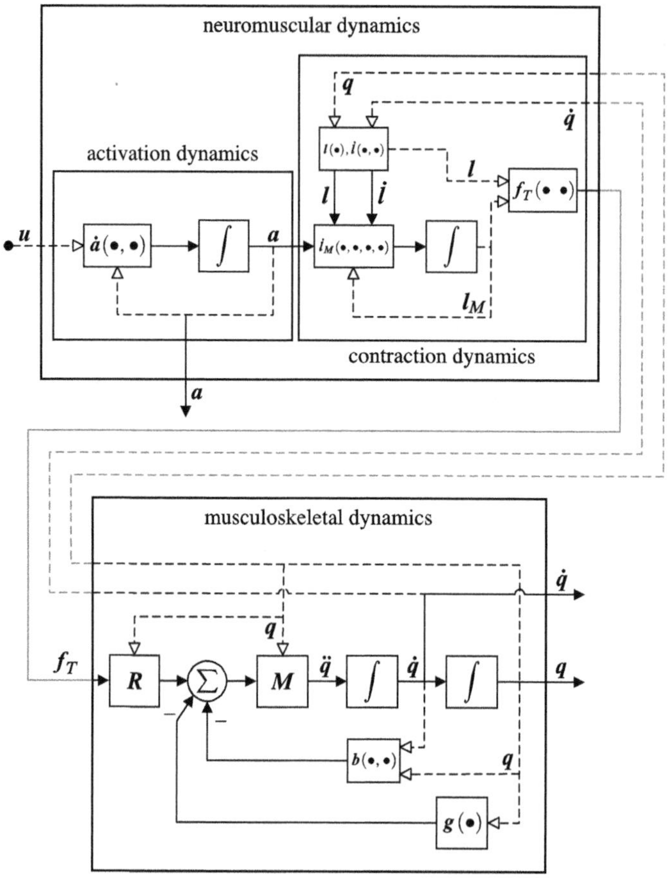

Figure 9.2 Neuromuscular and musculoskeletal system (feed-forward path). Neural excitations provide input to the activation dynamics. Output of the activation dynamics provides input to the contraction dynamics. Output of the contraction dynamics provides input to the musculoskeletal dynamics through the tendon forces.

where the internal states are $\boldsymbol{l}_M$ and $\boldsymbol{a}$. The tendon force can be written in terms of the states as

$$\boldsymbol{f}_T = \boldsymbol{f}_T(\boldsymbol{l}(\boldsymbol{q}), \boldsymbol{l}_M). \tag{9.12}$$

The musculoskeletal equation (9.4) and neuromuscular equations (9.10), (9.11), and (9.12) can be represented in block diagram form as depicted in Figure 9.2.

9.1.3 Stiff Tendon Model

If we make the assumption that the tendon is infinitely stiff, l_M is no longer an independent state but rather is algebraically related to the overall musculotendon length, $l(\boldsymbol{q})$. This provides a useful simplification to the full musculoskeletal dynamics described in the previous sections. We can define the muscle saturation force, f_S, as the active muscle force at full activation ($a = 1$). That is, $f_S(\boldsymbol{q}, \dot{\boldsymbol{q}}) \triangleq f_A(\boldsymbol{q}, \dot{\boldsymbol{q}}, 1)$. We can also define the muscle force-activation gain, K_f, as the magnitude of force generation in the muscle per change in unit activation. Thus,

$$K_f \triangleq \frac{\partial f_T}{\partial a} = f_S \cos\alpha. \tag{9.13}$$

For a system of muscles, the muscle force-activation gain matrix is then

$$\boldsymbol{K}_f = \begin{pmatrix} f_{S_1}\cos\alpha_1 & & \mathbf{0} \\ & \ddots & \\ \mathbf{0} & & f_{S_r}\cos\alpha_r \end{pmatrix}, \tag{9.14}$$

and the muscle forces can now be expressed in terms of this gain:

$$\boldsymbol{f_T}(\boldsymbol{q}, \dot{\boldsymbol{q}}, \boldsymbol{a}) = \boldsymbol{f_P}(\boldsymbol{q}, \dot{\boldsymbol{q}}) + \boldsymbol{K}_f(\boldsymbol{q}, \dot{\boldsymbol{q}})\boldsymbol{a}, \tag{9.15}$$

where

$$\boldsymbol{f_P} = \begin{pmatrix} f_{P_1}\cos\alpha_1 & \cdots & f_{P_r}\cos\alpha_r \end{pmatrix}^T. \tag{9.16}$$

As with the musculotendon forces, it is useful to express our generalized forces using a gain relationship. To this end, we will define the muscle torque-activation gain matrix:

$$\boldsymbol{K}_\tau \triangleq \boldsymbol{R}\boldsymbol{K}_f. \tag{9.17}$$

The generalized forces can now be expressed in terms of this gain:

$$\boldsymbol{\tau}(\boldsymbol{q}, \dot{\boldsymbol{q}}, \boldsymbol{a}) = \boldsymbol{\tau_P}(\boldsymbol{q}, \dot{\boldsymbol{q}}) + \boldsymbol{K}_\tau(\boldsymbol{q}, \dot{\boldsymbol{q}})\boldsymbol{a}, \tag{9.18}$$

where

$$\boldsymbol{\tau_P} = \boldsymbol{R}\boldsymbol{f_P}. \tag{9.19}$$

The overall neuromusculoskeletal dynamics can then be expressed as

$$\dot{\boldsymbol{a}} = \begin{pmatrix} \tau(u_1, a_1) & 0 & 0 \\ 0 & \ddots & 0 \\ 0 & 0 & \tau(u_r, a_r) \end{pmatrix}^{-1} (\boldsymbol{u} - \boldsymbol{a}) \tag{9.20}$$

$$\boldsymbol{K}_\tau(\boldsymbol{q}, \dot{\boldsymbol{q}})\boldsymbol{a} = \boldsymbol{M}(\boldsymbol{q})\ddot{\boldsymbol{q}} + \boldsymbol{b}(\boldsymbol{q}, \dot{\boldsymbol{q}}) + \boldsymbol{g}(\boldsymbol{q}) - \boldsymbol{\tau_P}(\boldsymbol{q}, \dot{\boldsymbol{q}}). \tag{9.21}$$

Example: Consider the $n = 3$ degree-of-freedom biomechanical model of Figure 9.3. The arm kinematics are shown in Table 9.1. The superscript t refers to the torso, h to the humerus, u to the ulna, r to the radius, and m to the hand frame of reference. The terms, ${}^t_h\boldsymbol{Q}$, ${}^t_u\boldsymbol{Q}$, ${}^t_r\boldsymbol{Q}$, and ${}^t_m\boldsymbol{Q}$ are the rotation matrices of the humerus, ulna, radius, and hand, respectively, in the torso frame. The terms ${}^t\boldsymbol{r}_h$, ${}^t\boldsymbol{r}_u$, ${}^t\boldsymbol{r}_r$, and ${}^t\boldsymbol{r}_m$ are the positions of the proximal ends of the humerus, ulna, radius, and hand, respectively, in the torso frame, and $\boldsymbol{d}_1$, $\boldsymbol{d}_2$, $\boldsymbol{d}_3$, and $\boldsymbol{d}_4$ are fixed translation vectors.

Table 9.1 Kinematics for a 3 degree-of-freedom human arm model

Translation	Rotation
Humerus	
${}^{t}\boldsymbol{r}_h = \boldsymbol{d}_1$	${}^{t}_{h}\boldsymbol{Q} = \boldsymbol{Q}_z(q_1)$
Ulna	
${}^{t}\boldsymbol{r}_u = {}^{t}\boldsymbol{r}_h + {}^{t}_{h}\boldsymbol{Q}\boldsymbol{d}_2$	${}^{t}_{u}\boldsymbol{Q} = {}^{t}_{h}\boldsymbol{Q}\boldsymbol{Q}_z(q_2)$
Radius	
${}^{t}\boldsymbol{r}_r = {}^{t}\boldsymbol{r}_u + {}^{t}_{u}\boldsymbol{Q}\boldsymbol{d}_3$	${}^{t}_{r}\boldsymbol{Q} = {}^{t}_{u}\boldsymbol{Q}$
Hand	
${}^{t}\boldsymbol{r}_m = {}^{t}\boldsymbol{r}_r + {}^{t}_{u}\boldsymbol{Q}\boldsymbol{d}_4$	${}^{t}_{m}\boldsymbol{Q} = {}^{t}_{u}\boldsymbol{Q}\boldsymbol{Q}_z(q_3)$

The positions of the centers of mass of the bodies are given by

$$ {}^{t}\boldsymbol{r}_{G_h} = {}^{t}\boldsymbol{r}_h + {}^{t}_{h}\boldsymbol{Q}\,{}^{h}\boldsymbol{d}_{\overrightarrow{hG_h}}, \tag{9.22} $$

$$ {}^{t}\boldsymbol{r}_{G_u} = {}^{t}\boldsymbol{r}_u + {}^{t}_{u}\boldsymbol{Q}\,{}^{u}\boldsymbol{d}_{\overrightarrow{uG_u}}, \tag{9.23} $$

$$ {}^{t}\boldsymbol{r}_{G_r} = {}^{t}\boldsymbol{r}_r + {}^{t}_{r}\boldsymbol{Q}\,{}^{r}\boldsymbol{d}_{\overrightarrow{rG_r}}, \tag{9.24} $$

$$ {}^{t}\boldsymbol{r}_{G_m} = {}^{t}\boldsymbol{r}_m + {}^{t}_{m}\boldsymbol{Q}\,{}^{m}\boldsymbol{d}_{\overrightarrow{mG_m}}. \tag{9.25} $$

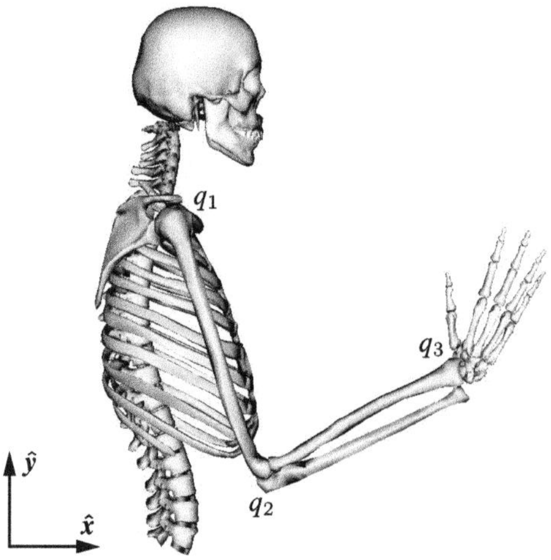

Figure 9.3 A simplified model of the human arm consisting of three generalized coordinates.

The translational velocity Jacobians are given by

$$^{t}\boldsymbol{\Gamma}_{G_h} = \frac{\partial\, ^{t}\boldsymbol{r}_{G_h}}{\partial \boldsymbol{q}}, \tag{9.26}$$

$$^{t}\boldsymbol{\Gamma}_{G_u} = \frac{\partial\, ^{t}\boldsymbol{r}_{G_u}}{\partial \boldsymbol{q}}, \tag{9.27}$$

$$^{t}\boldsymbol{\Gamma}_{G_r} = \frac{\partial\, ^{t}\boldsymbol{r}_{G_r}}{\partial \boldsymbol{q}}, \tag{9.28}$$

$$^{t}\boldsymbol{\Gamma}_{G_m} = \frac{\partial\, ^{t}\boldsymbol{r}_{G_m}}{\partial \boldsymbol{q}}. \tag{9.29}$$

For simplicity we will take the rotational inertias of the bodies to be *zero*. The dynamic terms are then

$$\boldsymbol{M}(\boldsymbol{q}) = M_h\, ^{t}\boldsymbol{\Gamma}^T_{G_h}\, ^{t}\boldsymbol{\Gamma}_{G_h} + M_u\, ^{t}\boldsymbol{\Gamma}^T_{G_u}\, ^{t}\boldsymbol{\Gamma}_{G_u} + M_r\, ^{t}\boldsymbol{\Gamma}^T_{G_r}\, ^{t}\boldsymbol{\Gamma}_{G_r} + M_m\, ^{t}\boldsymbol{\Gamma}^T_{G_m}\, ^{t}\boldsymbol{\Gamma}_{G_m}, \tag{9.30}$$

$$\boldsymbol{b}(\boldsymbol{q}, \dot{\boldsymbol{q}}) = \dot{\boldsymbol{M}}\dot{\boldsymbol{q}} - \frac{1}{2}\begin{pmatrix} \dot{\boldsymbol{q}}^T \frac{\partial \boldsymbol{M}}{\partial q_1}\dot{\boldsymbol{q}} \\ \dot{\boldsymbol{q}}^T \frac{\partial \boldsymbol{M}}{\partial q_2}\dot{\boldsymbol{q}} \\ \dot{\boldsymbol{q}}^T \frac{\partial \boldsymbol{M}}{\partial q_3}\dot{\boldsymbol{q}} \end{pmatrix}, \tag{9.31}$$

$$\boldsymbol{g}(\boldsymbol{q}) = (M_h\, ^{t}\boldsymbol{\Gamma}^T_{G_h} + M_u\, ^{t}\boldsymbol{\Gamma}^T_{G_u} + M_r\, ^{t}\boldsymbol{\Gamma}^T_{G_r} + M_m\, ^{t}\boldsymbol{\Gamma}^T_{G_m}) g\hat{\boldsymbol{e}}_{0_3}. \tag{9.32}$$

For the musculotendon units we will employ a stiff tendon model. We will also make two additional simplifications. We will take the pennation angle, α, and the passive force in the muscle, f_P , to be *zero*. Thus,

$$\boldsymbol{f}_T = \boldsymbol{f}_A = \underbrace{\begin{pmatrix} f_{S_1} & & \boldsymbol{0} \\ & \ddots & \\ \boldsymbol{0} & & f_{S_r} \end{pmatrix}}_{\boldsymbol{K}_f} \boldsymbol{a}. \tag{9.33}$$

The muscles for the arm are shown in Figure 9.4, and the muscle properties are summarized in Table 9.2. The muscle attachment and force-length data were taken from the study of Holzbaur, Murray, and Delp (2005). The saturation muscle force for a given muscle will be modeled as

$$f_S = f = f_o e^{-5\left(\frac{l_M - l_{M_o}}{l_{M_o}}\right)^2}\left[\operatorname{erf}\left(2\frac{\dot{l}_M}{v_{M_o}}\right) + 1\right], \tag{9.34}$$

where the terms f_o, l_{M_o}, and v_{M_o} are the maximum isometric force, optimal fiber length, and maximum muscle fiber contraction rate, respectively. It is noted that in the stiff tendon model,

$$l_M = l(\boldsymbol{q}) - l_{T_o} \quad \text{and} \quad \dot{l}_M = \dot{l}(\boldsymbol{q}, \dot{\boldsymbol{q}}), \tag{9.35}$$

where l_{T_o} is a constant tendon slack length.

Table 9.2 Maximum isometric forces, f_o, optimal fiber lengths, l_{M_o}, tendon slack lengths, l_{T_o}, and maximum contraction rate, v_{M_o}, for the 24 muscles used in the model of Figure 9.4

Muscle	Joints Spanned	f_o (N)	l_{M_o} (cm)	l_{T_o} (cm)	v_{M_o} (cm/s)	Input
DELT2	shoulder	1142.60	10.78	10.95	100	u_1
DELT3	shoulder	259.88	13.67	3.80	100	u_2
SUPSP	shoulder	487.82	6.82	3.95	100	u_3
INFSP	shoulder	1210.84	7.55	3.08	100	u_4
SUBSC	shoulder	1377.81	8.73	3.30	100	u_5
TMAJ	shoulder	425.39	16.24	2.0	100	u_6
PECM1	shoulder	364.41	14.42	0.28	100	u_7
PECM2	shoulder	515.41	13.85	8.90	100	u_8
CORB	shoulder	242.46	9.32	9.70	100	u_9
TLONG	shoulder, elbow	798.52	13.40	14.30	100	u_{10}
TLAT	elbow	624.30	11.38	9.80	100	u_{11}
TMED	elbow	624.30	11.38	9.08	100	u_{12}
BLONG	shoulder, elbow	624.30	11.57	27.23	100	u_{13}
BSHORT	shoulder, elbow	435.56	13.21	19.23	100	u_{14}
BRA	elbow	987.26	8.58	5.35	100	u_{15}
BRD	elbow	261.33	17.26	13.30	100	u_{16}
ECRL	elbow, wrist	304.89	8.10	22.40	100	u_{17}
ECRB	elbow, wrist	100.52	5.85	22.23	100	u_{18}
PT	elbow, wrist	566.22	4.92	9.80	100	u_{19}
EDCM	elbow, wrist	35.32	7.24	33.50	100	u_{20}
EIP	wrist	21.70	5.89	18.60	100	u_{21}
EPL	wrist	39.46	5.40	22.05	100	u_{22}
FDSM	wrist	91.03	7.49	29.50	100	u_{23}
FDPM	wrist	81.65	8.35	29.30	100	u_{24}

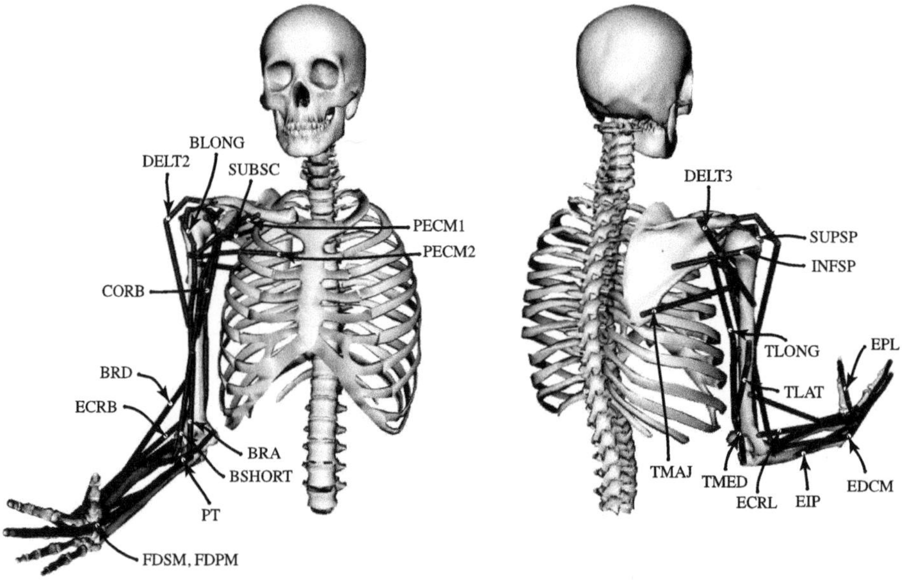

Figure 9.4 The simplified model of the human arm actuated by 24 muscles.

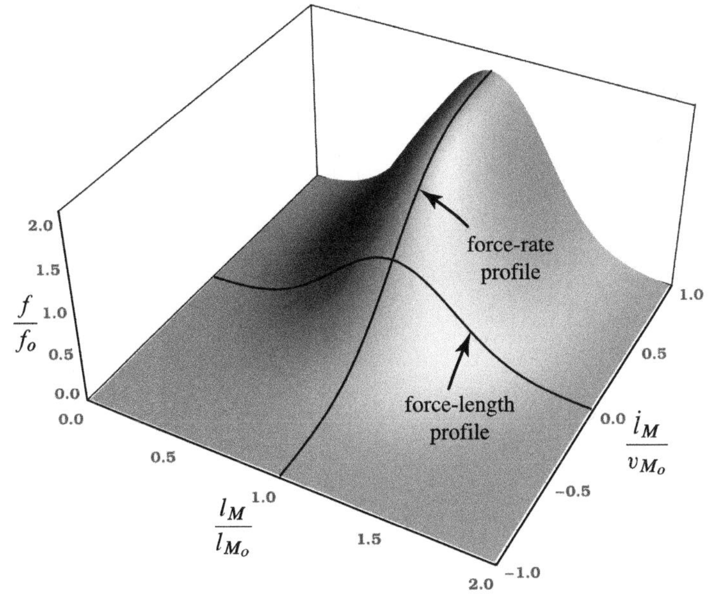

Figure 9.5 Muscle force-length-rate surface at full activation. The terms f_o, l_{M_o}, and v_{M_o} are the maximum isometric force, optimal fiber length, and maximum contraction rate, respectively. These are used as normalizing constants.

Normalizing (9.34) yields

$$\frac{f}{f_o} = e^{-5\left(\frac{l_M}{l_{M_o}}-1\right)^2}\left[\operatorname{erf}\left(2\frac{\dot{l}_M}{v_{M_o}}\right)+1\right]. \tag{9.36}$$

A plot of this normalized function is shown in Figure 9.5. Finally, the activation dynamics will be modeled as

$$\dot{\boldsymbol{a}} = \frac{1}{\tau}(\boldsymbol{u} - \boldsymbol{a}), \tag{9.37}$$

where an activation time constant, τ, of 20 ms is used. The overall neuromusculoskeletal system dynamics can then be expressed as

$$\dot{\boldsymbol{a}} = \frac{1}{\tau}(\boldsymbol{u} - \boldsymbol{a}) \tag{9.38}$$

$$\boldsymbol{K}_\tau(\boldsymbol{q}, \dot{\boldsymbol{q}})\boldsymbol{a} = \boldsymbol{M}(\boldsymbol{q})\ddot{\boldsymbol{q}} + \boldsymbol{b}(\boldsymbol{q}, \dot{\boldsymbol{q}}) + \boldsymbol{g}(\boldsymbol{q}). \tag{9.39}$$

We can now simulate the dynamics of this system. For our neural input we will use a predetermined time history, $\boldsymbol{u}(t)$, computed to produce a specific trajectory. Plots of the time histories of the hand trajectory, muscle excitations, and activations are shown in Figures 9.6 and 9.7.

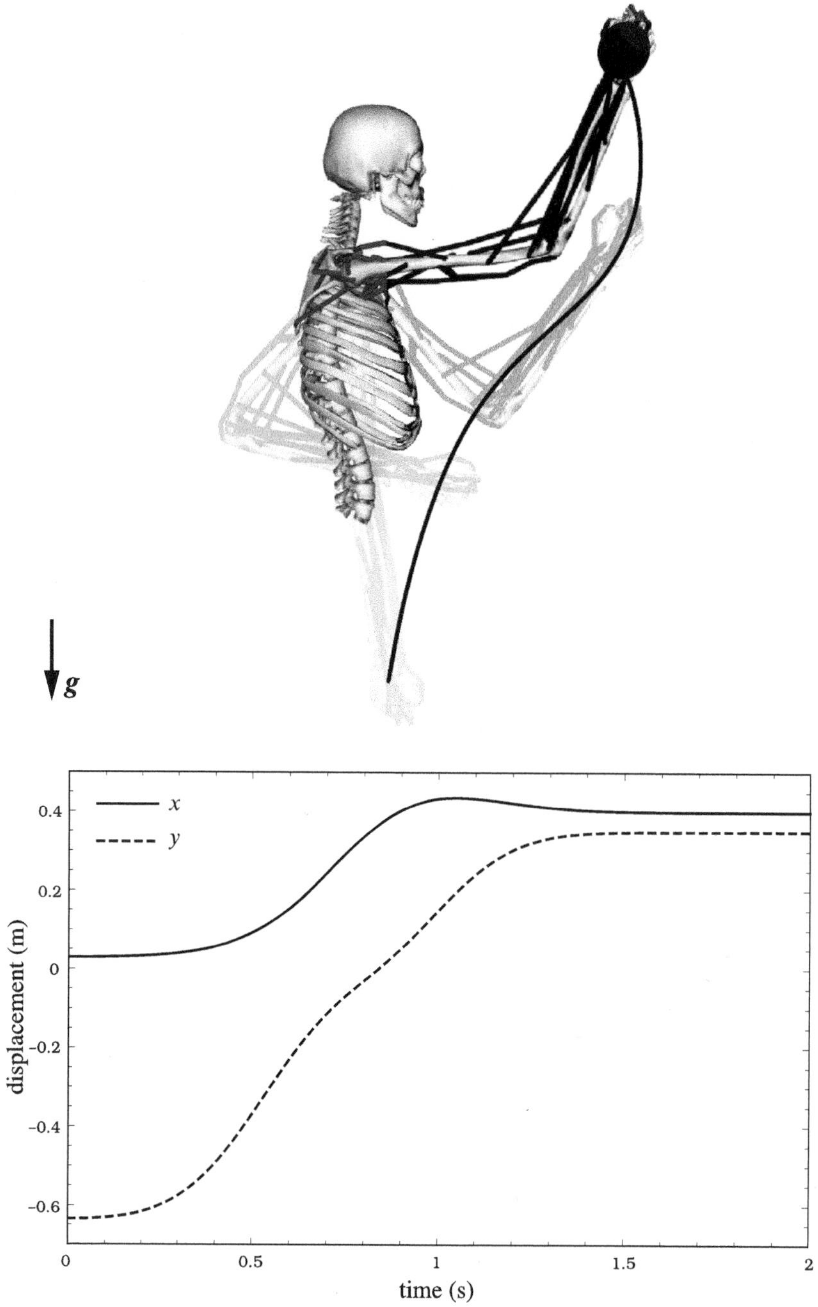

Figure 9.6 (Top) Animation frames from the simulation of a simplified model of the human arm actuated by 24 muscles. (Bottom) Time history of Cartesian coordinates of the hand, $\mathbf{x}$, following a specified trajectory.

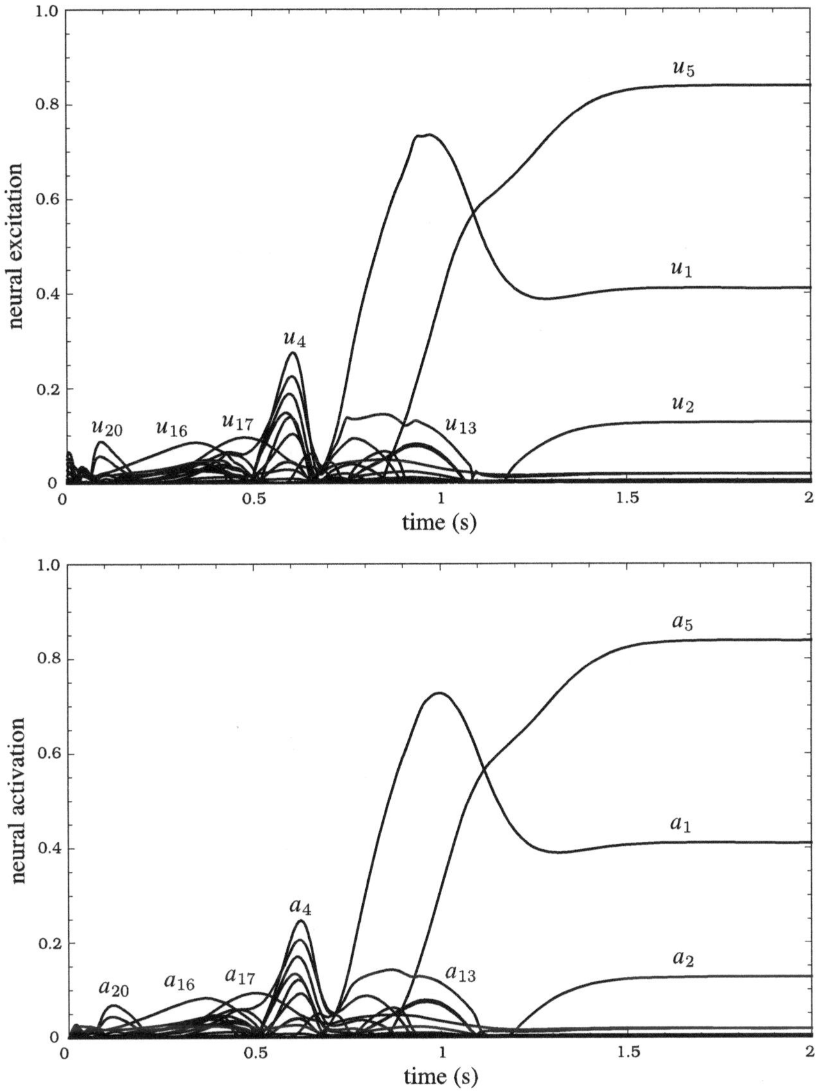

Figure 9.7 (Top) Time history of neural excitations, $\boldsymbol{u}$ (input). (Bottom) Time history of muscle activations, $\boldsymbol{a}$.

9.2 Constrained Dynamics of Biomechanical Systems

The biomechanical study of human motion as well as the design of anthropomorphic robotic mechanisms require faithful representations of human skeletal kinematics. In recent years, there has been a proliferation of humanoid robotic systems. Human skeletal kinematics has been modeled in these systems at a basic level, but some important aspects have been oversimplified or overlooked. While the representation of any skeletal joint as an ideal revolute or spherical joint is only an approximation, it is typically an acceptable one. Exceptions to this include the knee joint, which does not rotate about an absolute center but translates, as well, during knee extension (Delp et al. 1990). This added complexity, along with the presence of the patella (kneecap), has a significant influence on the generation of muscle moments about the knee. This is an important consideration if one wishes to simulate the human knee or emulate it in a humanoid robot that is to be driven by artificial muscles or cables. Thus, properly addressing the complexity of skeletal kinematics is important for both biomechanical simulations and anthropomorphic robot design.

9.2.1 Muscle-Based Actuation of Constrained Systems

In biomechanical simulations it is desirable to actuate the constrained biomechanical system using a set of musculotendon actuators. By using either a stiff tendon model or a steady state evaluation of the musculotendon forces, we can express $\boldsymbol{f}_T = \boldsymbol{f}_T(\boldsymbol{q}, \dot{\boldsymbol{q}}, \boldsymbol{a})$. In either case the joint moments induced by these musculotendon forces are

$$\boldsymbol{\tau} = \boldsymbol{R}(\boldsymbol{q})\boldsymbol{f}_T. \tag{9.40}$$

The constrained equation of motion can thus be expressed in terms of muscle actuation:

$$\boldsymbol{R}\boldsymbol{f}_T(\boldsymbol{q}, \dot{\boldsymbol{q}}, \boldsymbol{a}) + \boldsymbol{\Phi}^T\boldsymbol{\lambda} = \boldsymbol{M}\ddot{\boldsymbol{q}} + \boldsymbol{b} + \boldsymbol{g}. \tag{9.41}$$

Premultiplying (9.41) by $\bar{\boldsymbol{J}}^T$ gives the operational space form

$$\boldsymbol{T}(\boldsymbol{q})\boldsymbol{f}_T(\boldsymbol{q}, \dot{\boldsymbol{q}}, \boldsymbol{a}) + \begin{pmatrix} \boldsymbol{0} \\ \boldsymbol{\lambda} \end{pmatrix} = \check{\boldsymbol{\Lambda}} \begin{pmatrix} \ddot{\boldsymbol{x}} \\ \boldsymbol{0} \end{pmatrix} + \check{\boldsymbol{\mu}} + \check{\boldsymbol{p}}, \tag{9.42}$$

where $\boldsymbol{T}(\boldsymbol{q}) = \bar{\boldsymbol{J}}^T \boldsymbol{R} \in \mathbb{R}^{(m_T+m_C)\times r}$. Our task motion equation can then be expressed as

$$\boldsymbol{T}_1\boldsymbol{f}_T(\boldsymbol{q}, \dot{\boldsymbol{q}}, \boldsymbol{a}) = \check{\boldsymbol{\Lambda}}_{11}\ddot{\boldsymbol{x}} + \check{\boldsymbol{\mu}}_1 + \check{\boldsymbol{p}}_1, \tag{9.43}$$

and the equation corresponding to constraint forces can be expressed as

$$\boldsymbol{T}_2\boldsymbol{f}_T(\boldsymbol{q}, \dot{\boldsymbol{q}}, \boldsymbol{a}) + \boldsymbol{\lambda} = \check{\boldsymbol{\Lambda}}_{21}\ddot{\boldsymbol{x}} + \check{\boldsymbol{\mu}}_2 + \check{\boldsymbol{p}}_2, \tag{9.44}$$

where $\boldsymbol{T}_1$ is the $m_T \times r$ submatrix of $\boldsymbol{T}$ and $\boldsymbol{T}_2$ is the $m_C \times r$ submatrix of $\boldsymbol{T}$.

Example: Perhaps the most kinematically complicated system in the human skeletal system is the shoulder complex. While the purpose of the shoulder complex is to produce spherical articulation of the humerus, the resultant motion does not exclusively involve motion of the glenohumeral joint (see Figure 9.8).

The shoulder girdle, which comprises the clavicle and scapula, connects the glenohumeral joint to the torso and produces some of the motion associated with the overall articulation of the humerus. While this motion is small compared to the glenohumeral motion, its impact on overall arm function is significant (Klopčar and Lenarčič 2001;

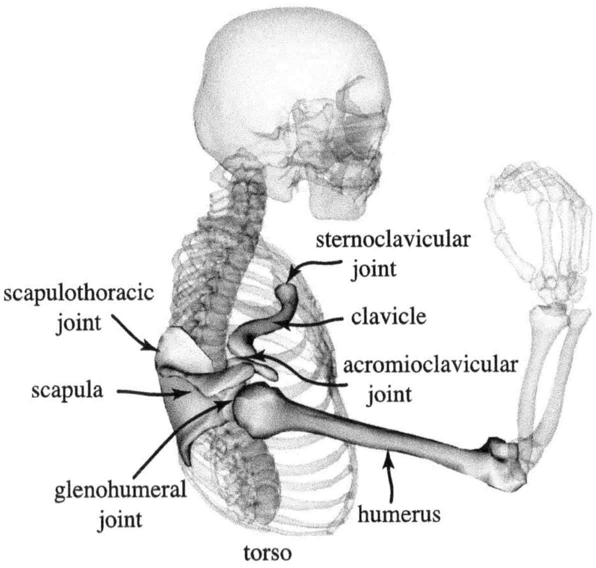

Figure 9.8 Various constituents of the shoulder complex including the scapula, clavicle, and humerus. The glenohumeral joint produces spherical motion of the humerus. The shoulder girdle attaches the glenohumeral joint to the torso and influences the resultant motion of the humerus through scapulothoracic, sternoclavicular, and acromioclavicular motion.

Lenarčič and Parenti-Castelli 2000). This impact is associated not only with the influence of the shoulder girdle on the skeletal kinematics of the shoulder complex but also with its influence on the routing and performance of muscles spanning the shoulder. As a consequence, shoulder kinematics is tightly coupled to the behavior of muscles spanning the shoulder. In turn, the action of these muscles (moments induced about the joints) influences the overall musculoskeletal dynamics of the shoulder. This coupling is illustrated in Figure 9.2. Muscle activation, $\boldsymbol{a}$, causes force generation in the muscles. Force generation, $\boldsymbol{f}_T$, is dependant on muscle length, contraction rate, and musculotendon length ($\boldsymbol{l}_M$, $\dot{\boldsymbol{l}}_M$, and $\boldsymbol{l}$), which in turn are dependant on skeletal configuration ($\boldsymbol{q}$ and $\dot{\boldsymbol{q}}$). The muscle-induced joint moments, $\boldsymbol{\tau}$, are dependent on these muscle forces as well as on muscle moment arms, $\boldsymbol{R}$ (which again are dependent on skeletal kinematics). Finally, these joint moments influence the multibody dynamics of the skeletal system.

We address the upper extremity model of Holzbaur et al. (2005) in this example. The model consists of a shoulder complex as well as a lower arm model. Holzbaur et al. implemented their model in the SIMM environment (Delp and Loan 2000) and, later, in OpenSim (Delp et al. 2007; Saul et al. 2015), where a minimal set of three generalized coordinates was chosen to describe the configuration of the shoulder complex. Since a minimal set of coordinates as employed in Holzbaur et al. (2005) the constraints that model the shoulder girdle are implicitly handled. Thus, all motions of the shoulder girdle are dependent on the three glenohumeral rotation coordinates. These are elevation plane, h_1, elevation angle, h_2, and shoulder rotation, h_3.

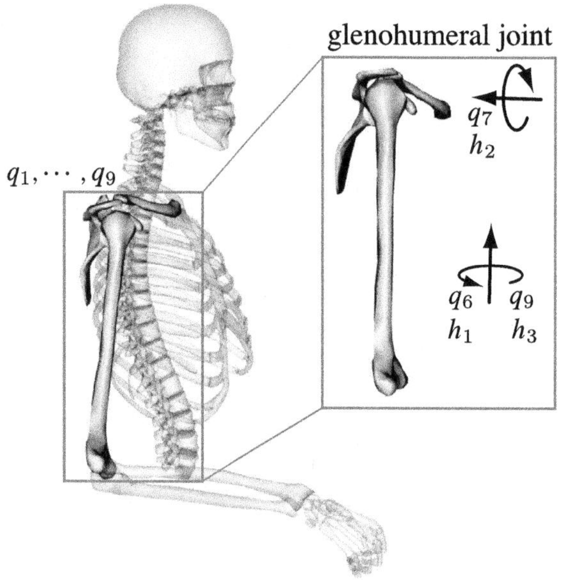

Figure 9.9 Parameterization of a shoulder model using a nonminimal set coordinates, q_6, q_7, and q_9, corresponding to the glenohumeral rotations, h_1, h_2, and h_3, in the model of Holzbaur et al. (2005). Five holonomic constraints couple the movement of the shoulder girdle with the glenohumeral rotations.

The constrained movement of the shoulder girdle was determined from the shoulder rhythm regression analysis of de Groot and Brand (2001). The model obtained from this regression analysis was shown to fit well for an independent set of shoulder motions and on a different set of subjects than was used for the regression analysis (de Groot and Brand 2001). The model of de Groot and Brand is considered to be superior in predicting shoulder motion to a simple unconstrained model that only reflects glenohumeral rotation.

For the purposes of formulating the dynamics it is often preferable to use a nonminimal, but standardized, set of generalized coordinates that are amenable to numerical formulation. Additionally, it is preferable to use a parameterization that preserves the physical meaning of the generalized forces as torques about individual joints. Often, when using a minimal set of coordinates, this is not the case, since a single generalized coordinate may influence multiple joint rotations, as in the parameterization of Holzbaur et al. (2005).

For these reasons we will reparameterize the model of Holzbaur et al. (2005) to include a total of $n = 9$ generalized coordinates to describe the unconstrained configuration of the shoulder (De Sapio, Holzbaur, and Khatib 2006). As shown in Figure 9.9, the coordinates q_6, q_7, and q_9 correspond to the independent coordinates for the shoulder complex used in Holzbaur et al. (2005): elevation plane, elevation angle, and shoulder rotation, respectively.

The shoulder kinematics associated with this nonminimal set of generalized coordinates is shown in Table 9.3. The superscript t refers to the torso, c to the clavicle, s to the

Table 9.3 Shoulder kinematics using a nonminimal set of coordinates

Translation	Rotation
Clavicle	
${}^{t}\boldsymbol{r}_c = \boldsymbol{d}_1$	${}^{t}_{c}\boldsymbol{Q} = \boldsymbol{Q}_1(q_1)\boldsymbol{Q}_2(q_2)$
Scapula	
${}^{t}\boldsymbol{r}_s = {}^{t}\boldsymbol{r}_c + {}^{t}_{c}\boldsymbol{Q}\boldsymbol{d}_2$	${}^{t}_{s}\boldsymbol{Q} = \boldsymbol{Q}_3(q_3)\boldsymbol{Q}_4(q_4)\boldsymbol{Q}_5(q_5)$
Humerus	
${}^{t}\boldsymbol{r}_h = {}^{t}\boldsymbol{r}_s + {}^{t}_{s}\boldsymbol{Q}\boldsymbol{d}_3$	${}^{t}_{h}\boldsymbol{Q} = \boldsymbol{Q}_6(q_6)\boldsymbol{Q}_7(q_7)\boldsymbol{Q}_6(q_8)\boldsymbol{Q}_6(q_9)$

scapula, and h to the humerus frame of reference. The terms ${}^{t}_{c}\boldsymbol{Q}$, ${}^{t}_{s}\boldsymbol{Q}$, and ${}^{t}_{h}\boldsymbol{Q}$ are the rotation matrices of the clavicle, scapula, and humerus, respectively, in the torso frame, and $\boldsymbol{Q}_1, \ldots, \boldsymbol{Q}_7$ are rotation matrices associated with spins about successive local coordinate axes, where the arguments identify the spin angles. The terms ${}^{t}\boldsymbol{r}_c$, ${}^{t}\boldsymbol{r}_s$, and ${}^{t}\boldsymbol{r}_h$ are the positions of the proximal ends of the clavicle, scapula, and humerus, respectively, in the torso frame, and $\boldsymbol{d}_1$, $\boldsymbol{d}_2$, and $\boldsymbol{d}_3$ are fixed translation vectors. The constraint constants, $\boldsymbol{b}$, associated with the dependency on humerus elevation plane, q_6, and $\boldsymbol{c}$, associated with the dependency on humerus elevation angle, q_7, were obtained from the regression analysis of de Groot and Brand (2001). They are

$$\boldsymbol{b} = \begin{pmatrix} 0.120 & -0.046 & 0.140 & -0.079 & -0.028 \end{pmatrix} \tag{9.45}$$

$$\boldsymbol{c} = \begin{pmatrix} -0.242 & 0.123 & -0.049 & 0.396 & 0.184 \end{pmatrix}. \tag{9.46}$$

Five holonomic constraints need to be imposed to properly constrain the motion of the shoulder girdle. With an additional constraint at the glenohumeral joint, we have a total of $m_C = 6$ constraints. This yields $p = n - m_C = 3$ degrees of kinematic freedom. Since this framework does not limit the dependent coordinates to functions of only a single independent coordinate, as in the case of the SIMM model (Holzbaur et al., 2005), we can implement the complete set of shoulder rhythm constraints (de Groot and Brand 2001) for our analysis. These constraint equations, $\boldsymbol{\phi}(\boldsymbol{q}) = \boldsymbol{0}$, are given by

$$\boldsymbol{\phi}(\boldsymbol{q}) = \begin{pmatrix} q_1 - b_1 q_6 - c_1 q_7 \\ q_2 - b_2 q_6 - c_2 q_7 \\ q_3 - b_3 q_6 - c_3 q_7 \\ q_4 - b_4 q_6 - c_4 q_7 \\ q_5 - b_5 q_6 - c_5 q_7 \\ q_8 + q_6 \end{pmatrix} = \boldsymbol{0}. \tag{9.47}$$

Taking the derivative yields

$$\dot{\boldsymbol{\phi}} = \begin{pmatrix} \dot{q}_1 - b_1\dot{q}_6 - c_1\dot{q}_7 \\ \dot{q}_2 - b_2\dot{q}_6 - c_2\dot{q}_7 \\ \dot{q}_3 - b_3\dot{q}_6 - c_3\dot{q}_7 \\ \dot{q}_4 - b_4\dot{q}_6 - c_4\dot{q}_7 \\ \dot{q}_5 - b_5\dot{q}_6 - c_5\dot{q}_7 \\ \dot{q}_8 + \dot{q}_6 \end{pmatrix} = \boldsymbol{\Phi}\dot{\boldsymbol{q}} = \mathbf{0}. \tag{9.48}$$

So,

$$\boldsymbol{\Phi}\dot{\boldsymbol{q}} = \begin{pmatrix} 1 & 0 & 0 & 0 & 0 & -b_1 & -c_1 & 0 & 0 \\ 0 & 1 & 0 & 0 & 0 & -b_2 & -c_2 & 0 & 0 \\ 0 & 0 & 1 & 0 & 0 & -b_3 & -c_3 & 0 & 0 \\ 0 & 0 & 0 & 1 & 0 & -b_4 & -c_4 & 0 & 0 \\ 0 & 0 & 0 & 0 & 1 & -b_5 & -c_5 & 0 & 0 \\ 0 & 0 & 0 & 0 & 0 & 1 & 0 & 1 & 0 \end{pmatrix} \begin{pmatrix} \dot{q}_1 \\ \vdots \\ \vdots \\ \dot{q}_9 \end{pmatrix} = \mathbf{0}, \tag{9.49}$$

where

$$\boldsymbol{\Phi} = \begin{pmatrix} 1 & 0 & 0 & 0 & 0 & -b_1 & -c_1 & 0 & 0 \\ 0 & 1 & 0 & 0 & 0 & -b_2 & -c_2 & 0 & 0 \\ 0 & 0 & 1 & 0 & 0 & -b_3 & -c_3 & 0 & 0 \\ 0 & 0 & 0 & 1 & 0 & -b_4 & -c_4 & 0 & 0 \\ 0 & 0 & 0 & 0 & 1 & -b_5 & -c_5 & 0 & 0 \\ 0 & 0 & 0 & 0 & 0 & 1 & 0 & 1 & 0 \end{pmatrix}. \tag{9.50}$$

The positions of the centers of mass of the bodies are given by

$${}^{t}\boldsymbol{r}_{G_c} = {}^{t}\boldsymbol{r}_{c} + {}^{t}_{c}\boldsymbol{Q}\,{}^{c}\boldsymbol{d}_{\overrightarrow{cG_c}}, \tag{9.51}$$

$${}^{t}\boldsymbol{r}_{G_s} = {}^{t}\boldsymbol{r}_{s} + {}^{t}_{s}\boldsymbol{Q}\,{}^{s}\boldsymbol{d}_{\overrightarrow{sG_s}}, \tag{9.52}$$

$${}^{t}\boldsymbol{r}_{G_h} = {}^{t}\boldsymbol{r}_{h} + {}^{t}_{h}\boldsymbol{Q}\,{}^{h}\boldsymbol{d}_{\overrightarrow{hG_h}}. \tag{9.53}$$

We can compute the angular velocities in the local frame by noting that

$${}^{c}\boldsymbol{\Omega}_{c} = {}^{t}_{c}\boldsymbol{Q}^{T}\,{}^{t}_{c}\dot{\boldsymbol{Q}}, \tag{9.54}$$

$${}^{s}\boldsymbol{\Omega}_{s} = {}^{t}_{s}\boldsymbol{Q}^{T}\,{}^{t}_{s}\dot{\boldsymbol{Q}}, \tag{9.55}$$

$${}^{h}\boldsymbol{\Omega}_{h} = {}^{t}_{h}\boldsymbol{Q}^{T}\,{}^{t}_{h}\dot{\boldsymbol{Q}}. \tag{9.56}$$

The translational velocity Jacobians are given by

$$^t\boldsymbol{\Gamma}_{G_c} = \frac{\partial^t \boldsymbol{r}_{G_c}}{\partial \boldsymbol{q}}, \tag{9.57}$$

$$^t\boldsymbol{\Gamma}_{G_s} = \frac{\partial^t \boldsymbol{r}_{G_s}}{\partial \boldsymbol{q}}, \tag{9.58}$$

$$^t\boldsymbol{\Gamma}_{G_h} = \frac{\partial^t \boldsymbol{r}_{G_h}}{\partial \boldsymbol{q}}, \tag{9.59}$$

and the angular velocity Jacobians are given by

$$^c\boldsymbol{\Pi}_c = \frac{\partial^c \boldsymbol{\omega}_c}{\partial \dot{\boldsymbol{q}}}, \tag{9.60}$$

$$^s\boldsymbol{\Pi}_s = \frac{\partial^s \boldsymbol{\omega}_s}{\partial \dot{\boldsymbol{q}}}, \tag{9.61}$$

$$^h\boldsymbol{\Pi}_h = \frac{\partial^h \boldsymbol{\omega}_h}{\partial \dot{\boldsymbol{q}}}. \tag{9.62}$$

The dynamic terms are then

$$\begin{aligned} \boldsymbol{M}(\boldsymbol{q}) &= M_c{}^t\boldsymbol{\Gamma}_{G_c}^T {}^t\boldsymbol{\Gamma}_{G_c} + M_s{}^t\boldsymbol{\Gamma}_{G_s}^T {}^t\boldsymbol{\Gamma}_{G_s} + M_h{}^t\boldsymbol{\Gamma}_{G_h}^T {}^t\boldsymbol{\Gamma}_{G_h} \\ &+ {}^c\boldsymbol{\Pi}_c^T {}^c\boldsymbol{I}_c^{G_c c}\boldsymbol{\Pi}_c + {}^s\boldsymbol{\Pi}_s^T {}^s\boldsymbol{I}_s^{G_s s}\boldsymbol{\Pi}_s + {}^h\boldsymbol{\Pi}_h^T {}^h\boldsymbol{I}_h^{G_h h}\boldsymbol{\Pi}_h, \end{aligned} \tag{9.63}$$

$$\boldsymbol{b}(\boldsymbol{q}, \dot{\boldsymbol{q}}) = \dot{\boldsymbol{M}}\dot{\boldsymbol{q}} - \frac{1}{2}\begin{pmatrix} \dot{\boldsymbol{q}}^T \frac{\partial \boldsymbol{M}}{\partial q_1}\dot{\boldsymbol{q}} \\ \vdots \\ \dot{\boldsymbol{q}}^T \frac{\partial \boldsymbol{M}}{\partial q_9}\dot{\boldsymbol{q}} \end{pmatrix}, \tag{9.64}$$

$$\boldsymbol{g}(\boldsymbol{q}) = (M_c{}^t\boldsymbol{\Gamma}_{G_c}^T + M_s{}^t\boldsymbol{\Gamma}_{G_s}^T + M_h{}^t\boldsymbol{\Gamma}_{G_h}^T) g\hat{\boldsymbol{e}}_{0_3}. \tag{9.65}$$

The constrained equation of motion is

$$\boldsymbol{\tau} + \boldsymbol{\Phi}^T\boldsymbol{\lambda} = \boldsymbol{M}\ddot{\boldsymbol{q}} + \boldsymbol{b} + \boldsymbol{g}, \tag{9.66}$$

or, expressed in terms of muscle actuation,

$$\boldsymbol{R}\boldsymbol{f}_T(\boldsymbol{q}, \dot{\boldsymbol{q}}, \boldsymbol{a}) + \boldsymbol{\Phi}^T\boldsymbol{\lambda} = \boldsymbol{M}\ddot{\boldsymbol{q}} + \boldsymbol{b} + \boldsymbol{g}. \tag{9.67}$$

Using the configuration space properties computed, we can formulate the task space dynamics. We will specify the task to be the orientation angles of the humerus (see Figure 9.9):

$$\mathbf{x} = \begin{pmatrix} q_6 & q_7 & q_9 \end{pmatrix}^T. \tag{9.68}$$

The task Jacobian is

$$\boldsymbol{J}(\boldsymbol{q}) = \frac{\partial \mathbf{x}}{\partial \boldsymbol{q}} = \begin{pmatrix} 0 & 0 & 0 & 0 & 0 & 1 & 0 & 0 & 0 \\ 0 & 0 & 0 & 0 & 0 & 0 & 1 & 0 & 0 \\ 0 & 0 & 0 & 0 & 0 & 0 & 0 & 0 & 1 \end{pmatrix}. \tag{9.69}$$

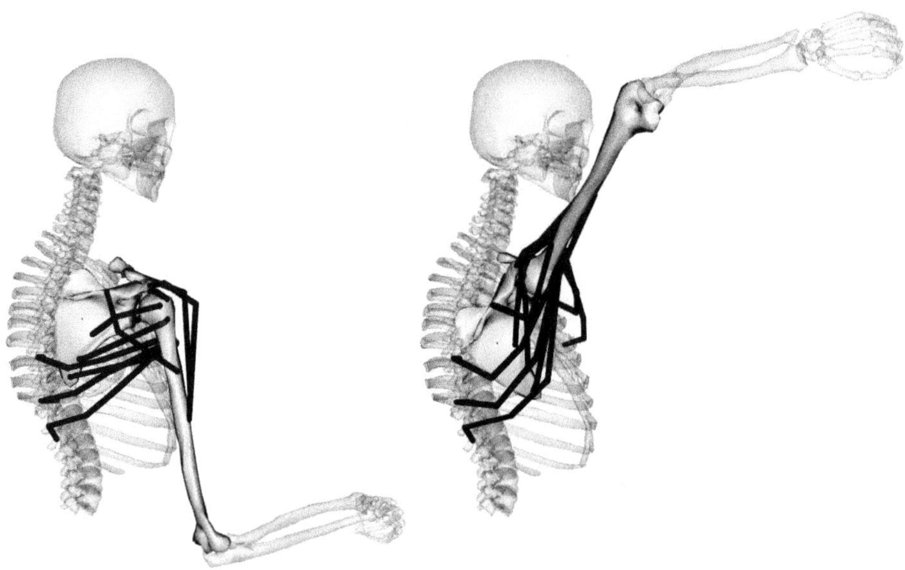

Figure 9.10 Muscle paths spanning the shoulder complex. A shoulder abduction sequence is shown. The movement of the scapula, a constituent of the shoulder girdle, can be observed. This movement influences the paths of the muscle-tendon units wrapping over the scapula. Since the muscle moment arms are determined from the muscle-tendon path data, the motion of the shoulder girdle influences the moment arms about the glenohumeral joint.

Using $\boldsymbol{J}(\boldsymbol{q})$, as well as $\boldsymbol{\Phi}(\boldsymbol{q})$, $\boldsymbol{M}(\boldsymbol{q})$, $\boldsymbol{b}(\boldsymbol{q}, \dot{\boldsymbol{q}})$, and $\boldsymbol{g}(\boldsymbol{q})$, we can compute the task/constraint space terms,

$$\check{\boldsymbol{\Lambda}}(\boldsymbol{q}) = \begin{pmatrix} \boldsymbol{J}\boldsymbol{M}^{-1}\boldsymbol{J}^T & \boldsymbol{J}\boldsymbol{M}^{-1}\boldsymbol{\Phi}^T \\ \boldsymbol{\Phi}\boldsymbol{M}^{-1}\boldsymbol{J}^T & \boldsymbol{\Phi}\boldsymbol{M}^{-1}\boldsymbol{\Phi}^T \end{pmatrix}^{-1}, \tag{9.70}$$

$$\check{\boldsymbol{\mu}}(\boldsymbol{q}, \dot{\boldsymbol{q}}) = \check{\boldsymbol{\Lambda}} \begin{pmatrix} \boldsymbol{J}\boldsymbol{M}^{-1}\boldsymbol{b} - \dot{\boldsymbol{J}}\dot{\boldsymbol{q}} \\ \boldsymbol{\Phi}\boldsymbol{M}^{-1}\boldsymbol{b} - \dot{\boldsymbol{\Phi}}\dot{\boldsymbol{q}} \end{pmatrix}, \tag{9.71}$$

$$\check{\boldsymbol{p}}(\boldsymbol{q}) = \check{\boldsymbol{\Lambda}} \begin{pmatrix} \boldsymbol{J}\boldsymbol{M}^{-1}\boldsymbol{g} \\ \boldsymbol{\Phi}\boldsymbol{M}^{-1}\boldsymbol{g} \end{pmatrix}. \tag{9.72}$$

As in the previous examples, since these terms involve matrix inverses, it is preferable to compute them numerically at each time step within the numerical integration loop rather than computing them symbolically.

Our task/constraint space equation of motion for the dynamics at the task point is

$$\boldsymbol{f} = \check{\boldsymbol{\Lambda}}_{11}\ddot{\boldsymbol{x}} + \check{\boldsymbol{\mu}}_1 + \check{\boldsymbol{p}}_1, \tag{9.73}$$

and the equation corresponding to constraint forces in the system is

$$\boldsymbol{f}_c + \boldsymbol{\lambda} = \check{\boldsymbol{\Lambda}}_{21}\ddot{\boldsymbol{x}} + \check{\boldsymbol{\mu}}_2 + \check{\boldsymbol{p}}_2. \tag{9.74}$$

The muscle-actuated task motion equation is

$$\boldsymbol{T}_1\boldsymbol{f}_T(\boldsymbol{q}, \dot{\boldsymbol{q}}, \boldsymbol{a}) = \check{\boldsymbol{\Lambda}}_{11}\ddot{\boldsymbol{x}} + \check{\boldsymbol{\mu}}_1 + \check{\boldsymbol{p}}_1, \tag{9.75}$$

and the equation corresponding to constraint forces can be expressed as

$$T_2 f_T(q, \dot{q}, a) + \lambda = \check{\Lambda}_{21}\ddot{x} + \check{\mu}_2 + \check{p}_2. \tag{9.76}$$

The dynamics of the constrained shoulder model addressed here, which involves kinematic coupling between the humerus, scapula, and clavicle, differs from a simple unconstrained shoulder model. The constrained model also differs from a simple model in the degree to which the system of muscles is able to generate control forces for a given motion task. This is due to the influence of the constrained motion between the humerus, scapula, and clavicle on the muscle forces and muscle moment arms about the glenohumeral joint (see Figure 9.10).

An example of this is shown in Figures 9.11 and 9.12. Predicted muscle moment arms, muscle forces, and moment-generating capacities for the deltoid muscles are compared for the simple and constrained shoulder models. The muscle path and force-length data were taken from the study of Holzbaur et al. (2005). In the constrained shoulder model the motions of the scapula and clavicle are highly coupled to humerus elevation angle (q_7 coordinate), whereas in the simple shoulder model the motions of the scapula and clavicle are not coupled to glenohumeral motion. The paths of the deltoid muscles are affected by the constrained motion of the humerus, scapula, and clavicle. This results in significant differences in moment arms predicted by the two models, with the constrained model often generating moment arms of substantially larger magnitude than the simple model.

Additionally, the predicted isometric muscle forces (computed at full activation) generated by the two models differ. The resulting moment-generating capacities of the constrained model are often substantially larger in magnitude than the simple model. This implies that the simple model, which excludes the constrained shoulder girdle motion, typically underestimates the moment-generating capacities of muscles that span the shoulder, since Holzbaur et al. (2005) demonstrated correlation between predicted and experimental moment-generating capacities for the constrained model. This is critical in various application areas involving the study and synthesis of human movement (De Sapio, Khatib, and Delp 2005).

Example: The purpose of the biomechanical human shoulder model discussed in the previous section is to simulate physiological shoulder motion and musculotendon routing. As such, it is ultimately intended to be actuated in a physiological manner, that is, by a system of musculotendon actuators that simulate skeletal muscle. For robotic applications, a mechanical analog of the human shoulder may be sought. If this mechanical analog is to be actuated by artificial muscles or cables, it is desirable to emulate human shoulder kinematics and muscle routing to reproduce the human-like action of muscles (moment arms) about the joints.

Standard robotic actuation may be adapted to a humanoid robotic shoulder complex rather than actuation that attempts to emulate musculoskeletal physiology. This offers the advantages of human shoulder kinematics without requiring complicated actuation. An example of such a system is the shoulder complex proposed by Lenarčič and

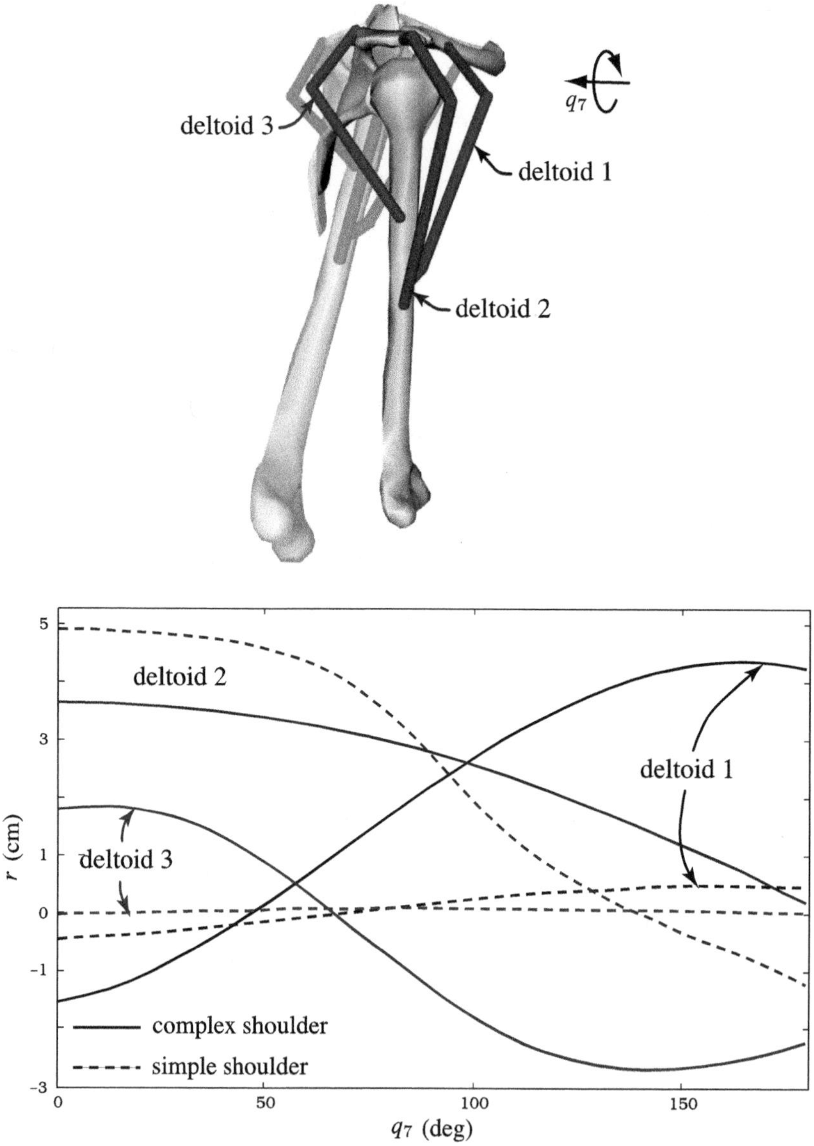

Figure 9.11 (Top) Muscle paths for the deltoid muscles. Elevation angle, q_7 is varied. (Bottom) Muscle moment arms for the deltoid muscles, as predicted by the constrained and simple shoulder models. The constrained model typically generates moment arms of substantially larger magnitude than those of the simple model.

colleagues (Lenarčič and Parenti-Castelli 2000; Lenarčič and Stanišić 2003). Their mechanism consists of a parallel-serial kinematic structure with four actuated prismatic joints (parallel part) and three actuated revolute joints (serial part).

The parallel part, consisting of four extensible legs attached to a moveable platform, acts as the shoulder girdle that supports the glenohumeral joint. As such, its

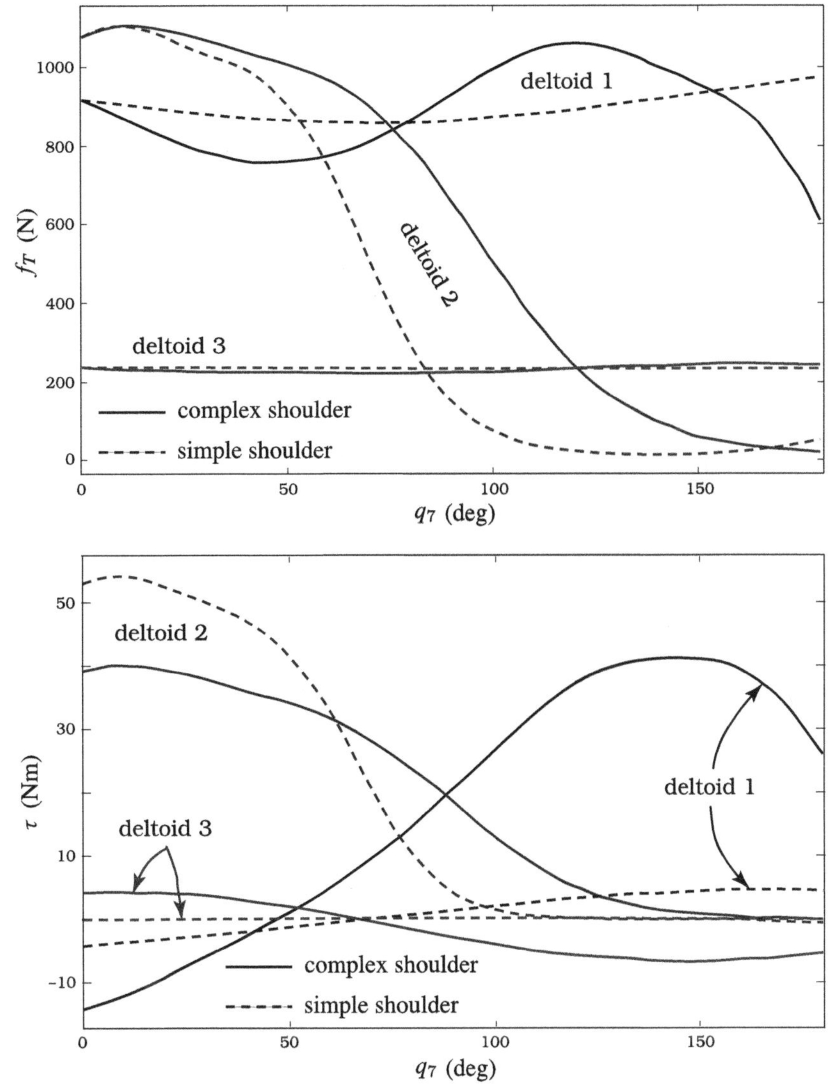

Figure 9.12 (Top) Muscle forces for the deltoid muscles, as predicted by the constrained and simple shoulder models. (Bottom) Moment-generating capacities for the deltoid muscles, as predicted by the constrained and simple shoulder models. The moment-generating capacities associated with the constrained model are typically larger in magnitude than those associated with the simple model.

design is intended to emulate the functionality of the scapula and clavicle attached to a fixed torso and connected by the scapulothoracic, sternoclavicular, and acromioclavicular joints. A parallel kinematic structure was chosen because of the need for high stiffness and precision in orienting the attached glenohumeral joint (Lenarčič and Parenti-Castelli 2000). The serial part consists of a spherical glenohumeral joint

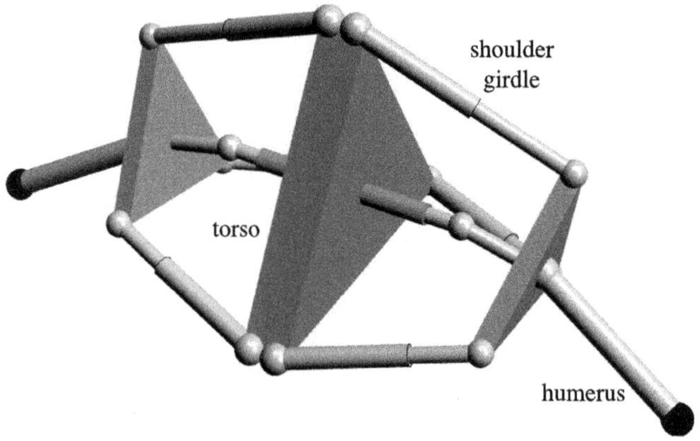

Figure 9.13 The parallel and serial parts composing the humanoid shoulder complex proposed by Lenarčič et al. For each shoulder a total of 16 generalized coordinates are employed ($n = 16$). These are constrained by 9 holonomic constraints ($m_C = 9$), yielding 7 degrees of freedom ($p = 7$).

attaching the humerus link to the shoulder girdle. A bilateral version of this is shown in Figure 9.13.

Lenarčič et al. presented detailed kinematic analyses of their shoulder complex design. For our purposes we will present a kinematic parameterization and constraint definition suitable for use in our constraint-based control framework. The system is partitioned into four serial chains with a total of 16 generalized coordinates defined as shown in Figure 9.14.

Holonomic loop constraints need to be imposed that reflect the connection of the extensible legs to the moveable platform. These constraints are expressed as

$$\boldsymbol{r}_{l_i} = \boldsymbol{r}_{p_i} \qquad \text{for } i = 1, \ldots, 3, \tag{9.77}$$

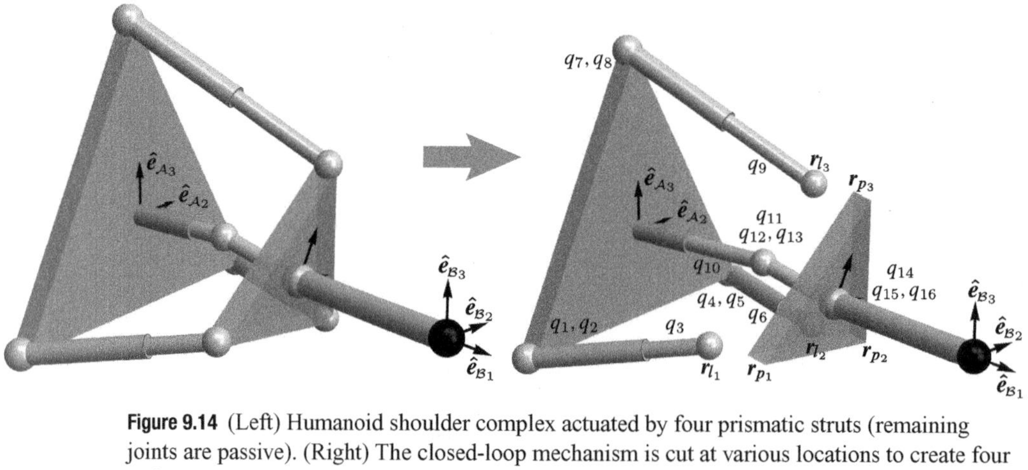

Figure 9.14 (Left) Humanoid shoulder complex actuated by four prismatic struts (remaining joints are passive). (Right) The closed-loop mechanism is cut at various locations to create four serial subsystems described by the set of generalized coordinates, $q_1, \ldots, q_{24}$.

where $\boldsymbol{r}_{l_i}$ is the terminal point of the ith strut subsystem that connects to $\boldsymbol{r}_{p_i}$, the ith position on the platform subsystem. In vector form we have

$$\boldsymbol{\phi}(\boldsymbol{q}) = \begin{pmatrix} \boldsymbol{r}_{l_1} - \boldsymbol{r}_{p_1} \\ \boldsymbol{r}_{l_2} - \boldsymbol{r}_{p_2} \\ \boldsymbol{r}_{l_3} - \boldsymbol{r}_{p_3} \end{pmatrix} = \boldsymbol{0}. \tag{9.78}$$

With these constraints the system possesses $p = n - m_C = 16 - 9 = 7$ degrees of freedom (4 for the parallel part and 3 for the serial part). Taking the derivative yields

$$\dot{\boldsymbol{\phi}} = \begin{pmatrix} \dot{\boldsymbol{r}}_{l_1} - \dot{\boldsymbol{r}}_{p_1} \\ \dot{\boldsymbol{r}}_{l_2} - \dot{\boldsymbol{r}}_{p_2} \\ \dot{\boldsymbol{r}}_{l_3} - \dot{\boldsymbol{r}}_{p_3} \end{pmatrix} = \boldsymbol{\Phi}\dot{\boldsymbol{q}} = \boldsymbol{0}, \tag{9.79}$$

where

$$\dot{\boldsymbol{r}}_{l_i} = \boldsymbol{\Gamma}_{l_i} \begin{pmatrix} \dot{q}_{3i-2} \\ \dot{q}_{3i-1} \\ \dot{q}_{3i} \end{pmatrix} \quad \text{and} \quad \dot{\boldsymbol{r}}_{p_i} = \boldsymbol{\Gamma}_{p_i} \begin{pmatrix} \dot{q}_{10} \\ \vdots \\ \dot{q}_{13} \end{pmatrix}, \quad \text{for } i = 1, \ldots, 3. \tag{9.80}$$

The terms $\boldsymbol{\Gamma}_{l_i}$ and $\boldsymbol{\Gamma}_{p_i}$ are the corresponding Jacobians of $\boldsymbol{r}_{l_i}$ and $\boldsymbol{r}_{p_i}$, respectively. So,

$$\boldsymbol{\Phi}\dot{\boldsymbol{q}} = \begin{pmatrix} \boldsymbol{\Gamma}_{l_1} & \boldsymbol{0} & \boldsymbol{0} & -\boldsymbol{\Gamma}_{p_1} & \boldsymbol{0} \\ \boldsymbol{0} & \boldsymbol{\Gamma}_{l_2} & \boldsymbol{0} & -\boldsymbol{\Gamma}_{p_2} & \boldsymbol{0} \\ \boldsymbol{0} & \boldsymbol{0} & \boldsymbol{\Gamma}_{l_3} & -\boldsymbol{\Gamma}_{p_3} & \boldsymbol{0} \end{pmatrix} \begin{pmatrix} \dot{q}_1 \\ \vdots \\ \dot{q}_{13} \\ \vdots \\ \dot{q}_{16} \end{pmatrix} = \boldsymbol{0}, \tag{9.81}$$

where

$$\boldsymbol{\Phi} = \begin{pmatrix} \boldsymbol{\Gamma}_{l_1} & \boldsymbol{0} & \boldsymbol{0} & -\boldsymbol{\Gamma}_{p_1} & \boldsymbol{0} \\ \boldsymbol{0} & \boldsymbol{\Gamma}_{l_2} & \boldsymbol{0} & -\boldsymbol{\Gamma}_{p_2} & \boldsymbol{0} \\ \boldsymbol{0} & \boldsymbol{0} & \boldsymbol{\Gamma}_{l_3} & -\boldsymbol{\Gamma}_{p_3} & \boldsymbol{0} \end{pmatrix}. \tag{9.82}$$

The constraint forces, $\boldsymbol{\lambda}$, are shown in Figure 9.15.

The unconstrained equations of motion for the three strut serial subsystems are as follows:

$$\begin{pmatrix} 0 \\ 0 \\ \tau_i \end{pmatrix} = \boldsymbol{M}_{l_i} \begin{pmatrix} \ddot{q}_{3i-2} \\ \ddot{q}_{3i-1} \\ \ddot{q}_{3i} \end{pmatrix} + \boldsymbol{b}_{l_i}(q_{3i-2}, \ldots, \dot{q}_{3i}) + \boldsymbol{g}_{l_i}(q_{3i-2}, q_{3i-1}, \ddot{q}_{3i}), \quad \text{for } i = 1, \ldots, 3, \tag{9.83}$$

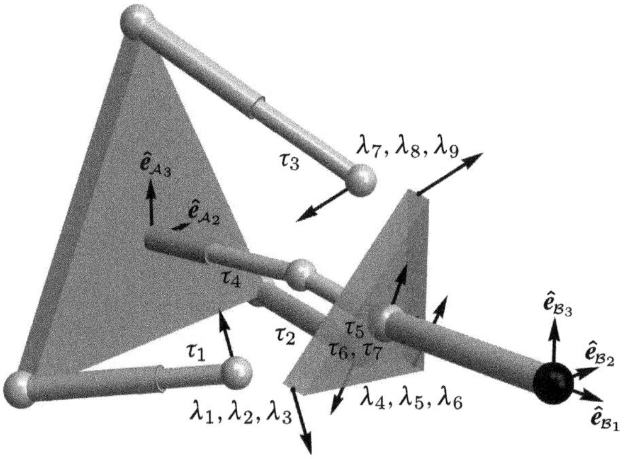

Figure 9.15 Constraint forces associated with loop closures in the humanoid shoulder complex.

where

$$\boldsymbol{M}_{l_i} = M_{l_i}\boldsymbol{\Gamma}_{G_{l_i}}^T\boldsymbol{\Gamma}_{G_{l_i}} + \boldsymbol{\Pi}_{l_i}^T\boldsymbol{I}_{l_i}^{G_i}\boldsymbol{\Pi}_{l_i}, \tag{9.84}$$

$$\boldsymbol{b}_{l_i} = \dot{\boldsymbol{M}}_{l_i}\dot{\boldsymbol{q}} - \frac{1}{2}\begin{pmatrix} \dot{\boldsymbol{q}}^T\frac{\partial \boldsymbol{M}_{l_i}}{\partial q_1}\dot{\boldsymbol{q}} \\ \vdots \\ \dot{\boldsymbol{q}}^T\frac{\partial \boldsymbol{M}_{l_i}}{\partial q_n}\dot{\boldsymbol{q}} \end{pmatrix}, \tag{9.85}$$

$$\boldsymbol{g}_{l_i} = M_{l_i}\boldsymbol{\Gamma}_{G_{l_i}}^T\hat{\boldsymbol{e}}_{0_3}, \tag{9.86}$$

$$\text{for } i = 1, \ldots, 3.$$

The terms $\boldsymbol{\Gamma}_{G_{l_i}}$ and M_{l_i} are the center of mass Jacobian and mass, respectively, of the ith strut. For the platform serial subsystem (including the 3 degree-of-freedom end link), we have

$$\begin{pmatrix} \tau_4 \\ \boldsymbol{0} \\ \tau_5 \\ \tau_6 \\ \tau_7 \end{pmatrix} = \boldsymbol{M}_p\begin{pmatrix} \ddot{q}_{10} \\ \vdots \\ \ddot{q}_{16} \end{pmatrix} + \boldsymbol{b}_p(q_{10}, \ldots, \dot{q}_{16}) + \boldsymbol{g}_p(q_{10}, \ldots, q_{16}), \tag{9.87}$$

where

$$\boldsymbol{M}_p = \sum_{i=1}^{3}\left(M_{p_i}\boldsymbol{\Gamma}_{G_{p_i}}^T\boldsymbol{\Gamma}_{G_{p_i}} + \boldsymbol{\Pi}_{p_i}^T\boldsymbol{I}_{p_i}^{G_i}\boldsymbol{\Pi}_{p_i}\right), \tag{9.88}$$

$$\boldsymbol{b}_p = \dot{\boldsymbol{M}}_p\dot{\boldsymbol{q}} - \frac{1}{2}\begin{pmatrix} \dot{\boldsymbol{q}}^T\frac{\partial \boldsymbol{M}_p}{\partial q_1}\dot{\boldsymbol{q}} \\ \vdots \\ \dot{\boldsymbol{q}}^T\frac{\partial \boldsymbol{M}_p}{\partial q_n}\dot{\boldsymbol{q}} \end{pmatrix}, \tag{9.89}$$

$$\boldsymbol{g}_p = \sum_{i=1}^{3} M_{p_i}\boldsymbol{\Gamma}_{G_{p_i}}^T\hat{\boldsymbol{e}}_{0_3}, \tag{9.90}$$

$$\text{for } i = 1, \ldots, 3.$$

The terms $\boldsymbol{\Gamma}_{G_{p_i}}$ and M_{p_i} are the center of mass Jacobian and mass, respectively, of the ith link in the platform chain serial subsystem.

The entire unconstrained system is described by

$$\begin{pmatrix} \mathbf{0} \\ \tau_1 \\ \mathbf{0} \\ \tau_2 \\ \mathbf{0} \\ \tau_3 \\ \tau_4 \\ \mathbf{0} \\ \tau_5 \\ \tau_6 \\ \tau_7 \end{pmatrix} = \begin{pmatrix} \boldsymbol{M}_{l_1} & \mathbf{0} & \mathbf{0} & \mathbf{0} \\ \mathbf{0} & \boldsymbol{M}_{l_2} & \mathbf{0} & \mathbf{0} \\ \mathbf{0} & \mathbf{0} & \boldsymbol{M}_{l_3} & \mathbf{0} \\ \mathbf{0} & \mathbf{0} & \mathbf{0} & \boldsymbol{M}_p \end{pmatrix} \begin{pmatrix} \ddot{q}_1 \\ \ddot{q}_2 \\ \vdots \\ \ddot{q}_{16} \end{pmatrix} + \begin{pmatrix} \boldsymbol{b}_{l_1} \\ \boldsymbol{b}_{l_2} \\ \boldsymbol{b}_{l_3} \\ \boldsymbol{b}_p \end{pmatrix} + \begin{pmatrix} \boldsymbol{g}_{l_1} \\ \boldsymbol{g}_{l_2} \\ \boldsymbol{g}_{l_3} \\ \boldsymbol{g}_p \end{pmatrix}, \tag{9.91}$$

and the constrained system is

$$\begin{pmatrix} \mathbf{0} \\ \tau_1 \\ \mathbf{0} \\ \tau_2 \\ \mathbf{0} \\ \tau_3 \\ \tau_4 \\ \mathbf{0} \\ \tau_5 \\ \tau_6 \\ \tau_7 \end{pmatrix} = \begin{pmatrix} \boldsymbol{M}_{l_1} & \mathbf{0} & \mathbf{0} & \mathbf{0} \\ \mathbf{0} & \boldsymbol{M}_{l_2} & \mathbf{0} & \mathbf{0} \\ \mathbf{0} & \mathbf{0} & \boldsymbol{M}_{l_3} & \mathbf{0} \\ \mathbf{0} & \mathbf{0} & \mathbf{0} & \boldsymbol{M}_p \end{pmatrix} \begin{pmatrix} \ddot{q}_1 \\ \ddot{q}_2 \\ \vdots \\ \ddot{q}_{16} \end{pmatrix} + \begin{pmatrix} \boldsymbol{b}_{l_1} \\ \boldsymbol{b}_{l_2} \\ \boldsymbol{b}_{l_3} \\ \boldsymbol{b}_p \end{pmatrix} + \begin{pmatrix} \boldsymbol{g}_{l_1} \\ \boldsymbol{g}_{l_2} \\ \boldsymbol{g}_{l_3} \\ \boldsymbol{g}_p \end{pmatrix}$$

$$+ \begin{pmatrix} \boldsymbol{\Gamma}_{l_1}^T & \mathbf{0} & \mathbf{0} \\ \mathbf{0} & \boldsymbol{\Gamma}_{l_2}^T & \mathbf{0} \\ \mathbf{0} & \mathbf{0} & \boldsymbol{\Gamma}_{l_3}^T \\ -\boldsymbol{\Gamma}_{p_1}^T & -\boldsymbol{\Gamma}_{p_2}^T & -\boldsymbol{\Gamma}_{p_3}^T \\ \mathbf{0} & \mathbf{0} & \mathbf{0} \end{pmatrix} \begin{pmatrix} \lambda_1 \\ \vdots \\ \lambda_9 \end{pmatrix}, \tag{9.92}$$

where the dimensions of the terms are

$$\boldsymbol{M} \in \mathbb{R}^{n \times n}, \qquad \boldsymbol{b}, \boldsymbol{g}, \boldsymbol{\tau} \in \mathbb{R}^n, \qquad \boldsymbol{\Phi} \in \mathbb{R}^{m \times n}, \qquad \boldsymbol{\lambda} \in \mathbb{R}^m, \tag{9.93}$$

and $n = 16$ and $m = 9$.

Using the configuration space properties computed, we can formulate the task space dynamics. We will specify the task to be the endpoint, $\boldsymbol{r}_E$, of the serial chain attached to

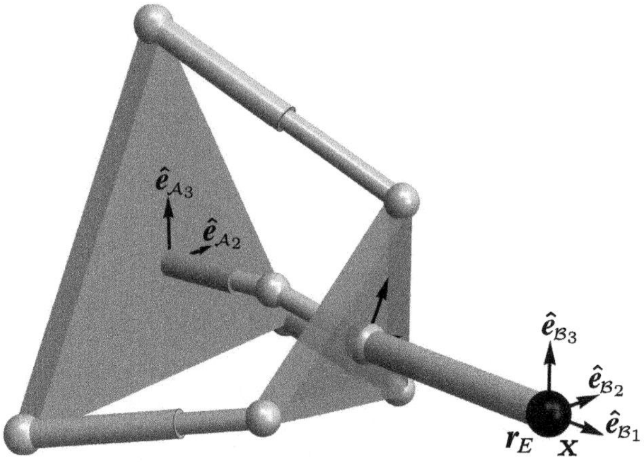

Figure 9.16 Humanoid shoulder complex showing task point.

the moving platform (see Figure 9.16):

$$\mathbf{x} = {}^0\boldsymbol{r}_E. \tag{9.94}$$

The task Jacobian is

$$\boldsymbol{J}(\boldsymbol{q}) = \frac{\partial \mathbf{x}}{\partial \boldsymbol{q}}. \tag{9.95}$$

Using $\boldsymbol{J}(\boldsymbol{q})$, as well as $\boldsymbol{\Phi}(\boldsymbol{q})$, $\boldsymbol{M}(\boldsymbol{q})$, $\boldsymbol{b}(\boldsymbol{q}, \dot{\boldsymbol{q}})$, and $\mathbf{g}(\boldsymbol{q})$, we can compute the task/constraint space terms:

$$\check{\boldsymbol{\Lambda}}(\boldsymbol{q}) = \begin{pmatrix} \boldsymbol{J}\boldsymbol{M}^{-1}\boldsymbol{J}^T & \boldsymbol{J}\boldsymbol{M}^{-1}\boldsymbol{\Phi}^T \\ \boldsymbol{\Phi}\boldsymbol{M}^{-1}\boldsymbol{J}^T & \boldsymbol{\Phi}\boldsymbol{M}^{-1}\boldsymbol{\Phi}^T \end{pmatrix}^{-1}, \tag{9.96}$$

$$\check{\boldsymbol{\mu}}(\boldsymbol{q}, \dot{\boldsymbol{q}}) = \check{\boldsymbol{\Lambda}} \begin{pmatrix} \boldsymbol{J}\boldsymbol{M}^{-1}\boldsymbol{b} - \dot{\boldsymbol{J}}\dot{\boldsymbol{q}} \\ \boldsymbol{\Phi}\boldsymbol{M}^{-1}\boldsymbol{b} - \dot{\boldsymbol{\Phi}}\dot{\boldsymbol{q}} \end{pmatrix}, \tag{9.97}$$

$$\check{\boldsymbol{p}}(\boldsymbol{q}) = \check{\boldsymbol{\Lambda}} \begin{pmatrix} \boldsymbol{J}\boldsymbol{M}^{-1}\boldsymbol{g} \\ \boldsymbol{\Phi}\boldsymbol{M}^{-1}\boldsymbol{g} \end{pmatrix}. \tag{9.98}$$

As in the previous examples, since these terms involve matrix inverses, it is preferable to compute them numerically at each time step within the numerical integration loop rather than computing them symbolically.

Our task/constraint space equation of motion for the dynamics at the task point is

$$\boldsymbol{f} = \check{\boldsymbol{\Lambda}}_{11}\ddot{\mathbf{x}} + \check{\boldsymbol{\mu}}_1 + \check{\boldsymbol{p}}_1, \tag{9.99}$$

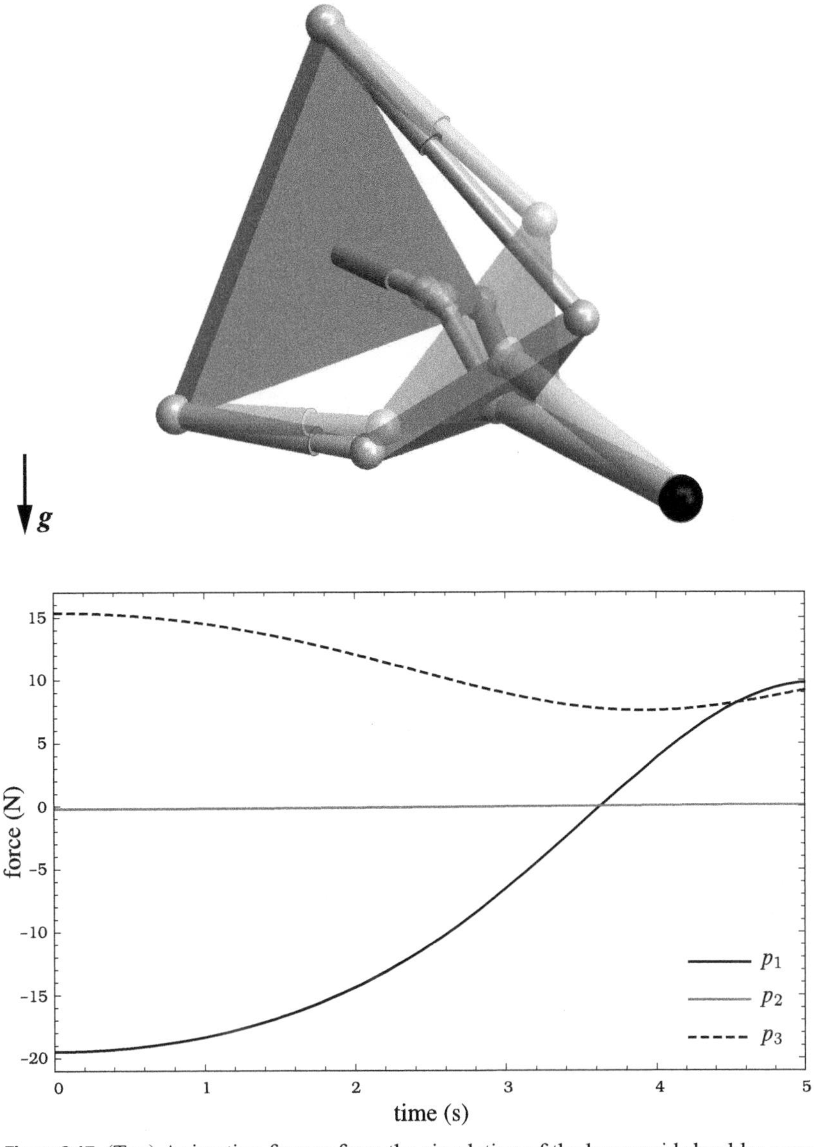

Figure 9.17 (Top) Animation frames from the simulation of the humanoid shoulder complex parallel shoulder mechanism. (Bottom) Time history of the humanoid shoulder complex task space gravity vector shoulder mechanism task space gravity vector.

and the equation corresponding to constraint forces in the system is

$$\boldsymbol{f}_c + \boldsymbol{\lambda} = \check{\boldsymbol{\Lambda}}_{21}\ddot{\boldsymbol{x}} + \check{\boldsymbol{\mu}}_2 + \check{\boldsymbol{p}}_2. \tag{9.100}$$

We will specify a task space force at the task point to cancel out the task/constraint space centrifugal and Coriolis force, and the task/constraint space gravity force. That is,

$$\boldsymbol{f} = \check{\boldsymbol{\mu}}_1 + \check{\boldsymbol{p}}_1. \tag{9.101}$$

This will result in the following dynamics:

$$\check{\mathbf{\Lambda}}_{11}\ddot{\mathbf{x}} = \mathbf{0}. \tag{9.102}$$

Therefore, the task point will have *zero* acceleration. Given *zero* velocity initial conditions, the task point will not move. The system will, however, move in the null space.

Results from simulation of the system under gravity are shown in Figures 9.17.

10 Software for Analytical Dynamics

This chapter provides a brief and very selective survey of some software useful for solving problems in analytical dynamics. I have chosen to include only software intended for academic and industrial use that has been validated for modeling accuracy. Consequently, I have omitted the many dynamics engines used for video games and entertainment purposes.

10.1 General Purpose Mathematical Software

A number of powerful tools for mathematical analysis exist. Two of the most popular are Mathematica by Wolfram Research Inc. and MATLAB by The MathWorks Inc. They provide capabilities for both symbolic manipulation of expressions (i.e., computer algebra systems) and numerical computation and evaluation of expressions. While the basic Mathematica and MATLAB applications do not provide high-level programming abstractions specific to multibody system modeling and analysis, they are extremely useful for the low-level symbolic and numerical processing required in solving problems in analytical dynamics (see Figure 10.1 for a screenshot of Mathematica). This has a benefit in the learning process, as they require the student to understand the central concepts of analytical dynamics and to set up the solution procedure in a detailed manner, while alleviating much of the tedium of the algebraic and numerical work.

10.1.1 Packages and Extensions for Multibody Dynamics

Wolfram Research and MathWorks both provide domain-specific extensions for analyzing multibody systems. This gives the user access to high-level abstractions for modeling, simulating, and analyzing multibody systems. These high-level abstractions can be accessed through a programming language or through a diagrammatic interface supporting both acausal and causal modeling. Acausal modeling involves component-based schematic representations of the physical system, whereas causal modeling involves block diagram representations of the underlying mathematical description of the system.

Simscape Multibody (formerly SimMechanics) by The MathWorks Inc. allows users to model multibody systems using a library of blocks associated with bodies, joints, constraints, spatial transformations, force elements, and sensors (see Figure 10.2 for

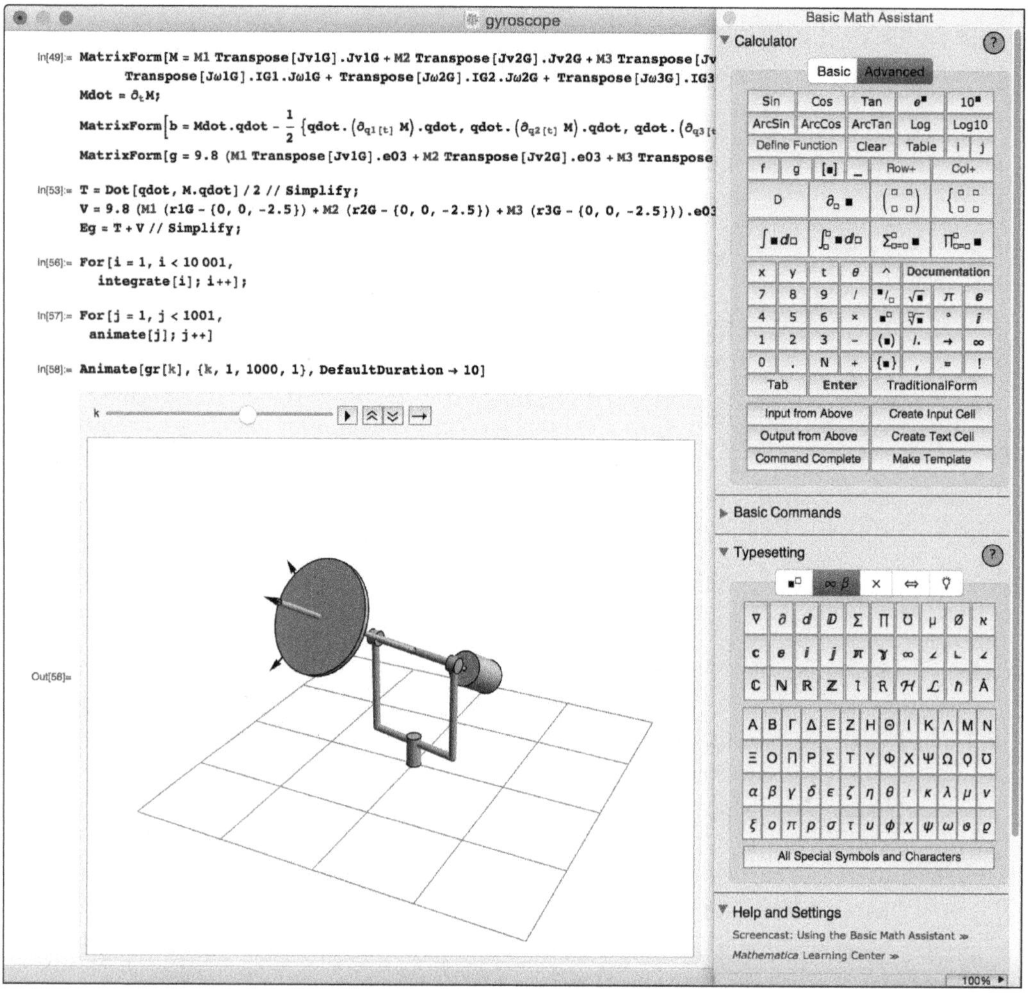

Figure 10.1 Mathematica by Wolfram Research Inc. offers general symbolic and numerical capabilities suitable for modeling, analysis, and simulation of multibody systems. A gimballed gyroscope is modeled here. Reprinted with permission of Wolfram Research Inc. Mathematica® is a registered trademark of Wolfram Research Inc.

a screenshot of Simscape Multibody). Entire CAD assemblies can be imported along with associated geometry, inertial properties, joint parameters, and constraints. Simscape Multibody then formulates and integrates the equations of motion for the multibody system. Animations are generated during simulation to allow visualization of the motion of the system.

Simscape Multibody also allows design and simulation of control systems in Simulink, which uses a causal block diagram interface for control system modeling. Simscape Multibody supports automatic C code generation for model deployment in other simulation environments, including hardware-in-the-loop (HIL) systems. In addition to the block diagrammatic user interface, the object-oriented Simscape

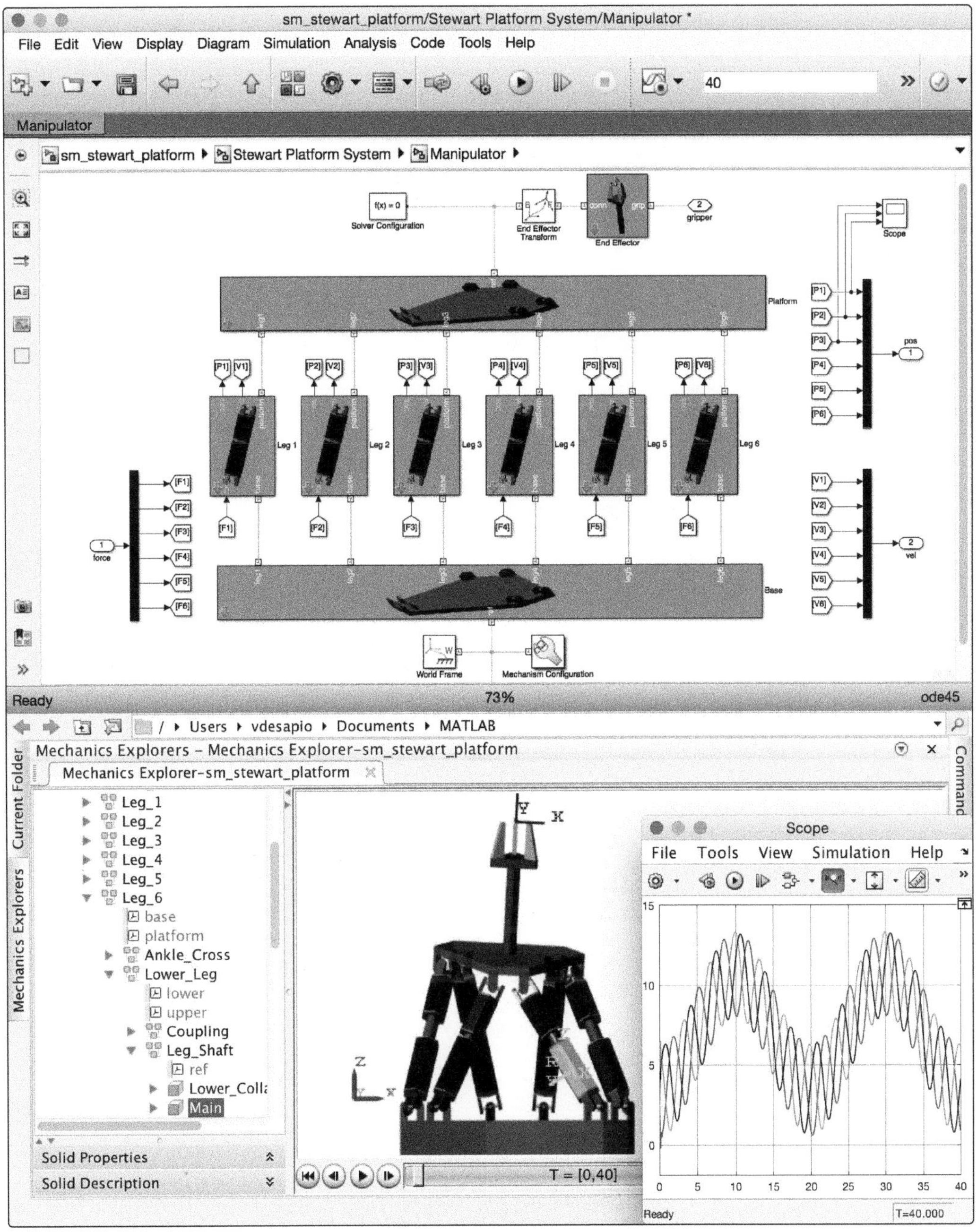

Figure 10.2 Simscape Multibody by The MathWorks Inc. provides a multibody simulation environment for 3D mechanical systems using blocks representing bodies, joints, constraints, force elements, and sensors. Reprinted with permission of The MathWorks Inc. Simscape™ Multibody™ is a trademark of The MathWorks Inc.

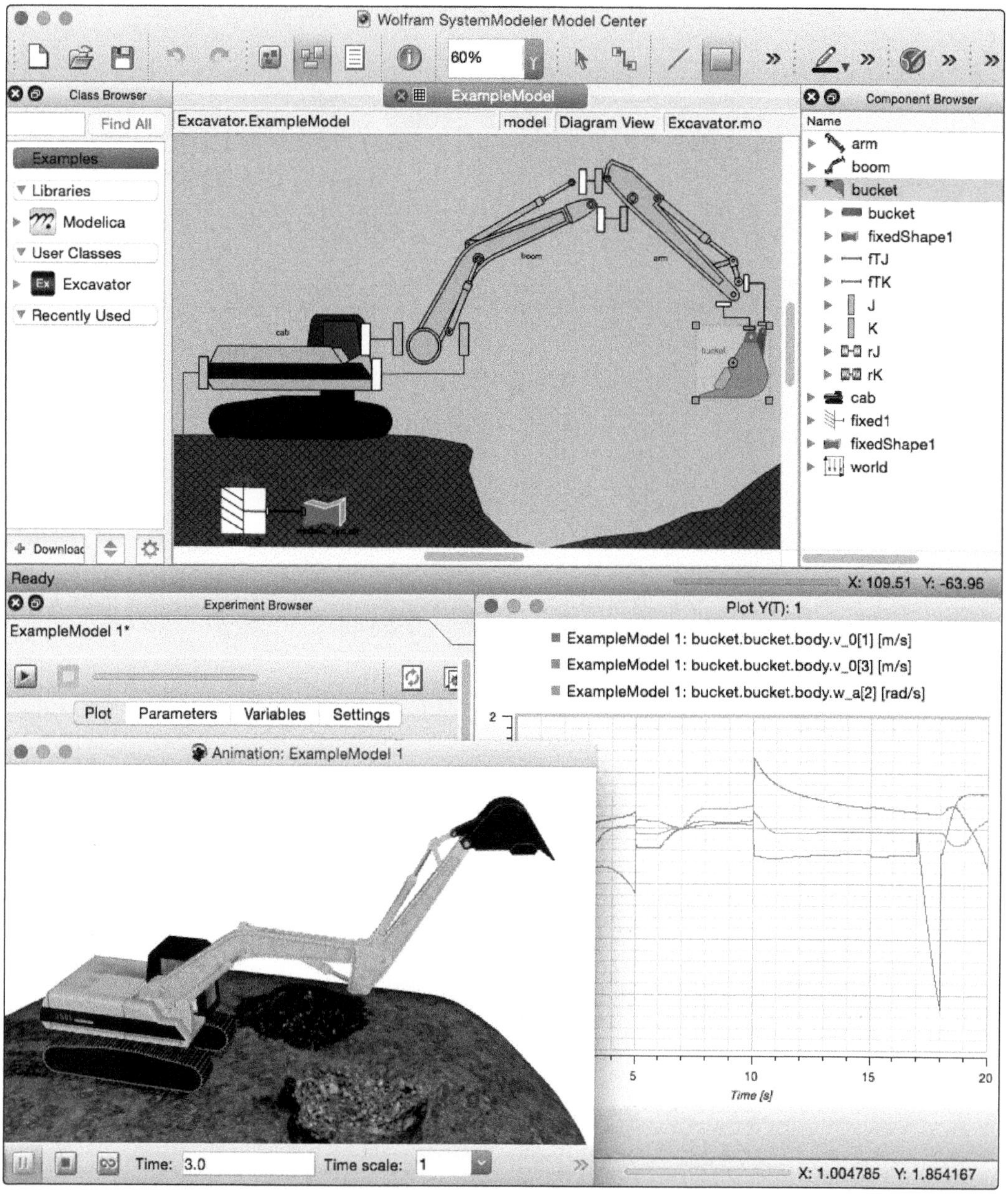

Figure 10.3 SystemModeler by Wolfram Research Inc. offers drag-and-drop component-based modeling, analysis, and simulation of multibody systems. Reprinted with permission of Wolfram Research Inc. SystemModeler® is a registered trademark of Wolfram Research Inc.

language builds on the MATLAB programming language with features specific to physical modeling.

SystemModeler by Wolfram Research provides multidomain modeling, simulation, and analysis capabilities (see Figure 10.3 for a screenshot of SystemModeler).

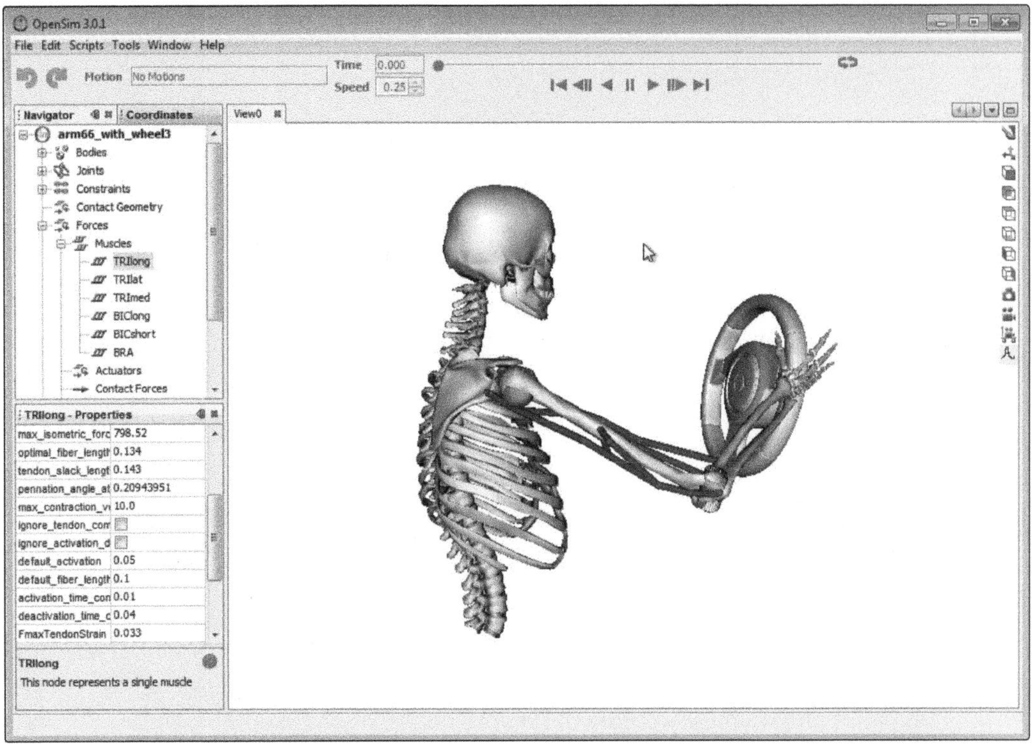

Figure 10.4 OpenSim biomechanical simulation software incorporating Simbody. Simbody's solver was derived from the NIH Internal Variable Dynamics Module (IVM) for molecular dynamics, itself based on the spatial operator algebra.

Like Simscape Multibody, SystemModeler allows users to create hierarchical component-based models that are consistent with the physical topology of the system. Models are translated into optimized equations of motion. Built-in numerical solvers handle systems with events and discontinuous behavior. As with Simscape, animations are generated during simulation to allow visualization of the motion of the system.

SystemModeler incorporates Modelica, a modeling language. Modelica is a nonproprietary, object-oriented, equation-based language to conveniently model complex physical systems .

Neweul-M^2, developed at the University of Stuttgart, is a software package for the analysis of multibody systems. It uses the MATLAB Symbolic Math Toolbox to generate the equations of motion for multibody systems in symbolic form. The formulation of the equations of motion is based on the Newton-Euler formalism and d'Alembert's and Jourdain's Principles. Currently constraints are restricted to holonomic ones. A command line interface and a graphical user interface are provided. Since the equations of motion are computed in symbolic form, they can be exported into the C programming language or used in a Simulink S-function.

10.2 Dedicated Multibody Dynamics Software

10.2.1 Commerical Software

Adams (Automated Dynamic Analysis of Mechanical Systems) by MSC Software Corporation is a multibody dynamics simulation suite consisting of Adams, Adams Machinery, Adams Car, Adams Flexible Multibody Systems, and Adams MaxFlex. CAD assemblies can be imported from many major CAD systems. Multibody systems are created using libraries of joints and constraints.

Simpack by SIMULIA (a division of Dassault Systèmes) is another dedicated multibody simulation environment used primarily in the automotive, railway, engine, wind turbine, power transmission, and aerospace industries. Simpack is specialized to address high-degree-of-freedom models with flexible bodies and impact between bodies.

10.2.2 Open Source Software

Simbody was developed as part of SimBios, the NIH Center for Biomedical Computation at Stanford University (see Figure 10.4 for a screenshot of OpenSim, which uses the Simbody engine). It has been used for coarse-grained molecular models as well as large-scale mechanical models, such as neuromusculoskeletal models of human gait, and robotic systems. Simbody is provided as an open source, object-oriented C++ application programming interface (API). It was derived from the public domain NIH Internal Variable Dynamics Module (IVM) for molecular dynamics, itself based on the spatial operator algebra (Rodriguez, Jain, and Kreutz-Delgado 1991) developed at NASA's Jet Propulsion Laboratory (JPL).

Appendix
Inclusion of Flexible Bodies

The previous chapters have detailed the dynamics of constrained multibody systems where the bodies are rigid. These methods can be extended to handle constrained systems involving a mix of flexible and rigid bodies. The method of absolute nodal coordinates described by Shabana (1998) has a demonstrated efficacy in application to flexible/rigid multibody systems. Using this method, the flexible subsystems are described using finite element nodal coordinates with respect to an absolute global coordinate system. For the rigid-body subsystems we can choose generalized coordinates that describe relative joint motion. Coupling the flexible and rigid subsystems results in a constrained dynamical system with graph topology.

For conciseness, in the following sections, the Einstein summation convention will be used. This specifies that summation is implied over all values of the index when an index variable appears twice in a single term. Consequently, the summation symbol can be omitted in these cases. For example,

$$A_{ij}\hat{\boldsymbol{e}}_i \otimes \hat{\boldsymbol{e}}_j = \sum_{i=1}^{3}\sum_{j=1}^{3} A_{ij}\hat{\boldsymbol{e}}_i \otimes \hat{\boldsymbol{e}}_j. \tag{A.1}$$

A.1 Continuum Kinematics

Continuum kinematics refers to the study of the motion of a continuum without regard to the causes of that motion. Unlike the kinematics of discrete systems of points and rigid bodies, the kinematics of continuous systems involve infinitely divisible deformable material elements.

A.1.1 Deformation

A *deformation* is formally defined as a diffeomorphism between a reference manifold, M_o, and a deformed manifold, M_d. That is, $\{\boldsymbol{x}|M_o \rightarrow M_d\}$:

$$\boldsymbol{x} = \boldsymbol{x}(\boldsymbol{X}, t), \tag{A.2}$$

where $\boldsymbol{x}$ represents the spatial field coordinates and $\boldsymbol{X}$ represents the material coordinates. We can represent the inversion of this as

$$\boldsymbol{X} = \boldsymbol{X}(\boldsymbol{x}, t). \tag{A.3}$$

The displacement, $\boldsymbol{u}(\boldsymbol{X}, t)$, of a material point is the difference between the deformed position vector, $\boldsymbol{x}(\boldsymbol{X}, t)$, and the reference position vector, $\boldsymbol{X}$. The material description of displacement is

$$\boldsymbol{u}(\boldsymbol{X}, t) = \boldsymbol{x}(\boldsymbol{X}, t) - \boldsymbol{X}. \tag{A.4}$$

Using (A.3), we can express the spatial description as

$$\boldsymbol{u}(\boldsymbol{x}, t) = \boldsymbol{x} - \boldsymbol{X}(\boldsymbol{x}, t). \tag{A.5}$$

The material description of the material velocity is given by

$$\boldsymbol{v}(\boldsymbol{X}, t) = \frac{\partial \boldsymbol{u}(\boldsymbol{X}, t)}{\partial t}, \tag{A.6}$$

and the material description of the material acceleration is given by

$$\boldsymbol{a}(\boldsymbol{X}, t) = \frac{\partial \boldsymbol{v}(\boldsymbol{X}, t)}{\partial t} = \frac{\partial^2 \boldsymbol{u}(\boldsymbol{X}, t)}{\partial t^2}. \tag{A.7}$$

Using (A.3), the spatial descriptions of the material velocity and material acceleration are

$$\boldsymbol{v} = \boldsymbol{v}(\boldsymbol{x}, t) \tag{A.8}$$

and

$$\boldsymbol{a} = \boldsymbol{a}(\boldsymbol{x}, t). \tag{A.9}$$

We can also arrive at (A.9) by taking the material derivative of the spatial description of velocity,

$$\boldsymbol{a}(\boldsymbol{x}, t) = \frac{D\boldsymbol{v}(\boldsymbol{x}, t)}{Dt} = \frac{\partial \boldsymbol{v}(\boldsymbol{x}, t)}{\partial t} + \boldsymbol{v}\nabla[\boldsymbol{v}(\boldsymbol{x}, t)], \tag{A.10}$$

where

$$\hat{\boldsymbol{a}}(\boldsymbol{x}, t) = \frac{\partial \boldsymbol{v}(\boldsymbol{x}, t)}{\partial t} \tag{A.11}$$

is the spatial description of the spatial acceleration vector.

A.2 Continuum Dynamics

Continuum dynamics builds upon continuum kinematics by adding descriptions of the physical causes of motion. This includes both physical laws relating force or stress to acceleration (Newton's law) and constitutive equations relating stress to material strain or strain rate. Given a finite element discretization, we can assemble the equations of motion for a flexible-body system.

The absolute nodal coordinate method (Shabana 1998) is useful for dealing with systems involving large rigid body motions and/or large deformations. Kübler, Eberhard,

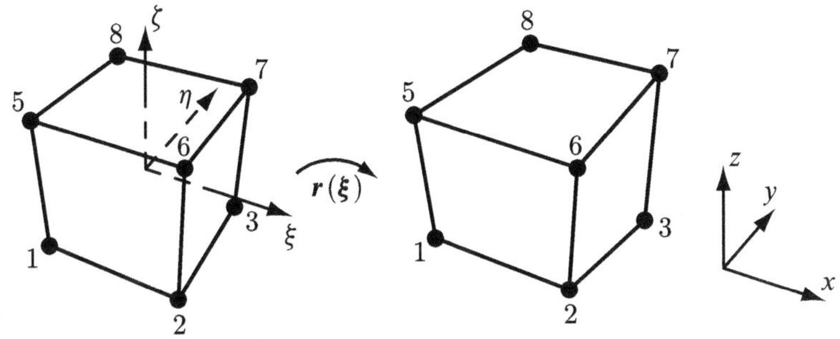

Figure A.1 Eight-node hexahedral element and coordinates of the intrinsic element parameter space, ξ, η, and ζ.

and Geisler (2003a, 2003b) provide an excellent description of the absolute nodal coordinate method with regard to flexible/rigid multibody systems. A brief review is presented here.

For a given flexible body subsystem a Lagrangian, or material, description is chosen that relates all quantities to the reference configuration, $\boldsymbol{X}$, of the system with domain Ω_o. Using a nonlinear finite element approach, a set of shape or interpolating functions, $\{N_1, \ldots, N_s\}$, associated with a particular finite element discretization of s nodes can be chosen. In the case of eight-node isoparametric hexahedral elements (see Figure A.1), the interpolating functions are given as (Hughes 2000)

$$N_a(\boldsymbol{\xi}) = \frac{1}{8}(1 + \xi_a \xi)(1 + \eta_a \eta)(1 + \zeta_a \zeta), \quad \text{for } a = 1, 2, \ldots, 8, \tag{A.12}$$

where $\boldsymbol{\xi} = (\xi \quad \eta \quad \zeta)^T$ are the coordinates of the intrinsic element parameter space. Defining the shape matrix, $\boldsymbol{N} \in \mathbb{R}^{3\times 3s}$,

$$\boldsymbol{N} \triangleq \begin{pmatrix} N_1 & 0 & 0 & \cdots & N_s & 0 & 0 \\ 0 & N_1 & 0 & \cdots & 0 & N_s & 0 \\ 0 & 0 & N_1 & \cdots & 0 & 0 & N_s \end{pmatrix}, \tag{A.13}$$

the displacement field, $\boldsymbol{u}(\boldsymbol{X}, t) \in \mathbb{R}^3$, is given by $\boldsymbol{u} = \boldsymbol{N}\boldsymbol{d}$, where $\boldsymbol{d} \in \mathbb{R}^{3s}$ is the vector of nodal displacements. The current material configuration is then given by $\boldsymbol{x}(\boldsymbol{X}, t) = \boldsymbol{X} + \boldsymbol{u}(\boldsymbol{X}, t)$. The weak form statement (summation convention applied) of the flexible-body dynamics problem is

$$\delta d_j (M_{jk} \ddot{d}_k + k_j + g_j) = 0, \tag{A.14}$$

where

$$M_{jk} = \int_{\Omega_o} \rho_o N_{ij} N_{ik} dV, \tag{A.15}$$

$$k_j = \int_{\Omega_o} \frac{\partial N_{ij}}{\partial X_k} F_{il} S_{lk} dV = \int_{\Omega_o} \frac{\partial N_{ij}}{\partial X_k} \left(\delta_{il} + \frac{\partial N_{il}}{\partial X_k} d_k \right) S_{lk} dV, \tag{A.16}$$

$$g_j = -\int_{\Omega_o} \rho_o N_{ij} b_i dV - \int_{\partial\Omega_o} N_{ij} p_i dA. \tag{A.17}$$

The term ρ_o is the material density field in the reference configuration, F_{il} is the deformation gradient , b_i are the body forces (e.g., gravity), p_i are the surface tractions, δ_{il} is the Kronecker delta, and S_{lk} is the second Piola-Kirchoff stress tensor. The system can thus be stated as

$$\boldsymbol{M}\ddot{\boldsymbol{d}} + \boldsymbol{k} + \boldsymbol{g} = \boldsymbol{0}, \tag{A.18}$$

or more generally as

$$\boldsymbol{M}\ddot{\boldsymbol{d}} + \boldsymbol{k} + \boldsymbol{g} = \boldsymbol{f}, \tag{A.19}$$

where $\boldsymbol{f} \in \mathbb{R}^{3s}$ is a vector of external forces applied at the nodes. The terms $\boldsymbol{M} \in \mathbb{R}^{3s \times 3s}$, $\boldsymbol{k} \in \mathbb{R}^{3s}$, and $\boldsymbol{g} \in \mathbb{R}^{3s}$ are the mass matrix, stiffness vector, and body/surface force vector, respectively. It is noted that due to integration with respect to the reference configuration, Ω_o, the mass matrix, $\boldsymbol{M}$, is constant. The stress tensor, $\boldsymbol{S}$, is highly nonlinear, however.

Different constitutive models (Belytschko et al., 2013) can be incorporated into (A.16) to provide a material specific means for computing $\boldsymbol{S}$. Generally speaking, a constitutive model can express $\boldsymbol{S} = \boldsymbol{S}(\boldsymbol{E}, \ldots)$, where $\boldsymbol{E}$ is the Green-Lagrange strain tensor and the ellipses represent functional dependence on rate terms (as in the case of viscous behavior). This allows us to relate the stress tensor to the displacements, since

$$E_{lk} = \frac{1}{2}(F_{il} F_{ik} - \delta_{lk}) \tag{A.20}$$

and

$$F_{ij} = \delta_{ij} + \frac{\partial N_{ij}}{\partial X_k} d_k. \tag{A.21}$$

The constitutive relationships thus provide a means of evaluating $\boldsymbol{k}$ based on the system states, $\boldsymbol{d}$ and $\dot{\boldsymbol{d}}$. We will not consider any detailed constitutive models here but will rather focus on the mathematical structure of (A.19).

A.3 Subsystem Assembly

We can assemble a set of rigid and flexible subsystems into a single constrained system. First, we specify the dynamics of a set of y unconstrained rigid-body subsystems,

$$\begin{aligned} \boldsymbol{\tau}_1 &= \boldsymbol{M}_{r_1}\ddot{\boldsymbol{q}}_1 + \boldsymbol{b}_1 + \boldsymbol{g}_{r_1}, \\ &\vdots \\ \boldsymbol{\tau}_y &= \boldsymbol{M}_{r_y}\ddot{\boldsymbol{q}}_y + \boldsymbol{b}_y + \boldsymbol{g}_{r_y}. \end{aligned} \tag{A.22}$$

Next, we specify the dynamics of a set of z flexible subsystems that have been discretized using absolute nodal coordinates,

$$\begin{aligned} \boldsymbol{f}_1 &= \boldsymbol{M}_{f_1}\ddot{\boldsymbol{d}}_1 + \boldsymbol{k}_1 + \boldsymbol{g}_{f_1}, \\ &\vdots \\ \boldsymbol{f}_z &= \boldsymbol{M}_{f_z}\ddot{\boldsymbol{d}}_z + \boldsymbol{k}_z + \boldsymbol{g}_{f_z}. \end{aligned} \tag{A.23}$$

The sets of equations given by (A.22) and (A.23) can be assembled into a single system equation of the form

$$\boldsymbol{\tau} = \boldsymbol{M}\ddot{\boldsymbol{q}} + \boldsymbol{b} + \boldsymbol{g}, \tag{A.24}$$

where

$$\boldsymbol{M} = \text{diag}(\boldsymbol{M}_{r_1}, \ldots, \boldsymbol{M}_{r_y}, \boldsymbol{M}_{f_1}, \ldots, \boldsymbol{M}_{f_z}), \tag{A.25}$$

$$\boldsymbol{\tau} = \left(\boldsymbol{\tau}_1^T \cdots \boldsymbol{\tau}_y^T \; \boldsymbol{f}_1^T \cdots \boldsymbol{f}_z^T\right)^T, \tag{A.26}$$

$$\ddot{\boldsymbol{q}} = \left(\ddot{\boldsymbol{q}}_1^T \cdots \ddot{\boldsymbol{q}}_y^T \; \ddot{\boldsymbol{d}}_1^T \cdots \ddot{\boldsymbol{d}}_z^T\right)^T, \tag{A.27}$$

$$\boldsymbol{b} = \left(\boldsymbol{b}_1^T \cdots \boldsymbol{b}_y^T \; \boldsymbol{k}_1^T \cdots \boldsymbol{k}_z^T\right)^T, \tag{A.28}$$

$$\boldsymbol{g} = \left(\boldsymbol{g}_{r_1}^T \cdots \boldsymbol{g}_{r_y}^T \; \boldsymbol{g}_{f_1}^T \cdots \boldsymbol{g}_{f_z}^T\right)^T. \tag{A.29}$$

We have $\boldsymbol{M} \in \mathbb{R}^{n \times n}$ and $\boldsymbol{\tau}, \ddot{\boldsymbol{q}}, \boldsymbol{b}, \boldsymbol{g} \in \mathbb{R}^n$, where

$$n = \sum_{i=1}^{y} n_{r_i} + \sum_{i=1}^{z} n_{f_i} \tag{A.30}$$

$$n_{f_i} = 3s_i. \tag{A.31}$$

Imposing a set of m_C holonomic constraint equations to establish system connectivity yields the familiar equation

$$\boldsymbol{\tau} = \boldsymbol{M}\ddot{\boldsymbol{q}} + \boldsymbol{b} + \boldsymbol{g} - \boldsymbol{\Phi}^T\boldsymbol{\lambda}, \tag{A.32}$$

subject to

$$\boldsymbol{\Phi}\ddot{\boldsymbol{q}} + \dot{\boldsymbol{\Phi}}\dot{\boldsymbol{q}} = \boldsymbol{0}. \tag{A.33}$$

References

Appell, P. (1900), Sur une forme générale des équations de la dynamique, *Journal für die Reine und Angewandte Mathematik* **121**, 310–319.

Baumgarte, J. (1972), Stabilization of constraints and integrals of motion in dynamical systems, *Computer Methods in Applied Mechanics and Engineering* **1**(1), 1–16.

Belytschko, T., Liu, W. K., Moran, B. and Elkhodary, K. (2013), *Nonlinear Finite Elements for Continua and Structures*, John Wiley.

Blajer, W. (1997), A geometric unification of constrained system dynamics, *Multibody System Dynamics* **1**(1), 3–21.

Clavel, R. (1991), *Conception d' un robot parallèle rapide à 4 degrés de liberté*, École Polytechnique Fédérale de Lausanne (EPFL).

de Groot, J. H. and Brand, R. (2001), A three-dimensional regression model of the shoulder rhythm, *IEEE Transactions on Biomedical Engineering* **16**, 735–743.

De Sapio, V., Holzbaur, K. and Khatib, O. (2006), The control of kinematically constrained shoulder complexes: Physiological and humanoid examples, *in Proceedings of the 2006 IEEE International Conference on Robotics and Automatio*, Vol. 1–10, IEEE, pp. 2952–2959.

De Sapio, V. and Khatib, O. (2005), Operational space control of multibody systems with explicit holonomic constraints, *in Proceedings of the 2005 IEEE International Conference on Robotics and Automation*, Vol. 1–4, IEEE, pp. 2961–2967.

De Sapio, V., Khatib, O. and Delp, S. (2005), Simulating the task-level control of human motion: A methodology and framework for implementation, *The Visual Computer* **21**(5), 289–302.

De Sapio, V., Khatib, O. and Delp, S. (2006), Task-level approaches for the control of constrained multibody systems, *Multibody System Dynamics* **16**(1), 73–102.

Delp, S. L., Anderson, F. C., Arnold, A. S., Loan, P., Habib, A., John, C. T., Guendelman, E. and Thelen, D. G. (2007), Opensim: Open-source software to create and analyze dynamic simulations of movement, *IEEE Transactions* Biomedical Engineering **54**(11), 1940–1950.

Delp, S. L. and Loan, J. P. (2000), A computational framework for simulating and analyzing human and animal movement, *IEEE Computing in Science and Engineering* **2**(5), 46–55.

Delp, S. L., Loan, J. P., Zajac, F. E., Topp, E. L. and Rosen, J. M. (1990), An interactive graphics-based model of the lower extremity to study orthopaedic surgical procedures, *IEEE Transactions on Biomedical Engineering* **37**, 757–767.

Denavit, J. and Hartenberg, R. (1955), A kinematic notation for lower-pair mechanisms based on matrices, *Journal of Applied Mechanics* **23**, 215–221.

Dirac, P. A. M. (1958), Generalized hamiltonian dynamics, *Proceedings of the Royal Society of London, Series A* **246**, 326–332.

Do Carmo, M. P. (1976), *Differential Geometry of Curves and Surfaces*, Prentice Hall.

Dugas, R. (1988), *A History of Mechanics*, Dover.

Flannery, M. R. (2005), The enigma of nonholonomic constraints, *American Journal of Physics* **73**(3), 265–272.

Gauss, K. F. (1829), Über ein neues allgemeines grundgesetz der mechanik [On a new fundamental law of mechanics], *Journal für die Reine und Angewandte Mathematik* **4**, 232–235.

Gibbs, J. W. (1879), On the fundamental formulae of dynamics, *American Journal of Mathematics* **2**(1), 49–64.

Goldstein, H., Poole, C. and Safko, J. (2002), *Classical Mechanics*, Addison Wesley.

Hertz, H. (1894), *Die prinzipien der mechanik in neuem zusammenhange dargestellt*, Barth.

Holzbaur, K. R. S., Murray, W. M. and Delp, S. L. (2005), A model of the upper extremity for simulating musculoskeletal surgery and analyzing neuromuscular control, *Annals of Biomedical Engineering* **33**(6), 829–840.

Hughes, T. (2000), *The Finite Element Method: Linear Static and Dynamic Finite Element Analysis*, Dover.

Huston, R. L., Liu, C. Q. and Li, F. (2003), Equivalent control of constrained multibody systems, *Multibody System Dynamics* **10**(3), 313–321.

Jourdain, P. E. B. (1909), Note on an analogue of Gauss' principle of least constraint, *Quarterly Journal of Pure and Applied Mathematics* **40**, 153–157.

Jungnickel, U. (1994), Differential-algebraic equations in Riemannian spaces and applications to multibody system dynamics, *ZAMM* **74**(9), 409–415.

Kane, T. R. (1961), Dynamics of nonholonomic systems, *Journal of Applied Mechanics* **28**(4), 574–578.

Khatib, O. (1987), A unified approach to motion and force control of robot manipulators: The operational space formulation, *International Journal of Robotics Research* **3**(1), 43–53.

Khatib, O. (1995), Inertial properties in robotic manipulation: An object level framework, *International Journal of Robotics Research* **14**(1), 19–36.

Klopčar, N. and Lenarčič, J. (2001), Biomechanical considerations on the design of a humanoid shoulder girdle, *in Proceedings of the 2001 IEEE/ASME International Conference on Advanced Intelligent Mechatronics*, Vol. 1, IEEE, pp. 255–259.

Kübler, L., Eberhard, P. and Geisler, J. (2003a), Flexible multibody systems with large deformations and nonlinear structural damping using absolute nodal coordinates, *Nonlinear Dynamics* **34**(1–2), 31–52.

Kübler, L., Eberhard, P. and Geisler, J. (2003b), Flexible multibody systems with large deformations using absolute nodal coordinates for isoparametric solid brick elements, *in Proceedings of the 2003 ASME Design Engineering Technical Conference*, ASME, pp. 1–10.

Lanczos, C. (1986), *The Variational Principles of Mechanics*, Dover.

Lenarčič, J., Stanišić, M. M. and Parenti-Castelli, V. (2000), Kinematic design of a humanoid robotic shoulder complex, *in Proceedings of the 2000 IEEE International Conference on Robotics and Automation*, Vol. 1, IEEE, pp. 27–32.

Lenarčič, J. and Stanišić, M. M. (2003), A humanoid shoulder complex and the humeral pointing kinematics, *IEEE Transactions on Robotics and Automation* **19**(3), 499–506.

Naudet, J., Lefeber, D., Daerden, F. and Terze, Z. (2003), Forward dynamics of open-loop multibody mechanisms using an efficient recursive algorithm based on canonical momenta, *Multibody System Dynamics* **10**(1), 45–59.

Papastavridis, J. G. (2002), *Analytical Mechanics: A Comprehensive Treatise on the Dynamics of Constrained Systems for Engineers, Physicists, and Mathematicians*, Oxford University Press.

Rodriguez, G., Jain, A. and Kreutz-Delgado, K. (1991), A spatial operator algebra for manipulator modeling and control, *International Journal of Robotics Research* **10**(4), 371–381.

Russakow, J., Khatib, O. and Rock, S. M. (1995), Extended operational space formulation for serial-to-parallel chain (branching) manipulators, *in Proceedings of the 1995 International Conference on Robotics and Automation*, Vol. 1, IEEE, pp. 1056–1061.

Saul, K. R., Hu, X., Goehler, C. M., Vidt, M. E., Daly, M., Velisar, A. and Murray, W. M. (2015), Benchmarking of dynamic simulation predictions in two software platforms using an upper limb musculoskeletal model, *Computer Methods in Biomechanics and Biomedical Engineering* **18**(13), 1445–1458.

Schutte, L. M. (1992), Using musculsokeletal models to explore strategies for improving performance in electrical stimulation-induced leg cycle ergometry, PhD thesis, Stanford University.

Shabana, A. (1998), Computer implementation of the absolute nodal coordinate formulation for flexible multibody dynamics, *Nonlinear Dynamics* **16**(3), 293–306.

Udwadia, F. and Kalaba, R. E. (1992), A new perspective on constrained motion, *in Proceedings: Mathematical and Physical Sciences*, Vol. 439, Royal Society of London, pp. 407–410.

Vujanovic, B. D. and Atanackovic, T. M. (2004), *An Introduction to Modern Variational Techniques in Mechanics and Engineering*, Springer Science & Business Media.

Wehage and Haug, E. J. (1982), Generalized coordinate partitioning for dimension reduction in analysis of constrained dynamic systems, *Journal of Mechanical Design* **104**(1), 247–255.

Zajac, F. E. (1989), Critical reviews in biomedical engineering, *in* J. R. Bourne, ed., *Muscle and Tendon: Properties, Models, Scaling, and Application to Biomechanics and Motor Control*, CRC Press, pp. 359–411.

Zajac, F. E. (1993), Muscle coordination of movement: A perspective, *Journal of Biomechanics* **26**, 109–124.

Index